Quantum Field Theory

Kazuhiko Nishijima

Quantum Field Theory

By Academician Prof. Kazuhiko Nishijima - A Classic in Theoretical Physics

Masud Chaichian • Anca Tureanu
Editors

Springer

Author
Kazuhiko Nishijima
University of Tokyo
Tokyo, Japan

Editors
Masud Chaichian
University of Helsinki
Helsinki, Finland

Anca Tureanu
University of Helsinki
Helsinki, Finland

Translated by
Yuki Sato
Tokuyama College
National Institute of Technology
Yamaguchi, Japan

ISBN 978-94-024-2192-7 ISBN 978-94-024-2190-3 (eBook)
https://doi.org/10.1007/978-94-024-2190-3

Originally published by Kinokuniya Company Ltd., Tokyo, Japan, 1987

Japanese original edition BA NO RIRON by Kazuhiko Nishijima published in 1987 by Kinokuniya Company Ltd., Tokyo

Back Cover: Photo of Prof. Nishijima, 2008 (courtesy of late Mrs. Hideko Nishijima)

This Springer imprint is published by the registered company Springer Nature B.V.
The registered company address is: Van Godewijckstraat 30, 3311 GX Dordrecht, The Netherlands

Foreword

In the Preface, Professors Masud Chaichian and Anca Tureanu describe in vivid detail how the English version of this book on quantum field theory by the great Japanese physicist Kazuhiko Nishijma evolved as a labor of love on their parts. Nishijima enjoyed visiting Helsinki and Finland (Who doesn't? I certainly do.) regularly and was often seen consulting a physics book in Japanese, of which he was actually the author. (I can tell you from personal experience that authors like to read their own books; I certainly find my books ridiculously clear and easy to understand.) Chaichian and Tureanu urged Nishijima to have the book translated into English. After Nishijima's death in 2009, they found a translator, had it proofread, improved the English, arranged for a publisher, and invited one of the most eminent theoretical physicists of our times, Professor Yoichiro Nambu, to write a Foreword.

After Nambu's untimely death in 2015, they asked me to step in, perhaps because I have also written a textbook on quantum field theory. Any reader qualified to read this book knows full well that I am far from the level of achievements attained by Yoichiro Nambu, for whom I have the greatest admiration as a person and as a physicist. Nevertheless, out of respect for Nishijma, Nambu, Chaichian, and Tureanu, I agreed to do so.

I believe that Dick Feynman said that learning physics is something like painting a house. (In case you have no experience painting houses, one coat is never enough. Only the most shoddy house painters would do that.) John Wheeler told me something similar: one should not imbibe physics like sipping water from a cup, but more like drinking from a fire hydrant. Anybody who tells you that he or she could learn quantum field theory by reading a single textbook is not only kidding you, but also himself or herself. During my four years in graduate school, I attended Julian Schwinger's course on quantum field theory four times, not because I was especially stupid, but because quantum field theory is a profound subject. In this spirit, I strongly urge students, and also physics professors who claim that they understand quantum field theory, to read this book in addition to whatever other books they have read or possibly in parallel with other books they are reading. I also find that this book treats some topics that are not discussed in detail, or omitted entirely, in more recent books on quantum field theory.

Santa Barbara, CA, USA
July 2022

Anthony Zee

Preface to the English Edition

During several years of our collaboration with Prof. Kazuhiko Nishijma, we had noticed that always he was carrying a book in Japanese, which he consulted from time to time on different subjects. Prof. Nishijima used to visit us at the University of Helsinki regularly in the years 1985–2008. During one of his visits, he finally left a copy of that book with us in Helsinki to keep it for him for his future use. Not knowing which book it was, we asked about it and he explained that it was a book on Quantum Field Theory he had written. Upon our question whether it will be translated into English, Prof. Nishijima said that it was his wish to translate the book himself.

After Prof. Nishijima passed away in Tokyo on 15 February 2009, the book and Prof. Nishijima's wish sadly came to our mind and we reached a wishful thought of getting the book by all means translated into English as a tribute to Prof. Nishijima. Prof. Yoichiro Nambu, the 2008 laureate of the Nobel Prize in Physics, knew Prof. Nishijima very well and in the early 1960's was nominating him for the Nobel Prize for the discovery of the strangeness quantum number. Knowing this, we discussed the plan with Prof. Nambu, and he supported the idea with great enthusiasm, mentioning that he knew the book well and that he would like to write a foreword for its English edition. However, he also mentioned about the well-known difficulty of translating from Japanese to any other language of any group of English kind, and that it was for this reason that Nishijima wanted to translate his book himself.

We contacted Springer-Verlag about publishing the translated book and they consulted Prof. Nambu for a review. Prof. Nambu wrote a very strong evaluation, mentioning the transparency of the book and, in several parts, the originality of the presentations of the subject. The next task was to find a person who could translate the book, being a physicist with a good grasp of English, especially as regards the terminology used in the book. It was only thanks to the efforts of Prof. Nambu, who advertised the project in Japan, asking for volunteers, that a translator was found. Eventually, a young researcher in particle physics at Nagoya University, Dr. Yuki Sato, wrote to us mentioning that he would like to take the task merely as a sign of respect to Prof. Nishijima, one of Japan's greatest physicists who made significant contributions to theoretical physics.

The translation of the book by Dr. Yuki Sato, together with us the undersigned as editors going through it and checking the text and all the formulas again and again, as well as adding the reference list, took a few years and meanwhile Prof. Nambu passed away. Instead, Prof. Anthony Zee from University of California, Santa Barbara, most kindly accepted to write the foreword for the book.

After the translation of the book, another professional native English translator and a physicist by education, Mr. Stephen Lyle, went through the whole translated text and performed a more complete polishing of the text into a fine English language. We express our utmost thanks to Mr. Stephen Lyle for his devotion to perfection. We would like to thank Dr. Ramon Khanna, executive editor, and Ms. Christina Fehling, editorial assistant, at Springer Nature for their kind support and advice during the whole publication process, and also for their unlimited patience with several delays in our preparing the English edition of the book to send for print, with the aim that it will become a genuine tribute to Prof. Nishijima.

During the preparation of the English edition of the book we have received the encouragement of several physicists in Japan, to whom we are most grateful. Our special gratitude goes to Prof. Misao Sasaki from Kyoto University, and Prof. Kei-Ichi Kondo from Chiba University, for their constant support and much help during several years.

Helsinki, Finland
July 2022

Moshe M. Chaichian
Anca Tureanu

Preface of the Author

One of the important concepts in quantum mechanics is the dual particle-wave nature of matter. The same object is treated as a particle in one case and as a wave in another case, which is not the case in classical physics. This was one of the launch pads of quantum mechanics. Its typical example is light: in classical physics it was mainly described by the wave theory, although there existed also Newton's particle theory. However, once Einstein's photon hypothesis was proposed at the beginning of this century,[1] one could no longer claim that light is a wave or a particle. In this way quantum mechanics made its appearance. In its framework, the electron, which is thoroughly treated as a particle in the classical theory, is always described as a wave. With that the following problem comes up: How can we introduce the particle-wave duality in the same theory? Quantum field theory answers this question.

Quantum field theory was developed no later than the birth of quantum mechanics. It has experienced many changes up to the present. In the early stage, studies were entirely restricted to the area of electrodynamics, and turned out to be quite successful. Nevertheless, early quantum field theory reached a deadlock called the *problem of divergences*. However, with the advent of the renormalisation theory after the war, quantum electrodynamics based on perturbation theory reached completion. On the other hand, after pions were created by accelerators, the struggles of field theory started again. The reasons was that the perturbative approach can not be used for strong interactions, and one encountered the difficulty of extending the renormalisation theory to an approximation method beyond the perturbation theory. As a consequence, field theory was stagnant for a while; instead, the S-matrix theory based on dispersion relations became pre-eminent. In addition, after hundreds of "elementary" particles were created by accelerators, it became necessary to answer the question: What on earth are fundamental particles? And this is how the quark

This is the only Preface written by Prof. Nishijima in Japanese to the 1st edition published in 1987 in Tokyo; the 8th edition with extended chapters was published in 1997 in Tokyo.

[1] "This century" means the twentieth century.

model appeared. On the other hand, from the studies of weak interactions, gauge theories were introduced. This book covers a survey of quantum field theory related to the developments above.

After the emergence of the gauge theory, there have been quite many developments toward new directions. For example, the study of monopoles, solitons, and instantons related to the topological nature of classical gauge theories; lattice gauge theory, in which gauge invariance is given the leading role to the detriment of Lorentz invariance; applications of grand unified theories to cosmology; more recently, supersymmetric theories, supergravity theories and superstring theories, and so on. These theories are developing; each of them is sufficiently vast already as to require an independent book. I myself do not yet understand them very well, so that I have omitted them in this book, considering that more adequate people will write about them.

Although I published "Fundamental Particles" (1963) [1] and "Fields and Particles" (1969) [2] with the W.A. Benjamin Corporation as text books for elementary particle theory and quantum field theory, the contents of those books are by now outdated, and the topics to be emphasized have also changed in time. However, as fas as the fundamentally invariable parts are concerned, this book is based on the two books above. Additionally, since I have had a reluctance to write a book in the Mathematical Library in ignorance of mathematics, I have decided to write the book based not on the mathematical logic but on the physical logic, on the advice of Professor Seizo Ito. It has been long time since the request to write this book, but Professor Ito has patiently waited, and I would like to express my appreciation for him.

In the 61 year of the Showa era (1986), at Rakuhoku,

Tokyo, Japan Kazuhiko Nishijima

Contents

Chapter 1
Elementary Particle Theory and Field Theory

In a broad sense, field theory can be defined as quantum mechanics with infinite degrees of freedom. This broad definition includes condensed matter physics, which treats non-relativistic many-body problems. However, in this textbook, we discuss the formalism of field theory in a narrow sense, namely, in the framework of relativistic quantum mechanics. The objects of field theory in this narrow sense are the so-called *elementary particles*, and the academic discipline that studies the properties of elementary particles is called *elementary particle theory*. In a way, we can thus say that field theory provides a "grammar" for describing the properties of elementary particles. Since the two theories were developed together, they cannot be clearly separated. Elementary particle theory is concerned with the types of elementary particles that can exist and the types of interactions that they undergo. In each case, there is a phenomenological level and a fundamental level. For instance, atoms, which had been thought to be indivisible, in fact consist of atomic nuclei and electrons, while the atomic nuclei themselves consist of protons and neutrons. Thus, the solution to the problem of what the truly fundamental particles are has changed over time. Correspondingly, the solution to the problem of what the fundamental interactions are has also changed. We do not know if there will ever be an end to the changing viewpoints. However, what has been established so far is that particles called *quarks* and *leptons* seem to be the fundamental particles, while the fundamental interactions are assumed to be gauge interactions. We will discuss later how concepts such as quarks and gauge interactions arised. For now, we shall just give an overview of the developments of elementary particle theory at the phenomenological level.

The concept of field is a generic one used to refer to dynamical variables defined at each point in space. Examples of fields in macroscopic physics are the temperature and the product of the density and the stream velocity of a fluid. In microphysics, only very few such dynamical variables actually survive at the classical level, and perhaps the only candidate is the electromagnetic field. After the birth of quantum mechanics a crucial question was how to apply quantum

K. Nishijima, *Quantum Field Theory*,
https://doi.org/10.1007/978-94-024-2190-3_1

mechanics to the electromagnetic field. As mentioned in the foreword, although light is described classically as an electromagnetic wave, at the microscopic level, it also exhibits particle-like features, as expressed in Einstein's photon hypothesis. Therefore, it is essential that the quantum mechanical description of the electromagnetic field should contain such particle-like characteristics. This line of research was pursued by Dirac in [3]. In 1928, Dirac [4] also rewrote the non-relativistic Schrödinger equation as a relativistic equation. This led to the development of quantum electrodynamics. In 1928, Heisenberg and Pauli [5, 6] formulated quantum electrodynamics in a systematic way, and this provided the starting point for field theory.

Dirac's relativistic theory of electrons predicted the existence of a positron, the antiparticle of the electron. Its existence was confirmed by Anderson in 1932 [7]. Similarly, the existence of the anti-proton, the antiparticle of the proton, was also expected. However, we had to wait until 1955 for confirmation of its existence [8]. On the other hand, again in nuclear physics, the discovery of the neutron by Chadwick in [9, 10] was of fundamental importance, leading to the picture of atomic nuclei consisting of protons and neutrons. Yet, the *birth of elementary particle theory* originated in Yukawa's meson theory.

1.1 Classification of Interactions and Yukawa's Theory

Current theories such as grand unified theories or the theory of supergravity start by recognizing that there exist a variety of interactions with different strengths. Among these interactions, the best known, apart from gravity, is the electromagnetic interaction which, as mentioned before, has been studied since the birth of quantum mechanics. However, from the early days it was known that there were lots of phenomena that could not be explained by the electromagnetic interaction alone. Beta-decay is an example. The lifetime of a radioactive atomic nucleus is longer than that of states excited electromagnetically. This implies that the interaction causing beta-decay is much weaker than the electromagnetic interaction. In fact, in order to explain the continuous energy spectrum of electrons emitted in beta-decay, Pauli introduced the neutrino hypothesis [11, 12]. Pauli had a very convenient excuse for proposing such a new particle, namely, saving the conservation laws of energy, momentum, and angular momentum.

Fermi substantiated Pauli's suggestion by proposing the *four-fermion interaction* [13] to describe beta-decay in field theory. This was the second interaction discovered after the electromagnetic interaction. It meant that field theory now had to deal with electromagnetic and weak interactions.

After the detection of the neutron in 1932 [9, 10], it gradually became clear that an atomic nucleus consists of protons and neutrons. It was now essential to understand the nature of the so-called *nuclear force* binding protons and neutrons to form atomic nuclei. Since the neutron is electrically neutral, the nuclear force cannot

Table 1.1 Elementary particles and their interactions

Elementary particles	Interactions
ν	Weak
e	Weak + electromagnetic
n, p	Weak + electromagnetic + strong
γ	Electromagnetic (carrier)

be the electromagnetic force. Just after the proposal of Fermi's theory, Ivanenko and Tamm calculated the nuclear force by assuming that it was produced by exchange of an electron–neutrino pair via Fermi's interaction [14–16]. It turned out that the calculated force was too weak to be at the origin of the nuclear force.

Consequently, the nuclear force had to be of an entirely different kind from the electromagnetic and the weak interactions. It was also understood that various particles have their own roles and take part only in certain interactions. This is shown in Table 1.1. The symbols for the fundamental particles are ν (neutrino), e (electron), n (neutron), p (proton), and γ (photon), while "weak", "electromagnetic", and "strong" stand for the weak interaction, the electromagnetic interaction, and the strong interaction (the nuclear force), respectively. The proton and the neutron are collectively dubbed *nucleons* in the sense of elements of an atomic nucleus.

In 1935, Yukawa put forward the *meson theory*, reflecting the principle that *all forces are mediated by fields* [17]. Since the electromagnetic interaction is mediated by the electromagnetic field, the strong interaction and the weak interaction must be mediated by fields, too. Hence, Yukawa introduced the *meson field* as the mediator of the strong interaction. Just as quantizing the electromagnetic field yields the light quantum (photon), it turns out that quantizing any field yields corresponding quanta describing their particle-like aspects. The quantum corresponding to the meson field is an elementary particle called a *meson*. One of Yukawa's most important contributions is that he clarified the relationship between the mass of a quantum and the range of action of a force. In general, the range of a force mediated by some field is given by the Compton wavelength of its quantum. For instance, the range of the electromagnetic interaction mediated by a massless photon is infinite. Since the quantum of the meson field mediates the nuclear force with a range of around 10^{-13} cm, the meson mass must be about 200 times heavier than the electron mass. This mass is midway between the nucleon and electron masses, hence the name meson. Moreover, from a phenomenological standpoint, the quantum of the field mediating the weak interaction described by Fermi's model at a single point must be very massive. Yukawa thought that both the strong nuclear force and the weak interaction might be mediated by the meson field. Shortly afterwards, Klein suggested that a different field should be introduced for the weak interaction [18]. In any case, at that time, such ideas were mere hypotheses. In order to prove them, mesons had to be produced. However, since no accelerator had enough energy to create them, the only way to detect them was by studying cosmic rays. Historically, Yukawa's particle was discovered after some bumps and detours.

1.2 The Muon as the First Member of the Second Generation

Theorists predict the existence of experimentally unknown particles, while experimentalists detect sometimes particles which theorists have not predicted. The relation between Yukawa's particle and the muon can be described in this way. Using a modern term, we can say that the world surrounding us consists of elementary particles of the first generation. This is because elementary particles of the second generation are in general unstable, except for the neutrino. Thus, the discoveries of elementary particles of the second generation always gave rise to confusion. Elementary particles such as the muon, the strange quark, and the charm quark are in the second generation. The muon was discovered first.

In 1937, a particle with mass about 100 MeV, the muon, was detected in cosmic rays [19]. For almost a decade afterwards, this particle was thought to be Yukawa's particle.

In 1945, Conversi, Pancini, and Piccioni found that the interaction between the muon and an atomic nucleus was almost as strong as the interaction involved in the muon's weak decay [20], whence they were of a similar (weak) nature. One consequence of this discovery was that the muon and Yukawa's particle, the pion, had to be different particles. One year later, Powell's group observed the following two-step decay process [21] on a photographic plate:

$$\pi^+ \to \mu^+ + \nu\,, \quad \mu^+ \to e^+ + \nu + \nu\,.$$

Here π, μ, and ν stand for the pion, the muon, and the neutrino, respectively. Although there are several different neutrinos, we do not distinguish them here.

A two-meson theory which admits the existence of two different mesons, the pion and the muon, was developed by Sakata, Inoue, and Tanigawa in [22, 23]. One year later, it was also proposed in the United States by Marshak and Bethe [24].

1.3 Quantum Electrodynamics

After the Second World War, quantum electrodynamics made amazing progress. Quantum electrodynamics, or QED, is the most typical field theory. By introducing a new weapon called *renormalization*, it became possible to calculate, within the framework of QED, the Lamb shift in a hydrogen atom, i.e., the energy shift between the $2s$ and $2p$ states, and the anomalous magnetic moment of the electron. The formulation of the covariant perturbative theory by Tomonaga, Schwinger, Feynman, Dyson, and others was instrumental in this success. At this point, QED achieved the position of being the most successful discipline in exact science. Moreover, the notion of renormalizability was closely connected to subsequent developments in field theory, and in particular to the discovery of gauge theories, as the most important guiding principle in field theories.

1.4 The Road from Pions to Hadrons

In 1948, in Berkeley, pions were created artificially for the first time [25]. Furthermore, in 1952, Fermi's group found a resonance state of the pion–nucleon system, which they named Δ [26]. This resonance appears in the reaction

$$\pi + \mathrm{N} \to \Delta \to \mathrm{N} + \pi ,$$

where N stands for a nucleon. The existence of resonances is characteristic of strong interactions of the pion–nucleon system at low energies.

It soon became clear that the covariant perturbative theory which was such a useful tool in QED had almost no utility for understanding strong interactions. This made theorists suspicious of the usefulness of field theory for strong interactions. In fact, determinations of the pion's spin or parity were based on a general invariance principle, and the covariant perturbative theory was completely useless for this purpose. Of course, a non-perturbative method was developed in field theory, but it was not successful because the prescriptions proposed by renormalization were intimately connected to the perturbation theory.

This situation meant that many preferred to investigate the pion–nucleon system without considering the details of strong interactions, but considering this system in its own right. One thing that came up was the idea of *isospin*. The concept of isospin invariance was introduced by Kemmer in 1938 to make a connection between the charge independence of the nuclear force and Yukawa's theory [27]. The isospin transformation is a symmetry in a virtual isospin space. It became an active field of study again when pions could be created in the laboratory. This led to the first two examples of a modern term, *flavour*.

Another idea that came up was the dispersion relation, which is deeply connected to causality in field theory. This gives the relation between the absorptive (imaginary) part of a scattering amplitude and its dispersive (real) part. The form of the dispersion relation does not depend on the details of the interactions. The important factors for dispersion relations are just the representations of spacetime and internal symmetries carried by the particle states, together with the type of mass spectrum the dynamical system possesses. However, the dispersion relations leave slight signatures of the original interactions. For instance, although the best known dispersion relation for the forward scattering of the pion–nucleon system includes terms with first-order poles with respect to some Lorentz invariant variables, the poles can be described by the renormalized coupling constant of Yukawa's interaction. In addition, the dispersion relation for electromagnetic form factors is given by the charges of particles adjusted by so-called *subtraction constants*. In each case, these constants are renormalized coupling constants in field theory.

Much can be learned from the dispersion relation method. First, comparing with perturbative calculations in field theory, we can know what type of renormalization is carried out preturbatively. Considering a given *S*-matrix element (a scattering amplitude, for example) at some order in perturbation theory, its absorption part can be expressed in terms of lower order S-matrix elements because of the unitarity of the S-matrix. In this evaluation, integrals have to be carried out in a finite phase-space, but divergences do not appear here. Next, we obtain the dispersion part of the S-matrix element from its absorption part using the dispersion formula. This corresponds to carrying out a *Hilbert transformation*, and this integral sometimes diverges. Such a divergence just corresponds to the one in field theory. If a divergence exists, we can remove it using a dispersion formula in which a subtraction has been made. These subtractions correspond to the renormalization in field theory, and the subtraction constant corresponds to the renormalized coupling constant. Thus in the dispersion theory, the divergence appears only in the dispersion part, not in the absorption part. Furthermore, the validity of the subtraction is not restricted to a perturbative approach.

We may ask ourselves whether the dynamics of elementary particles can be formulated in terms of the S-matrix, that is, whether or not we can determine the S-matrix by combining the unitarity and the dispersion relations of the S-matrix as mentioned above. In this case, the renormalization is automatically processed by introducing the dispersion formula in which subtractions have been made, thereby realizing a renormalization procedure which is independent of perturbation theory. In this formalism, stable particles which cause poles to occur in scattering amplitudes give the same dispersion relations no matter whether they are elementary particles or composite particles. In other words, what is important in the S-matrix theory is only the stability of particles, so it does not matter whether the particles are elementary or composite. This ended up having a great significance on subsequent developments in the elementary particle theory. The theory proposed by Chew and Low in 1956 is a typical example of the non-relativistic S-matrix theory [28]. The idea was developed further by the discovery of the Mandelstam representation in 1958 [29]. In this representation, scattering amplitudes can be written as double dispersion integrations, and they have crossing symmetries (we shall not touch on such symmetries here, but they will be explained later).

Note, however, that the dynamical S-matrix theory encounters two difficulties. The first is the impossibility or extreme difficulty in finding the complete set of dispersion formulas. The second difficulty is that the solution cannot be determined unambiguously, even under the approximation where we neglect many-particle states. This is related to the fact that we cannot tell the difference between elementary particles and composite particles in the S-matrix theory, as mentioned before. For example, for a given pair of dispersion formulas we can adopt the Lagrangian of Yukawa's theory or the Lagrangian of quantum chromodynamics (QCD).

The S-matrix theory itself, however, went in a different direction. From the study of the high-energy behaviour of scattering amplitudes with crossing symmetries, the Regge pole theory, or Regge trajectory theory, was formulated in 1960s [30]. In the 1950s, following the discovery of the antiproton, many hadrons were detected at the Bevatron using bubble chambers, and the Regge trajectory was applied to classify those hadrons. "Hadron" is the generic name for elementary particles participating in strong interactions. When we consider a two-body bound state, the

Regge trajectory is obtained by plotting the relation between the angular momentum J extended to a continuous variable and the mass squared of the state, M^2. In this method, allowing many hadrons with different quantum numbers except for their spins to follow the same trajectory clarifies the dynamical family relationships among them, and this of course can help to classify them. In particular, Chew and Frautschi succeeded in classifying many hadrons by considering that all the Regge trajectories are continuous lines with universal tangents [31]. As mentioned before, the Regge trajectory corresponds to a bound state, so the success of the Chew–Frautschi plot implied that all hadrons were composite rather than fundamental. The way we interpret this fact is a problem of fundamental importance for elementary particle theory, and we will discuss it in great depth later.

1.5 Strange Particles as Members of the Second Generation

Elementary particles which obey Fermi statistics are classified into hadrons, which engage in strong interactions, and leptons, which do not. The muon, which belongs to the second generation, is a lepton. We have already mentioned its detection and the confusion this created. Similarly, when second generation hadrons were detected, they caused the same confusion. For this reason, hadrons belonging to the second generation were first called *strange particles*. In connection with the detection of the muon, we mentioned that experimentalists sometimes detect particles that have not been predicted by theorists, and this is indeed what happened in the case of strange particles.

In 1947, Rochester and Butler detected two V-particles in a cloud chamber [32]. The V-particle was named after the shape of the track in the cloud chamber, indicating a particle decay. Although the observations continued for two more years, no new particle was discovered. In such a case, two things can happen: either we find nothing when we carry out additional tests or we suddenly detect many cases by improving experimental methods. In the case of V-particles or strange particles, the latter happened. The improved method was to detect V-particles from cosmic rays by climbing high mountains. In the early 1950s, experimental groups in Pasadena and Manchester observed dozens of V-particles on the White Mountains [33] and at the Pic du Midi in the Pyrenees [34], respectively. The biggest problem with the V-particle was to reconcile its frequent detections with its relatively long lifetime of about 10^{-10} seconds (long for an elementary particle). This problem was somewhat similar to the muon problem. With hindsight, this was because both particles belonged to the second generation. Many new particles were subsequently detected in cloud chambers and on photographic plates. For example, *hyperons* like Λ^0, $\Sigma^\pm$, and Θ^-, which are hadrons satisfying Fermi statistics, heavy K-mesons (also called θ- or τ-mesons), and so on.

In order to explain the discrepancy between the frequent detections and the long lifetime, it was suggested that V-particles or strange particles might be created in pairs. Such theories were proposed in 1951 by Nambu, Yamaguchi, and the author

[35, 36], by Oneda [37], and by Miyazawa [38], then by Pais in [39]. However, the pair-production theories were hard to demonstrate from observations of cosmic rays. What these models had in common was that strange particles were always produced in pairs through strong interactions, and the isolated strange particles decayed through weak interactions. This explained both the frequent detections and the long lifetime. However, some phenomena could not be explained by this hypothesis alone. For example, it was not understood why the decay processes $\Sigma^- \to \Lambda^0 + \pi^-$ and $\Sigma^- \to n + \pi^-$ did not occur through strong interactions. Nor indeed why the heavy mesons with positive charges were more common then those with negative charges both on plates and in cloud chambers. These questions could not be answered using the above hypothesis.

In 1953, the pair production of strange particles that had not been confirmed by cosmic rays was verified experimentally using an accelerator, the Cosmotron, at Brookhaven [40]. The concept of *strangeness* was introduced by Nakano and the author [41], and independently by Gell-Mann [42]. This concept is closely related to isospin. All the members of a given isospin multiplet have the same value of the strangeness. Further, strangeness is an additive quantum number. It is conserved in strong and electromagnetic interactions, but not in weak interactions. As we shall see later, parity is a quantum number with the same features. Moreover, the K-meson came to play an extremely important role in elementary particle theory. In the history of modern physics, we may say that the K-meson has played the most important role after the hydrogen atom. The concept of strangeness is closely related to the fact that the neutral K-meson, denoted by K^0, differs from its antiparticle $\bar{K}^0$. Furthermore, the non-conservation of parity in weak interactions was discovered through the diversity of K-meson decay patterns. The fact that the CP transformation, where C stands for charge conjugation and P for parity, is not conserved in weak interactions was discovered through the observation of certain decay patterns of the neutral K-mesons. As can be seen from this, the K-meson continued to provide many new ideas. To introduce a modern term, strangeness is a third example of flavour and it is carried by hadrons of the second generation. We shall see later that the symmetry group $SU(3)$ was proposed as a way of combining the isospin symmetry group $SU(2)$ with strangeness.

1.6 Non-conservation of Parity

The strange K-meson was given different names according to its decay patterns:

$$\theta \to 2\pi , \quad \tau \to 3\pi . \tag{1.1}$$

The question was whether θ and τ were the same particle or not. The improvement in the accuracy of measurements revealed that the masses and lifetimes of θ and τ coincided. In addition, it became clear that the two production rates were constant, and quite independent of the production process. This suggested that θ and τ were the same type of particle. However, from the analyses by Dalitz [43] and Fabri [44], this implied the non-conservation of parity in decay processes. The discrepancy between the identification of the particles and the conservation of parity was called the θ–τ puzzle. Under the circumstances, given that the non-conservation of parity might occur, not only in K-meson decays, but in all weak interactions, Lee and Yang proposed various methods of experimental validation [45]. The ^{60}Co experiment by Wu's group was the first to confirm the non-conservation of parity in beta decays [46]. In particular, it was shown that parity was maximally broken in decay processes which involved neutrinos. This led to the two-component theory of the neutrino put forward by Lee and Yang [47]. Since parity is a space-time symmetry, in contrast to strangeness, many physicists were shocked by the non-conservation of parity.

After this discovery, theoretical research was very active. Here we mention the most important development. Assuming that all forces are mediated by fields, we can determine the characteristics of the fields that mediate the weak interactions by identifying the specific types of phenomenological Fermi interaction. An important first step in this direction was the $V - A$ theory proposed in 1958 [48–50]. This theory treated the fields mediating weak interactions as vector fields. The quanta of these vector fields were called W bosons, where W here stands for "weak".

In this connection, in the late 1940s, the conserved vector current (CVC) hypothesis was put forward [49], based on the universality of the Fermi interaction. The idea was proposed by Marshak and Sudarshan [48], by Sakurai [50], and by Gerstein and Zeldovich [51], and further promoted by Gell-Mann and Feynman [49]. According to experiments, the coupling constants for vector couplings of Fermi interactions for various processes are the same, irrespective of the processes, in spite of corrections from strong interactions. This can be explained by assuming that a four-vector density expressing hadronic parts of the Fermi interaction is proportional to an isospin four-current density. This is the *CVC hypothesis*, which will be explained in more detail in the text. When describing phenomenological Fermi interactions as products of two Yukawa interactions mediated by W bosons, each Yukawa interaction is described as the product of the conserved isospin current density and the W field. It is suggestive to compare with the fact that the electromagnetic interaction is the product of the conserved current density and the four-potential of the electromagnetic field.

The idea of including the axial vector current, corresponding to A in $V - A$, developed into the *current algebra method* [52, 53], inspired by the fact that in quantum mechanics a set of conserved quantities forms an algebra. In Sect. 1.4, we presented the methods based on the dispersion formulas and on symmetries for dealing with the strong interactions, although without going into the details. The current algebra method is an extension of the latter. The three components of isospin satisfy the same commutation relations as the components of angular momentum, and they generate the $SU(2)$ algebra. Although this is the algebra formed by just the spatial integral of the time component of the V part of $V - A$, i.e., the algebra satisfied by isospin, if we take $V - A$ itself, we also get the $SU(2)$ algebra. Since the $V - A$ current corresponds to the left-handed current, which will be explained later, we write this algebra as $SU(2)_L$. Introducing the strangeness-changing current, this algebra is extended to $SU(3)_L$. Furthermore, introducing $SU(3)_R$ in the context of the $V - A$ theory, the current algebra is extended to $SU(3)_L \times SU(3)_R$. The first successful application of this algebra was the Adler–Weisberger formula derived in 1965 [54–56]. Then, in 1963, Cabbibo extended the concept of the universal Fermi interaction to the strangeness-changing interaction [57], and this was an important step on the path to the quark model.

1.7 Second Generation Neutrinos

The neutrino was first introduced by Pauli in 1933 [11, 12], and immediately used by Fermi to formulate the field theory of beta-decay [13]. The existence of this neutrino was demonstrated by Cowan, Reines, et al. in [58]. This was the first generation neutrino.

Research on several processes involving weak interactions suggested that there might be a lepton number conservation law and a corresponding selection rule. However, two different definitions were given for the lepton number. One taking e^-, μ^+, and ν as leptons was given by Konopinski and Mahmoud in [59], and another taking e^-, μ^-, and the left-handed (anti)neutrino as leptons was given by Lee and Yang in [47]. It seemed that neither definition was in conflict with experiment. Then, in 1957, assuming a four-component neutrino and putting the case that each conservation law holds, the author set out to investigate which theories were consistent [60]. Combining the above two types of conservation law in the right way, it turned out that the electron family and the muon family could be attributed separate conservation laws. In the electron family, the difference between the numbers of e^- and ν_L and the numbers of e^+ and $\bar{\nu}_R$ was conserved, and likewise in the muon family for the difference between the numbers of μ^- and $\bar{\nu}_L$ and the numbers of μ^+ and ν_R.

In short, there were two types of lepton number conservation law: one concerned the lepton number of the first generation and the other concerned the lepton number of the second generation. The subscripts L and R stand for left-handed and the right-handed, respectively, and both are two-component neutrinos. There are thus two types of two-component neutrino: one belongs to the electron family in the first generation and the other to the muon family in the second generation. A similar idea was also introduced by Schwinger [61].

The second generation neutrino hypothesis was confirmed by a group at Columbia University in 1962 using the AGS accelerator in Brookhaven [62].

1.8 Democratic and Aristocratic Hadrons—The Quark Model

In 1948, Fermi and Yang constructed a theory treating the pion as a bound state of a nucleon and an anti-nucleon [63]. They were motivated by the idea that, since there were so many so-called elementary particles, it would be better to consider some of them as composite particles. In 1955, including the strange Λ particle with a pair of fundamental particles, the proton and the neutron, Sakata extended the Fermi–Yang model for hadrons [64]. He put forward the idea that all hadrons were bound states of protons, neutrons, the Λ particles, and their antiparticles. Given that there might exist a symmetry between p, n, and Λ, Ikeda, Ogawa, and Ohnuki introduced the symmetry group $SU(3)$, and studied the types of representations the hadrons belonged to [65]. For mesons and the hadrons satisfying Bose statistics, Sakata's model achieved many successful results. The η meson found in 1961 [66] was one successful example of Sakata's predictions. However, his model could not give a good explanation for baryons. In any case, this was just one attempt to interpret hundreds of hadrons as bound states of the most fundamental particles.

In 1961, Gell-Mann [52, 67] and Ne′eman [68] independently extracted the concept of $SU(3)$ symmetry from Sakata's model. In Sakata's model, p, n, and Λ were assigned to the three-dimensional representation of $SU(3)$. On the other hand, including Σ and Θ with the particles in Sakata's model, they were able to assign these particles to the eight-dimensional representation. This was subsequently called the *eightfold way*. For mesons, there is no qualitative difference between Sakata's model and the eightfold way, because mesons constitute the eight-dimensional representation in both models.

The $SU(3)$ symmetry is not exact, but explicitly broken. This led to the relation called the *Gell-Mann–Okubo formula* [69, 70] among the masses of particles in one irreducible representation. The highlight of this development was the observation of Ω^- with a mass of 1972 MeV, while the mass predicted by the mass formula was 1983 MeV [71].

Let us return to the question of how to interpret the large number of hadrons discussed in Sect. 1.4. From the phenomenological point of view, hadrons can be categorized by the Chew–Frautschi plot [72] and the $SU(3)$ group. The success of this classification method implied that all the observed hadrons were composite particles. There were two possible ways to interpret this outcome: one was that no fundamental particle existed other than the already observed hadrons, and the other was that more fundamental particles existed, but that they had not yet been observed. These were referred to as *democratic hadrons* and *aristocratic hadrons*, respectively.

The former idea was put forward in Berkeley around 1960, especially by Chew's group [73]. In modern physics, there is a tendency always to seek more fundamental levels, such as molecules $\rightarrow$ atoms $\rightarrow$ nucleons $\rightarrow$ elementary particles, but Chew's group suggested cutting off this sequence. According to this idea, all hadrons should be treated on an equal footing. In other words, since no hadron is more fundamental than the others, each hadron from the lightest pion to the heaviest nucleon should be thought of as a bound state of other hadrons. In this case, each hadron's mass and spin should be determined by a self-consistent method. In this approach, there is no idea of a most fundamental field, so field theory is not suitable to describe it. In this sense, it is a pure S-matrix theory. In particular, the dispersion formula is suitable for this purpose because it holds whether the relevant particles are elementary or composite. However, a framework describing the complete dynamics in a mathematically rigorous way was never provided. This was the idea of a democracy of hadrons.

The other idea is that detectable hadrons are composite particles of more fundamental particles. We can say that this is the aristocracy of the hadron world, treating the more fundamental particles as aristocrats. In 1964, Gell-Mann and Zweig proposed the quark model for hadrons based on the $SU(3)$ group [74, 75]. In this model, quarks became the most fundamental particles. This can be considered as a refinement of Sakata's model. Rather than p, n, and Λ used in Sakata's model, the quark model introduces quarks with three different flavours, namely u, d, and s. Mesons are considered to be bound states of a quark and an antiquark, which is similar to the construction in Sakata's model, while baryons are considered to be bound states of three quarks. The possible representations are:

$$\text{mesons} \quad \mathbf{3} \otimes \mathbf{3} = \mathbf{1} \oplus \mathbf{8} ,$$

$$\text{baryons} \quad \mathbf{3} \otimes \mathbf{3} \otimes \mathbf{3} = \mathbf{1} \oplus \mathbf{8} \oplus \mathbf{8} \oplus \mathbf{10} .$$

One merit of this model is that all the representations appearing on the right-hand side are actually realized in nature. Furthermore, this model succeeded in explaining intrinsic features of hadrons such as the mass and the magnetic moment. However, Gell-Mann [67, 68] and Han and Nambu [76, 77] showed that, when this model is applied to the interpretation of real hadrons, in order to avoid a contradiction with the exclusion principle, an until then unknown degree of freedom had to be introduced. This degree of freedom was called *colour* to distinguish it from flavour.

Unlike the democratic theory of hadrons, the quark model can be easily included in field theory. Quarks and leptons are described by fundamental fields, and hadrons are considered to be composite particles. Further, the left-handed or right-handed current densities in the current algebras can be expressed in terms of very simple bilinear forms using the quark model. In other words, when writing weak interactions in Yukawa form, the hadron parts can be written as products of bilinear forms and W fields. This was a great advantage, recalling from above that expressing the left-handed and right-handed current densities as hadron fields required highly non-linear representations. Thus the quark model was highly promising from the standpoint of including weak interactions.

The charges of the three types of quark can be determined through the fact that the baryon consists of three quarks. The charge of u is $2e/3$, and the charges of d and s are $-e/3$, where e is the positron charge. It is a feature of the quark model that quarks must have fractional charges. However, although experimentalists have tried to detect such fractional charges, they have not been found yet. Therefore, the hypothesis was put forward that independent quarks could not be observed in principle, for some reason. This is the hypothesis known as *quark confinement.* If we accept this hypothesis, there is a sense in which Chew's idea of a democracy of hadrons can still be realized. This is because the S-matrix theory treats only observable hadrons. The dispersion relation is given for the S-matrix elements relevant to the processes involving only hadrons, while the unitarity condition is also expressed by states consisting only of hadrons, which means that the quark states do not appear.

However, the dynamics of hadrons is not determined within the framework of the S-matrix theory. To do this, we need to know the dynamics of gluons mediating strong interactions and quarks. We can say that such an application of the quark model to both strong and weak interactions has paved the way for theories of gauge fields mediating strong, electromagnetic, and weak interactions. We will discuss the theories of gauge fields in the latter half of this book. In this chapter, we went through the story of trial and error that eventually led to the theories of gauge fields. We have used many terms without giving strict definitions, but these will be explained step by step in the following chapters.

Chapter 2
Canonical Formalism and Quantum Mechanics

Quantum field theory is the quantum mechanics of systems with an infinite number of degrees of freedom, and it is constructed as the limit of systems with a finite number of degrees of freedom. After reviewing general features of quantum mechanics for systems with a finite number of degrees of freedom, we will attempt to extend the formalism.

2.1 Schrödinger's Picture and Heisenberg's Picture

In general, in quantum mechanics, the equations of motion are either Schrödinger's or Heisenberg's. Given the Hamiltonian of a system as a function of the set of canonically conjugate variables, $q_1, q_2, \ldots, q_f$ and $p_1, p_2, \ldots, p_f$, Schrödinger's equation is

$$\mathrm{i}\frac{\partial}{\partial t}\Psi(t) = H\left(q_r, \frac{1}{\mathrm{i}}\frac{\partial}{\partial q_r}\right)\Psi(t) . \tag{2.1}$$

$\Psi(t)$ is called a *probability amplitude* or *state vector*. This equation holds in the so-called *Schrödinger picture*, in which an operator $\mathscr{O}$ representing a physical quantity is explicitly independent of time. The expectation value of the dynamical quantity $\mathscr{O}$ at time t is

$$\langle \mathscr{O} \rangle = \int \Psi^*(t)\mathscr{O}\Psi(t)\mathrm{d}q , \quad \mathrm{d}q = \rho \mathrm{d}q_1 \mathrm{d}q_2 \ldots \mathrm{d}q_f , \tag{2.2}$$

where ρ stands for an appropriate density.

In this book, we use natural units, such that $\hbar = c = 1$, where h and c stand for Planck's constant and the speed of light, respectively. From the point of view of relativity, it is natural to take the same unit for time and space; the unit of time is

K. Nishijima, *Quantum Field Theory*,
https://doi.org/10.1007/978-94-024-2190-3_2

considered to be the time light takes to travel a unit length, for instance 1 cm. In addition, although the reduced Planck constant $\hbar = h/2\pi$, frequently appearing in quantum mechanics, is a very small unit from the perspective of the human scale, it is useful to choose this as unit because it always appears when physical quantities are expressed in the microscopic world. For example, angular momentum always appears in units of $\hbar$, and in natural units this is merely an integer or a half-integer. In the natural system, all the units of physical quantities are given as certain powers of length.

In Schrödinger's picture, all the time dependence is assigned to state vectors, while in Heisenberg's picture the time dependence is assigned to the operators corresponding to physical quantities. Since all the physical quantities are functions of the canonical variables, they can be determined using the equations of motion, which are Heisenberg's equations in our case:

$$\frac{\mathrm{d}q_r(t)}{\mathrm{d}t} = \mathrm{i}[H, q_r(t)] \ , \qquad \frac{\mathrm{d}p_r(t)}{\mathrm{d}t} = \mathrm{i}[H, p_r(t)] \ , \tag{2.3}$$

where the canonical variables must satisfy the canonical commutation relations

$$[q_r(t), p_s(t)] = \mathrm{i}\delta_{rs} \ , \quad [q_r(t), q_s(t)] = 0 \ , \quad [p_r(t), p_s(t)] = 0 \ . \tag{2.4}$$

In this case, the expectation value of $\mathscr{O}$ at time t, corresponding to (2.2), is given by

$$\langle \mathscr{O} \rangle = \int \Psi^* \mathscr{O}(t) \Psi \mathrm{d}q \ . \tag{2.5}$$

The two representations are related by the following unitary transformations:

$$\Psi(t) = \mathrm{e}^{-\mathrm{i}tH}\Psi \ , \quad \mathscr{O}(t) = \mathrm{e}^{\mathrm{i}tH}\mathscr{O}\mathrm{e}^{-\mathrm{i}tH}, \tag{2.6}$$

for a state vector and an operator, respectively. Taking appropriate care of the operator orderings, as a consequence of (2.4), we obtain for a function F of p and q the equations

$$\frac{\partial F}{\partial p_r} = \mathrm{i}[F, q_r] \ , \qquad \frac{\partial F}{\partial q_r} = -\mathrm{i}[F, p_r] \ . \tag{2.7}$$

Hence, (2.3) can be written in the canonical form

$$\frac{\mathrm{d}q_r(t)}{\mathrm{d}t} = \frac{\partial H}{\partial p_r(t)} \ , \qquad \frac{\mathrm{d}p_r(t)}{\mathrm{d}t} = -\frac{\partial H}{\partial q_r(t)} \ , \tag{2.8}$$

where the t-dependence has been made explicit in $p_r(t)$ and $q_r(t)$, because these operators are defined in the Heisenberg picture.

2.2 Hamilton's Principle

In classical dynamics, the equation of motion, called the *Euler–Lagrange equation*, is derived from the variation principle

$$\delta \int_{t_0}^{t_1} \mathrm{d}t L(q_r, \dot{q}_r) = 0 . \tag{2.9}$$

This is *Hamilton's principle*. Considering $q_r(t_0)$ and $q_r(t_1)$ as given and imposing the conditions

$$\delta q_r(t_0) = \delta q_r(t_1) = 0 , \tag{2.10}$$

the Euler–Lagrange equations are derived as

$$\frac{\mathrm{d}}{\mathrm{d}t}\left(\frac{\partial L}{\partial \dot{q}_r}\right) - \frac{\partial L}{\partial q_r} = 0 , \tag{2.11}$$

where $\dot{q}_r$ is the time derivative of q_r. To pass from here to the canonical form, it is enough to define the canonically conjugate momentum p_r as

$$p_r = \frac{\partial L}{\partial \dot{q}_r} . \tag{2.12}$$

When $\dot{q}_r$ can be written in terms of q and p, from this equation, we call L a *non-singular Lagrangian*. In this case, the Hamiltonian is defined by

$$H = \sum_{r=1}^{f} p_r \dot{q}_r - L . \tag{2.13}$$

2.3 Equivalence Between the Canonical Equations and Lagrange's Equations

In this section, we derive Lagrange's equations by combining the canonical equations and the canonical commutation relations. A combination of the canonical equations (2.3), the canonical commutation relations (2.4), and the definition of the

Hamiltonian (2.13) yield

$$\begin{aligned}\frac{\mathrm{d}q_r}{\mathrm{d}t} &= \mathrm{i}[H, q_r] \\ &= \mathrm{i}\left[\sum_{j=1}^{f} p_j \dot{q}_j - L, q_r\right] \\ &= \dot{q}_r + \mathrm{i}\sum p_j[\dot{q}_j, q_r] - \mathrm{i}[L, q_r]\,.\end{aligned}$$

Taking into account the fact that the q's commute, we use

$$[L, q_r] = \sum_j \frac{\partial L}{\partial \dot{q}_j}[\dot{q}_j, q_r] + \sum_j \frac{\partial L}{\partial q_j}[q_j, q_r] = \sum_j \frac{\partial L}{\partial \dot{q}_j}[\dot{q}_j, q_r]\,.$$

Inserting this in the previous equation, we find

$$\frac{\mathrm{d}q_r}{\mathrm{d}t} = \dot{q}_r + \mathrm{i}\sum_j \left(p_j - \frac{\partial L}{\partial \dot{q}_j}\right)[\dot{q}_j, q_r]\,.$$

This equation always holds as long as the identity (2.12) does. Starting from the other canonical equation, we find similarly

$$\begin{aligned}\frac{\mathrm{d}p_r}{\mathrm{d}t} &= \mathrm{i}[H, p_r] \\ &= \frac{\partial L}{\partial q_r} + \mathrm{i}\sum_j \left(p_j - \frac{\partial L}{\partial \dot{q}_j}\right)[\dot{q}_j, p_r]\,.\end{aligned}$$

Thus, if (2.13) holds,

$$\frac{\mathrm{d}p_r}{\mathrm{d}t} - \frac{\partial L}{\partial q_r} = \frac{\mathrm{d}}{\mathrm{d}t}\left(\frac{\partial L}{\partial \dot{q}_r}\right) - \frac{\partial L}{\partial q_r} = 0\,.$$

As a consequence, we can understand that the canonical equations are equivalent to Lagrange's equations as long as the canonical commutation relations hold. This fact will be used to quantize fields.

2.4 Equal-Time Canonical Commutation Relations

As we have seen in the previous section, the equal-time commutation relations (2.4) are the basic relations of quantum mechanics. We prove below that these relations hold at any time if they hold at some time $t = t_0$. First, from the canonical equations (2.3) and the Jacobi identity, we get the differential equation

$$\frac{\mathrm{d}}{\mathrm{d}t}[q_r(t), p_s(t)] = \mathrm{i}\big[H, [q_r(t), p_s(t)]\big] . \tag{2.14}$$

Solving this equation with the initial condition

$$[q_r(t_0), p_s(t_0)] = \mathrm{i}\delta_{rs} , \tag{2.15}$$

we obtain the following relation for arbitrary time t :

$$[q_r(t), p_s(t)] = \mathrm{i}\delta_{rs} . \tag{2.16}$$

Chapter 3
Quantization of Free Fields

The dynamical systems described in the previous chapter had a finite number of degrees of freedom f. How can we extend the theory to the limit $f \to \infty$? To investigate this problem, we start by implementing an easy extension of the theory and then generalizing it.

3.1 Field Theory Based on Canonical Formalism

First, we consider N systems with no interaction among them. The total Lagrangian is given by the sum of Lagrangians corresponding to each system, namely,

$$L = \sum_{n=1}^{N} L_n(q_n, \dot{q}_n) , \tag{3.1}$$

and Hamilton's principle, or the principle of least action, is given by

$$\delta \int_{t_0}^{t_1} \mathrm{d}t \sum_{n=1}^{N} L_n(q_n, \dot{q}_n) = 0 . \tag{3.2}$$

Then, treating N as a continuous infinity, we replace the discrete number n by the continuous parameter x. At the same time, we change the symbol of the dynamical variable from q_n to $\varphi(x)$. We call the dynamical variable, which is a function of continuous variables, a *field*. At the same time, designating a domain of integration, the principle of least action assumes the form

$$\delta \int \mathrm{d}t L = \delta \int \mathrm{d}t \left(\int \mathrm{d}x \mathscr{L}[\varphi(x), \dot{\varphi}(x)] \right) = 0 . \tag{3.3}$$

K. Nishijima, *Quantum Field Theory*,
https://doi.org/10.1007/978-94-024-2190-3_3

$\mathscr{L}$ is called a *Lagrangian density*. In the case when x stands for the spatial coordinates in three dimensions, $\mathrm{d}x$ denotes the volume element $\mathrm{d}^3x = \mathrm{d}x\,\mathrm{d}y\,\mathrm{d}z$. At the same time, $\mathrm{d}t\,\mathrm{d}^3x = \mathrm{d}^4x$ stands for the four-dimensional volume element in natural units. In this system, t is sometimes written as x_0. Thus, the principle of least action in three-dimensional space can be written as

$$\delta \int_{\Omega} \mathrm{d}^4x \mathscr{L}[\varphi(x), \dot{\varphi}(x)] = 0 , \tag{3.4}$$

where Ω stands for a certain four-dimensional domain and x denotes the four-dimensional coordinate. This is the principle of least action for the case when there are no interactions between fields at different points, but in this case we cannot describe the waves propagating through space. In relativistic field theory based on the principle of actions through a medium, a field at some point interacts only with its infinitesimal neighbours. Put another way, the Lagrangian density is a function of $\varphi(x)$, $\dot{\varphi}(x)$, and $\varphi(x + \mathrm{d}x)$. Then, using

$$\varphi(x + \mathrm{d}x) = \varphi(x) + \frac{\partial \varphi(x)}{\partial x_k}\mathrm{d}x_k , \quad k = 1, 2, 3 ,$$

the Lagrangian density becomes a function of φ and its time-like and space-like derivatives. Putting these together, we can write

$$\mathscr{L} = \mathscr{L}\left[\varphi(x), \frac{\partial \varphi(x)}{\partial x_\mu}\right], \quad \mu = 1, 2, 3, 4 , \tag{3.5}$$

where $x_4 = \mathrm{i}x_0 = \mathrm{i}t$. In general, there exist several fields, and we distinguish them by an additional label α as $\varphi_\alpha(x)$. Thus, the principle of least action can be written as

$$\delta \int \mathrm{d}^4x \mathscr{L}\left[\varphi_\alpha(x), \frac{\partial \varphi_\alpha(x)}{\partial x_\mu}\right] = 0 . \tag{3.6}$$

3.1.1 Canonical Commutation Relations

Let us consider how we can extend the canonical commutation relations for a system with a finite number of degrees of freedom to those for a system with an infinite number of degrees of freedom. First, writing the equal-time canonical commutation relations in a system with finite degrees of freedom as

$$\sum_s [q_r, p_s] A_s = \mathrm{i}A_r , \tag{3.7}$$

where A_r is an arbitrary quantity, we can make a dictionary to translate from discrete variables to continuous variables:

$$q_s \to \varphi_\alpha(x) \, ,$$

$$p_s = \frac{\partial L}{\partial \dot{q}_s} \to \pi_\alpha(x) = \frac{\partial}{\partial \dot{\varphi}(x)} \mathscr{L}\left[\varphi_\alpha(x), \frac{\partial \varphi_\alpha(x)}{\partial x_\mu}\right] ,$$

$$\sum_s \to \sum_\alpha \int \mathrm{d}^3 x \, .$$

Using these rules, we translate (3.7) to continuous variables. Since the condition is given at equal times, we assume $x_0 = x_0'$ in the following, and we have

$$\sum_\beta \int \mathrm{d}^3 x' \left[\varphi_\alpha(x), \pi_\beta(x')\right] f_\beta(x') = \mathrm{i} f_\alpha(x) \, . \tag{3.8}$$

Assuming that this holds for arbitrary $f_\beta(x')$, we get the canonical commutation relation in field theory:

$$\left[\varphi_\alpha(x), \pi_\beta(x')\right] = \mathrm{i}\delta_{\alpha\beta}\delta^3(x - x') \, , \tag{3.9}$$

where $x_0 = x_0'$, and

$$\delta^3(x - x') = \delta(x - x')\delta(y - y')\delta(z - z').$$

As we shall see later, commutation relations like this hold only for fields satisfying Bose statistics. For fields satisfying Fermi statistics, we need to replace the commutation relations by anti-commutation relations. We shall discuss this issue later. In addition, the relations corresponding to

$$[q_r, q_s] = [p_r, p_s] = 0 \tag{3.10}$$

are

$$\left[\varphi_\alpha(x), \varphi_\beta(x')\right] = \left[\pi_\alpha(x), \pi_\beta(x')\right] = 0 \, , \quad x_0 = x_0' \, . \tag{3.11}$$

The canonical quantization of fields as presented above was introduced by Heisenberg and Pauli.

3.1.2 Euler–Lagrange Equations

The equations of motion for fields are derived through the principle of least action

$$\delta \int_{\Omega} \mathrm{d}^4 x \mathscr{L} = 0 , \tag{3.12}$$

where φ_α is given in the four-dimensional domain Ω, such that, on the boundary surface $\partial\Omega$ of the domain, the variations of the field vanish, i.e., $\delta\varphi_\alpha = 0$. The variation considered here is $\delta\varphi_\alpha(x) = \varphi'_\alpha(x) - \varphi_\alpha(x)$, and a variation commutes with a derivative. In addition, from now on we shall write derivatives as follows:

$$\frac{\partial \varphi_\alpha(x)}{\partial x_\mu} = \partial_\mu \varphi_\alpha(x) = \varphi_{\alpha,\mu}(x) . \tag{3.13}$$

We start with the following equation:

$$\delta \int_{\Omega} \mathrm{d}^4 x \mathscr{L} = \int_{\Omega} \mathrm{d}^4 x \left[\frac{\partial \mathscr{L}(x)}{\partial \varphi_\alpha(x)} \delta\varphi_\alpha(x) + \frac{\partial \mathscr{L}(x)}{\partial \varphi_{\alpha,\mu}(x)} \delta\varphi_{\alpha,\mu}(x) \right] . \tag{3.14}$$

Using the commutativity of a variation and a derivative, viz.,

$$\delta\varphi_{\alpha,\mu}(x) = \partial_\mu \delta\varphi_\alpha(x) , \tag{3.15}$$

we perform a partial integral, using Gauss' theorem:

$$\int_{\Omega} \mathrm{d}^4 x \partial_\mu A_\mu = \int_{\partial\Omega} \mathrm{d}\sigma_\mu A_\mu , \tag{3.16}$$

where $\mathrm{d}\sigma_\mu = \mathrm{d}^4 x / \mathrm{d}x_\mu$. Thus,

$$\delta \int_{\Omega} \mathrm{d}^4 x \mathscr{L}(x) = \int_{\Omega} \mathrm{d}^4 x \left[\frac{\partial \mathscr{L}(x)}{\partial \varphi_\alpha(x)} - \frac{\partial}{\partial x_\mu} \left(\frac{\partial \mathscr{L}(x)}{\partial \varphi_{\alpha,\mu}(x)} \right) \right] \delta\varphi_\alpha(x)$$
$$+ \int_{\partial\Omega} \mathrm{d}\sigma_\mu \frac{\partial \mathscr{L}(x)}{\partial \varphi_{\alpha,\mu}(x)} \delta\varphi_\alpha(x) . \tag{3.17}$$

From the assumption that the variations of the field vanish on the boundary surface, the second term vanishes. Using the fact that the variations $\delta\varphi_\alpha$ are arbitrary, the first term leads to

$$[\mathscr{L}]_{\varphi_\alpha} = \frac{\partial \mathscr{L}}{\partial \varphi_\alpha} - \frac{\partial}{\partial x_\mu} \left(\frac{\partial \mathscr{L}}{\partial \varphi_{\alpha,\mu}} \right) = 0 . \tag{3.18}$$

These are the Euler–Lagrange equations. We call the left-hand side an *Euler derivative*.

Example: Klein–Gordon Equation

The following equation is the *Klein–Gordon equation*:

$$(\Box - m^2)\varphi = 0\,, \tag{3.19}$$

where $\Box = \partial_\mu^2 = \Delta - \partial_0^2$. We choose the Lagrangian density

$$\mathscr{L} = -\frac{1}{2}\left[(\partial_\mu\varphi)^2 + m^2\varphi^2\right]. \tag{3.20}$$

In the following we shall use the Pauli metric. That is to say, we define an inner product of vectors as

$$x \cdot y = \boldsymbol{x}\boldsymbol{y} - x_0 y_0\,. \tag{3.21}$$

Thus, the field equation derived from (3.20) becomes

$$[\mathscr{L}]_\varphi = -m^2\varphi + \partial_\mu^2\varphi = (\Box - m^2)\varphi = 0\,,$$

and (3.19) is reproduced. The kinetic term in the Lagrangian density is $\dot{\varphi}^2/2$ due to the overall minus sign, and since this term is positive, the energy in the whole system is positive-definite, which will be explained below.

3.1.3 Hamiltonian

To obtain the Hamiltonian for fields we again apply the translation from point-particle systems. We rewrite the defining identity of the Hamiltonian (2.13) using the rule given before:

$$H = \sum_\alpha \int \mathrm{d}^3x\, \pi_\alpha(x)\dot{\varphi}_\alpha(x) - \int \mathrm{d}^3x \mathscr{L}(x) \equiv \int \mathrm{d}^3x \mathscr{H}(x)\,, \tag{3.22}$$

$$\mathscr{H}(x) = \sum_\alpha \pi_\alpha(x)\dot{\varphi}_\alpha(x) - \mathscr{L}(x)\,, \tag{3.23}$$

where H is the Hamiltonian and $\mathscr{H}$ is the Hamiltonian density.

Example: Hamiltonian for Real Scalar Field

Rewriting the previous example (3.20),

$$\mathscr{L} = \frac{1}{2}\left[\dot{\varphi}^2 - (\nabla\varphi)^2 - m^2\varphi^2\right] . \tag{3.24}$$

From this, the canonical variable conjugate to φ is

$$\pi = \frac{\partial \mathscr{L}}{\partial \dot{\varphi}} = \dot{\varphi} . \tag{3.25}$$

Therefore,

$$\mathscr{H} = \pi\dot{\varphi} - \mathscr{L} = \frac{1}{2}\left[\pi^2 + (\nabla\varphi)^2 + m^2\varphi^2\right] . \tag{3.26}$$

This is clearly positive-definite.

As we have seen above, the canonical formalism for point-particle systems can be extended to field theory by using the simple translation rule.

3.2 Relativistic Generalization of the Canonical Equations

The canonical equation in quantum mechanics, viz.,

$$\frac{\mathrm{d}\mathscr{O}}{\mathrm{d}t} = \mathrm{i}[H, \mathscr{O}] , \tag{3.27}$$

holds directly for field theory. Since P_0 is the time component of the energy–momentum vector, writing $P_4 = \mathrm{i}P_0 = \mathrm{i}H$ and considering $\mathscr{O}(x)$ as a polynomial in the field operators, we can write (3.27) in the form

$$\left[P_4, \mathscr{O}(x)\right] = \mathrm{i}\frac{\partial \mathscr{O}(x)}{\partial x_4} . \tag{3.28}$$

If a theory is relativistically invariant, we can generalize this equation to

$$\left[\mathscr{O}(x), P_\mu\right] = \frac{1}{\mathrm{i}}\frac{\partial}{\partial x_\mu}\mathscr{O}(x) , \tag{3.29}$$

so there must be spatial components P_k ($k = 1, 2, 3$) satisfying this equation.

In fact, it can be shown that the proper choice for P_k is

$$P_k = -\int \mathrm{d}^3x \sum_\alpha \pi_\alpha(x)\partial_k\varphi_\alpha(x) . \tag{3.30}$$

As an example, taking φ_α or π_α as $\mathscr{O}$, we find

$$\left[\varphi_\alpha(x), P_k\right] = -\mathrm{i}\int \mathrm{d}^3x'\delta^3(x-x')\partial_k\varphi_\alpha(x') = \frac{1}{\mathrm{i}}\frac{\partial}{\partial x_k}\varphi_\alpha(x)$$

and

$$\left[\pi_\alpha(x), P_k\right] = \mathrm{i}\int \mathrm{d}^3x'\pi_\alpha(x')\frac{\partial}{\partial x'_k}\delta^3(x-x') = \frac{1}{\mathrm{i}}\frac{\partial}{\partial x_k}\pi_\alpha(x)\ .$$

In this way, we obtain the four-dimensional energy–momentum vector P_μ. These four components commute with each other:

$$[P_\mu, P_\nu] = 0\ . \tag{3.31}$$

Now we sandwich a general operator $\mathscr{O}(x)$ between the eigenstates $|a\rangle$ and $|b\rangle$ of P_μ, which satisfy

$$P_\mu|a\rangle = p_\mu|a\rangle\ , \quad P_\mu|b\rangle = q_\mu|b\rangle. \tag{3.32}$$

This gives

$$\frac{\partial}{\partial x_\mu}\langle a|\mathscr{O}(x)|b\rangle = \mathrm{i}\langle a|[\mathscr{O}(x), P_\mu]|b\rangle = -\mathrm{i}(p_\mu - q_\mu)\langle a|\mathscr{O}(x)|b\rangle\ . \tag{3.33}$$

The solution of this differential equation is

$$\langle a|\mathscr{O}(x)|b\rangle = \langle a|\mathscr{O}(0)|b\rangle \mathrm{e}^{-\mathrm{i}(p-q)\cdot x}\ , \tag{3.34}$$

and the x-dependence of $\mathscr{O}(x)$ can thus be determined. This relation is important and will be used often later.

3.3 Quantization of the Real Scalar Field

Since the real scalar field has already appeared as an example, we now proceed to quantize it. First, we consider a non-interacting case. To choose the Lagrangian density $\mathscr{L}$, we take into account the following conditions:

1. $\mathscr{L}$ is invariant under the Lorentz transformations.
2. The Hamiltonian density $\mathscr{H}$ calculated from $\mathscr{L}$ is positive-definite.
3. In the non-interacting case, the fields satisfy the Klein–Gordon equation.

We now examine condition 1 above in detail. It is enough that $\mathscr{L}$ should be invariant only under the proper Lorentz transformations, i.e., the transformations generated by accumulating infinitesimal Lorentz transformations. Regarding condition 3, if we replace p_μ by the operator $-\mathrm{i}\partial/\partial x_\mu$ in the Einstein relation

$$p^2 + m^2 = p_\mu^2 + m^2 = 0 \tag{3.35}$$

and apply the result to a field, we obtain the Klein–Gordon equation. This is the condition associating a given free field to a particle of given mass.

The Lagrangian density (3.20) is a scalar, i.e., Lorentz invariant. In addition, the Hamiltonian density (3.26) calculated from it is positive-definite. The field equation derived using the principle of least action becomes the Klein–Gordon equation itself. Thus we can say that (3.20) is the correct Lagrangian density. The quantity canonically conjugate to φ is the time derivative of ϕ, as given by (3.25).

The fields are quantized using the equal-time commutation relations:

$$[\varphi(x), \varphi(x')] = [\pi(x), \pi(x')] = 0 \tag{3.36}$$

and

$$[\varphi(x), \pi(x')] = \mathrm{i}\delta^3(x - x') \ , \tag{3.37}$$

where $x_0 = x'_0$.

Next, we put the scalar field system inside a cube with volume $V = L^3$ and impose a periodic condition on it. Then we expand $\varphi(x)$ in a Fourier series. In this box, the set of functions

$$\left\{\mathrm{e}^{\mathrm{i}\boldsymbol{p}\boldsymbol{x}}\right\} \tag{3.38}$$

forms a complete set, where $p_j = n_j(2\pi/L)$ and the n_j are integers, so we get the expansion

$$\varphi(x) = \frac{1}{\sqrt{V}} \sum_{\boldsymbol{p}} \mathrm{e}^{\mathrm{i}\boldsymbol{p}\boldsymbol{x}} c(\boldsymbol{p}, t) \ . \tag{3.39}$$

Substituting this expansion into the Klein–Gordon equation, we get the equation for the coefficient functions $c(\boldsymbol{p}, t)$:

$$\ddot{c}(\boldsymbol{p}, t) = -(\boldsymbol{p}^2 + m^2)c(\boldsymbol{p}, t) \ . \tag{3.40}$$

Then, setting $p_0 = \sqrt{\boldsymbol{p}^2 + m^2}$, the general solution for the equation above is

$$c(\boldsymbol{p}, t) = c_1(\boldsymbol{p})\mathrm{e}^{-\mathrm{i}p_0 t} + c_2(\boldsymbol{p})\mathrm{e}^{\mathrm{i}p_0 t} \ . \tag{3.41}$$

Therefore,

$$\varphi(x) = \frac{1}{\sqrt{V}} \sum_{p} \mathrm{e}^{\mathrm{i}\boldsymbol{px}} \big[c_1(\boldsymbol{p}) \mathrm{e}^{-\mathrm{i}p_0 t} + c_2(\boldsymbol{p}) \mathrm{e}^{\mathrm{i}p_0 t} \big]$$

$$= \frac{1}{\sqrt{V}} \sum_{p} \big[\mathrm{e}^{\mathrm{i}\boldsymbol{px} - \mathrm{i}p_0 t} c_1(\boldsymbol{p}) + \mathrm{e}^{-\mathrm{i}\boldsymbol{px} + \mathrm{i}p_0 t} c_2(-\boldsymbol{p}) \big] . \tag{3.42}$$

In addition, φ is real or Hermitian as an operator, so we have

$$c_2(\boldsymbol{p}) = c_1^{\dagger}(-\boldsymbol{p}) . \tag{3.43}$$

From now on we shall simply write c_1 as c and use the notation $p \cdot x = \boldsymbol{px} - p_0 t$. Therefore, φ and π are

$$\varphi(x) = \frac{1}{\sqrt{V}} \sum_{p} \big[\mathrm{e}^{\mathrm{i}p \cdot x} c(\boldsymbol{p}) + \mathrm{e}^{-\mathrm{i}p \cdot x} c^{\dagger}(\boldsymbol{p}) \big] \tag{3.44}$$

and

$$\pi(x) = \frac{1}{\sqrt{V}} \sum_{p} \big[-\mathrm{i}p_0 \mathrm{e}^{\mathrm{i}p \cdot x} c(\boldsymbol{p}) + \mathrm{i}p_0 \mathrm{e}^{-\mathrm{i}p \cdot x} c^{\dagger}(\boldsymbol{p}) \big] . \tag{3.45}$$

Substituting these expansions into the equal-time commutation relations (3.36), we find

$$[c(\boldsymbol{p}), c(\boldsymbol{q})] = \big[c^{\dagger}(\boldsymbol{p}), c^{\dagger}(\boldsymbol{q}) \big] = 0 , \tag{3.46}$$

$$\big[c(\boldsymbol{p}), c^{\dagger}(\boldsymbol{q}) \big] + \big[c^{\dagger}(-\boldsymbol{p}), c(-\boldsymbol{q}) \big] = 0 . \tag{3.47}$$

Next, we rewrite (3.37) as

$$[\varphi(x), \pi(x')] = \frac{\mathrm{i}}{V} \sum_{p} \mathrm{e}^{\mathrm{i}\boldsymbol{p}(\boldsymbol{x}-\boldsymbol{x}')} ,$$

where $x_0' = x_0$. Substituting the expansions (3.44) and (3.45) into this equation, we have

$$\mathrm{i}p_0 \big[c(\boldsymbol{p}), c^{\dagger}(\boldsymbol{q}) \big] - \mathrm{i}p_0 \big[c^{\dagger}(-\boldsymbol{p}), c(-\boldsymbol{q}) \big] = \mathrm{i}\delta_{\boldsymbol{p},\boldsymbol{q}} ,$$

or

$$\big[c(\boldsymbol{p}), c^{\dagger}(\boldsymbol{q}) \big] - \big[c^{\dagger}(-\boldsymbol{p}), c(-\boldsymbol{q}) \big] = \frac{1}{p_0} \delta_{\boldsymbol{p},\boldsymbol{q}} . \tag{3.48}$$

Combining this equation and (3.47) yields

$$\left[c(\boldsymbol{p}), c^\dagger(\boldsymbol{q})\right] = \frac{1}{2p_0}\delta_{\boldsymbol{p},\boldsymbol{q}} \ . \tag{3.49}$$

Then, instead of $c(\boldsymbol{p})$, we define $a(\boldsymbol{p})$ by

$$c(\boldsymbol{p}) = \frac{1}{\sqrt{2p_0}} a(\boldsymbol{p}) \ , \quad c^\dagger(\boldsymbol{p}) = \frac{1}{\sqrt{2p_0}} a^\dagger(\boldsymbol{p}) \ . \tag{3.50}$$

The commutation relations simplify to

$$[a(\boldsymbol{p}), a(\boldsymbol{q})] = \left[a^\dagger(\boldsymbol{p}), a^\dagger(\boldsymbol{q})\right] = 0 \tag{3.51}$$

and

$$\left[a(\boldsymbol{p}), a^\dagger(\boldsymbol{q})\right] = \delta_{\boldsymbol{p},\boldsymbol{q}} \ . \tag{3.52}$$

Substituting (3.50) into (3.44), we obtain

$$\varphi(x) = \sum_{\boldsymbol{p}} \frac{1}{\sqrt{2p_0 V}} \left[\mathrm{e}^{\mathrm{i}p\cdot x} a(\boldsymbol{p}) + \mathrm{e}^{-\mathrm{i}p\cdot x} a^\dagger(\boldsymbol{p})\right] \ . \tag{3.53}$$

In order to interpret the operators a and $a^\dagger$, we represent the energy and the momentum in a total system by a and $a^\dagger$:

$$H = \int_V \mathrm{d}^3x \, \mathscr{H}(x) = \frac{1}{2} \sum_{\boldsymbol{p}} p_0 \left[a(\boldsymbol{p}) a^\dagger(\boldsymbol{p}) + a^\dagger(\boldsymbol{p}) a(\boldsymbol{p})\right] , \tag{3.54}$$

$$P_k = -\int_V \mathrm{d}^3x \, \pi(x) \partial_k \varphi(x) = \frac{1}{2} \sum_{\boldsymbol{p}} p_k \left[a(\boldsymbol{p}) a^\dagger(\boldsymbol{p}) + a^\dagger(\boldsymbol{p}) a(\boldsymbol{p})\right] . \tag{3.55}$$

Then we define the following so-called number density operator $n(\boldsymbol{p})$:

$$n(\boldsymbol{p}) = a^\dagger(\boldsymbol{p}) a(\boldsymbol{p}). \tag{3.56}$$

Since (3.52) implies $a(\boldsymbol{p}) a^\dagger(\boldsymbol{p}) = n(\boldsymbol{p}) + 1$, (3.54) and (3.55) become

$$H = \sum_{\boldsymbol{p}} p_0 \left[n(\boldsymbol{p}) + \frac{1}{2}\right] , \quad P_k = \sum_{\boldsymbol{p}} p_k \left[n(\boldsymbol{p}) + \frac{1}{2}\right] . \tag{3.57}$$

What we would like to emphasize here is that the eigenvalues of $n(\boldsymbol{p})$ (which can be written as n by fixing $\boldsymbol{p}$) become non-negative integers like 0, 1, 2, ... To prove this, we start with $[a, a^\dagger] = 1$. Therefore,

$$[n, a] = -a\ , \quad [n, a^\dagger] = a^\dagger\ . \tag{3.58}$$

Here we describe the eigenstate of n as $|n'\rangle$. n' is the eigenvalue of the operator n, satisfying

$$n|n'\rangle = n'|n'\rangle\ . \tag{3.59}$$

Then, from (3.58),

$$\langle n'|[n, a^\dagger]|n''\rangle = (n' - n'')\langle n'|a^\dagger|n''\rangle = \langle n'|a^\dagger|n''\rangle\ .$$

Thus, the matrix elements of $a^\dagger$ are nonzero only if $n' = n'' + 1$. In other words, $a^\dagger$ is the operator raising the eigenvalues of n by 1. At the same time, a is the operator lowering the eigenvalues of n by 1. Therefore, the eigenvalues of n for the states connected by $a^\dagger$ or a become

$$\ldots,\ n_1 - 2\ ,\ n_1 - 1\ ,\ n_1\ ,\ n_1 + 1\ ,\ n_1 + 2\ , \ldots\ , \tag{3.60}$$

if we start with a given eigenvalue n_1. According to the definition (3.56), the eigenvalues of n cannot be negative, so there must be a minimum value for the series above. Denoting it by n_0, (3.60) becomes

$$n_0\ ,\ n_0 + 1\ ,\ n_0 + 2\ ,\ \ldots\ . \tag{3.61}$$

We find

$$\begin{aligned}\langle n'|n|n'\rangle &= \sum_{n''}\langle n'|a^\dagger|n''\rangle\langle n''|a|n'\rangle \\ &= \langle n'|a^\dagger|n' - 1\rangle\langle n' - 1|a|n'\rangle \\ &= \left|\langle n' - 1|a|n'\rangle\right|^2 = n'\ .\end{aligned} \tag{3.62}$$

Therefore, we can choose the undetermined phase factor to be

$$\langle n' - 1|a|n'\rangle = \langle n'|a^\dagger|n' - 1\rangle = \sqrt{n'}\ . \tag{3.63}$$

To find n_0, we impose the condition

$$a|n_0\rangle = 0\ . \tag{3.64}$$

Comparing it with (3.63), we can immediately find

$$n_0 = 0 . \tag{3.65}$$

This completely determines the eigenvalues of n. This n is called a *particle number operator*.

Going back to (3.57),

$$H = \sum_{p} p_0 n(\boldsymbol{p}) + \text{c-number} , \quad P_k = \sum_{p} p_k n(\boldsymbol{p}) + \text{c-number} . \tag{3.66}$$

We drop the c-numbers for two reasons:

1. The choice of operator orderings when going from classical mechanics to quantum mechanics is not unique. For instance, in (3.54), if we go to quantum mechanics after rewriting $aa^\dagger$ as $a^\dagger a$ within the framework of classical mechanics, then these c-numbers do not appear.
2. Only the differences between the eigenvalues of these operators can actually be observed.

If we drop the c-numbers, unifying the equations (3.66), we obtain

$$P_\mu = \sum_{p} p_\mu n(\boldsymbol{p}) . \tag{3.67}$$

The interpretation of this is that $n(\boldsymbol{p})$ stands for the number of particles with momentum $\boldsymbol{p}$ and energy p_0. This interpretation is possible because, as mentioned before, the eigenvalues of $n(\boldsymbol{p})$ are non-negative integers. Thus, although we have considered the scalar field as a wave satisfying the wave equation, its particle-like features have shown up as a consequence of imposing the canonical commutation relations to quantize it. This is nothing but an extension of Einstein's photon hypothesis to the scalar field. In addition, as is clear from (3.35), m stands for the mass of the quantum of the scalar field. As we call n the particle number operator, a and $a^\dagger$ are called the *annihilation operator* and *creation operator*, respectively.

3.4 Quantization of the Complex Scalar Field

Complex scalar fields are quantized in a similar manner to real scalar fields. The first method is to quantize a complex scalar field by decomposing it into two real scalar fields. We choose the Lagrangian density

$$\mathscr{L} = -\left(\partial_\mu \varphi^\dagger \cdot \partial_\mu \varphi + m^2 \varphi^\dagger \varphi\right) , \tag{3.68}$$

where $\varphi^\dagger$ is the operator Hermitian conjugate to φ. In classical theory, φ^* is used as the complex conjugate of φ. However, we use the notation $\varphi^\dagger$ because φ becomes an operator through quantization. Then, making the change of variables

$$\varphi = \frac{1}{\sqrt{2}}(\varphi_1 - \mathrm{i}\varphi_2)\,, \quad \varphi^\dagger = \frac{1}{\sqrt{2}}(\varphi_1 + \mathrm{i}\varphi_2)\,, \tag{3.69}$$

the Lagrangian density can be rewritten in the form

$$\mathscr{L} = -\frac{1}{2}\left[(\partial_\mu\varphi_1)^2 + m^2\varphi_1^2\right] - \frac{1}{2}\left[(\partial_\mu\varphi_2)^2 + m^2\varphi_2^2\right]. \tag{3.70}$$

Since these are two real scalar fields, the corresponding quantization method is the same as in the previous section.

The second method is to quantize complex scalar fields without decomposing them into real scalar fields. To begin with, we define the canonical conjugate quantities of φ and $\varphi^\dagger$ as follows:

$$\pi^\dagger = \frac{\partial\mathscr{L}}{\partial\dot{\varphi}}\,, \quad \pi = \frac{\partial\mathscr{L}}{\partial\dot{\varphi}^\dagger}\,. \tag{3.71}$$

As a result, we have

$$\pi^\dagger = \dot{\varphi}^\dagger\,, \quad \pi = \dot{\varphi}\,. \tag{3.72}$$

The equal-time commutation relations that do not vanish and the Fourier expansions are given by

$$\left[\varphi(x), \pi^\dagger(x')\right] = \left[\varphi^\dagger(x), \pi(x')\right] = \mathrm{i}\delta^3(x - x')\,, \quad \text{where } x_0 = x_0', \tag{3.73}$$

$$\varphi(x) = \sum_{\boldsymbol{p}} \frac{1}{\sqrt{2p_0V}}\left[a(\boldsymbol{p})\mathrm{e}^{\mathrm{i}p\cdot x} + b^\dagger(\boldsymbol{p})\mathrm{e}^{-\mathrm{i}p\cdot x}\right], \tag{3.74}$$

and

$$\varphi^\dagger(x) = \sum_{\boldsymbol{p}} \frac{1}{\sqrt{2p_0V}}\left[b(\boldsymbol{p})\mathrm{e}^{\mathrm{i}p\cdot x} + a^\dagger(\boldsymbol{p})\mathrm{e}^{-\mathrm{i}p\cdot x}\right]. \tag{3.75}$$

In this way, when quantizing the complex scalar fields, we need two types of operators, a and b. Expressing the equal-time commutation relations in terms of a and b and repeating the calculation in the previous section, we obtain the following equations:

$$\left[a(\boldsymbol{p}), a^\dagger(\boldsymbol{q})\right] = \left[b(\boldsymbol{p}), b^\dagger(\boldsymbol{q})\right] = \delta_{\boldsymbol{p},\boldsymbol{q}}\,, \tag{3.76}$$

where other commutators such as $[a(\boldsymbol{p}), a(\boldsymbol{q})]$ all vanish. Therefore, introducing

$$n_+(\boldsymbol{p}) = a^\dagger(\boldsymbol{p})a(\boldsymbol{p}) \,, \quad n_-(\boldsymbol{p}) = b^\dagger(\boldsymbol{p})b(\boldsymbol{p}) \,, \tag{3.77}$$

both operators become particle number operators. They both correspond to quanta with mass m, but there is a quantum number that distinguishes between the two types of quanta, like the electric charge, for example. π^+ and π^-, or K^0 and $\bar{K}^0$, are examples of particle–antiparticle pairs. As for the latter, it is convenient to use a real scalar representation, and the corresponding particles are K_1^0 and K_2^0. The two types of quanta, being complex conjugates, are said to be *antiparticles* of each other. This non-interacting theory is invariant under the interchange of the two fields, and the interchange can be expressed as a unitary transformation. Writing it as $\mathscr{C}$, we have

$$\mathscr{C}^{-1}\varphi(x)\mathscr{C} = \mathrm{e}^{\mathrm{i}\alpha}\varphi^\dagger(x) \,, \quad \mathscr{C}^{-1}\varphi^\dagger(x)\mathscr{C} = \mathrm{e}^{-\mathrm{i}\alpha}\varphi(x) \,, \tag{3.78}$$

where α is a real number. We call this transformation *charge conjugation*. Since it commutes with P_μ, a particle and its antiparticle have the same mass. By changing the phase of φ in a suitable way in the above equation, we can choose the phase factor $\mathrm{e}^{\mathrm{i}\alpha}$ to be ± 1.

3.5 Dirac Equation

Many particles with spin 1/2 obey Dirac's equation. Here we will discuss only the properties relevant to quantization.[1]

To translate the non-relativistic Schrödinger equation into the relativistic Dirac equation, the following conditions are used:

1. The wave function satisfies the Klein–Gordon equation.
2. A probability density ρ and a current density of probability $\boldsymbol{j}$ are defined as quadratic forms of the wave function, satisfying the following equation of continuity:

$$\frac{\partial \rho}{\partial t} + \operatorname{div} \boldsymbol{j} = 0 \,, \quad \rho \geq 0 \,. \tag{3.79}$$

[1] The properties of the Dirac equation are discussed in detail in my Japanese book [78].

To obtain Pauli's two-component wave function in a non-relativistic approximation, the wave function must be multicomponent. We write this as follows:

$$\psi = \begin{pmatrix} \psi_1 \\ \psi_2 \\ \vdots \end{pmatrix} . \tag{3.80}$$

Therefore, ρ should be written as a quadratic form in ψ or, more precisely, as the following bilinear form in ψ and $\psi^\dagger$:

$$\rho = \sum_{r,s} \psi_r^* a_{rs} \psi_s \ . \tag{3.81}$$

In order for this to be positive-definite, the coefficients a_{rs} should form a positive-definite Hermitian matrix that can be diagonalized by a suitable unitary transformation. Furthermore, since we can rewrite it as a unit matrix by a scale transformation, without loss of generality, we can choose the following form from the beginning:

$$\rho = \sum_r \psi_r^* \psi_r = \psi^\dagger \psi \ . \tag{3.82}$$

Here, $\psi^\dagger$ is the Hermitian conjugate of ψ having the form of a $1 \times n$ matrix.

In terms of the Hamiltonian H, the equation of motion can be written in the form

$$\mathrm{i}\frac{\partial}{\partial t}\psi = H\psi \ , \quad \text{or} \quad \left(\mathrm{i}\frac{\partial}{\partial t} - H\right)\psi = 0 \ . \tag{3.83}$$

From this equation we can obtain the second order differential equation with respect to time:

$$\left(\frac{\partial^2}{\partial t^2} + H^2\right)\psi = 0. \tag{3.84}$$

This coincides with the Klein–Gordon equation if we set

$$H^2 = m^2 + \boldsymbol{p}^2 = m^2 - \nabla^2 \ . \tag{3.85}$$

Assuming that H is linear in p, the most general form becomes

$$H = \boldsymbol{\alpha} \cdot \boldsymbol{p} + m\beta = \alpha_x p_x + \alpha_y p_y + \alpha_z p_z + m\beta \ , \tag{3.86}$$

where $\boldsymbol{\alpha}$ and β are suitable matrices. Thus, by using the matrix elements of $\boldsymbol{\alpha}$ and β, the equation of motion becomes

$$\mathrm{i}\frac{\partial}{\partial t}\psi_r = \sum_s \left(\boldsymbol{\alpha}_{rs}\frac{1}{\mathrm{i}}\cdot\nabla + m\beta_{rs}\right)\psi_s \ . \tag{3.87}$$

We now determine the properties of α and β:

1. Since H is Hermitian, $\boldsymbol{\alpha}$ and β are Hermitian matrices.
2. H must satisfy (3.85).

Substituting (3.86) into (3.85), we arrive at the following relations:

$$\alpha_x^2 = \alpha_y^2 = \alpha_z^2 = \beta^2 = \mathbf{1} \quad \text{(unit matrix)} \ ,$$

$$\alpha_x\alpha_y + \alpha_y\alpha_x = \ldots = 0 \ , \qquad \alpha_x\beta + \beta\alpha_x = \ldots = 0 \ .$$

Combining the above relations, we obtain the alternative form

$$\alpha_k\alpha_l + \alpha_l\alpha_k = 2\delta_{kl}\mathbf{1} \ , \quad \alpha_k\beta + \beta\alpha_k = 0 \ , \quad \beta^2 = \mathbf{1} \ . \tag{3.88}$$

Next, we derive the equation of continuity. To this end, we set

$$\rho = \psi^\dagger\psi \ , \quad \boldsymbol{j} = \psi^\dagger\boldsymbol{\alpha}\psi \ . \tag{3.89}$$

Taking the complex conjugate of (3.87) and using the hermiticity of $\boldsymbol{\alpha}$ and β, we find

$$\mathrm{i}\frac{\partial}{\partial t}\psi_r^* = \sum_s \left(\frac{1}{\mathrm{i}}\nabla\psi_s^*\cdot\boldsymbol{\alpha}_{sr} - m\psi_s^*\beta_{sr}\right) \ . \tag{3.90}$$

Combining (3.87) and (3.90), the equation of continuity can be derived:

$$\begin{aligned}
\mathrm{i}\frac{\partial\rho}{\partial t} &= \mathrm{i}\sum_r\left(\frac{\partial\psi_r^*}{\partial t}\psi_r + \psi_r^*\frac{\partial\psi_r}{\partial t}\right) \\
&= \sum_{s,r}\left(\frac{1}{\mathrm{i}}\nabla\psi_s^*\cdot\boldsymbol{\alpha}_{sr}\psi_r - m\psi_s^*\beta_{sr}\psi_r\right) + \sum_{r,s}\left(\psi_r^*\boldsymbol{\alpha}_{rs}\frac{1}{\mathrm{i}}\cdot\nabla\psi_s + m\psi_r^*\beta_{rs}\psi_s\right) \\
&= -\mathrm{i}\nabla\psi^\dagger\cdot\boldsymbol{\alpha}\psi - \mathrm{i}\psi^\dagger\boldsymbol{\alpha}\cdot\nabla\psi \\
&= -\mathrm{i}\,\mathrm{div}(\psi^\dagger\boldsymbol{\alpha}\psi) \\
&= -\mathrm{i}\,\mathrm{div}\boldsymbol{j} \ .
\end{aligned}$$

Pauli Representation

To satisfy the commutation relations (3.88), the dimension of the matrices of $\boldsymbol{\alpha}$ and β must be at least four. In order to derive explicit representations, Pauli introduced the four-dimensional σ-matrices and the ρ-matrix:

$$\sigma_j = \begin{pmatrix} \boldsymbol{\sigma}_j & \mathbf{0} \\ \mathbf{0} & \boldsymbol{\sigma}_j \end{pmatrix}, \quad j = 1, 2, 3 , \tag{3.91}$$

where $\boldsymbol{\sigma}_j$ are the two-dimensional Pauli matrices, $\mathbf{0}$ is the two-dimensional zero matrix, and σ_j are four-dimensional Pauli matrices. It is clear that the relations among them are

$$\sigma_1\sigma_2 = -\sigma_2\sigma_1 = \mathrm{i}\sigma_3 , \quad \text{and its cyclic permutations}, \tag{3.92}$$

$$\sigma_1^2 = \sigma_2^2 = \sigma_3^2 = \mathbf{1} \quad \text{(four-dimensional unit matrix)} . \tag{3.93}$$

Then the ρ-matrix is defined by

$$\rho_1 = \begin{pmatrix} \mathbf{0} & \boldsymbol{I} \\ \boldsymbol{I} & \mathbf{0} \end{pmatrix}, \quad \rho_2 = \begin{pmatrix} \mathbf{0} & -\mathrm{i}\boldsymbol{I} \\ \mathrm{i}\boldsymbol{I} & \mathbf{0} \end{pmatrix}, \quad \rho_3 = \begin{pmatrix} \boldsymbol{I} & \mathbf{0} \\ \mathbf{0} & -\boldsymbol{I} \end{pmatrix}, \tag{3.94}$$

where $\boldsymbol{I}$ is the two-dimensional unit matrix. The properties (3.92) and (3.93) also hold for ρ:

$$\rho_1\rho_2 = -\rho_2\rho_1 = \mathrm{i}\rho_3 , \quad \text{and its cyclic permutations}, \tag{3.95}$$

$$\rho_1^2 = \rho_2^2 = \rho_3^2 = \mathbf{1} . \tag{3.96}$$

The three σ-matrices and the three ρ-matrices commute with each other. It is easily shown that $[\sigma_i, \rho_j] = 0\ (i, j = 1, 2, 3)$.

Then, in Pauli's representation, $\boldsymbol{\alpha}$ and β are given by

$$\alpha_j = \rho_1\sigma_j = \begin{pmatrix} \mathbf{0} & \boldsymbol{\sigma}_j \\ \boldsymbol{\sigma}_j & \mathbf{0} \end{pmatrix}, \quad \beta = \rho_3 . \tag{3.97}$$

$\boldsymbol{\alpha}$ and β constructed in this way clearly satisfy the conditions (3.88).

Weyl Equation

So far the discussion about the Dirac equation has been based on the assumption that $m \neq 0$. Although in that case the wave function has four components, when $m = 0$ we can make do with two components. In this case, since

$$H = \boldsymbol{\alpha} \cdot \boldsymbol{p} , \tag{3.98}$$

β is not needed and the required commutation relation is only the first equation in (3.88), i.e., the commutation relations among the α's. There are two-dimensional representations satisfying this, namely,

$$\alpha_k = \pm\boldsymbol{\sigma}_k \ . \tag{3.99}$$

Corresponding to this representation, the equation of motion becomes the two-component equation

$$\mathrm{i}\frac{\partial}{\partial t}\psi = \pm\boldsymbol{\sigma}\cdot\boldsymbol{p}\psi \ . \tag{3.100}$$

This is called *Weyl's equation*. At first, this equation was not taken seriously because it is not compatible with the law of conservation of parity. However, after the proposal of the non-conservation of parity by Lee and Yang in [45], it attracted a lot of attention. Nowadays, it is used as the equation for massless neutrinos. The connection with the non-conservation of parity will be mentioned in the next section.

3.6 Relativistic Transformations of Dirac's Wave Function

Field operators generally possess definite properties under the action of the Lorentz transformations. For example, under the Lorentz transformation

$$x_\mu \to x'_\mu = \sum_\nu a_{\mu\nu}x_\nu \ , \tag{3.101}$$

a scalar field transforms in such a way that

$$\varphi'(x') = \varphi(x) \ . \tag{3.102}$$

We have changed φ to φ' because the functional form of the field changes upon a transformation of the coordinates. In other words, in the case of the scalar field, the change of variable and the change of functional form cancel each other, and as a whole the field is invariant. Whether we write x or x', both labels do in fact stand for the same point. On the other hand, when we consider a vector field $A_\mu(x)$, its transformation property is

$$A_\mu(x) \to A'_\mu(x') = \sum_\nu a_{\mu\nu}A_\nu(x) \ . \tag{3.103}$$

Here, the problem is to see how the four-component Dirac field ψ transforms under Lorentz transformations. We assume the following linear transformation:

$$\psi \rightarrow \psi' = S\psi \,. \tag{3.104}$$

More precisely,

$$\psi_r(x) \rightarrow \psi'_r(x') = \sum_s S_{rs}\psi_s(x) \,. \tag{3.105}$$

We define the matrix S so that this transformation is compatible with the Dirac equation. Of course, S must be one of the representations of the Lorentz transformation (3.101). If such a matrix S exists, then the Dirac equation has exactly the same form in every inertial system, i.e., it is Lorentz covariant.

In order to determine the transformation, we first rewrite Dirac's equation. Multiplying

$$\left(-\mathrm{i}\frac{\partial}{\partial t} + \frac{1}{\mathrm{i}}\boldsymbol{\alpha}\cdot\nabla + m\beta\right)\psi = 0$$

from the left by β, we obtain

$$\left(\beta\frac{\partial}{\partial x_4} + \frac{1}{\mathrm{i}}\beta\boldsymbol{\alpha}\cdot\nabla + m\right)\psi = 0 \,. \tag{3.106}$$

Then we define the γ-matrices by

$$\gamma_j = \frac{1}{\mathrm{i}}\beta\alpha_j \,, \quad j = 1, 2, 3 \,, \qquad \gamma_4 = \beta \,. \tag{3.107}$$

The four γ-matrices are Hermitian, and the commutation relations (3.88) can be expressed in the following unified form:

$$\gamma_\mu\gamma_\nu + \gamma_\nu\gamma_\mu = 2\delta_{\mu\nu}\mathbf{1} \,, \quad \mu, \nu = 1, 2, 3, 4 \,. \tag{3.108}$$

Then, the Dirac equation can be written as

$$\left(\gamma_\mu\frac{\partial}{\partial x_\mu} + m\right)\psi = 0 \,. \tag{3.109}$$

It remains to prove that, if we start with the equation above and determine S properly, then the equation

$$\left(\gamma_\mu\frac{\partial}{\partial x'_\mu} + m\right)\psi' = 0 \tag{3.110}$$

will also hold. First, we consider the Lorentz transformation

$$x'_\mu = \sum_\nu a_{\mu\nu} x_\nu \ , \quad \sum_\mu a_{\mu\rho} a_{\mu\sigma} = \delta_{\rho\sigma} \ ,$$

$$x_\nu = \sum_\mu a_{\mu\nu} x'_\mu \ , \quad \sum_\mu a_{\rho\mu} a_{\sigma\mu} = \delta_{\rho\sigma} \ . \tag{3.111}$$

Here, $(a_{\mu\nu})$ is an orthogonal matrix, and a_{j4} and a_{4j} $(j = 1, 2, 3)$ are pure imaginary numbers. Substituting

$$\frac{\partial}{\partial x'_\mu} = \sum_\nu \frac{\partial x_\nu}{\partial x'_\mu} \frac{\partial}{\partial x_\nu} = \sum_\nu a_{\mu\nu} \frac{\partial}{\partial x_\nu}$$

into (3.110) and using (3.104), we obtain

$$\sum_\mu \sum_\nu \gamma_\mu S a_{\mu\nu} \frac{\partial}{\partial x_\nu} \psi + m S \psi = 0 \ .$$

Multiplying from the left by S^{-1}, we have

$$\sum_\mu \sum_\nu (S^{-1} \gamma_\mu S) a_{\mu\nu} \frac{\partial}{\partial x_\nu} \psi + m \psi = 0 \ .$$

In order for this to coincide with (3.109), we need to require

$$\sum_\mu (S^{-1} \gamma_\mu S) a_{\mu\nu} = \gamma_\nu \ , \quad \text{or} \quad S^{-1} \gamma_\mu S = \sum_\nu a_{\mu\nu} \gamma_\nu \ . \tag{3.112}$$

This is one of the conditions for determining S. This S is a representation of the Lorentz transformation, as will become clear from the following discussion.

Assuming that, for another Lorentz transformation $(b_{\mu\nu})$, the relation

$$T^{-1} \gamma_\mu T = \sum_\nu b_{\mu\nu} \gamma_\nu$$

holds as well as the relation between S and $(a_{\mu\nu})$, we immediately obtain

$$(ST)^{-1} \gamma_\mu (ST) = \sum_\sigma \Big(\sum_\nu a_{\mu\nu} b_{\nu\sigma} \Big) \gamma_\sigma \ .$$

Thus, if S and T correspond to a and b, respectively, then ST corresponds to ab. In this sense, S and T are representations of the Lorentz transformations a and b.

Next, we derive another condition which S must satisfy. The complex conjugate of Dirac's equation is

$$\left[(\gamma_\mu)^*_{rs}\frac{\partial}{\partial x^*_\mu} + m\delta_{rs}\right]\psi^*_s = 0\,, \tag{3.113}$$

where we sum over the index s. The space-like components of x_μ are real numbers, and the time-like component is a pure imaginary number. In addition, the γ-matrices are Hermitian. Taking all this into account, we obtain

$$\sum_{j=1}^{3}\frac{\partial\psi^*_s}{\partial x_j}(\gamma_j)_{sr} - \frac{\partial\psi^*_s}{\partial x_4}(\gamma_4)_{sr} + m\psi^*_r = 0\,.$$

Then, multiplying by γ_4 from the right and using its anti-commutativity with γ_1, γ_2, and γ_3, we obtain

$$-\frac{\partial\bar{\psi}}{\partial x_\mu}\gamma_\mu + m\bar{\psi} = 0\,, \tag{3.114}$$

where

$$\bar{\psi} = \psi^\dagger\gamma_4\,. \tag{3.115}$$

In order for (3.114) to hold in any inertial system, $\bar{\psi}$ has to transform as

$$\bar{\psi} \to \bar{\psi}' = \bar{\psi}S^{-1}\,. \tag{3.116}$$

We have considered $\bar{\psi}$ instead of $\psi^\dagger$ because the transformation property is easier to handle. However, since ψ and $\bar{\psi}$ are not independent, we have to impose a restriction on S to make (3.104) and (3.114) compatible. Using the Hermitian conjugate of (3.104), we find

$$\bar{\psi}' = \psi'^\dagger\gamma_4 = \psi^\dagger S^\dagger\gamma_4 = \bar{\psi}\gamma_4 S^\dagger\gamma_4\,.$$

Comparing this equation with (3.116), we arrive at the condition

$$\gamma_4 S^\dagger\gamma_4 = S^{-1}\,, \quad \text{or} \quad S^\dagger\gamma_4 = \gamma_4 S^{-1}\,. \tag{3.117}$$

It should be emphasized that $S^\dagger$ and S^{-1} cannot be treated as equal because S is not necessarily unitary. Then the two conditions determining S are (3.112) and (3.117). In the following, we shall determine the matrix S for an infinitesimal Lorentz transformation.

We write down the infinitesimal Lorentz transformation as

$$x'_\mu = x_\mu + \sum_\nu \epsilon_{\mu\nu} x_\nu \ , \qquad \epsilon_{\mu\nu} = -\epsilon_{\nu\mu} \ \text{infinitesimal} . \tag{3.118}$$

Here $\epsilon_{\mu\nu}$ is a pure imaginary number if either μ or ν is 4. Corresponding to (3.118), we assume S to have the form

$$S = \mathbf{1} + \frac{1}{2} \sum_\mu \sum_\nu \epsilon_{\mu\nu} T_{\mu\nu} \ , \qquad T_{\mu\nu} = -T_{\nu\mu} \ , \tag{3.119}$$

where $T_{\mu\nu}$ are 4×4 matrices produced by γ-matrices. Substituting the equation above into (3.112), the linear term in ϵ is

$$\begin{aligned} \left[\gamma_\mu, \frac{1}{2} \sum_\lambda \sum_\nu \epsilon_{\lambda\nu} T_{\lambda\nu} \right] &= \sum_\nu \epsilon_{\mu\nu} \gamma_\nu \\ &= \frac{1}{2} \sum_\lambda \sum_\nu \epsilon_{\lambda\nu} (\delta_{\lambda\mu} \gamma_\nu - \delta_{\mu\nu} \gamma_\lambda) \ . \end{aligned}$$

Comparing the anti-symmetrized coefficients of the anti-symmetric tensor $\epsilon_{\lambda\nu}$, we deduce that

$$\gamma_\mu T_{\lambda\nu} - T_{\lambda\nu} \gamma_\mu = \delta_{\lambda\mu} \gamma_{nu} - \delta_{\nu\mu} \gamma_\lambda \ . \tag{3.120}$$

This is the condition for determining the anti-symmetric tensor $T_{\lambda\nu}$. The tensor which satisfies this equation is

$$T_{\lambda\nu} = \frac{1}{4} (\gamma_\lambda \gamma_\nu - \gamma_\nu \gamma_\lambda) \ . \tag{3.121}$$

It can be checked that this equation satisfies not only (3.112), but also (3.117).

Substituting (3.119) into (3.117) yields

$$\sum_\mu \sum_\nu \epsilon^*_{\mu\nu} T^\dagger_{\mu\nu} = - \sum_\mu \sum_\nu \epsilon_{\mu\nu} \gamma_4 T_{\mu\nu} \gamma_4 \ . \tag{3.122}$$

We separate the cases when μ and ν contain 4 or not:

$$\epsilon_{ik} T^\dagger_{ik} = -\epsilon_{ik} \gamma_4 T_{ik} \gamma_4 = -\epsilon_{ik} T_{ik} \ ,$$

$$-\epsilon_{4k} T^\dagger_{4k} = -\epsilon_{4k} \gamma_4 T_{4k} \gamma_4 = \epsilon_{4k} T_{4k} \ .$$

Since both cases turn out to give

$$T_{\mu\nu}^{\dagger} = -T_{\mu\nu} , \tag{3.123}$$

we see that the condition can be satisfied. Thus it has been proven that for the infinitesimal Lorentz transformation, $T_{\mu\nu}$, i.e., the matrix S, exists. In addition, this tells us that

$$\mathrm{Tr}(T_{\mu\nu}) = 0 . \tag{3.124}$$

It also gives the following restriction on S:

$$\det S = 1 . \tag{3.125}$$

Generalizing, it can be shown that S exists for the proper Lorentz transformations generated by infinitesimal Lorentz transformations. Actually, (3.125) also holds for finite transformations.[2]

Space Inversion

One example of a special Lorentz transformation that does not belong to the proper Lorentz transformations is space inversion. In this case, S must be determined separately. Space inversion is defined by

$$x_j' = -x_j , \quad j = 1, 2, 3 , \quad x_4' = x_4 . \tag{3.126}$$

The corresponding conditions which S must satisfy are

$$\gamma_j S = -S\gamma_j , \quad \gamma_4 S = S\gamma_4 , \quad S^{\dagger}\gamma_4 = \gamma_4 S^{-1} . \tag{3.127}$$

Up to an undetermined phase factor, the solution is

$$S = \gamma_4 . \tag{3.128}$$

Thus we can say that Dirac's equation preserves parity. However, this statement does not hold for the two-component Weyl equation. We shall discuss this below.

Relativistic Covariance of Weyl's Equation

Although in Sect. 3.5 we used boldface to distinguish Pauli's matrix in two dimensions from the one in four dimensions, from now on we shall use boldface only for three-dimensional vectors and normal characters for their components. In the following, we shall prove the relativistic covariance of the Weyl equation

[2] To derive the explicit form of S corresponding to finite transformations, the reader is referred to [45].

similarly to that of the Dirac equation. We start with Weyl's equation (3.100):

$$\mathrm{i}\frac{\partial}{\partial t}\psi = \pm\frac{1}{\mathrm{i}}\boldsymbol{\sigma}\cdot\nabla\psi \ . \tag{3.129}$$

The positive sign corresponds to the right-handed neutrino and the negative one to the left-handed neutrino, as will be explained in the next section. Then we introduce the matrices Γ_μ:

$$\Gamma_j = \sigma_j \ , \quad j = 1, 2, 3 \ , \quad \Gamma_4 = \pm\mathrm{i} \ . \tag{3.130}$$

Using the Γ-matrix, Weyl's equation can be written in the form

$$\Gamma_\mu \frac{\partial \psi}{\partial x_\mu} = 0 \ , \qquad \frac{\partial \psi^\dagger}{\partial x_\mu}\Gamma_\mu = 0 \ . \tag{3.131}$$

Thus, the Lorentz transformation of the Weyl spinors becomes

$$\psi \to \psi' = S\psi \ , \quad \psi^\dagger \to \psi'^\dagger = \psi^\dagger S^\dagger \ . \tag{3.132}$$

The big difference with Dirac's equation is that in Weyl's equation there is no mass term. The covariance of the two equations in (3.131) (in other words, the condition that they hold in the same form in any inertial system) can be expressed as follows:

$$S^\dagger \Gamma_\mu S = \sum_\nu a_{\mu\nu}\Gamma_\nu \ . \tag{3.133}$$

Assuming that S has the form (3.119) for the infinitesimal Lorentz transformation (3.118),

$$\frac{1}{2}\epsilon^*_{\mu\nu}T^\dagger_{\mu\nu}\Gamma_\lambda + \frac{1}{2}\Gamma_\lambda\epsilon_{\mu\nu}T_{\mu\nu} = \epsilon_{\lambda\rho}\Gamma_\rho = \frac{1}{2}\epsilon_{\mu\nu}(\delta_{\mu\lambda}\Gamma_\nu - \delta_{\nu\lambda}\Gamma_\mu) \ . \tag{3.134}$$

From this equation, $T_{\mu\nu}$ can be obtained immediately:

$$T_{12} = \pm T_{34} = \frac{\mathrm{i}}{2}\sigma_3 \ , \quad T_{23} = \pm T_{14} = \frac{1}{2}\sigma_1 \ , \quad T_{31} = \pm T_{24} = \frac{\mathrm{i}}{2}\sigma_2 \ . \tag{3.135}$$

Again we find the condition

$$\mathrm{Tr}(T_{\mu\nu}) = 0 \ . \tag{3.136}$$

In this case the matrix S is unimodular, i.e.,

$$\det S = 1 \ . \tag{3.137}$$

For the space inversion (3.126), S must satisfy the following four equations:

$$S^\dagger \sigma_j S = -\sigma_j\ , \quad j = 1, 2, 3\ , \quad S^\dagger S = \mathbf{1}\ . \tag{3.138}$$

However, there exists no two-dimensional matrix satisfying these four equations. Thus, Weyl's equation does not maintain its form under space inversion. In other words, this equation is not defined with respect to space inversion. Hence, parity is not conserved. For this reason Weyl's equation was long considered unphysical. However, once the non-conservation of parity proposed by Lee and Yang [45] had been confirmed experimentally, Weyl's equation began to bask in the limelight as the equation of the massless neutrino. The theory of neutrinos described by this equation is called the *two-component theory of the neutrino*.

3.7 Solutions of the Free Dirac Equation

To quantize the Dirac field, we must first derive the complete set of solutions of the free Dirac equation. In Pauli's representation, the Hamiltonian is

$$H = \boldsymbol{\alpha} \cdot \boldsymbol{p} + \beta m = \rho_1 \boldsymbol{\sigma} \cdot \boldsymbol{p} + \rho_3 m\ , \tag{3.139}$$

with the ρ-matrices defined by (3.94). We begin by deriving the invariant quantities for the Dirac equation, in other words, the conserved quantities:

$$[H, H] = 0\ , \quad [p, H] = 0\ . \tag{3.140}$$

These equations express conservation of energy and momentum. Next, we consider the angular momentum:

$$\boldsymbol{L} = \boldsymbol{r} \times \boldsymbol{p}\ . \tag{3.141}$$

We note that

$$[L_3, \boldsymbol{\alpha} \cdot \boldsymbol{p} + \beta m] = \boldsymbol{\alpha} \cdot [L_3, \boldsymbol{p}] = \mathrm{i}(\alpha_1 p_2 - \alpha_2 p_1) = \mathrm{i}(\boldsymbol{\alpha} \times \boldsymbol{p})_3\ .$$

Generalizing this equation we find that the orbital angular momentum is not a conserved quantity:

$$[\boldsymbol{L}, H] = \mathrm{i}\boldsymbol{\alpha} \times \boldsymbol{p} = \mathrm{i}\rho_1 \boldsymbol{\sigma} \times \boldsymbol{p}. \tag{3.142}$$

The same happens for spins:

$$[\sigma_3, \boldsymbol{\alpha} \cdot \boldsymbol{p} + \beta m] = \rho_1 [\sigma_3, \boldsymbol{\sigma}] \cdot \boldsymbol{p} = 2\mathrm{i}\rho_1(\sigma_2 p_1 - \sigma_1 p_2) = -2\mathrm{i}\rho_1(\boldsymbol{\sigma} \times \boldsymbol{p})_3\ ,$$

or

$$[\boldsymbol{\sigma}, H] = -2\mathrm{i}\rho_1 \boldsymbol{\sigma} \times \boldsymbol{p} \ . \tag{3.143}$$

Combining (3.142) and (3.143), we find that the total angular momentum is a conserved quantity:

$$[\boldsymbol{J}, H] = 0 \ , \quad \text{where} \ \boldsymbol{J} = \boldsymbol{L} + \frac{1}{2}\boldsymbol{\sigma} \ . \tag{3.144}$$

In contrast with the non-relativistic theory, even for the free particle, the spin and angular momentum are not conserved separately. Only their sum is conserved.

The *helicity* is another important conserved quantity. That is,

$$h = \boldsymbol{\sigma} \cdot \boldsymbol{p}/p \ , \quad \text{or} \ \lambda = h/2 \ , \tag{3.145}$$

is conserved, and the eigenvalues of h are ± 1. To be precise, λ is the helicity, but here we call h helicity as well.

The motion of a free Dirac particle is determined by specifying its momentum and helicity. The wave function with the given momentum p takes the form

$$\psi(x) = \frac{1}{\sqrt{V}} \mathrm{e}^{\mathrm{i}\boldsymbol{p}\cdot\boldsymbol{x} - \mathrm{i}Et} u(p) \ . \tag{3.146}$$

Here, V is the volume of the whole system within which the particle is confined, and $u(p)$ is a four-component spinor.

In contrast with the scalar and the vector, the spinor is a physical quantity obeying the transformation rule (3.104). Substituting (3.146) into the Dirac equation, we find

$$Eu(p) = (\rho_1 \boldsymbol{\sigma} \cdot \boldsymbol{p} + \rho_3 m)u(p) \equiv H(p)u(p) \ . \tag{3.147}$$

In order to obtain the energy eigenvalue E, we diagonalize $H(p)$ using a unitary matrix U. We use the formula

$$\exp\left(-\frac{\mathrm{i}}{2}a\rho_2\right)\rho_3 \exp\left(\frac{\mathrm{i}}{2}a\rho_2\right) = \rho_3 \cos a + \rho_1 \sin a \ . \tag{3.148}$$

Then we write $H(p)$ as

$$\rho_1 \boldsymbol{\sigma} \cdot \boldsymbol{p} + \rho_3 m = A(\rho_3 \cos a + \rho_1 \sin a) \ ,$$

where

$$A = \sqrt{(\boldsymbol{\sigma} \cdot \boldsymbol{p})^2 + m^2} = \sqrt{p^2 + m^2} \ , \quad \tan a = \boldsymbol{\sigma} \cdot \boldsymbol{p}/m \ , \tag{3.149}$$

with p standing for the absolute value of the vector $\boldsymbol{p}$. Then, the matrix U is

$$U = \exp\left(\frac{\mathrm{i}}{2}a\rho_2\right) = \exp\left(\frac{\mathrm{i}}{2}\rho_2 \tan^{-1}\frac{\boldsymbol{\sigma}\cdot\boldsymbol{p}}{m}\right) . \tag{3.150}$$

The diagonal form of $H(\boldsymbol{p})$ is

$$UH(\boldsymbol{p})U^{-1} = \sqrt{p^2+m^2}\rho_3 . \tag{3.151}$$

Using the formula

$$\tan^{-1}\frac{\boldsymbol{\sigma}\cdot\boldsymbol{p}}{m} = \frac{\boldsymbol{\sigma}\cdot\boldsymbol{p}}{m}\frac{\tan^{-1}(p/m)}{p/m} ,$$

we put U in the form

$$U = \frac{1}{\sqrt{2A}}\left(\sqrt{A+m} + \mathrm{i}\rho_2\frac{\boldsymbol{\sigma}\cdot\boldsymbol{p}}{\sqrt{A+m}}\right) . \tag{3.152}$$

Next, we introduce the spinor $v(\boldsymbol{p}) = Uu(\boldsymbol{p})$, such that

$$Ev(\boldsymbol{p}) = \sqrt{p^2+m^2}\rho_3 v(\boldsymbol{p}) . \tag{3.153}$$

Since the eigenvalues of ρ_3 are ± 1, the eigenvalues E turn out to be

$$E = \pm\sqrt{p^2+m^2} . \tag{3.154}$$

Because each eigenvalue is doubly degenerate, we distinguish the solutions by the helicity:

1) $E = \sqrt{p^2+m^2}, \quad h = 1$ 2) $E = \sqrt{p^2+m^2}, \quad h = -1$

$$v(\boldsymbol{p}) = \begin{pmatrix} \cos\theta/2 \\ \mathrm{e}^{\mathrm{i}\varphi}\sin\theta/2 \\ 0 \\ 0 \end{pmatrix}, \qquad v(\boldsymbol{p}) = \begin{pmatrix} -\mathrm{e}^{\mathrm{i}\varphi}\sin\theta/2 \\ \cos\theta/2 \\ 0 \\ 0 \end{pmatrix},$$

3) $E = -\sqrt{p^2+m^2}, \quad h = 1$ 4) $E = -\sqrt{p^2+m^2}, \quad h = -1$

$$v(\boldsymbol{p}) = \begin{pmatrix} 0 \\ 0 \\ \cos\theta/2 \\ \mathrm{e}^{\mathrm{i}\varphi}\sin\theta/2 \end{pmatrix}, \qquad v(\boldsymbol{p}) = \begin{pmatrix} 0 \\ 0 \\ -\mathrm{e}^{\mathrm{i}\varphi}\sin\theta/2 \\ \cos\theta/2 \end{pmatrix}.$$

This completes the solution of the free Dirac equation. The wave function can be normalized by requiring

$$\int_V \mathrm{d}^3x\, \psi^\dagger(x)\psi(x) = 1 \,. \tag{3.155}$$

Note that there are positive and negative energy eigenvalues. The interpretation of negative energy solutions can be correctly given only after quantization. In addition, the Dirac particle does not obey Bose statistics like the spin-0 scalar particle. It obeys Fermi statistics, being a spin-$1/2$ particle. What kind of quantization must be chosen to satisfy the statement above? This is the subject of the next section.

Relation Between Dirac's Equation and Weyl's Equation

Weyl's equation is obtained from Dirac's equation without the mass term:

$$H\psi = \boldsymbol{\alpha}\cdot\boldsymbol{p}\psi = E\psi \,. \tag{3.156}$$

We introduce the following Hermitian matrices:

$$\sigma_{\mu\nu} = \frac{1}{2\mathrm{i}}(\gamma_\mu\gamma_\nu - \gamma_\nu\gamma_\mu) = \frac{2}{\mathrm{i}}T_{\mu\nu} \,, \tag{3.157}$$

$$\gamma_5 = \gamma_1\gamma_2\gamma_3\gamma_4 \,. \tag{3.158}$$

The matrices σ_{23}, σ_{31}, and σ_{12} are equivalent to σ_1, σ_2, and σ_3 in (3.91). Then we can write

$$\boldsymbol{\alpha}\cdot\boldsymbol{p} = -\gamma_5\boldsymbol{\sigma}\cdot\boldsymbol{p} \,, \tag{3.159}$$

where $\boldsymbol{\sigma}$ stands for the three-dimensional vector given by the three components σ_1, σ_2, and σ_3. This matrix can be thought of as being either two-dimensional or four-dimensional, as we choose. We then decompose the wave function in the following way:

$$\psi = \psi_{\mathrm{L}} + \psi_{\mathrm{R}} = \frac{1+\gamma_5}{2}\psi + \frac{1-\gamma_5}{2}\psi \,. \tag{3.160}$$

In the case of a massless particle, Dirac's equation (3.156) reduces to the two equations

$$E\psi_{\mathrm{L}} = -\boldsymbol{\sigma}\cdot\boldsymbol{p}\psi_{\mathrm{L}} \,, \quad E\psi_{\mathrm{R}} = \boldsymbol{\sigma}\cdot\boldsymbol{p}\psi_{\mathrm{R}} \,. \tag{3.161}$$

By restricting to the positive energy solution, we find that, since $E = p$, ψ_{L} corresponds to the case $h = -1$, called the *left-handed* wave function, with spin antiparallel to its momentum, while ψ_{R} corresponds to $h = 1$, called the *right-handed* wave function, with spin parallel to the momentum. In the Weyl representation, the matrix γ_5 is diagonal:

$$\gamma_5 = \begin{pmatrix} 1 & & & \\ & 1 & & \\ & & -1 & \\ & & & -1 \end{pmatrix} . \tag{3.162}$$

Hence, ψ_L and ψ_R become the following two-component spinors:

$$\psi_L = \begin{pmatrix} \times \\ \times \\ 0 \\ 0 \end{pmatrix} , \quad \psi_R = \begin{pmatrix} 0 \\ 0 \\ \times \\ \times \end{pmatrix} , \tag{3.163}$$

where $\times$ stands for the components which are in general not zero. This completes the derivation of the two-component Weyl equation. However, as we have seen above, it can be expressed in a four-component form.

As mentioned before, the neutrino can be considered to obey Weyl's equation. If this is true, is the sign of its helicity positive or negative? In other words, is the neutrino left-handed or right-handed? The helicity of the neutrino was measured by Goldhaber, Grozins, and Sunyar in [79]. The process they used is an elementary reaction called *K-capture* induced by essentially the same interaction as β-decay, namely

$$\mathrm{e}^- + \mathrm{p} \to \mathrm{n} + \nu . \tag{3.164}$$

Both p and n are nucleons in an atomic nucleus. The following two-stage process takes place:

$$\mathrm{e}^- + \underset{J=0}{\mathrm{A}} \longrightarrow \underset{J=1}{\mathrm{B}^*} + \nu , \quad \underset{J=1}{\mathrm{B}^*} \longrightarrow \underset{J=0}{\mathrm{B}} + \gamma , \tag{3.165}$$

where A is a spin-0 nucleus. Absorbing the electron on its K-shell according to (3.164), it becomes the spin-1 excited atomic nucleus B^* and simultaneously emits a neutrino. Then B^* reaches its ground state B by emitting a γ-ray. B is again in a spin-0 state. If there exists an atomic nucleus which undergoes such a chain of reactions, the helicity of the neutrino can be determined in the following way.

First, note that, when they are emitted in opposite directions, the neutrino ν and the γ particle carry the maximum energy, thus satisfying the resonant scattering condition. We will check the conservation of angular momentum in this case. The z-axis is chosen as the direction in which ν is radiated, and we consider J_z before and after the interaction.

If ν is left-handed, so that its spin is antiparallel to the z-direction, i.e., $-1/2$, we have the following possibilities for J_z:

$$\begin{array}{lccccccc} & e^- & +\,A & \longrightarrow B^* & +\;\nu & \longrightarrow \gamma & +\,B & +\;\nu \\ J_z: & \frac{1}{2} & 0 & 1 & -\frac{1}{2} & 1 & 0 & -\frac{1}{2} \\ J_z: & -\frac{1}{2} & 0 & 0 & -\frac{1}{2} & \underset{\text{forbidden}}{0} & 0 & -\frac{1}{2} \end{array}$$

In the analysis of the conservation law above, the case where the electron has $J_z = -1/2$ is forbidden, because the z component of the spin of the photon emitted in the $-z$ direction cannot be zero. This is a feature of massless particles, which will be explained later. Therefore, the only allowed case is when the photon is emitted in the $-z$ direction with $J_z = 1$, i.e., when the photon is left-handed (meaning that it has left-circular polarization). Thus, if ν is left-handed, the photon emitted with the maximum energy will also be left-handed. Similarly, if ν is right-handed, then the photon will also be right-handed.

The atomic nuclei used in the actual experiments were

$$A = {}^{152}_{63}\mathrm{Eu}(0^-)\,, \quad B = {}^{152}_{62}\mathrm{Sm}(0^+)\,, \quad B^* = {}^{152}_{62}\mathrm{Sm}(1^-)\,, \tag{3.166}$$

where the numbers 0 or 1 in brackets represent the spins of the nuclei and the plus and minus signs indicate the parity. By measuring the polarization of the photons with the highest energies, it was deduced that the neutrino is *left-handed*.

3.8 Quantization of the Dirac Field

To quantize the Dirac field, we start from the Lagrangian density in the form

$$\mathscr{L} = -\bar{\psi}\left(\gamma_\mu \frac{\partial}{\partial x_\mu} + m\right)\psi\,, \tag{3.167}$$

and derive the Dirac equation. We must treat ψ and $\bar{\psi}$ as independent quantities when deriving the field equation from this Lagrangian density using the variational principle. As a consequence, the following equations are obtained:

$$\left(\gamma_\mu \frac{\partial}{\partial x_\mu} + m\right)\psi = 0\,, \quad -\frac{\partial \bar{\psi}}{\partial x_\mu}\gamma_\mu + m\bar{\psi} = 0\,. \tag{3.168}$$

In addition, the quantity canonically conjugate to ψ_r, denoted $\pi_r^\dagger$, is

$$\pi_r^\dagger = \frac{\partial \mathscr{L}}{\partial \dot{\psi}_r} = \mathrm{i}(\bar{\psi}\gamma_4)_r = \mathrm{i}\psi_r^\dagger\,. \tag{3.169}$$

The four-dimensional continuity equation related to the law of conservation of probability, which we have discussed before, is

$$\partial_\mu j_\mu = 0 \, , \quad \text{where} \quad j_\mu = \mathrm{i}\bar{\psi}\gamma_\mu\psi \, . \tag{3.170}$$

Combining the equations (3.104), (3.116), and (3.112), it is easily shown that j_μ transforms as a four-vector under Lorentz transformations. In addition, it will be explained later that the canonical algebra for Dirac fields obeying Fermi statistics must be chosen as the following anti-commutation relation instead of (3.9):

$$\left[\psi_r(x), \pi_s^\dagger(x')\right]_+ = \mathrm{i}\delta_{rs}\delta^3(x - x') \, , \quad \text{where} \quad x_0 = x_0' \, . \tag{3.171}$$

Here, $[\ ,\ ]_+$ denotes the *anti-commutator*, sometimes written as $\{\ ,\ \}$, and defined by

$$[A, B]_+ = \{A, B\} = AB + BA \, . \tag{3.172}$$

Sometimes, the commutator $[\ ,\ ]$ is denoted, in turn, by $[\ ,\ ]_-$.

For $x_0 = x_0'$, we can write the equal-time anti-commutators in the form

$$\left[\psi_r(x), \psi_s^\dagger(x')\right]_+ = \delta_{rs}\delta^3(x - x') \, ,$$

$$\left[\psi_r(x), \psi_s(x')\right]_+ = \left[\psi_r^\dagger(x), \psi_s^\dagger\right]_+ = 0 \, . \tag{3.173}$$

Then, the Hamiltonian density becomes

$$\begin{aligned} \mathscr{H} &= \sum_r \pi_r^\dagger \dot{\psi}_r - \mathscr{L} = \mathrm{i}\bar{\psi}\gamma_4\dot{\psi} + \bar{\psi}(\gamma_\mu\partial_\mu + m)\psi \\ &= \psi^\dagger(\boldsymbol{\alpha}\cdot\boldsymbol{p} + \beta m)\psi \, , \quad \text{where} \quad \boldsymbol{p} = -\mathrm{i}\nabla \, . \end{aligned} \tag{3.174}$$

Note the appearance of the Dirac Hamiltonian sandwiched between ψ and $\psi^\dagger$. The Hamiltonian itself is given by the space integration:

$$H = \int \mathrm{d}^3x \mathscr{H}(x) = \int \mathrm{d}^3x \psi^\dagger(x)\left(\frac{1}{\mathrm{i}}\boldsymbol{\alpha}\cdot\nabla + \beta m\right)\psi(x) \, . \tag{3.175}$$

This completes the application of the canonical formalism.

To quantize, we introduce the Fourier expansion

$$\psi_r(x) = \frac{1}{\sqrt{V}}\sum_{\boldsymbol{q}}\left[\mathrm{e}^{\mathrm{i}\boldsymbol{q}\cdot\boldsymbol{x} - \mathrm{i}E_q x_0}\sum_h u_r^{(+)}(\boldsymbol{q}, h)a_+(\boldsymbol{q}, h) + \mathrm{e}^{\mathrm{i}\boldsymbol{q}\cdot\boldsymbol{x} + \mathrm{i}E_q x_0}\sum_h u_r^{(-)}(\boldsymbol{q}, h)a_-(\boldsymbol{q}, h)\right] , \tag{3.176}$$

where $E_q = \sqrt{q^2 + m^2}$, $q = |\boldsymbol{q}|$, and $u^{(+)}$ and $u^{(-)}$ are the positive and negative energy solutions, respectively, satisfying the equations

$$(\boldsymbol{\alpha} \cdot \boldsymbol{q} + \beta m) u^{(\pm)}(\boldsymbol{q}, h) = \pm E_q u^{(\pm)}(\boldsymbol{q}, h) \ . \tag{3.177}$$

Substituting the Fourier expansion above into the Hamiltonian (3.175), we obtain

$$H = \sum_{\boldsymbol{q}} E_q \left[\sum_h a_+^\dagger(\boldsymbol{q}, h) a_+(\boldsymbol{q}, h) - \sum_h a_-^\dagger(\boldsymbol{q}, h) a_-(\boldsymbol{q}, h) \right] . \tag{3.178}$$

In addition, the following conserved quantity (which is in fact proportional to the electric charge) is also important:

$$N = \int \mathrm{d}^3 x j_0(x) = \sum_{\boldsymbol{q}} \sum_h \left[a_+^\dagger(\boldsymbol{q}, h) a_+(\boldsymbol{q}, h) + a_-^\dagger(\boldsymbol{q}, h) a_-(\boldsymbol{q}, h) \right] . \tag{3.179}$$

Next, substituting the Fourier expansion (3.176) into (3.173), we find

$$\begin{aligned} \left[a_+(\boldsymbol{q}, h), a_+^\dagger(\boldsymbol{q}', h') \right]_+ &= \delta_{h,h'} \delta_{\boldsymbol{q},\boldsymbol{q}'} \ , \\ \left[a_-(\boldsymbol{q}, h), a_-^\dagger(\boldsymbol{q}', h') \right]_+ &= \delta_{h,h'} \delta_{\boldsymbol{q},\boldsymbol{q}'} \ , \\ \text{other anti-commutators} &= 0 \ . \end{aligned} \tag{3.180}$$

Looking at the expression (3.178) for the Hamiltonian, the energy in the system is apparently not positive-definite.

Although the question regarding the negative energy was originally considered as a drawback of the theory, Dirac was able to get round it by assuming that the vacuum is the state filled completely with negative energy states, taking advantage of the fact that the electron obeys Fermi statistics. However, this idea is unnatural if we think of, say, the electric field created by the infinite number of electrons in the negative energy states. A much clearer solution is achieved by quantizing the field. To explain this, we use the following trick:

$$-\sum_h a_-^\dagger(\boldsymbol{q}, h) a_-(\boldsymbol{q}, h) = \sum_h a_-(\boldsymbol{q}, h) a_-^\dagger(\boldsymbol{q}, h) - 2 \ . \tag{3.181}$$

Neglecting the constant 2, as we did in (3.66), the Hamiltonian becomes positive-definite:

$$H = \sum_{\boldsymbol{q}} E_q \left[\sum_h a_+^\dagger(\boldsymbol{q}, h) a_+(\boldsymbol{q}, h) + \sum_h a_-(\boldsymbol{q}, h) a_-^\dagger(\boldsymbol{q}, h) \right] . \tag{3.182}$$

Although the term we neglected is negative infinity, it is still a constant, viz.,

$$-2\sum_{q} E_q \,. \tag{3.183}$$

This corresponds to neglecting the total energy of the electrons in the negative energy states in Dirac's hole theory. Let us reconstruct the interpretation in the hole theory by reinterpreting the operators. $a^{\dagger}{}_{+}$ and a_+ are the operators creating and annihilating, respectively, the electron with positive energy (the Dirac particle), while $a_-^{\dagger}$ and a_- are the operators creating and annihilating, respectively, particles with negative energy. However, creating the electron with negative energy is equivalent to annihilating a particle with positive energy. In addition, creating some particle with an additive quantum number is equivalent to annihilating another with the opposite quantum number. Here we consider that creating the electron with negative energy is equivalent to annihilating the positive-energy particle with the opposite quantum numbers. We make the replacement

$$\begin{aligned} a_+(\boldsymbol{q},h) &\to a(\boldsymbol{q},h)\,, \quad a_+^{\dagger}(\boldsymbol{q},h) \to a^{\dagger}(\boldsymbol{q},h)\,, \\ a_-(\boldsymbol{q},h) &\to b^{\dagger}(-\boldsymbol{q},h)\,, \quad a_-^{\dagger}(\boldsymbol{q},h) \to b(-\boldsymbol{q},h)\,. \end{aligned} \tag{3.184}$$

The helicity is a product of the additive quantum numbers $\boldsymbol{p}$ and $\boldsymbol{\sigma}$, so its sign is unchanged.

Since the particle whose additive quantum numbers all have opposite sign to the electron has a positive electric charge, it is called a *positron*. We define the operators

$$N_+(\boldsymbol{q},h) = a^{\dagger}(\boldsymbol{q},h)a(\boldsymbol{q},h)\,, \quad N_-(\boldsymbol{q},h) = b^{\dagger}(\boldsymbol{q},h)b(\boldsymbol{q},h)\,. \tag{3.185}$$

Hence, H and N can be written as

$$H = \sum_{\boldsymbol{q}}\sum_{h} E_q\big[N_+(\boldsymbol{q},h) + N_-(\boldsymbol{q},h)\big]\,, \tag{3.186}$$

$$N = \sum_{\boldsymbol{q}}\sum_{h} \big[N_+(\boldsymbol{q},h) - N_-(\boldsymbol{q},h)\big]\,, \tag{3.187}$$

where b and $b^{\dagger}$ are the creation and annihilation operators of the positron, respectively. $N_{\pm}$ are the particle number operators and, unlike the case of Bose statistics, their eigenvalues can only take the values 0 and 1, a feature of Fermi statistics. This can be shown using the anti-commutation relations (3.180). Fixing $\boldsymbol{q}$ and h,

$$N^2 = a^{\dagger}aa^{\dagger}a = a^{\dagger}(1 - a^{\dagger}a)a = a^{\dagger}a - (a^{\dagger})^2a^2 = a^{\dagger}a = N\,,$$

i.e., from the anti-commutation relations, we find

$$\left[N_{\pm}(\boldsymbol{q},h)\right]^2 = N_{\pm}(\boldsymbol{q},h) \,. \tag{3.188}$$

This means that the only permissible eigenvalues are 0 and 1, reflecting Pauli's exclusion principle, namely that one state can be occupied by at most one particle. For this reason, the field of particles obeying Fermi statistics must be quantized using anti-commutation relations.

Using the new operators, the Fourier expansion of the Dirac field can be written as

$$\psi_r(x) = \frac{1}{\sqrt{V}} \sum_{\boldsymbol{q}} \sum_{h} \left[\mathrm{e}^{\mathrm{i}q\cdot x} u_r(q,h) a(q,h) + \mathrm{e}^{-\mathrm{i}q\cdot x} v_r(q,h) b^{\dagger}(q,h) \right] , \tag{3.189}$$

where $q\cdot x = \boldsymbol{qx} - E_q x_0$ is the four-dimensional scalar product, and we have simply written $\boldsymbol{q}$ in a or b as q. In addition, we have used the notation

$$u_r(q,h) = u_r^{(+)}(\boldsymbol{q},h) \,, \quad v_r(q,h) = u_r^{(-)}(-\boldsymbol{q},h) \,. \tag{3.190}$$

Casimir Operator

When experiments are carried out, very often only the momentum and energy are measured, and not the spin or the helicity. In such cases, we must sum over the spin states, and for this purpose, it is useful to define the so-called *Casimir operator*. We shall write the electron and the positron, or more generally the particle state with four-momentum p and the antiparticle state with four-momentum $\bar{p}$, as $|p,h\rangle$ and $|\bar{p},h\rangle$, respectively. Here h denotes the helicity. In this case, from the Fourier expansion, we obtain

$$\langle 0|\psi(x)|p,h\rangle = \frac{1}{\sqrt{V}} \mathrm{e}^{\mathrm{i}p\cdot x} u(p,h) \,, \tag{3.191a}$$

$$\langle p,h|\bar{\psi}(x)|0\rangle = \frac{1}{\sqrt{V}} \mathrm{e}^{-\mathrm{i}p\cdot x} \bar{u}(p,h) \,, \tag{3.191b}$$

where $|0\rangle$ stands for the vacuum state. For the antiparticle state, we find

$$\langle 0|\bar{\psi}(x)|\bar{p},h\rangle = \frac{1}{\sqrt{V}} \mathrm{e}^{\mathrm{i}\bar{p}\cdot x} \bar{v}(\bar{p},h) \,, \tag{3.192a}$$

$$\langle \bar{p},h|\psi(x)|0\rangle = \frac{1}{\sqrt{V}} \mathrm{e}^{-\mathrm{i}\bar{p}\cdot x} v(\bar{p},h) \,. \tag{3.192b}$$

From the orthogonality of the solutions of (3.147) and the normalization of the solution based on (3.155), we have

$$\sum_s \left[\sum_{h'} u_r^{(+)}(\boldsymbol{p}, h') u_s^{(+)*}(\boldsymbol{p}, h') \right] u_s^{(+)}(\boldsymbol{p}, h') = u_r^{(+)}(\boldsymbol{p}, h) ,$$
$$\sum_s \left[\sum_{h'} u_r^{(+)}(\boldsymbol{p}, h') u_s^{(+)*}(\boldsymbol{p}, h') \right] u_s^{(-)}(\boldsymbol{p}, h') = 0 . \tag{3.193}$$

These relations tell us that the term in the brackets is equivalent to the projection operator which selects the positive energy solution. Hence,

$$\sum_{h'} u_r^{(+)}(\boldsymbol{p}, h') u_s^{(+)*}(\boldsymbol{p}, h') = \left(\frac{H(p) + E_p}{2E_p} \right)_{rs} = \left(\frac{\boldsymbol{\alpha} \cdot \boldsymbol{p} + \beta m + E_p}{2E_p} \right)_{rs} . \tag{3.194}$$

We rewrite the above equation as

$$\sum_h u_r(p, h) \bar{u}_s(p, h) = \left(\frac{-\mathrm{i} p \cdot \gamma + m}{2E_p} \right)_{rs} . \tag{3.195}$$

Similarly, for the negative energy solution,

$$\sum_{h'} u_r^{(-)}(-\boldsymbol{p}, h') u_s^{(-)*}(-\boldsymbol{p}, h') = \left(\frac{-\boldsymbol{\alpha} \cdot \boldsymbol{p} + \beta m - E_p}{-2E_p} \right)_{rs} , \tag{3.196}$$

i.e.,

$$\sum_h v_r(\bar{p}, h) \bar{v}_s(\bar{p}, h) = \left(\frac{-\mathrm{i} \bar{p} \cdot \gamma - m}{2E_p} \right)_{rs} . \tag{3.197}$$

This projection operator is called Casimir operator. Using this operator we can perform all the calculations without using an explicit representation of the Dirac spinor.

3.9 Charge Conjugation

In the quantization of the complex scalar field and the Dirac field, the concept of particle and antiparticle arises. We will now focus on the symmetry between them. Such a symmetry always exists in the case of free fields. Since exchanging a particle with an antiparticle means a flip in the sign of the electric charge, such a transformation is called *charge conjugation* and, following (3.78), it is denoted by $\mathscr{C}$. For the Dirac field, the $\mathscr{C}$-operation interchanges ψ and $\bar{\psi}$, i.e., in general,

$$\psi_r(x) \to \mathscr{C}^{-1}\psi_r(x)\mathscr{C} = C_{rs}\bar{\psi}_s(x) \, , \tag{3.198}$$

where C is a 4×4 matrix. In addition, we have summed over the spinor index s. In the following, we will study the properties of the matrix C.

First, we consider that four-momentum and spin or helicity are unchanged when interchanging a particle and an antiparticle. From this property and (3.187), we obtain

$$\mathscr{C}^{-1}N\mathscr{C} = -N \, . \tag{3.199}$$

Thus, the same equation holds for the density as well:

$$\mathscr{C}^{-1}j_\mu\mathscr{C} = -j_\mu \, . \tag{3.200}$$

In quantized theories, it is more convenient to define $j_\mu(x)$ by

$$j_\mu(x) = \mathrm{i}\bar{\psi}(x)\gamma_\mu\psi(x) \to \frac{\mathrm{i}}{2}\left[\bar{\psi}(x), \gamma_\mu\psi(x)\right] = \frac{\mathrm{i}}{2}(\gamma_\mu)_{rs}\left[\bar{\psi}_r(x), \psi_s(x)\right] . \tag{3.201}$$

With this redefinition, the vacuum expectation value of j_μ vanishes.

Then we write the Dirac equation (3.168) as

$$(\gamma \cdot \partial + m)\psi = 0 \, , \quad (\gamma^{\mathrm{T}} \cdot \partial - m)\bar{\psi}^{\mathrm{T}} = 0 \, , \tag{3.202}$$

where γ^{T} stands for the transpose of the matrix γ_μ. If Dirac's equation does not change form under the charge conjugation operation, then

$$(\gamma \cdot \partial + m)\mathscr{C}^{-1}\psi\mathscr{C} = (\gamma \cdot \partial + m)C\bar{\psi}^{\mathrm{T}} = 0 \, .$$

Consequently,

$$(C^{-1}\gamma_\mu C \cdot \partial_\mu + m)\bar{\psi}^{\mathrm{T}} = 0 \, . \tag{3.203}$$

This equation is equivalent to the second equation in (3.202) if

$$C^{-1}\gamma_\mu C = -\gamma_\mu^{\mathrm{T}} \, . \tag{3.204}$$

Clearly, the matrices $-\gamma_\mu^{\mathrm{T}}$ satisfy the commutation relations (3.108), just like γ_μ. We skip the details, but given an explicit representation of the γ-matrices, the one obtained by replacing γ_μ by $-\gamma_\mu^{\mathrm{T}}$ is also a representation of the Dirac matrices and it is equivalent to the original one. As a result, we can choose the matrix C as a unitary matrix. In addition, from (3.204), we have

$$C\gamma_\mu^{\mathrm{T}}C^{-1} = -\gamma_\mu \,. \tag{3.205}$$

The equations derived by transposing (3.204) are

$$C^{\mathrm{T}}\gamma_\mu^{\mathrm{T}}\left(C^{\mathrm{T}}\right)^{-1} = -\gamma_\mu \,,$$

$$\left(C^{\mathrm{T}}\right)^{-1}\gamma_\mu C^{\mathrm{T}} = -\gamma_\mu^{\mathrm{T}} \,. \tag{3.206}$$

Substituting the right-hand side of the latter relation into the left-hand side of (3.205), we find

$$\left(C^{\mathrm{T}}C^{-1}\right)\gamma_\mu\left(C^{\mathrm{T}}C^{-1}\right)^{-1} = \gamma_\mu \,. \tag{3.207}$$

All 4×4 matrices can be expressed as polynomials of the four γ-matrices. The equation above shows that $C^{\mathrm{T}}C^{-1}$ commutes with all the γ-matrices, whence it commutes with all the 4×4 matrices. Such a matrix must therefore be proportional to the unit matrix:

$$C^{\mathrm{T}}C^{-1} = a\mathbf{1} \,. \tag{3.208}$$

The coefficient a does not depend on which representation we choose. This can be proved as follows.

Considering a unitary matrix S, we change the representation of γ as follows:

$$\gamma_\mu{}' = S\gamma_\mu S^{-1} \,. \tag{3.209}$$

In this case, solving (3.204) in the new representation, the matrix C' will be given by

$$C' = SCS^{\mathrm{T}} \,. \tag{3.210}$$

Therefore,

$$(C')^{\mathrm{T}}(C')^{-1} = SC^{\mathrm{T}}S^{\mathrm{T}}(S^{\mathrm{T}})^{-1}C^{-1}S^{-1} = a\mathbf{1} = C^{\mathrm{T}}C^{-1} \,.$$

Since the coefficient a is shared by all the equivalent representations, it is enough to determine it for one representation, say the Pauli representation. In this representation,

$$\gamma_\mu^{\mathrm{T}} = \gamma_\mu \,, \quad \mu = 2, 4 \,, \qquad \gamma_\mu^{\mathrm{T}} = -\gamma_\mu \,, \quad \mu = 1, 3 \,, \tag{3.211}$$

and (3.204) is satisfied by choosing $C = \gamma_2\gamma_4$. In this case,

$$C^{\mathrm{T}} = (\gamma_2\gamma_4)^{\mathrm{T}} = \gamma_4{}^{\mathrm{T}}\gamma_2{}^{\mathrm{T}} = \gamma_4\gamma_2 = -\gamma_2\gamma_4 = -C\,,$$

$$C^{\mathrm{T}}C^{-1} = -1\,. \tag{3.212}$$

Thus, we have found that $a = -1$ and the matrix C is anti-symmetric.

The charge conjugation transformation of $\bar{\psi}$ can be derived from the charge conjugation of ψ. Although we skip the intermediate calculations, using (3.198), (3.204), the unitarity of C, and the hermiticity of γ_μ, we find

$$\mathscr{C}^{-1}\bar{\psi}_r\mathscr{C} = (C^{-1})_{rs}\psi_s(x)\,. \tag{3.213}$$

Charge Conjugation of the Bilinear Forms

Since the γ-matrices are four-dimensional matrices, there exist 16 linearly independent polynomials that can be created using them. Denoting a four-dimensional matrix by O, we consider the bilinear form

$$\left[\bar{\psi}(x), O\psi(x)\right] = O_{rs}\left[\bar{\psi}_r(x), \psi_s(x)\right]. \tag{3.214}$$

From the transformation property of this bilinear form under Lorentz transformations, the matrix O can be categorized into the following five types:

$$\begin{aligned}&1\ (\text{scalar})\,, \quad \gamma_\mu\ (\text{vector})\,, \quad \sigma_{\mu\nu}\ (\text{anti-symmetric tensor})\,,\\ &\gamma_\mu\gamma_5\ (\text{axial vector})\,, \quad \gamma_5\ (\text{pseudo scalar})\,.\end{aligned} \tag{3.215}$$

The number of matrices in all categories is $1+4+6+4+1 = 16$, so they exhaust the linearly independent 4×4 matrices.

We now examine the transformation properties of the bilinear forms (3.214) under charge conjugation. Combining (3.198) and (3.213),

$$\mathscr{C}^{-1}\left[\bar{\psi}(x), O\psi(x)\right]\mathscr{C} = \left[\bar{\psi}(x), O'\psi(x)\right], \tag{3.216}$$

where

$$O' = \left(C^{-1}OC\right)^{\mathrm{T}} = CO^{\mathrm{T}}C^{-1} \equiv \epsilon O\,. \tag{3.217}$$

Calculating ϵ for the various cases using (3.205), we obtain Table 3.1.

Table 3.1 Sign changes under charge conjugation

O	1	γ_μ	$\sigma_{\mu\nu}$	$\gamma_\mu\gamma_5$	γ_5
ϵ	1	-1	-1	1	1

3.10 Quantization of the Complex Vector Field

So far we have studied the quantization of the spin-0 scalar field and the spin-1/2 spinor field, and in the following we will examine the quantization of a vector field. A real vector field and a complex vector field do not differ essentially, so we consider the latter.

We denote a complex vector field and its conjugate by φ_μ and $\varphi_\mu^\dagger$, respectively. As quantized operators, they are Hermitian conjugates of each other. When using the Pauli metric convention, as we do now, the i attributed to quantum theory changes its sign under the Hermitian conjugation operator $\dagger$, but the one attributed to relativity theory does not change its sign. Therefore, φ_j and $\varphi_j^\dagger$ are complex conjugates for $j = 1, 2, 3$, but when we write

$$\varphi_4 = \mathrm{i}\varphi_0 \,, \quad \varphi_4^\dagger = \mathrm{i}\varphi_0^\dagger \,, \tag{3.218}$$

it turns out that φ_0 and $\varphi_0^\dagger$ are complex conjugates, and φ_4 and $\varphi_4^\dagger$ are not. Bearing this in mind, we will discuss the conditions that a free vector field must satisfy:

$$K\varphi_\mu \equiv (\Box - m^2)\varphi_\mu = 0 \,, \tag{3.219}$$

$$\partial_\mu \varphi_\mu = 0 \,. \tag{3.220}$$

The first is the Klein–Gordon equation for each component. The second is called the *irreducibility condition*. If this four-divergence does not vanish, it represents a scalar field, and then φ_μ represents a field describing a mixture of spin-0 and spin-1 particles.

Let us derive the Lagrangian density satisfying the two conditions above. If we try the form

$$\mathscr{L} = -\left(\partial_\nu \varphi_\mu^\dagger \cdot \partial_\nu \varphi_\mu + m^2 \varphi_\mu^\dagger \cdot \varphi_\mu\right) ,$$

by analogy with the scalar field Lagrangian, we get into trouble, because although (3.219) holds, (3.220) does not. In addition, the Hamiltonian density derived from this Lagrangian density is not positive-definite, as can be seen here:

$$\begin{aligned}\mathscr{H} &= \pi_\mu^\dagger \pi_\mu + \nabla\varphi_\mu^\dagger \cdot \nabla\varphi_\mu + m^2 \varphi_\mu^\dagger \varphi_\mu \\ &= \sum_{j=1}^{3} \left(\pi_j^\dagger \pi_j + \nabla\varphi_j^\dagger \cdot \nabla\varphi_j + m^2 \varphi_j^\dagger \varphi_j\right) - \left(\pi_0^\dagger \pi_0 + \nabla\varphi_0^\dagger \cdot \nabla\varphi_0 + m^2 \varphi_0^\dagger \varphi_0\right) .\end{aligned}$$

We have to find another form, and we will try the one suggested by the Lagrangian of the electromagnetic field:

$$\mathscr{L} = -\frac{1}{2}(\partial_\mu \varphi_\nu^\dagger - \partial_\nu \varphi_\mu^\dagger)(\partial_\mu \varphi_\nu - \partial_\nu \varphi_\mu) - m^2 \varphi_\mu^\dagger \varphi_\mu \,. \tag{3.221}$$

Additionally, we define the field strength $F_{\mu\nu}$ by

$$F_{\mu\nu} = \partial_\mu \varphi_\nu - \partial_\nu \varphi_\mu \ . \tag{3.222}$$

The Euler–Lagrange equation derived from the variational principle is

$$\frac{\partial \mathscr{L}}{\partial \varphi_\nu^\dagger} - \frac{\partial}{\partial x_\mu}\left(\frac{\partial \mathscr{L}}{\partial \varphi_{\nu,\mu}^\dagger}\right) = -m^2 \varphi_\nu + \partial_\mu F_{\mu\nu} = 0 \ ,$$

whence

$$\partial_\mu F_{\mu\nu} = m^2 \varphi_\nu \ , \quad \partial_\mu F_{\mu\nu}^\dagger = m^2 \varphi_\nu^\dagger \ . \tag{3.223}$$

In the following, we assume $m \neq 0$. Since $F_{\mu\nu}$ is an anti-symmetric tensor,

$$m^2 \partial_\nu \varphi_\nu = \partial_\nu \partial_\mu F_{\mu\nu} = 0 \ .$$

Thus, from the assumption that $m \neq 0$, we can derive (3.220). Additionally, using (3.220) in (3.223), we find that (3.219) is also satisfied:

$$\partial_\mu(\partial_\mu \varphi_\nu - \partial_\nu \varphi_\mu) = \Box \varphi_\nu = m^2 \varphi_\nu \ .$$

Next, we define π as follows:

$$\pi_\nu^\dagger = \frac{\partial \mathscr{L}}{\partial \dot{\varphi}_\nu} = \frac{1}{\mathrm{i}}(\partial_\nu \varphi_4^\dagger - \partial_4 \varphi_\nu^\dagger) = \frac{1}{\mathrm{i}} F_{\nu 4}^\dagger \ ,$$

$$\pi_\nu = \frac{\partial \mathscr{L}}{\partial \dot{\varphi}_\nu^\dagger} = \frac{1}{\mathrm{i}}(\partial_\nu \varphi_4 - \partial_4 \varphi_\nu) = \frac{1}{\mathrm{i}} F_{\nu 4} \ . \tag{3.224}$$

Here we encounter a formal difficulty, i.e.,

$$\pi_4 = \pi_4^\dagger = 0 \ . \tag{3.225}$$

It turns out that we cannot define the canonically conjugate momentum for the fourth component. However, noticing that the fourth component is not independent of the other three components, it becomes clear that this is not a fundamental difficulty. From the equation of motion (3.223) and the definition of the canonically conjugate momentum (3.224), we see how the fourth component can be expressed in terms of the other components:

$$\varphi_4 = \frac{\mathrm{i}}{m^2} \operatorname{div} \boldsymbol{\pi} \ , \quad \varphi_4^\dagger = \frac{\mathrm{i}}{m^2} \operatorname{div} \boldsymbol{\pi}^\dagger \ . \tag{3.226}$$

In addition, the time derivative of φ_j can be written in terms of the three components π_j as follows:

$$\dot{\varphi}_j = \pi_j + \mathrm{i}\frac{\partial \varphi_4}{\partial x_j} , \quad \dot{\varphi}_j^\dagger = \pi_j^\dagger + \mathrm{i}\frac{\partial \varphi_4^\dagger}{\partial x_j} . \tag{3.227}$$

Thus, considering only the three space components as independent quantities, the Hamiltonian density can be written as

$$\mathscr{H} = \sum_{j=1}^{3} (\pi_j^\dagger \dot{\varphi}_j + \pi_j \dot{\varphi}_j^\dagger) - \mathscr{L} . \tag{3.228}$$

However, to construct the Hamiltonian we use

$$H = \int \mathrm{d}^3 x \mathscr{H}(x) = \int \mathrm{d}^3 x \mathscr{H}'(x) , \tag{3.229}$$

where $\mathscr{H}'$ is defined by adding a three-dimensional divergence term to $\mathscr{H}$:

$$\mathscr{H}' = \mathscr{H} - \mathrm{i}\,\mathrm{div}(\boldsymbol{\pi}^\dagger \varphi_4 + \boldsymbol{\pi} \varphi_4^\dagger) . \tag{3.230}$$

The advantage of using $\mathscr{H}'$ is that it makes manifest the positivity of the Hamiltonian:

$$\mathscr{H}' = \boldsymbol{\pi}^\dagger \cdot \boldsymbol{\pi} + \frac{1}{m^2}(\mathrm{div}\,\boldsymbol{\pi}^\dagger)(\mathrm{div}\,\boldsymbol{\pi}) + m^2 \boldsymbol{\varphi}^\dagger \cdot \boldsymbol{\varphi} + (\mathrm{rot}\,\boldsymbol{\varphi}^\dagger)\cdot(\mathrm{rot}\,\boldsymbol{\varphi}) . \tag{3.231}$$

Canonical Quantization and Canonical Equations

Starting from the Lagrangian density (3.221), we introduce the following equal-time commutation relations:

$$\left[\pi_j^\dagger(x), \varphi_k(x')\right] = \left[\pi_j(x), \varphi_k^\dagger(x')\right] = -\mathrm{i}\delta_{jk}\delta^3(x - x') , \quad \text{where } x_0 = x_0' . \tag{3.232}$$

All other commutators are set to zero. Then we derive the canonical equations:

$$\begin{aligned}\dot{\varphi}_j(x) &= \mathrm{i}\int \mathrm{d}^3 x' \left[\mathscr{H}(x'), \varphi_j(x)\right] \\ &= \int \mathrm{d}^3 x' \left[\pi_j(x')\delta^3(x - x') + \frac{1}{m^2}\mathrm{div}\,\boldsymbol{\pi}(x')\frac{\partial}{\partial x_j'}\delta^3(x - x')\right] ,\end{aligned}$$

whence

$$\dot{\boldsymbol{\varphi}}(x) = \boldsymbol{\pi}(x) - \frac{1}{m^2}\mathrm{grad}(\mathrm{div}\,\boldsymbol{\pi}) . \tag{3.233}$$

Similarly, we obtain

$$\dot{\boldsymbol{\pi}}(x) = -m^2\boldsymbol{\varphi}(x) - \mathrm{rot}\,\mathrm{rot}\,\boldsymbol{\varphi}(x) . \tag{3.234}$$

Combining the two equations above and using (3.226), we can derive the equation of motion (3.219) for φ_μ. In addition, differentiating (3.226) with respect to time and combining the result with equation (3.224), we obtain the condition (3.220). Thus, the canonical equations (3.233) and (3.234) are equivalent to the original equations.

We can also derive the important equal-time commutation relation

$$\begin{aligned}[\dot{\varphi}_j(x), \varphi_k^\dagger(x')] &= \left[\pi_j(x) - \frac{1}{m^2}\frac{\partial}{\partial x_j}\mathrm{div}\,\boldsymbol{\pi}(x), \varphi_k^\dagger(x')\right] \\ &= -\mathrm{i}\left(\delta_{jk} - \frac{1}{m^2}\frac{\partial^2}{\partial x_j \partial x_k}\delta^3(x - x')\right) .\end{aligned} \tag{3.235}$$

Fourier Expansion

Similarly to what we have done so far, we introduce the Fourier expansion for vector fields:

$$\varphi_j(x) = \sum_{\boldsymbol{p}} \frac{1}{\sqrt{2p_0 V}}\left[\mathrm{e}^{\mathrm{i}p\cdot x}a_j(\boldsymbol{p}) + \mathrm{e}^{-\mathrm{i}p\cdot x}b_j^\dagger(\boldsymbol{p})\right] ,$$

$$\varphi_j^\dagger(x) = \sum_{\boldsymbol{p}} \frac{1}{\sqrt{2p_0 V}}\left[\mathrm{e}^{\mathrm{i}p\cdot x}b_j(\boldsymbol{p}) + \mathrm{e}^{-\mathrm{i}p\cdot x}a_j^\dagger(\boldsymbol{p})\right] , \quad j = 1, 2, 3 . \tag{3.236}$$

We can write the Fourier expansions for φ_4 and $\varphi_4^\dagger$ in terms of a_j and b_j above because they can be expressed in terms of the other components, but here we skip this.

Following the same method as before, we can determine the commutation relations between coefficients. Here we only give the result:

$$\left[a_j(\boldsymbol{p}), a_k^\dagger(\boldsymbol{q})\right] = \left[b_j(\boldsymbol{p}), b_k^\dagger(\boldsymbol{q})\right] = \delta_{\boldsymbol{p},\boldsymbol{q}}\left(\delta_{jk} + \frac{p_j p_k}{m^2}\right) . \tag{3.237}$$

All other commutators are zero. Then, including the fourth component, we rewrite the Fourier expansions:

$$\varphi_\mu(x) = \sum_{\boldsymbol{p}} \frac{1}{\sqrt{2p_0V}} \left[\mathrm{e}^{\mathrm{i}p\cdot x} a_\mu(\boldsymbol{p}) + \mathrm{e}^{-\mathrm{i}p\cdot x} b_\mu^\dagger(\boldsymbol{p}) \right] ,$$

$$\varphi_\mu^\dagger(x) = \sum_{\boldsymbol{p}} \frac{1}{\sqrt{2p_0V}} \left[\mathrm{e}^{\mathrm{i}p\cdot x} b_\mu(\boldsymbol{p}) + \mathrm{e}^{-\mathrm{i}p\cdot x} a_\mu^\dagger(\boldsymbol{p}) \right] . \tag{3.238}$$

Using the irreducibility condition (3.220), we find that

$$p_\mu a_\mu(\boldsymbol{p}) = p_\mu a_\mu^\dagger(\boldsymbol{p}) = p_\mu b_\mu(\boldsymbol{p}) = p_\mu b_\mu^\dagger(\boldsymbol{p}) = 0 . \tag{3.239}$$

From these relations, it turns out that the zeroth component of a and b can be expressed in terms of the other three components:

$$a_0(\boldsymbol{p}) = \frac{p_j}{p_0} a_j(\boldsymbol{p}) , \quad \text{and similar equations.} \tag{3.240}$$

Combining (3.237) with the above equation,

$$\left[a_\mu(\boldsymbol{p}), a_\nu^\dagger(\boldsymbol{q}) \right] = \left[b_\mu(\boldsymbol{p}), b_\nu^\dagger(\boldsymbol{q}) \right] = \delta_{\boldsymbol{p},\boldsymbol{q}} \left(\delta_{\mu\nu} + \frac{p_\mu p_\nu}{m^2} \right) . \tag{3.241}$$

All other commutation relations are again zero.

We choose three unit vectors orthogonal to each other as

$$\boldsymbol{e}_1 = \boldsymbol{p}/p , \quad \boldsymbol{e}_2 , \quad \boldsymbol{e}_3 = \boldsymbol{e}_1 \times \boldsymbol{e}_2 , \tag{3.242}$$

and we expand $\boldsymbol{a}(\boldsymbol{p})$ as

$$\boldsymbol{a}(\boldsymbol{p}) = \boldsymbol{e}_1 a^{(1)}(\boldsymbol{p}) + \boldsymbol{e}_2 a^{(2)}(\boldsymbol{p}) + \boldsymbol{e}_3 a^{(3)}(\boldsymbol{p}) . \tag{3.243}$$

In this new coordinate system,

$$\left[a^{(1)}(\boldsymbol{p}), a^{(1)\dagger}(\boldsymbol{p}) \right] = 1 + \frac{p^2}{m^2} = \frac{p_0^2}{m^2} ,$$

$$\left[a^{(2)}(\boldsymbol{p}), a^{(2)\dagger}(\boldsymbol{p}) \right] = \left[a^{(3)}(\boldsymbol{p}), a^{(3)\dagger}(\boldsymbol{p}) \right] = 1 , \tag{3.244}$$

where $a^{(1)}$ corresponds to the longitudinal component and $a^{(2)}$ and $a^{(3)}$ to the transverse components. Here we change the definition of $a^{(j)}(\boldsymbol{p})$. Instead of (3.243), we set

$$\boldsymbol{a}(\boldsymbol{p}) = \boldsymbol{e}_1 \left(\frac{p_0}{m} \right) a^{(1)}(\boldsymbol{p}) + \boldsymbol{e}_2 a^{(2)}(\boldsymbol{p}) + \boldsymbol{e}_3 a^{(3)}(\boldsymbol{p}) . \tag{3.245}$$

Consequently, this new operator satisfies the simple commutation relation

$$\left[a^{(j)}(\boldsymbol{p}), a^{(k)}(\boldsymbol{p})\right] = \delta_{jk} \,. \tag{3.246}$$

Furthermore, using this new operator, the energy and momentum of the field are

$$H = \sum_{p} \sum_{j} p_0 \left[n_+^{(j)}(\boldsymbol{p}) + n_-^{i}(\boldsymbol{p}) \right] , \tag{3.247}$$

$$P_k = \sum_{p} \sum_{j} p_k \left[n_+^{(j)}(\boldsymbol{p}) + n_-^{(j)}(\boldsymbol{p}) \right] , \tag{3.248}$$

$$n_+^{(j)}(\boldsymbol{p}) = a^{(j)\dagger}(\boldsymbol{p}) a^{(j)}(\boldsymbol{p}) \,, \quad n_-^{(j)}(\boldsymbol{p}) = b^{(j)\dagger}(\boldsymbol{p}) b^{(j)}(\boldsymbol{p}) \,, \tag{3.249}$$

where we have assumed that the momentum $\boldsymbol{P}$ is given by interpreting α as the three components of the vector field in (3.30). n_+ and n_- are the particle number operators. Their eigenvalues are non-negative integers, $0, 1, 2, \ldots$, reflecting the Bose statistics.

In addition, the operation of charge conjugation connects $\varphi_\mu(x)$ and $\varphi_\mu^\dagger(x)$ by a relation similar to (3.78).

In this chapter, considering the fields with spin 0, 1/2, and 1 as quantum-mechanical operators, we have quantized them by using the canonical commutation relations. A field satisfying the so-called *wave equation* as equation of motion can be considered as a classical wave. However, as a consequence of the quantization, the eigenvalues of its energy and momentum coincide with those for particles of the same mass, which implies that fields possess particle-like aspects. Intuitively speaking, in the free state of motion, wave-like aspects described by the wave equation show up strongly, and when the state of motion is changed via interactions, the change in the four-momentum has particle-like aspects.

The dual particle–wave nature arising in quantum mechanics can be understood only after quantizing fields.

Chapter 4
Invariant Functions and Quantization of Free Fields

In the last chapter, we quantized free fields using the canonical quantization procedure. However, we did this by fixing a time axis, and we should ask whether the result is independent of this choice. To answer this question, we need to study the unequal-time commutation relations. By introducing several kinds of Lorentz-invariant Green functions, we will show that free fields can be quantized without fixing a direction of time.

4.1 Unequal-Time Commutation Relations for Real Scalar Fields

Using the previously obtained Fourier expansion for the real scalar field, one can derive the commutation relations among two fields at unequal times. Using (3.51), (3.52), and (3.53),

$$
\begin{aligned}
[\varphi(x), \varphi(y)] &= \sum_{\boldsymbol{p}} \sum_{\boldsymbol{q}} \frac{1}{\sqrt{2p_0V}} \frac{1}{\sqrt{2q_0V}} \Big(\mathrm{e}^{\mathrm{i}p\cdot x - \mathrm{i}q\cdot y}[a(\boldsymbol{p}), a^\dagger(\boldsymbol{q})] \\
&\qquad + \mathrm{e}^{-\mathrm{i}p\cdot x + \mathrm{i}q\cdot y}[a^\dagger(\boldsymbol{p}), a(\boldsymbol{q})] \Big) \\
&= \frac{1}{2p_0V} \left[\mathrm{e}^{\mathrm{i}p\cdot(x-y)} - \mathrm{e}^{-\mathrm{i}p\cdot(x-y)} \right] . \qquad (4.1)
\end{aligned}
$$

K. Nishijima, *Quantum Field Theory*,
https://doi.org/10.1007/978-94-024-2190-3_4

It is clear from (3.38) that the possible values of the momentum in a cube with fixed volume V are quantized by periodic conditions. Then, when V or L is large, the sum over momenta is replaced by an integral, i.e.,

$$\sum_{\boldsymbol{p}} \to \frac{V}{(2\pi)^3} \int \mathrm{d}^3 p \,. \tag{4.2}$$

Making this replacement in (4.1), the right-hand side becomes

$$\frac{1}{(2\pi)^3} \int \frac{\mathrm{d}^3 p}{2p_0} \left[\mathrm{e}^{\mathrm{i}p\cdot(x-y)} - \mathrm{e}^{-\mathrm{i}p\cdot(x-y)} \right] \equiv \mathrm{i}\Delta(x-y) \,. \tag{4.3}$$

This function was introduced by Pauli and Jordan [80], but the sign is different from the original, i.e.,

$$\Delta(x) = -\frac{\mathrm{i}}{(2\pi)^3} \int \frac{\mathrm{d}^3 p}{2p_0} \left(\mathrm{e}^{\mathrm{i}p\cdot x} - \mathrm{e}^{-\mathrm{i}p\cdot x} \right) . \tag{4.4}$$

In the following, we list some properties of this function.

1. $\Delta(x)$ is a Lorentz-invariant function, i.e., for the proper Lorentz transformation,

$$x_\mu \to x'_\mu = \sum_\nu a_{\mu\nu} x_\nu \,,$$

the following equality holds:

$$\Delta(x') = \Delta(x) \,. \tag{4.5}$$

Proof $\delta(p^2 + m^2)$ is clearly Lorentz invariant, and using the formula

$$\delta(ab) = |b|^{-1}\delta(a) + |a|^{-1}\delta(b) \,,$$

one finds

$$\delta(p^2 + m^2) = \frac{1}{2\sqrt{\boldsymbol{p}^2 + m^2}} \left[\delta\left(p_0 - \sqrt{\boldsymbol{p}^2 + m^2} \right) + \delta\left(p_0 + \sqrt{\boldsymbol{p}^2 + m^2} \right) \right] . \tag{4.6}$$

The first term is nonzero in the region $p_0 \geq m$ and the second in the region $p_0 \leq -m$. Since these two regions do not overlap, each term is invariant under the proper Lorentz transformation. We now introduce the following discontinuous functions:

$$\theta(p_0) = \begin{cases} 1 \text{ for } p_0 > 0 , \\ 0 \text{ for } p_0 < 0 , \end{cases} \tag{4.7}$$

$$\epsilon(p_0) = \theta(p_0) - \theta(-p_0) = p_0/|p_0| . \tag{4.8}$$

From the above, we note that both $\theta(p_0)\delta(p^2 + m^2)$ and $\epsilon(p_0)\delta(p^2 + m^2)$ are Lorentz invariant, whence the following function is also Lorentz invariant:

$$\int \mathrm{d}^4 p \epsilon(p_0)\delta(p^2 + m^2)\mathrm{e}^{\mathrm{i}p\cdot x}$$
$$= \frac{1}{2} \int \frac{\mathrm{d}p_0 \mathrm{d}^3 p}{\sqrt{\boldsymbol{p}^2 + m^2}} \left[\delta\left(p_0 - \sqrt{\boldsymbol{p}^2 + m^2}\right) - \delta\left(p_0 + \sqrt{\boldsymbol{p}^2 + m^2}\right) \right] \mathrm{e}^{\mathrm{i}\boldsymbol{p}\boldsymbol{x} - \mathrm{i}p_0 x_0} .$$

Integrating this equation over p_0 and inverting the integration variable according to $\boldsymbol{p} \to -\boldsymbol{p}$ in the second term, the above equation becomes

$$\int \mathrm{d}^4 p \epsilon(p_0)\delta(p^2 + m^2)\mathrm{e}^{\mathrm{i}p\cdot x} = \int \frac{\mathrm{d}^3 p}{2p_0} \left(\mathrm{e}^{\mathrm{i}p\cdot x} - \mathrm{e}^{-\mathrm{i}p\cdot x}\right) .$$

In this equation, p_0 stands for $\sqrt{\boldsymbol{p}^2 + m^2}$, and so is positive-definite. This equation coincides with (4.4) up to the coefficient. Therefore, $\Delta(x)$ is Lorentz invariant.

2. $\Delta(x)$ is an odd function, i.e.,

$$\Delta(-x) = -\Delta(x) . \tag{4.9}$$

3. If x is a space-like vector, i.e., $\boldsymbol{x}^2 > x_0^2$, then one has

$$\Delta(x) = 0 . \tag{4.10}$$

Proof If x is a space-like vector, then there exists a coordinate system with $x'_0 = 0$. However, the four-vector $x' = (\boldsymbol{x}', 0)$ can be transformed into $(-\boldsymbol{x}', 0)$ by a spatial rotation. This rotation is, of course, a Lorentz transformation. Thus, from properties (1) and (2),

$$\Delta(x) = \Delta(x') = \Delta(-x') = -\Delta(x') = 0 .$$

This tells us that $\varphi(x)$ and $\varphi(y)$ commute if x and y are spatially separated. According to Einstein, no action can ever be transmitted faster than the speed of light, whence observations at two spatially separated points can never interfere with each other. According to quantum mechanics, two physical quantities that can be observed without interfering with each other or getting in each other's way are said to be commutative. This consequence is thus a quantum mechanical expression of Einstein causality. It is called *local commutativity* or *microscopic causality*. In addition, the Lorentz invariance of $\Delta(x)$ means that the canonical quantization with a fixed choice of time axis is in fact independent of this choice of time axis. Thus, the commutation relation

$$[\varphi(x), \varphi(y)] = \mathrm{i}\Delta(x - y) \tag{4.11}$$

does not in effect depend on the way we choose the time axis.

4. $\Delta(x)$ satisfies the Klein–Gordon equation:

$$K_x \Delta(x) = (\Box_x - m^2)\Delta(x) = 0. \tag{4.12}$$

Proof We have

$$(\Box_x - m^2)[\varphi(x), \varphi(0)] = [(\Box_x - m^2)\varphi(x), \varphi(0)] = 0\,.$$

5. For $x_0 = 0$,

$$\Delta(x) = 0\,, \quad \frac{\partial}{\partial x_0}\Delta(x) = -\delta^3(x)\,. \tag{4.13}$$

Proof The first equation is a special case of property (3). The second can be obtained from the canonical commutation relation:

$$[\dot{\varphi}(x), \varphi(0)] = -\delta^3(x)\,, \quad \text{where } x_0 = 0\,.$$

It can be also derived from the Fourier representation of $\Delta(x)$.

These are the properties of the Jordan–Pauli invariant Δ-function. When the mass m vanishes, we write $\Delta(x)$ as $D(x)$.

4.2 Various Invariant Functions

In the previous section, we defined the invariant Δ-function using the commutation relation. Next, we introduce several kinds of Lorentz-invariant function that are closely associated with it.

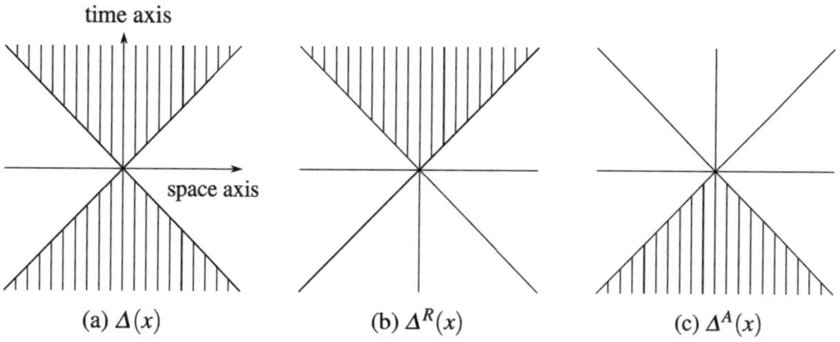

Fig. 4.1 Regions where the Δ-function is nonzero

First, since $\Delta(x)$ vanishes for space-like x, it is nonzero only in the future and past light cones. We define the retarded Δ-function Δ^{R} and the advanced Δ-function Δ^{A} as follows:

$$\Delta^{\mathrm{R}}(x) = -\theta(x_0)\Delta(x) = \begin{cases} -\Delta(x) & \text{when } x_0 > 0\,, \\ 0 & \text{when } x_0 < 0\,, \end{cases} \tag{4.14}$$

$$\Delta^{\mathrm{A}}(x) = \theta(-x_0)\Delta(x) = \begin{cases} 0 & \text{when } x_0 > 0\,, \\ \Delta(x) & \text{when } x_0 < 0\,. \end{cases} \tag{4.15}$$

The regions where these functions are not zero are the shaded areas in Fig. 4.1. Clearly, we have

$$\Delta(x) = \Delta^{\mathrm{A}}(x) - \Delta^{\mathrm{R}}(x)\,, \quad \Delta^{\mathrm{A}}(x) = \Delta^{\mathrm{R}}(-x)\,. \tag{4.16}$$

Next, we define the even function $\bar{\Delta}$ by summing Δ^{A} and $\Delta^{\mathrm{R}}(x)$:

$$\bar{\Delta}(x) = \frac{1}{2}\left[\Delta^{\mathrm{A}}(x) + \Delta^{\mathrm{R}}(x)\right] = -\frac{1}{2}\epsilon(x_0)\Delta(x)\,. \tag{4.17}$$

The above functions have been derived by decomposing space-time into the appropriate regions. Furthermore, we can also decompose the function Δ in the momentum space in terms of positive and negative frequency modes:

$$\Delta(x) = \Delta^{(+)}(x) + \Delta^{(-)}(x)\,, \tag{4.18}$$

$$\mathrm{i}\Delta^{(+)}(x) = \frac{1}{(2\pi)^3}\int \frac{\mathrm{d}^3 p}{2p_0}\mathrm{e}^{\mathrm{i}p\cdot x} = \frac{1}{(2\pi)^3}\int \mathrm{d}^4 p\,\theta(p_0)\delta(p^2+m^2)\mathrm{e}^{\mathrm{i}p\cdot x}\,, \tag{4.19}$$

$$
\begin{aligned}
-\mathrm{i}\Delta^{(-)}(x) &= \frac{1}{(2\pi)^3}\int \frac{\mathrm{d}^3 p}{2p_0}\mathrm{e}^{-\mathrm{i}p\cdot x} \\
&= \frac{1}{(2\pi)^3}\int \mathrm{d}^4 p\theta(-p_0)\delta(p^2+m^2)\mathrm{e}^{\mathrm{i}p\cdot x} .
\end{aligned}
\tag{4.20}
$$

Therefore,

$$
\Delta^{(+)}(-x) = -\Delta^{(-)}(x)\,, \quad \Delta^{(-)}(-x) = -\Delta^{(+)}(x)\,. \tag{4.21}
$$

Next, we introduce the even function $\Delta^{(1)}$:

$$
\Delta^{(1)}(x) = \mathrm{i}\big[\Delta^{(+)}(x) - \Delta^{(-)}(x)\big] = \frac{1}{(2\pi)^3}\int \mathrm{d}^4 p\delta(p^2+m^2)\mathrm{e}^{\mathrm{i}p\cdot x} . \tag{4.22}
$$

From these, we obtain the following equations:

$$
\Delta^{(+)}(x) = \frac{1}{2}\big[\Delta(x) - \mathrm{i}\Delta^{(1)}(x)\big]\,, \quad \Delta^{(-)}(x) = \frac{1}{2}\big[\Delta(x) + \mathrm{i}\Delta^{(1)}(x)\big]\,. \tag{4.23}
$$

We observe that $\Delta^{(+)}$ and $\Delta^{(-)}$ satisfy the Klein–Gordon equation since $\Delta(x)$ and $\Delta^{(1)}(x)$ do. Furthermore,

$$
K_x\Delta^{\mathrm{R}}(x) = K_x\Delta^{\mathrm{A}}(x) = K_x\bar{\Delta}(x) = -\delta^4(x)\,. \tag{4.24}
$$

To prove these equations, we can use

$$
\Delta^{\mathrm{R}}(x) = \bar{\Delta}(x) - \frac{1}{2}\Delta(x)\,, \quad \Delta^{\mathrm{A}}(x) = \bar{\Delta}(x) + \frac{1}{2}\Delta(x)\,, \tag{4.25}
$$

and then we only need to prove the last equation in (4.24). For this purpose, we derive the Fourier expression of $\bar{\Delta}(x)$. Denoting Cauchy's principal value by P, we can write

$$
\epsilon(x_0) = \frac{x_0}{|x_0|} = \frac{2}{\pi}\int_0^\infty \frac{\mathrm{d}\tau}{\tau}\sin(\tau x_0) = \frac{1}{\mathrm{i}\pi}\mathrm{P}\int_{-\infty}^{\infty}\frac{\mathrm{d}\tau}{\tau}\mathrm{e}^{\mathrm{i}\tau x_0}\,. \tag{4.26}
$$

Therefore,

$$
\begin{aligned}
\bar{\Delta}(x) &= \frac{\mathrm{i}}{2\pi}\mathrm{P}\int\frac{\mathrm{d}\tau}{\tau}\mathrm{e}^{\mathrm{i}\tau x_0}\frac{(-\mathrm{i})}{(2\pi)^3}\int \mathrm{d}^4 k\epsilon(k_0)\delta(k^2+m^2)\mathrm{e}^{\mathrm{i}k\cdot x} \\
&= \frac{1}{(2\pi)^4}\int \mathrm{d}^4 p\;\mathrm{e}^{\mathrm{i}p\cdot x}\mathrm{P}\int\frac{\mathrm{d}\tau}{\tau}\delta\big[\boldsymbol{p}^2 - (p_0+\tau)^2 + m^2\big]\frac{p_0+\tau}{|p_0+\tau|} \\
&= \frac{1}{(2\pi)^4}\mathrm{P}\int\frac{\mathrm{d}^4 p}{p^2+m^2}\mathrm{e}^{\mathrm{i}p\cdot x}\,,
\end{aligned}
\tag{4.27}
$$

where we have taken $\boldsymbol{p} = \boldsymbol{k}$ and $p_0 = k_0 - \tau$. Thus,

$$K_x \bar{\Delta}(x) = -\frac{1}{(2\pi)^4} \int \mathrm{d}^4 p \, \mathrm{e}^{\mathrm{i}p\cdot x} = -\delta^4(x) , \tag{4.28}$$

which proves (4.24).

Finally, we introduce Feynman's function Δ_F, also known as Stueckelberg's causal function. This is defined by

$$\begin{aligned} \Delta_\mathrm{F} &= \frac{1}{2}\Delta^{(1)}(x) - \mathrm{i}\bar{\Delta}(x) , \\ &= \mathrm{i}\theta(x_0)\Delta^{(+)}(x) + \mathrm{i}\theta(-x_0)\delta^{(+)}(-x) , \\ &= \frac{-\mathrm{i}}{(2\pi)^4} \int \mathrm{d}^4 p \, \mathrm{e}^{\mathrm{i}p\cdot x} \left[\frac{\mathrm{P}}{p^2 + m^2} + \mathrm{i}\pi\delta(p^2 + m^2) \right] , \\ &= \frac{-\mathrm{i}}{(2\pi)^4} \lim_{\epsilon \to +0} \int \mathrm{d}^4 p \, \mathrm{e}^{\mathrm{i}p\cdot x} \frac{1}{p^2 + m^2 - \mathrm{i}\epsilon} . \end{aligned} \tag{4.29}$$

From now on, we will omit the limit symbol, taking it as understood. Clearly, this function satisfies

$$K_x \Delta_\mathrm{F}(x) = \mathrm{i}\delta^4(x) . \tag{4.30}$$

4.3 Unequal-Time Commutation Relations of Free Fields

Up to now, we have investigated the equal-time commutation relations associated with the canonical quantization. In fact, using the invariant function we can obtain the unequal-time commutation relation directly.

For the real scalar field, as mentioned in Sect. 4.1, we have

$$[\varphi(x), \varphi(y)] = \mathrm{i}\Delta(x - y) . \tag{4.31}$$

Similarly, for the complex scalar field,

$$[\varphi(x), \varphi^\dagger(y)] = [\varphi^\dagger(x), \varphi(y)] = \mathrm{i}\Delta(x - y) , \tag{4.32}$$

$$[\varphi(x), \varphi(y)] = [\varphi^\dagger(x), \varphi^\dagger(y)] = 0 . \tag{4.33}$$

Next, we derive the unequal-time anti-commutation relation for spinor fields. We solve this as an initial value problem of a differential equation, i.e., we start with the equation

$$\left(\gamma_\mu \frac{\partial}{\partial x_\mu} + m \right) \{\psi(x), \bar{\psi}(y)\} = 0 . \tag{4.34}$$

As an initial condition, we use the equal-time anti-commutation relation

$$\{\psi_r(x), \bar{\psi}_s(y)\} = (\gamma_4)_{rs}\delta^3(x-y)\ , \quad \text{when } x_0 = y_0\ . \tag{4.35}$$

The solution of (4.34) is given uniquely by

$$\{\psi_r(x), \bar{\psi}_s(y)\} = -\mathrm{i}\left(\gamma_\mu \frac{\partial}{\partial x_\mu} - m\right)_{rs} \Delta(x-y) = \mathrm{i}S_{rs}(x-y)\ . \tag{4.36}$$

Setting $x_0 = y_0$ in this equation, the right-hand side becomes

$$-\mathrm{i}\gamma_4 \frac{\partial}{\partial x_4}\Delta(x-y) = \gamma_4\delta^3(x-y)\ ,$$

which gives (4.35). Thus, in this case the anti-commutation relations are

$$\begin{aligned} \{\psi_r(x), \bar{\psi}_s(y)\} &= \mathrm{i}S_{rs}(x-y)\ , \\ \{\psi_r(x), \psi_s(y)\} &= \{\bar{\psi}_r(x), \bar{\psi}_s(y)\} = 0\ . \end{aligned} \tag{4.37}$$

Next, we consider the complex vector field. In this case, the commutator satisfies the equations

$$K_x[\varphi_\mu(x), \varphi_\nu^\dagger(y)] = 0\ , \tag{4.38}$$

$$\frac{\partial}{\partial x_\mu}[\varphi_\mu(x), \varphi_\nu^\dagger(y)] = 0\ . \tag{4.39}$$

To begin with, we consider only the space components. The time component will be analyzed using the second equation. The initial condition is given by the equal-time commutation relations:

$$[\varphi_j(x), \varphi_k^\dagger(y)] = 0\ , \quad [\dot{\varphi}_j(x), \varphi_k^\dagger(y)] = -\mathrm{i}\left(\delta_{jk} - \frac{1}{m^2}\frac{\partial^2}{\partial x_j \partial x_k}\right)\delta^3(x-y)\ , \tag{4.40}$$

where $x_0 = y_0$. The solution which satisfies (4.38), (4.39), and (4.40) is

$$\begin{aligned} [\varphi_\mu(x), \varphi_\nu^\dagger(y)] &= \mathrm{i}\left(\delta_{\mu\nu} - \mathrm{i}\frac{1}{m^2}\frac{\partial^2}{\partial x_\mu \partial x_\nu}\right)\Delta(x-y)\ , \\ [\varphi_\mu(x), \varphi_\nu(y)] &= [\varphi_\mu^\dagger(x), \varphi_\nu^\dagger(y)] = 0\ . \end{aligned} \tag{4.41}$$

This implies that, although canonical quantization gives an initial condition to the above differential equation by fixing a time axis, the commutators obtained in the end are independent of the way we choose a specific inertial system. Thus, the method of canonical quantization yields a Lorentz-invariant quantization, at least for free fields.

In Sect. 2.3, we showed that the canonical equation and the Euler–Lagrange equation coincide as long as the canonical commutation relations hold. Conversely, if we assume that the canonical equation coincides with the Euler–Lagrange equation, does this restrict the commutation relations? Furthermore, can we derive the commutation relations without specifying a time axis as we did in the method of canonical quantization? Using these ideas, we shall now quantize free fields.

4.4 Generalities of the Quantization of Free Fields

As we mentioned in the previous section, from the requirement that the canonical equation and Euler's equation must coincide, we can derive the commutation relations for free fields. In this case, an important assumption is that all free fields satisfy the Klein–Gordon equation as a consequence of the field equation.

Assume that the Lagrangian density has the form

$$\mathscr{L} = \mathscr{L}\left[\varphi_\alpha, \varphi_\alpha^\dagger, \varphi_{\alpha,\mu}, \varphi_{\alpha,\mu}^\dagger\right], \tag{4.42}$$

where φ_α is a complex field. We will deal with the changes needed to handle a real field whenever necessary. In the case of free fields, $\mathscr{L}$ is a bilinear form in φ, $\varphi^\dagger$, and their derivatives. The Euler–Lagrange equation can therefore be written in the form

$$[\mathscr{L}]_{\varphi_\alpha^\dagger} = D_{\alpha\beta}(\partial)\varphi_\beta = 0\,, \tag{4.43}$$

where D is a matrix which includes differential operators. In the case of the complex scalar field, we have

$$[\mathscr{L}]_{\varphi^\dagger} = K\varphi = (\Box - m^2)\varphi = 0\,, \tag{4.44}$$

so in this case the one-dimensional matrix D is given by

$$D(\partial) = K = \Box - m^2\,. \tag{4.45}$$

As a perturbation, we add to the Lagrangian density of the above free field an interaction term between $\varphi_\alpha(x)$ and an external complex field $Q(x)$, and assume equivalence of the canonical equation and the Euler–Lagrange equation:

$$\mathscr{L}(x) \to \mathscr{L}(x) - Q(x)\varphi_\alpha^\dagger(x)\varphi_\alpha(x)\,, \tag{4.46}$$

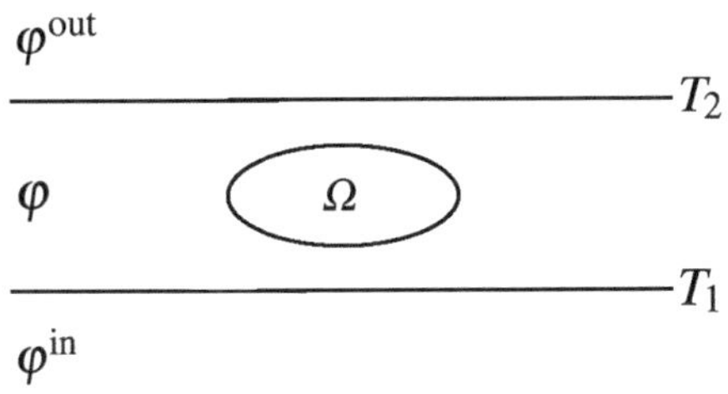

Fig. 4.2 Relation between T_1, T_2, and Ω

where we sum over α. From the variational principle,

$$D_{\alpha\beta}(\partial)\varphi_\beta - Q\varphi_\alpha = 0\,. \tag{4.47}$$

If φ is a real scalar field, then Q can be replaced by $2Q$ in the above equation. We will return to this difference later.

Next, we assume that Q is nonzero only in a certain finite space-time region Ω, while

$$Q = 0 \;\text{ when } t > T_2 \text{ or } t < T_1\,. \tag{4.48}$$

We denote φ_α before T_1 by $\varphi_\alpha^{\text{in}}$, and φ_α after T_2 by $\varphi_\alpha^{\text{out}}$. This anteroposterior relation is illustrated in Fig. 4.2. Since the external field vanishes before T_1 and after T_2, we obtain

$$D_{\alpha\beta}(\partial)\varphi_\beta^{\text{in}} = 0 \quad \text{for } t < T_1\,, \tag{4.49}$$

$$D_{\alpha\beta}(\partial)\varphi_\beta^{\text{out}} = 0 \quad \text{for } t > T_2\,. \tag{4.50}$$

If Q is zero in the whole of space-time, then of course we obtain

$$\varphi_\alpha = \varphi_\alpha^{\text{in}} = \varphi_\alpha^{\text{out}}\,.$$

We next try to solve for φ_α in terms of φ_α^{in}, and set

$$\varphi_\alpha = \varphi_\alpha^{\text{in}} + \varphi'_\alpha\,. \tag{4.51}$$

It is clear that $\varphi'_\alpha = 0$ for $t < T_1$. Applying $D_{\alpha\beta}$ to the above equation, we find

$$D_{\alpha\beta}(\partial)\varphi_\beta = D_{\alpha\beta}(\partial)\varphi_\beta^{\text{in}} + D_{\alpha\beta}(\partial)\varphi'_\beta = Q\varphi_\alpha\,.$$

Therefore, from (4.49),

$$D_{\alpha\beta}(\partial)\varphi'_\beta = Q\varphi_\alpha\,. \tag{4.52}$$

Since we have the boundary condition that φ' is 0 in the region $t < T_1$, we can integrate the above equation using the retarded Δ-function, i.e., defining Δ^{R} by

$$D_{\alpha\beta}(\partial)\Delta^{\mathrm{R}}_{\beta\gamma}(x) = -\delta_{\alpha\gamma}\delta^4(x)\,, \quad \Delta^{\mathrm{R}}_{\alpha\beta} = 0 \;\text{ when } x_0 < 0\,, \tag{4.53}$$

the quantity φ'_α can be written in the form

$$\varphi'_\alpha(x) = -\int_\Omega \mathrm{d}^4x'\,\Delta^{\mathrm{R}}_{\alpha\beta}(x-x')Q(x')\varphi_\beta(x')\,. \tag{4.54}$$

Thus, we find that φ'_α vanishes in the region $x_0 < T_1$ thanks to the properties of Δ^{R} and Ω.

Rewriting (4.51) for φ, we obtain the integral equation

$$\varphi_\alpha(x) = \varphi^{\mathrm{in}}_\alpha - \int_\Omega \mathrm{d}^4x'\,\Delta^{\mathrm{R}}_{\alpha\beta}(x-x')Q(x')\varphi_\beta(x')\,. \tag{4.55}$$

In addition, using $\varphi^{\mathrm{out}}_\alpha$ and Δ^{A} defined by changing the boundary condition in (4.53), we have

$$\varphi_\alpha(x) = \varphi^{\mathrm{out}}_\alpha(x) - \int_\Omega \mathrm{d}^4x'\,\Delta^{\mathrm{A}}_{\alpha\beta}(x-x')Q(x')\varphi_\beta(x')\,. \tag{4.56}$$

The above are consequences of the Euler–Lagrange equation.

We now discuss the canonical equation. To begin with, the Hamiltonian density is

$$\mathscr{H} = \sum_\alpha\left(\frac{\partial\mathscr{L}}{\partial\dot\varphi_\alpha}\dot\varphi_\alpha + \frac{\partial\mathscr{L}}{\partial\dot\varphi^\dagger_\alpha}\dot\varphi^\dagger_\alpha\right) - \mathscr{L}\,. \tag{4.57}$$

Introducing Q, the modification corresponding to (4.46) leads to

$$\mathscr{H}(x) \to \mathscr{H}(x) + Q(x)\varphi^\dagger_\alpha(x)\varphi_\alpha(x) \equiv \mathscr{H}(x) + \mathscr{H}'(x)\,. \tag{4.58}$$

The Hamiltonian of this system is

$$H = \int_{x_0=t} \mathrm{d}^3x\big[\mathscr{H}(x) + \mathscr{H}'(x)\big] \equiv \mathscr{H}(x) + \mathscr{H}'(x)\,. \tag{4.59}$$

When $Q(x)$ is nonzero, H depends explicitly on t. When a physical quantity F depends on t both implicitly via canonical variables and explicitly through some external field, its total derivative is given by

$$\frac{\mathrm{d}F}{\mathrm{d}t} = \frac{\partial F}{\partial(t)} + \mathrm{i}[H(t), F]\,, \tag{4.60}$$

where the first term on the right-hand side is the derivative with respect to the explicit t-dependence through the external field.

In particular, for a field, we have

$$\frac{\partial \varphi_\alpha(x)}{\partial x_0} = \mathrm{i}\big[H(x_0), \varphi_\alpha(x)\big] . \tag{4.61}$$

Since $Q = 0$ for $x_0 < T$, we obtain

$$\frac{\partial \varphi_\alpha^{\mathrm{in}}(x)}{\partial x_0} = \mathrm{i}\big[H_0[\varphi_\alpha^{\mathrm{in}}(x), \ldots], \varphi_\alpha^{\mathrm{in}}(x)\big] , \tag{4.62}$$

where $H_0[\cdots]$ stands for a functional of $\varphi_\alpha^{\mathrm{in}}(x)$. When Q is nonzero, we take φ_α as follows (by analogy with the super multi-time theory to be discussed later):

$$\varphi_\alpha(x) = U(x_0)^{-1}\varphi_\alpha^{\mathrm{in}}(x)U(x_0) . \tag{4.63}$$

The reason why we adopt this form is that φ_α and $\varphi_\alpha^{\mathrm{in}}$ satisfy the same equal-time commutation relation. Since φ_α is equal to $\varphi_\alpha^{\mathrm{in}}$ for $x_0 < T_1$,

$$U = 1 \quad \text{when } x_0 < T_1 . \tag{4.64}$$

Substituting (4.63) into (4.61), the right-hand side becomes

$$\begin{aligned} &\mathrm{i}U(x_0)^{-1}\big[H_0[\varphi_\alpha^{\mathrm{in}}, \ldots], \varphi_\alpha^{\mathrm{in}}(x)\big]U(x_0) + \mathrm{i}U(x_0)^{-1}\big[H'[\varphi_\alpha^{\mathrm{in}}, \ldots], \varphi_\alpha^{\mathrm{in}}(x)\big]U(x_0) \\ &\quad = U(x_0)^{-1}\frac{\partial \varphi_\alpha^{\mathrm{in}}(x)}{\partial x_0}U(x_0) + \mathrm{i}U(x_0)^{-1}\big[H'[\varphi_\alpha^{\mathrm{in}}(x), \ldots], \varphi_\alpha^{\mathrm{in}}(x)\big]U(x_0) . \end{aligned} \tag{4.65}$$

On the other hand, the left-hand side becomes

$$\begin{aligned} \frac{\partial}{\partial x_0}\big[U(x_0)^{-1}\varphi_\alpha^{\mathrm{in}}(x)U(x_0)\big] &= U(x_0)^{-1}\frac{\partial \varphi_\alpha^{\mathrm{in}}(x)}{\partial x_0}U(x_0) + U(x_0)^{-1}\varphi_\alpha^{\mathrm{in}}(x)\frac{\partial U(x_0)}{\partial x_0} \\ &= -U(x_0)^{-1}\frac{\partial U(x_0)}{\partial x_0}U(x_0)^{-1}\varphi_\alpha^{\mathrm{in}}(x)U(x_0) . \end{aligned} \tag{4.66}$$

In deriving this equation, we have used the following formula:

$$\frac{\partial}{\partial x_0}U(x_0)^{-1} = -U(x_0)^{-1}\frac{\partial U(x_0)}{\partial x_0}U(x_0)^{-1} .$$

Equating (4.65) and (4.66), we obtain

$$\left[\mathrm{i}H'\big[\varphi_\alpha^{\mathrm{in}}(x), \ldots\big] + \frac{\partial U(x_0)}{\partial x_0}U(x_0)^{-1}, \varphi_\alpha^{\mathrm{in}}(x)\right] = 0 . \tag{4.67}$$

This equation can be satisfied by setting the first factor in the commutator equal to zero, viz.,

$$\mathrm{i}\frac{\partial U(x_0)}{\partial x_0} = H'\big[\varphi_\alpha^{\mathrm{in}}(x), \ldots\big]U(x_0)\,, \tag{4.68}$$

where

$$H'\big[\varphi_\alpha^{\mathrm{in}}(x), \ldots\big] = \int \mathrm{d}^3x\, Q(x)\varphi_\alpha^{\dagger\mathrm{in}}(x)\varphi_\alpha^{\mathrm{in}}(x)\,. \tag{4.69}$$

As a consequence, we have shown that, if $U(x_0)$ satisfies (4.68), $\varphi_\alpha(x)$ can be written in the form (4.63). The boundary condition which $U(x_0)$ must satisfy is (4.64). Then, taking into account (4.64) and integrating (4.68), we obtain the integral equation

$$U(x_0) = 1 - \mathrm{i}\int_{-\infty}^{x_0} \mathrm{d}^4x'\, Q(x')\varphi_\beta^{\dagger\mathrm{in}}(x')\varphi_\beta^{\mathrm{in}}(x')U(x_0')\,. \tag{4.70}$$

We now compare the two integral equations (4.55) and (4.70). Assuming that Q is small, we consider the power series expansion with respect to Q, keeping terms up to the second order in Q. The solutions of the two integral equations are

$$\varphi_\alpha(x) = \varphi_\alpha^{\mathrm{in}}(x) - \int_\Omega \mathrm{d}^4x'\, \Delta_{\alpha\beta}^{\mathrm{R}}(x-x')Q(x')\varphi_\beta^{\mathrm{in}}(x') + \text{second order terms in } Q\,, \tag{4.71}$$

$$U(x_0) = 1 - \mathrm{i}\int_{-\infty}^{x_0} \mathrm{d}^4x'\, Q(x')\varphi_\beta^{\dagger\mathrm{in}}(x')\varphi_\beta^{\mathrm{in}}(x') + \text{second order terms in } Q\,. \tag{4.72}$$

In the same approximation,

$$\begin{aligned}\varphi_\alpha(x) &= U(x_0)^{-1}\varphi_\alpha^{\mathrm{in}}(x)U(x_0) \qquad (4.73)\\ &= \varphi_\alpha^{\mathrm{in}}(x) - \mathrm{i}\int_{-\infty}^{\infty} \mathrm{d}^4x'\theta(x_0-x_0')\big[\varphi_\alpha^{\mathrm{in}}(x), \varphi_\beta^{\dagger\mathrm{in}}(x')\varphi_\beta^{\mathrm{in}}(x')\big]Q(x') + \cdots\,.\end{aligned}$$

Requiring that (4.71) coincide with (4.73),

$$\theta(x_0-x_0')\big[\varphi_\alpha^{\mathrm{in}}(x), \varphi_\beta^{\dagger\mathrm{in}}(x')\varphi_\beta^{\mathrm{in}}(x')\big] = -\mathrm{i}\Delta_{\alpha\beta}^{\mathrm{R}}(x-x')\varphi_\beta^{\mathrm{in}}(x')\,. \tag{4.74}$$

Taking $Q \to 0$, $\varphi_\alpha^{\mathrm{in}}$ becomes φ_α, and we obtain for the free field

$$\theta(x_0-x_0')\big[\varphi_\alpha(x), \varphi_\beta^{\dagger}(x')\varphi_\beta(x')\big] = -\mathrm{i}\Delta_{\alpha\beta}^{\mathrm{R}}(x-x')\varphi_\beta(x')\,. \tag{4.75}$$

In a similar manner, we obtain

$$\theta(x_0' - x_0)\left[\varphi_\alpha(x), \varphi_\beta^\dagger(x')\varphi_\beta(x')\right] = \mathrm{i}\Delta_{\alpha\beta}^{\mathrm{A}}(x - x')\varphi_\beta(x') . \tag{4.76}$$

We define the Δ-function:

$$\Delta_{\alpha\beta}(x) = \Delta_{\alpha\beta}^{\mathrm{A}}(x) - \Delta_{\alpha\beta}^{\mathrm{R}}(x) . \tag{4.77}$$

Adding (4.75) and (4.76), we find the following commutation relation:

$$[\varphi_\alpha(x), \varphi_\beta^\dagger(x')\varphi_\beta(x')] = \mathrm{i}\Delta_{\alpha\beta}(x - x')\varphi_\beta(x') . \tag{4.78}$$

In the above, we considered a complex field. For real fields, a factor of 2 appears, as mentioned earlier:

$$\left[\varphi_\alpha(x), \varphi_\beta(x')\varphi_\beta(x')\right] = 2\mathrm{i}\Delta_{\alpha\beta}(x - x')\varphi_\beta(x') . \tag{4.79}$$

From the two equations above, we can derive the commutators among two fields. In the case of the complex field, the results are as follows:

1. Solution for Bose statistics:

$$\begin{aligned} \left[\varphi_\alpha(x), \varphi_\beta(x')\right] &= \left[\varphi_\alpha^\dagger(x), \varphi_\beta^\dagger(x')\right] = 0 , \\ \left[\varphi_\alpha(x), \varphi_\beta^\dagger(x')\right] &= \mathrm{i}\Delta_{\alpha\beta}(x - x') . \end{aligned} \tag{4.80}$$

2. Solution for Fermi statistics:

$$\begin{aligned} \left\{\varphi_\alpha(x), \varphi_\beta(x')\right\} &= \left\{\varphi_\alpha^\dagger(x), \varphi_\beta^\dagger(x')\right\} = 0 , \\ \left\{\varphi_\alpha(x), \varphi_\beta^\dagger(x)'\right\} &= \mathrm{i}\Delta_{\alpha\beta}(x - x') . \end{aligned} \tag{4.81}$$

Assuming Bose statistics for the real field,

$$\left[\varphi_\alpha(x), \varphi_\beta(x')\right] = \mathrm{i}\Delta_{\alpha\beta}(x - x') . \tag{4.82}$$

Regarding the applicability of Bose or Fermi statistics, we have the *spin–statistics theorem* due to Pauli [81, 82], which establishes a connection between spin and statistics. If we assume the Lorentz invariance of the theory, microscopic causality, and a positive-definite metric, particles with integer spin must obey Bose statistics, while particles with half-integer spin must obey Fermi statistics. We will discuss the notion of a positive-definite metric in the next chapter.

At the beginning of this section, we assumed that the fields satisfy the Klein–Gordon equation as a consequence of the field equation (4.43). This means that mathematically there exists a matrix of differential operators C satisfying

$$D_{\alpha\beta}(\partial)C_{\beta\gamma}(\partial) = C_{\alpha\beta}(\partial)D_{\beta\gamma}(\partial) = \delta_{\alpha\gamma}(\Box - m^2) . \tag{4.83}$$

Therefore, (4.53) and the corresponding solution for the equation of Δ^{A} are given by

$$\Delta^{\mathrm{R}}_{\alpha\beta}(x) = C_{\alpha\beta}(\partial)\Delta^{\mathrm{R}}(x) , \quad \Delta^{\mathrm{A}}_{\alpha\beta}(x) = C_{\alpha\beta}(\partial)\Delta^{\mathrm{A}}(x) . \tag{4.84}$$

Thus,

$$\Delta_{\alpha\beta}(x) = C_{\alpha\beta}(\partial)\Delta(x) . \tag{4.85}$$

Using this formula, free fields can be quantized directly.

One-Component Real Scalar Field In this case, it is clear that

$$C(\partial) = 1 , \quad \left[\varphi(x), \varphi(x')\right] = \mathrm{i}\Delta(x - x') . \tag{4.86}$$

Faddeev–Popov Ghost Field Suppose two real scalar fields c and $\bar{c}$ obey Fermi statistics and are described by the Lagrangian density

$$\mathscr{L} = \mathrm{i}\partial_\mu \bar{c} \cdot \partial_\mu c . \tag{4.87}$$

In this case,

$$D(\partial) = -\mathrm{i}\Box , \quad C(\partial) = \mathrm{i} . \tag{4.88}$$

The commutation relations are therefore

$$\left\{c(x), c(x')\right\} = \left\{\bar{c}(x), \bar{c}(x')\right\} = 0 , \quad \left\{c(x), \bar{c}(x')\right\} = -D(x - x') . \tag{4.89}$$

This case violates Pauli's theorem for spin and statistics, so at least one of the assumptions leading to that theorem must be relaxed. In fact, in this case we drop the assumption of a positive-definite metric and introduce an indefinite metric. This field is then referred to as a *ghost field*. As will be discussed later, it plays an important role in non-Abelian gauge field theory. In general, indefinite metrics arise in gauge field theories, as will be discussed in detail in the next chapter.

Complex Vector Field From the Lagrangian density (3.221),

$$[\mathscr{L}]_{\varphi^\dagger_\alpha} = \partial_\beta(\partial_\beta\varphi_\alpha - \partial_\alpha\varphi_\beta) - m^2\varphi_\alpha = D_{\alpha\beta}(\partial)\varphi_\beta . \tag{4.90}$$

Therefore,

$$D_{\alpha\beta}(\partial) = \delta_{\alpha\beta}(\Box - m^2) - \partial_\alpha \partial_\beta \,. \tag{4.91}$$

From this, we obtain C as follows:

$$C_{\alpha\beta}(\partial) = \delta_{\alpha\beta} - \frac{1}{m^2}\partial_\alpha \partial_\beta \,. \tag{4.92}$$

This is the well-known form, reproducing the equation (4.41).
Compared to the method of canonical quantization, it is clear from the example of the vector field that, using this approach, the unequal-time commutation relations can be obtained rather easily.

Chapter 5
Indefinite Metric and the Electromagnetic Field

One may say that the history of quantum field theory began with the quantization of the electromagnetic field. However, the electromagnetic field cannot be treated as a simple field because one feature of this vector field, which is in fact a *gauge field*, is an indefinite metric. We thus begin by discussing this metric. For more detail, the reader is referred to the book *Quantum Field Theory* by Noboru Nakanishi, written in Japanese [83].

5.1 Indefinite Metric

A key tool for formulating quantum mechanics is Hilbert space, which is a vector space. We start with a brief review of finite-dimensional vector spaces $\mathscr{V}$, which will be intuitive rather than mathematically rigorous:

1. Vectors form a group under addition.
2. Vectors can be multiplied by complex numbers and the distributive property holds.
3. The number of linearly independent vectors defines the dimension of the vector space.

In quantum physics, vectors are written in the form "bra" $\langle x|$ or "ket" $|x\rangle$. We define an inner product which is used to introduce a metric.

The inner product has the following properties:

$$\langle y|x\rangle = \langle x|y\rangle^* , \tag{5.1}$$

$$\langle y|ax + bx'\rangle = a\langle y|x\rangle + b\langle y|x'\rangle , \tag{5.2}$$

$$\langle ay + by'|x\rangle = a^*\langle y|x\rangle + b^*\langle y'|x\rangle , \tag{5.3}$$

K. Nishijima, *Quantum Field Theory*,
https://doi.org/10.1007/978-94-024-2190-3_5

where a and b stand for complex numbers. The third equation follows from the first two. In mathematics, (5.2) and (5.3) are often the other way round, i.e., a^* and b^* appear in (5.2), while a and b appear in (5.3). The inner product is said to be positive-definite if

$$\langle x|x\rangle \geq 0 \,. \tag{5.4}$$

Then, $\langle x|x\rangle = 0$ and $|0\rangle = 0$ are equivalent. In this case, the space is Euclidean. An indefinite-metric space is one in which we relax the condition (5.4). Then, (5.1) implies that $\langle x|x\rangle$ is real but not necessarily positive.

Linear Operators

If T is such that, for an arbitrary $|x\rangle \in \mathscr{V}$, $T|x\rangle \in \mathscr{V}$, then we call T an *operator on* $\mathscr{V}$. If T satisfies the linearity condition

$$T|ax + by\rangle = aT|x\rangle + bT|y\rangle \,, \tag{5.5}$$

it is called a *linear operator*.

Given a set of linearly independent vectors, i.e., a basis $\{e_j\}$, we can construct a representation of T. An arbitrary $|x\rangle \in \mathscr{V}$ can be uniquely expressed in the form

$$|x\rangle = \sum_j x_j|e_j\rangle \,. \tag{5.6}$$

We can then write $T|e_j\rangle$ in the form

$$T|e_j\rangle = \sum_k |e_k\rangle t_{kj} \,. \tag{5.7}$$

The matrix t with components t_{kj} is called a *representation of* T. Denoting the corresponding representation of S by s,

$$\begin{aligned} ST|e_j\rangle &= \sum_k S|e_k\rangle t_{kj} \\ &= \sum_k \sum_l |e_l\rangle s_{lk} t_{kj} \\ &= \sum_l |e_l\rangle \left(\sum_k s_{lk} t_{kj} \right) , \end{aligned} \tag{5.8}$$

whence the matrix st represents the operator ST.

A matrix u such that $\det u \neq 0$ can be used to transform to another basis:

$$|e'_j\rangle = \sum_k |e_k\rangle u_{kj} \,. \tag{5.9}$$

Denoting the representation of T in the new basis by $\tilde{t}$, we have

$$T|e'_j\rangle = \sum_k |e'_k\rangle \tilde{t}_{kj} \ . \tag{5.10}$$

The transformation law for the representation t is then given by the similarity transformation

$$\tilde{t} = u^{-1} t u \ . \tag{5.11}$$

Metric Matrix
Given a basis $\{e_j\}$, we define the metric matrix by

$$\eta_{mn} = \langle e_m | e_n \rangle \ . \tag{5.12}$$

From the property (5.1) of the inner product, we then have $\eta_{nm} = \eta^*_{mn}$, i.e., $\eta^\dagger = \eta$, which shows that η is a Hermitian matrix. Under the change of basis (5.9), this matrix transforms by

$$\tilde{\eta} = u^\dagger \eta u \ . \tag{5.13}$$

Note the difference with (5.11).

Degeneracy
If there exists a vector $|k\rangle$ satisfying $\langle l|k\rangle = 0$ for any vector $|l\rangle$ in $\mathscr{V}$, the vector space is said to be degenerate.

Hermitian Conjugate
For a given linear operator T and arbitrary vectors $|k\rangle$ and $|l\rangle$, the relation

$$\langle k|T^\dagger|l\rangle = \langle l|T|k\rangle^* \tag{5.14}$$

defines the Hermitian conjugate $T^\dagger$ of T. If the matrices representing T and $T^\dagger$ are s and t, respectively, then

$$\eta s = t^\dagger \eta \ . \tag{5.15}$$

If in particular $\mathscr{V}$ is non-degenerate and $\det \eta \neq 0$, then we obtain

$$s = \eta^{-1} t^\dagger \eta \ . \tag{5.16}$$

Thus, in general, the matrix representing a Hermitian operator, i.e., one with $T^\dagger = T$, will not be a Hermitian matrix. In an indefinite-metric space, we use the following terminology for operators:

$$\begin{aligned} &U^\dagger = U^{-1} && \text{unitary matrix}\,, \\ &P^2 = P && \text{projection operator}\,, \\ &K^n = 0 \ \ (K^{n-1} \neq 0) && \text{nilpotent operator}\,. \end{aligned}$$

In a non-degenerate vector space, the eigenvalues of η can be normalized to ± 1 by a suitable transformation of the basis.

5.2 Generalized Eigenstates

When the metric is indefinite, things happen that would not be expected with a positive-definite metric. In the Lorentz-invariant formulation of gauge field theories, which include the electromagnetic field as a typical case, the indefinite metric is inevitable, so we discuss its properties here.

If we have

$$T|k\rangle = \lambda|k\rangle\,, \tag{5.17}$$

then, as in the case of the positive-definite metric, $|k\rangle$ is called an eigenstate of T and λ its eigenvalue. In addition, if $\mathscr{W}$ is a subspace of $\mathscr{V}$ with the property

$$|k\rangle \in \mathscr{W} \implies T|k\rangle \in \mathscr{W}\,, \tag{5.18}$$

then $\mathscr{W}$ is called an *invariant subspace* of T. For any matrix t representing T, we can define

$$f(x) = \det(x\boldsymbol{I} - t)\,, \tag{5.19}$$

where $\boldsymbol{I}$ is the unit matrix in the space of the representation t. This polynomial in x is called the *eigen polynomial* or *characteristic polynomial* of T and is independent of the choice of basis.

Cayley–Hamilton Theorem This states that T satisfies its own characteristic polynomial, i.e.,

$$f(T) = 0\,. \tag{5.20}$$

Proof We define the adjugate matrix $\tilde{s}$ of the matrix s by

$$\tilde{s}_{mn} = \frac{\partial}{\partial s_{nm}} \det s \; . \tag{5.21}$$

Clearly,

$$\sum_m s_{nm} \tilde{s}_{ml} = \delta_{nl} \det s \; . \tag{5.22}$$

We define the primary expression $s_{nm}(x)$ by

$$s_{nm}(x) = x\delta_{nm} - t_{nm} \; . \tag{5.23}$$

Therefore,

$$\sum_m s_{nm}(T)\tilde{s}_{ml}(T) = \delta_{nl} f(T) \; , \tag{5.24}$$

$$T|m\rangle = \sum_n |n\rangle t_{nm} \; . \tag{5.25}$$

Rewriting (5.25),

$$\sum_n (\delta_{nm} T - t_{nm})|n\rangle = \sum_n s_{nm}(T)|n\rangle = 0 \; .$$

Thus,

$$\begin{aligned} \sum_{n,m} \tilde{s}_{ml}(T) s_{nm}(T)|n\rangle &= \sum_{n,m} s_{nm}(T)\tilde{s}_{ml}(T)|n\rangle \\ &= \sum_n \delta_{ln} f(T)|n\rangle \\ &= f(T)|l\rangle \\ &= 0 \; . \end{aligned} \tag{5.26}$$

Here $|l\rangle$ is an arbitrary basis vector. Since this equation holds for every element of the basis, this proves (5.20). ■

Next, we factorize $f(x)$. Considering $\lambda_i \neq \lambda_j$ for $i \neq j$, we obtain

$$f(x) = \prod_{j=1}^{r} (x - \lambda_j)^{N_j} \; , \quad \sum_{j=1}^{r} N_j = N \; , \tag{5.27}$$

where N is the dimension of the vector space. We know that $f(T) = 0$, and consider the polynomial with the lowest degree satisfying $g(T) = 0$:

$$g(x) = \prod_{j=1}^{r} (x - \lambda_j)^{n_j} \ , \quad 0 < n_j \leq N_j \ . \tag{5.28}$$

Decomposition Theorem The vector space $\mathscr{V}$ can be decomposed as

$$\mathscr{V} = \sum_{j=1}^{r} \bigoplus \mathscr{V}(\lambda_j) \ , \tag{5.29}$$

where $|k\rangle \in \mathscr{V}(\lambda_j)$ means that $(T - \lambda_j)^{n_j}|k\rangle = 0$.

Proof First, we define $g_1(x), g_2(x), \ldots, g_r(x)$ by

$$g_j(x) = \frac{g(x)}{(x - \lambda_j)^{n_j}} = \prod_{i \neq j} (x - \lambda_i)^{n_i} \ . \tag{5.30}$$

Since $g_1(x), g_2(x), \ldots, g_r(x)$ are disjoint, there exist polynomials $h_j(x)$ satisfying

$$\sum_{j=1}^{r} h_j(x) g_j(x) = 1 \ . \tag{5.31}$$

We now set

$$P_j(T) = h_j(T) g_j(T) \ , \quad \sum_{j=1}^{r} P_j(T) = 1 \ . \tag{5.32}$$

Additionally, if $i \neq j$, $g_i(x) g_j(x)$ can be divided by $g(x)$, so

$$g_i(T) g_j(T) = 0 \ .$$

Hence, if $i \neq j$, $P_i(T) P_j(T) = 0$. Multiplying (5.32) by $P_j(T)$,

$$P_j^2(T) = P_j(T) \ . \tag{5.33}$$

This shows that $P_j(T)$ is a projection operator, and $\mathscr{V}(\lambda_j)$ can be defined by $P_j(T)\mathscr{V} = \mathscr{V}(\lambda_j)$, which proves the decomposition theorem. ■

The orthogonality condition for eigenvectors is complicated in the case of an indefinite metric. Even if H is a Hermitian operator, it is not always true that its eigenvalues are real. If h is the matrix representation of H, (5.15) implies

$$\eta h = h^\dagger \eta = (\eta h)^\dagger \ . \tag{5.34}$$

Thus, the product ηh is a Hermitian matrix. Here we can choose η to satisfy $\eta^2 = \mathbf{1}$ and (5.13) allows us to select a proper basis when $\mathscr{V}$ is not degenerate. As a consequence,

$$\det(x\boldsymbol{I} - h) = \det\eta \cdot \det(x\eta - \eta h) \ , \tag{5.35}$$

which shows that the characteristic polynomial of H is then a real polynomial. Note again that the characteristic polynomial is independent of the choice of basis. The eigenvalues of H are thus either real numbers or complex conjugate pairs of complex numbers. In the following, we take H to be a Hermitian operator and consider the subspace $\mathscr{V}(\lambda)$ defined above.

First, considering the two vectors which belong to $\mathscr{V}(\lambda_j)$ and $\mathscr{V}(\lambda_k)$, i.e., the ket vectors $|e_j\rangle$ and $|e_k\rangle$, we have

$$(H - \lambda_j)^{n_j}|e_j\rangle = 0 \ , \quad \langle e_k|(H - \lambda_k^*)^{n_k} = 0 \ . \tag{5.36}$$

If $\lambda_j \neq \lambda^*{}_k$, then $(x - \lambda_j)^{n_j}$ and $(x - \lambda_k^*)^{n_k}$ are disjoint, so there exist polynomials $f(x)$ and $g(x)$ satisfying

$$f(x)(x - \lambda_j)^{n_j} + g(x)(x - \lambda_k^*)^{n_k} = 1 \ . \tag{5.37}$$

We thus obtain the orthogonality relation

$$\langle e_k|e_j\rangle = \langle e_k|f(H)(H - \lambda_j)^{n_j}|e_j\rangle + \langle e_k|(H - \lambda_k^*)^{n_k} g(H)|e_j\rangle = 0 \ . \tag{5.38}$$

For complex eigenvalues $\lambda_j \neq \lambda_j^*$, we then have

$$\langle e_j|e_j\rangle = 0 \ , \tag{5.39}$$

so the norm of this vector is zero.

If λ_j is a real number, there must exist a vector $|x_j\rangle$ in $\mathscr{V}(\lambda_j)$ such that

$$\langle x_j|(H - \lambda_j)^{n_j - 1}|x_j\rangle \neq 0 \ . \tag{5.40}$$

Otherwise, we would always have $(H - \lambda_j)^{n_j - 1} = 0$ on $\mathscr{V}(\lambda_j)$, and this would contradict the condition that n_j is the smallest such number. Therefore, the following ket vectors are linearly independent:

$$|x_j\rangle \ , \quad (H - \lambda_j)|x_j\rangle \ , \quad \ldots \ , \quad (H - \lambda_j)^{n_j - 1}|x_j\rangle \ . \tag{5.41}$$

Multipole State

For a real eigenvalue λ, when there is no $|x'\rangle$ such that $|x\rangle = (H - \lambda)|x'\rangle$ and

$$(H-\lambda)^m|x\rangle = 0\ , \quad \langle x|(H-\lambda)^{m-1}|x\rangle \neq 0 \quad (m \geq 2)\ , \tag{5.42}$$

we call $|x\rangle$ an *m th order multipole state*, where "state" refers to the quantum-mechanical state.

Here we normalize $|x\rangle$ so that

$$\langle x|(H-\lambda)^{m-1}|x\rangle = \epsilon\ , \quad \epsilon = \pm 1\ , \tag{5.43}$$

and then consider a linear combination of the linearly independent vectors in (5.41):

$$|z\rangle = |x\rangle + \sum_{k=1}^{m-1} c_k (H-\lambda)^k |x\rangle\ . \tag{5.44}$$

Choosing c_k suitably, we can arrange for the following equation to hold:

$$\langle z|(H-\lambda)^k|z\rangle = 0\ , \quad k = 0, 1, \ldots, m-2\ . \tag{5.45}$$

Combining (5.43) and (5.45), we have

$$\langle z|(H-\lambda)^k|z\rangle = \epsilon\delta_{k,m-1}\ . \tag{5.46}$$

We now construct the m linearly independent vectors in (5.41). Here we choose the set starting from $|z\rangle$ instead of $|x\rangle$, i.e.,

$$|z\rangle\ , \quad (H-\lambda)|z\rangle\ , \quad \ldots\ , \quad (H-\lambda)^{m-1}|z\rangle\ . \tag{5.47}$$

For this basis, η and the representation h of H are given by

$$\eta = \begin{pmatrix} \mathbf{0} & & \epsilon \\ & \cdot^{\cdot^{\cdot}} & \\ \epsilon & & \mathbf{0} \end{pmatrix}\ , \quad \eta^2 = 1\ , \tag{5.48}$$

$$h = \begin{pmatrix} \lambda & & & & \mathbf{0} \\ 1 & \lambda & & & \\ & 1 & \ddots & & \\ & & \ddots & \lambda & \\ \mathbf{0} & & & 1 & \lambda \end{pmatrix}\ , \quad h^\dagger = \eta h \eta\ . \tag{5.49}$$

In addition, the projection operator for such a subspace $\mathscr{V}$ is given by

$$P(\mathscr{V}) = \epsilon \sum_{k=0}^{m-1} (H-\lambda)^k |z\rangle\langle z|(H-\lambda)^{m-k-1} . \tag{5.50}$$

The above works when λ is real.

When λ is complex, we consider $\mathscr{V}(\lambda) \bigoplus \mathscr{V}(\lambda^*)$. Using η and h in (5.48) and (5.49), the metric matrix and the matrix representation of H become

$$\begin{pmatrix} 0 & \eta \\ \eta & 0 \end{pmatrix} \quad \text{and} \quad \begin{pmatrix} h & 0 \\ 0 & h^* \end{pmatrix} , \tag{5.51}$$

respectively.

Subspace Describing Physical States

In various kinds of physical theories, especially those involving gauge fields, the indefinite metric is a necessary feature of the mathematical formulation. However, our observable world can be described by a positive-definite metric. The norm of the state vector corresponding to the probability is always positive. We thus pick out a subspace $\mathscr{V}_{\text{phys}}$ from the state-vector space $\mathscr{V}$, viz.,

$$\mathscr{V}_{\text{phys}} \subset \mathscr{V} , \tag{5.52}$$

and interpret only the former as observable. An arbitrary vector $|x\rangle$ in this subspace has the property

$$\langle x|x\rangle \geq 0 . \tag{5.53}$$

The difference with (5.4) is that the equation $\langle x|x\rangle = 0$ does not necessarily require $|x\rangle = 0$, i.e., we accept that there can be a nonzero state vector with zero norm. Setting aside the issue of defining this subspace, what we can say is that, even if the parent space $\mathscr{V}$ is non-degenerate, this subspace is degenerate.

To begin with, consider a vector $|y\rangle$ with zero norm which belongs to this subspace, and another, arbitrary vector $|x\rangle$ in this subspace, i.e.,

$$|x\rangle, |y\rangle \in \mathscr{V}_{\text{phys}} , \quad \langle y|y\rangle = 0 . \tag{5.54}$$

Since $|x\rangle + a|y\rangle$ also belongs to this subspace, its norm satisfies

$$\langle x|x\rangle + 2\text{Re}(a\langle x|y\rangle) + |a|^2\langle y|y\rangle = \langle x|x\rangle + 2\text{Re}(a\langle x|y\rangle) \geq 0 .$$

In order for this inequality to hold for an arbitrary complex number a, we must have

$$\langle x|y\rangle = 0 . \tag{5.55}$$

It turns out that the subspace generated by $|x\rangle$ and $|y\rangle$ is degenerate. Indeed, the vector $|y\rangle$ with zero norm is orthogonal to all the vectors in this subspace.

We thus consider the following direct-product decomposition:

$$\mathscr{V}_{\text{phys}} = \mathscr{V}_+ \oplus \mathscr{V}_0 \,, \tag{5.56}$$

where all the vectors belonging to $\mathscr{V}_+$ have positive norm and the vectors belonging to $\mathscr{V}_0$ have zero norm. Note that this decomposition is not unique. Although $\mathscr{V}_0$ can be determined uniquely, $\mathscr{V}_+$ cannot. This is because if $|x\rangle$ has positive norm under the condition (5.54), then $|x\rangle + a|y\rangle$ has the same positive norm, so the part of the vectors in $\mathscr{V}_0$ becomes indeterminate. Consider $|x_i'\rangle = |x_i\rangle + |y_i\rangle$, where $|x_i\rangle$ is a positive-norm vector, $|y_i\rangle$ is a zero-norm vector, and both belong to $\mathscr{V}_{\text{phys}}$. Then, $\langle x_1'|x_2'\rangle = \langle x_1|x_2\rangle$, so if the metric is given, $\mathscr{V}_+$ cannot be determined uniquely. This is because $\mathscr{V}_{\text{phys}}$ is degenerate.

We now consider an operator T having $\mathscr{V}_{\text{phys}}$ as an invariant subspace:

$$T\mathscr{V}_{\text{phys}} \subset \mathscr{V}_{\text{phys}} \,. \tag{5.57}$$

It is easy to show that

$$\langle x_1'|T|x_2'\rangle = \langle x_1|T|x_2\rangle \,. \tag{5.58}$$

Therefore, with respect to such an operator, $|x_i'\rangle$ and $|x_i\rangle$ are equivalent, so, formally, we can write

$$|x_i'\rangle \equiv |x_i\rangle \pmod{\mathscr{V}_0} \,. \tag{5.59}$$

An operator satisfying (5.57) is said to be an observable. Then we consider the following quotient space:

$$\mathscr{H}_{\text{phys}} = \mathscr{V}_{\text{phys}}/\mathscr{V}_0 \,. \tag{5.60}$$

If we now express a vector belonging to an equivalence class as a ket vector $|x\rangle\rangle$, we have

$$\langle\langle x|x\rangle\rangle \geq 0 \,, \tag{5.61}$$

where equality holds only for

$$|x\rangle\rangle \equiv 0 \pmod{\mathscr{V}_0} \,.$$

In this sense, the quotient space $\mathscr{H}_{\text{phys}}$ has a positive-definite metric, which allows a physical interpretation. Regarding the observable T, the matrix element can be defined by

$$\langle\langle x_1|T|x_2\rangle\rangle = \langle x_1 + y_1|T|x_2 + y_2\rangle \ , \quad |y_1\rangle, |y_2\rangle \in \mathscr{V}_0 \ . \tag{5.62}$$

T is then viewed as an operator in $\mathscr{H}_{\text{phys}}$. Introducing a suitable operator Q, $\mathscr{V}_{\text{phys}}$ can be defined for practical purposes by

$$\mathscr{V}_{\text{phys}} = \left\{|x\rangle \,\middle|\, Q|x\rangle = 0\,,\ |x\rangle \in \mathscr{V}\right\} . \tag{5.63}$$

5.3 Free Electromagnetic Field in the Fermi Gauge

In order to quantize the electromagnetic field, we consider the Lagrangian density constructed by rewriting (3.221) in terms of a real vector field, setting the mass m to zero:

$$\mathscr{L} = -\frac{1}{4}F_{\mu\nu}F_{\mu\nu} \ , \quad \text{where} \quad F_{\mu\nu} = \partial_\mu A_\nu - \partial_\nu A_\mu \ . \tag{5.64}$$

In this case, the electric and magnetic fields are defined by

$$F_{4k} = -F_{k4} = \mathrm{i}E_k \ , \quad F_{ij} = -F_{ji} = \epsilon_{ijk}H_k \ , \tag{5.65}$$

where ϵ_{ijk} is 1, -1, or 0 depending on whether (ijk) is an even permutation of (123), an odd permutation of (123), or neither. Then (5.64) can be written in the form

$$\mathscr{L} = \frac{1}{2}(\boldsymbol{E}^2 - \boldsymbol{H}^2) \ . \tag{5.66}$$

Using the variational principle to derive the equation of motion from this Lagrangian density, we obtain the source-free Maxwell equation:

$$\partial_\mu F_{\mu\nu} = \Box A_\nu - \partial_\mu\partial_\nu A_\mu = 0 \ . \tag{5.67}$$

Given any solution A_μ, other solutions can be obtained by the gauge transformation

$$A_\mu \to A'_\mu = A_\mu + \partial_\mu \Lambda \ , \tag{5.68}$$

where Λ is an arbitrary real function of x_μ. The fact that the solution allows such a gauge transformation comes from the fact that the operator $D_{\mu\nu}(\partial)$ appearing previously has no inverse. Although we quantized the massive gauge field by expressing the fourth component in terms of the other components, we cannot do this when the mass is zero. That is, if $m \neq 0$, we obtain the field equation

$$\partial_\mu A_\mu = 0 , \tag{5.69}$$

but if $m = 0$, then we cannot derive this.

Moreover, the quantity π_j canonically conjugate to A_j is

$$\pi_j = \frac{\partial \mathscr{L}}{\partial \dot{A}_j} = \mathrm{i}F_{4j} = -E_j , \tag{5.70}$$

but there is clearly no way to define a quantity canonically conjugate to A_4. Here we have a novel situation, not encountered previously. One way to deal with it is to remove the fourth component via a gauge transformation and quantize only the remaining three components. However, with such a quantization method, Lorentz invariance is unclear, and it is better to find other ways where invariance is manifest. One such method is due to Fermi.

Fermi's Method

Fermi introduced the Lagrangian density

$$\mathscr{L} = -\frac{1}{4}F_{\mu\nu}F_{\mu\nu} - \frac{1}{2}(\partial_\mu A_\mu)^2 . \tag{5.71}$$

Because of the additional term, the canonical conjugate quantity can be defined even for the fourth component. In fact, instead of (5.70), we have

$$\pi_j = \mathrm{i}F_{4j} , \quad \pi_4 = \mathrm{i}(\partial_\mu A_\mu) . \tag{5.72}$$

In addition, taking the Euler derivative, the field equation becomes

$$[\mathscr{L}]_{A_\mu} = \frac{\partial \mathscr{L}}{\partial A_\mu} - \frac{\partial}{\partial x_\nu}\left(\frac{\partial \mathscr{L}}{\partial A_{\mu,\nu}}\right) = D_{\mu\nu}(\partial)A_\nu = 0 , \tag{5.73}$$

where

$$D_{\mu\nu}(\partial) = \delta_{\mu\nu}\Box . \tag{5.74}$$

Therefore, (5.73) turns out to be the massless Klein–Gordon equation, i.e.,

$$\Box A_\mu = 0 . \tag{5.75}$$

Next, we define the inverse C of the operator D such that

$$D_{\alpha\beta}C_{\beta\gamma} = \delta_{\alpha\gamma}\Box . \tag{5.76}$$

Clearly,

$$C_{\beta\gamma} = \delta_{\beta\gamma} \ . \tag{5.77}$$

The four-dimensional commutation relation is

$$\left[A_\alpha(x), A_\beta(y)\right] = \mathrm{i}C_{\alpha\beta} D(x-y) = \mathrm{i}\delta_{\alpha\beta} D(x-y) \ , \tag{5.78}$$

where the Δ-function for $m = 0$ is denoted by D. In this case,

$$D(x) = -\frac{1}{2\pi}\epsilon(x_0)\delta(x^2) \ , \quad D^{(1)}(x) = \frac{1}{2\pi^2}\mathrm{P}\frac{1}{x^2} \ . \tag{5.79}$$

From this commutation relation, we obtain the following equal-time canonical commutator:

$$\left[A_\mu(x), \pi_\nu(y)\right] = \mathrm{i}\delta_{\mu\nu}\delta^3(x-y) \ , \quad \text{where} \quad x_0 = y_0 \ . \tag{5.80}$$

To derive this, we used the definition of π_ν in (5.72) and the following property of the D-function (4.13):

$$\frac{\partial}{\partial x_0} D(x) = -\delta^3(x) \ , \quad \text{where} \quad x_0 = 0 \ . \tag{5.81}$$

Additionally, it is straightforward to show that

$$\left[\partial_\mu A_\mu(x), \partial_\nu A_\nu(y)\right] = -\mathrm{i}\Box_x D(x-y) = 0 \ . \tag{5.82}$$

The field equation here is the Klein–Gordon equation given by (5.75), and we require (5.69) to get Maxwell's equation. However, when $m \neq 0$, (5.69) cannot be derived from the field equation. The method introduced by Fermi involves restricting the observable physical states by imposing an additional condition. His condition for an observable state $|\Psi\rangle$ is

$$\langle\Psi|\partial_\mu A_\mu(x)|\Psi\rangle = 0 \ . \tag{5.83}$$

The set of such states forms a vector space. Since this condition is bilinear in $|\Psi\rangle$, we will introduce its linearization later on.

When the state is restricted in this way, Maxwell's equation holds in terms of expectation values:

$$\langle\Psi|\partial_\mu F_{\mu\nu}|\Psi\rangle = \langle\Psi|\Box A_\nu(x)|\Psi\rangle - \partial_\nu\langle\Psi|\partial_\mu A_\mu(x)|\Psi\rangle = 0 \ . \tag{5.84}$$

The above can be achieved using the following Lagrangian density, which differs from (5.71) only by a divergence:

$$\mathscr{L}' = -\frac{1}{2}(\partial_\nu A_\mu)(\partial_\nu A_\mu) \ . \tag{5.85}$$

The energy and the momentum of this system are given by

$$H = \frac{1}{2}\int d^3x \left[\dot{A}_\mu^2 + (\nabla A_\mu)^2\right] , \tag{5.86}$$

$$\boldsymbol{P} = -\int d^3x \ \dot{A}_\mu \nabla A_\mu \ . \tag{5.87}$$

These are common to both Lagrangians.

Fourier Expansion

First, we implement the Fourier expansion of A_μ, although in the following, we change the variable of the Fourier coefficient from the three-dimensional vector $\boldsymbol{k}$ to the four-dimensional vector k :

$$A_\mu(x) = \sum_{\boldsymbol{k}} \frac{1}{\sqrt{2k_0 V}}\left[e^{ik\cdot x} A_\mu(k) + e^{-ik\cdot x} A_\mu^\dagger(k)\right] . \tag{5.88}$$

We then expand the Fourier coefficients $A_\mu(k)$ in terms of polarization vectors:

$$A_\mu(k) = \sum_\lambda e_\mu^{(\lambda)} a(k,\lambda) \ , \quad A_\mu^\dagger(k) = \sum_\lambda e_\mu^{(\lambda)} a^\dagger(k,\lambda) \ . \tag{5.89}$$

Here we define four polarization vectors such that

$$\begin{aligned} &e_4^{(1)} = e_4^{(2)} = e_4^{(3)} = 0 \ , \quad e_4^{(4)} = 1 \ , \\ &\boldsymbol{e}^{(3)} = \boldsymbol{e}^{(1)} \times \boldsymbol{e}^{(2)} = \frac{\boldsymbol{k}}{k_0} \ , \quad e^{(1)} \cdot k = e^{(2)} \cdot k = 0 \ . \end{aligned} \tag{5.90}$$

Thus, $e^{(1)}$ and $e^{(2)}$ correspond to a transverse wave, $e^{(3)}$ to a longitudinal wave, and $e^{(4)}$ to a scalar wave. These four vectors form a complete orthogonal system:

$$e_\mu^{(\lambda)} e_\mu^{(\lambda')} = \delta_{\lambda\lambda'} \ , \quad e_\mu^{(\lambda)} e_\nu^{(\lambda)} = \delta_{\mu\nu} \ . \tag{5.91}$$

We can now write the Fourier expansion in the form

$$A_\mu(x) = \sum_{\boldsymbol{k}}\sum_\lambda \frac{1}{\sqrt{2k_0 V}}\left[e_\mu^{(\lambda)} a(k,\lambda) e^{ik\cdot x} + e_\mu^{(\lambda)} a^\dagger(k,\lambda) e^{-ik\cdot x}\right] . \tag{5.92}$$

In the Pauli metric, the adjoint operator $a^\dagger(k,\lambda)$ of $a(k,\lambda)$ is the Hermitian conjugate for $\lambda = 1,2,3$, while $\mathrm{i}a(k,4)$ and $\mathrm{i}a^\dagger(k,4)$ are Hermitian conjugates for $\lambda = 4$. The commutation relation is

$$\left[a(k,\lambda), a^\dagger(k',\lambda')\right] = \delta_{\lambda\lambda'}\delta_{\boldsymbol{k},\boldsymbol{k}'} \,. \tag{5.93}$$

The a's commute among themselves and so do the $a^\dagger$'s.

Up to c-numbers, the energy and momentum of the whole system are given by

$$H = \sum_{\boldsymbol{k}} k_0 \sum_{\lambda} a^\dagger(k,\lambda)a(k,\lambda) \,, \tag{5.94}$$

$$P_j = \sum_{\boldsymbol{k}} k_j \sum_{\lambda} a^\dagger(k,\lambda)a(k,\lambda) \,. \tag{5.95}$$

For $\lambda = 1,2,3$, the particle number operator is

$$n(k,\lambda) = a^\dagger(k,\lambda)a(k,\lambda) \,. \tag{5.96}$$

Incidentally, $a^\dagger(k,4)$ is not the Hermitian conjugate of $a(k,4)$, but a so-called *adjoint operator*. Setting

$$a(k,4) = \mathrm{i}a(k,0) \,, \quad a^\dagger(k,4) = \mathrm{i}a^\dagger(k,0) \,, \tag{5.97}$$

then $a^\dagger(k,0)$ becomes the Hermitian conjugate of $a(k,0)$. As is clear from the commutation relation

$$\left[a(k,0), a^\dagger(k,0)\right] = -1 \,, \tag{5.98}$$

the scalar wave has indefinite metric. If $a(k,0)$ and $a^\dagger(k,0)$ are interpreted as an annihilation operator and a creation operator, respectively, then for the vacuum $|0\rangle$, we obtain

$$a(k,0)|0\rangle = 0 \,. \tag{5.99}$$

We introduce the n-particle state of the scalar wave by

$$|n\rangle = \frac{1}{\sqrt{n!}}\left[a^\dagger(k,0)\right]^n|0\rangle \,. \tag{5.100}$$

Its norm is given by

$$\langle n|n\rangle = (-1)^n\langle 0|0\rangle = (-1)^n \,.$$

This shows why the indefinite metric turns up. The particle number operator of the scalar wave is now given by

$$n(k,0) = a^{\dagger}(k,0)a(k,0) \ . \tag{5.101}$$

When k is fixed, there are four polarization states and the metric matrix $\eta(k)$ corresponding to the four states is

$$\eta(k) = \begin{pmatrix} 1 & & & \\ & 1 & & \\ & & 1 & \\ & & & -1 \end{pmatrix} . \tag{5.102}$$

It is clear that the indefinite metric of the one-particle state for the scalar wave has its origin in the Minkowski metric.

5.4 Lorenz Condition and Physical State Space

In the previous section, we imposed the additional condition (5.83) to pick out physically observable states. It has been written in bilinear form, so we now seek a way to rewrite it as a linear form.

Since the free field can be decomposed into a positive frequency part $A_{\mu}^{(+)}(x)$ corresponding to the annihilation operators and a negative frequency part $A_{\mu}^{(-)}(x)$ corresponding to the creation operators, we can write

$$A_{\mu}(x) = A_{\mu}^{(+)}(x) + A_{\mu}^{(-)}(x) \ . \tag{5.103}$$

The vacuum $|0\rangle$ then satisfies

$$A_{\mu}^{(+)}(x)|0\rangle = 0 \ . \tag{5.104}$$

Restricting the system to the free electromagnetic field, it turns out that the state-vector space $\mathscr{V}$ can be created by applying $A_{\mu}^{(-)}(x)$ to the vacuum $|0\rangle$. Since the space $\mathscr{V}$ has indefinite metric, in order to obtain a physical interpretation, we must introduce the subspace $\mathscr{V}_{\text{phys}}$ discussed in Sect. 5.2.

One possibility here is to choose the Coulomb gauge, in which the scalar wave reproducing the negative term in the metric matrix is dropped at the outset, but what we would like is rather to eliminate the indefiniteness of the metric while maintaining the manifest covariance of the theory. This can be done by adopting the *Lorenz condition*. In (5.63), we mentioned a suitable operator Q to select $\mathscr{V}_{\text{phys}}$, but in the following we shall see that there may exist more than one operator with the property (5.63). Then we define the subspace $\mathscr{V}_{\text{phys}}$ as follows:

$$\mathscr{V}_{\text{phys}} = \left\{|\Phi\rangle \;\middle|\; \partial_\mu A_\mu^{(+)}(x)|\Phi\rangle = 0\,,\quad |\Phi\rangle \in \mathscr{V}\right\}\,, \tag{5.105}$$

where x runs over all the points in four-dimensional space-time. The extra condition appearing here, i.e.,

$$\partial_\mu A_\mu^{(+)}(x)|\Phi\rangle = 0\,, \tag{5.106}$$

is called the *Lorenz condition*. It is clear that a state which satisfies this condition also obeys (5.83), i.e.,

$$\langle\Phi|\partial_\mu A_\mu(x)|\Phi\rangle = 0\,.$$

What we need to prove first is that arbitrary vectors which belong to $\mathscr{V}_{\text{phys}}$ have nonzero norm, i.e.,

$$\langle\Phi|\Phi\rangle \geq 0\,. \tag{5.107}$$

From the Fourier expansion (5.92),

$$\partial_\mu A_\mu^{(+)}(x) = \mathrm{i}\sum_{\boldsymbol{k}} \frac{1}{\sqrt{2k_0 V}}\mathrm{e}^{\mathrm{i}k\cdot x}\sum_\lambda k_\mu e_\mu^{(\lambda)} a(\boldsymbol{k},\lambda)\,. \tag{5.108}$$

The sum over λ is

$$\sum_\lambda k_\mu e_\mu^{(\lambda)} a(\boldsymbol{k},\lambda) = |\boldsymbol{k}|\left[a(\boldsymbol{k},3) - a(\boldsymbol{k},0)\right]\,, \tag{5.109}$$

where $a(\boldsymbol{k},3)$ is the operator annihilating the longitudinal wave. We then introduce the following operators:

$$a(\boldsymbol{k}) = \frac{1}{\sqrt{2}}\left[a(\boldsymbol{k},3) - a(\boldsymbol{k},0)\right]\,,\quad b(\boldsymbol{k}) = \frac{1}{\sqrt{2}}\left[a(\boldsymbol{k},3) + a(\boldsymbol{k},0)\right]\,. \tag{5.110}$$

Therefore, the Lorenz condition ends up in the form

$$a(\boldsymbol{k})|\Phi\rangle = 0\,. \tag{5.111}$$

For a given $\boldsymbol{k}$, we obtain the commutation relations

$$\left[a, a^\dagger\right] = \left[b, b^\dagger\right] = 0\,,\quad \left[a, b^\dagger\right] = \left[b, a^\dagger\right] = 1\,. \tag{5.112}$$

These operators commute with $a(k,1)$, $a(k,2)$, $a^\dagger(k,1)$, and $a^\dagger(k,2)$. $\mathscr{V}$ can be generated by applying polynomials in $a^\dagger(k,1)$, $a^\dagger(k,2)$, $a^\dagger(k)$, and $b^\dagger(k)$ to the vacuum state. On the other hand, it is clear from (5.112) that $\mathscr{V}_{\text{phys}}$ can be generated by applying the operators $a^\dagger(k,1)$, $a^\dagger(k,2)$, and $a^\dagger(k)$ to $|0\rangle$. We expand the state $|\Phi_j\rangle$ belonging to $\mathscr{V}_{\text{phys}}$ in terms of the number of $a^\dagger$ operators:

$$|\Phi_j\rangle = |\Phi_0\rangle + |\Phi_1\rangle + |\Phi_2\rangle + \cdots , \tag{5.113}$$

where $|\Phi_j\rangle$ denotes the state containing transverse photons and j a-photons. The inner product of two physical states is

$$\langle \Phi' | \Phi \rangle = \langle \Phi_0' | \Phi_0 \rangle , \tag{5.114}$$

since the norm is zero unless the number of a-photons is zero. For example, using (5.112), the inner product of states $a^\dagger(k)|\Phi_{\text{T}}\rangle$ and $a^\dagger(k')|\Phi_{\text{T}}'\rangle$ containing single a-photons, while both $|\Phi_T\rangle$ and $|\Phi_T'\rangle$ contain only transverse photons, is

$$\langle \Phi_{\text{T}}' | a(k') a^\dagger(k) | \Phi_{\text{T}} \rangle = \langle \Phi_{\text{T}}' | a^\dagger(k) a(k') | \Phi_{\text{T}} \rangle = 0 .$$

This is because when the annihilation operator $a(k')$ acts on the state $|\Phi_{\text{T}}\rangle$, it destroys it. This is the reason why (5.114) holds. If Φ belongs to $\mathscr{V}_{\text{phys}}$, then (5.114) implies

$$\langle \Phi | \Phi \rangle = \langle \Phi_0 | \Phi_0 \rangle \geq 0 . \tag{5.115}$$

If this vanishes, it does not mean $|\Phi\rangle = 0$, but rather $|\Phi_0\rangle = 0$. In the decomposition (5.56), it turns out that $\mathscr{V}_+$ consists of state vectors containing only transverse photons, while $\mathscr{V}_0$ consists of state vectors containing a-photons. However, since a time axis is chosen in the decomposition (5.56), the vector space $\mathscr{V}_+$ is not Lorentz invariant.

Note, however, that the quotient space $\mathscr{H}_{\text{phys}}$ defined by (5.60) is Lorentz invariant. Moreover, (5.114) can also be written

$$\langle \Phi' | \Phi \rangle = \langle \Phi' | P(\mathscr{V}_+) | \Phi \rangle , \quad \text{where } |\Phi\rangle, \ |\Phi'\rangle \in \mathscr{V}_{\text{phys}} . \tag{5.116}$$

Here $P(\mathscr{V}_+)$ is the operator projecting onto the subspace $\mathscr{V}_+$. Although $\mathscr{V}_+$ is not Lorentz invariant, (5.116) is independent of the choice of $\mathscr{V}_+$, so this inner product is itself Lorentz invariant, and so too is $\mathscr{V}_{\text{phys}}$.

We now give some examples of observables, as defined in Sect. 5.2. For instance, the following are observables:

$$F_{\mu\nu}(x) , \quad \partial_\mu A_\mu(x) . \tag{5.117}$$

This is because, as is easily shown, we have

$$\left[F_{\mu\nu}(x), \partial_\rho A_\rho^{(+)}(y)\right] = 0\,, \quad \left[\partial_\mu A_\mu(x), \partial_\rho A_\rho^{(+)}(y)\right] = 0\,. \tag{5.118}$$

Thus, if $|\Phi\rangle$ belongs to $\mathscr{V}_{\text{phys}}$, then both $F_{\mu\nu}(x)|\Phi\rangle\rangle$ and $\partial_\mu A_\mu(x)|\Phi\rangle$ belong to $\mathscr{V}_{\text{phys}}$. Therefore, if $|\Phi\rangle$ and $|\Phi'\rangle$ belong to $\mathscr{V}_{\text{phys}}$, we have

$$\langle\Phi'|F_{\mu\nu}(x)F_{\rho\sigma}(y)|\Phi\rangle = \langle\Phi'|F_{\mu\nu}(x)P(\mathscr{V}_+)F_{\rho\sigma}(y)|\Phi\rangle\,. \tag{5.119}$$

In the presence of interactions, an S-matrix can be defined. This is also an observable, satisfying

$$\left[S, \partial_\rho A_\rho^{(+)}(x)\right] = \left[S, \partial_\rho A_\rho(x)\right] = 0\,. \tag{5.120}$$

5.5 Free Electromagnetic Field: Generalization of Gauge Choices

So far we have discussed the quantization of the electromagnetic field in the Fermi gauge. It would be easier if the story were self-contained in this gauge, but later, when we come to consider renormalization for instance, it is useful to have the gauge choice generalized in advance. Therefore, generalizing (5.71), we introduce the Lagrangian density

$$\mathscr{L} = -\frac{1}{4}F_{\mu\nu}F_{\mu\nu} - \frac{1}{2\alpha}(\partial_\mu A_\mu)^2\,. \tag{5.121}$$

The parameter α is called a *gauge parameter*. The value $\alpha = 1$ corresponds to the Fermi gauge. We then consider the Euler derivative:

$$[\mathscr{L}]_{A_\mu} = D_{\mu\nu}(\partial)A_\nu\,, \tag{5.122}$$

where

$$D_{\mu\nu}(\partial) = \delta_{\mu\nu}\Box + \frac{1-\alpha}{\alpha}\partial_\mu\partial_\nu\,. \tag{5.123}$$

In order to quantize the field, we define the differential operator C to be such that

$$C_{\mu\sigma}D_{\sigma\nu} = D_{\mu\sigma}C_{\sigma\nu} = \delta_{\mu\nu}\Box\,. \tag{5.124}$$

From this, it follows that

$$C_{\mu\nu}(\partial) = \delta_{\mu\nu} - (1-\alpha)\frac{\partial_\mu\partial_\nu}{\Box}\,. \tag{5.125}$$

One feature of the general gauge is that $\Box$ appears in the denominator, and this yields the dipole state or dipole ghost discussed in Sect. 5.2. The reason it is called a ghost is that it is related to the indefinite metric, as mentioned at the end of Sect. 4.4. Using this C, we obtain the commutation relation

$$\begin{aligned}\left[A_\mu(x), A_\nu(y)\right] &= \mathrm{i}C_{\mu\nu}(\partial)D(x-y) \\ &= \mathrm{i}\delta_{\mu\nu}D(x-y) + \mathrm{i}(1-\alpha)\partial_\mu\partial_\nu E(x-y)\ ,\end{aligned} \tag{5.126}$$

where E is defined by

$$\Box E(x) = -D(x)\ , \quad E(x) = -\frac{1}{8\pi}\epsilon(x_0)\theta(-x^2)\ . \tag{5.127}$$

This function can be also written in the form

$$E(x) = -\left[\frac{\partial}{\partial m^2}\Delta(x, m^2)\right]_{m^2=0}\ , \tag{5.128}$$

where, as for the function $\Delta(x)$ defined in Chap. 4, we have written the mass dependence m^2 explicitly as $\Delta(x, m^2)$. We will use this definition later, too.

Although the field equation is different from both Maxwell's equation and the Klein–Gordon equation, the latter can be obtained by imposing the Lorenz condition, which is the same as the Fermi gauge.

From the field equation $D_{\mu\nu}(\partial)A_\nu = 0$, we have

$$\partial_\mu D_{\mu\nu}(\partial)A_\nu = \frac{1}{\alpha}\Box\partial_\nu A_\nu = 0\ , \quad \Box(\partial_\nu A_\nu) = 0\ . \tag{5.129}$$

Combining the field equation and (5.129),

$$\Box D_{\mu\nu}(\partial)A_\nu = \Box^2 A_\mu + \frac{1-\alpha}{\alpha}\partial_\mu\Box(\partial_\nu A_\nu) = \Box^2 A_\mu = 0\ . \tag{5.130}$$

Note that $\Box A_\mu \neq 0$. This is clear from the commutation relation

$$\begin{aligned}\left[\Box A_\mu(x), A_\nu(y)\right] &= \mathrm{i}\Box C_{\mu\nu}(\partial)D(x-y) \\ &= -\mathrm{i}(1-\alpha)\partial_\mu\partial_\nu D(x-y) \neq 0\ ,\end{aligned} \tag{5.131}$$

where we have assumed that α is different from 1. Therefore,

$$\Box A_\mu \neq 0\ , \quad \Box^2 A_\mu = 0\ . \tag{5.132}$$

These two equations imply the existence of the dipole ghost, as we shall now show.

Dipole Ghost

Recall that multipole states with indefinite metric were defined by using a Hermitian operator. In the following, we choose as Hermitian operator the mass operator:

$$H = -P_\mu^2 \, . \tag{5.133}$$

Then it can be proven that the state $A_\alpha(x)|0\rangle$ includes a dipole state. To show this, we use (3.29) and $P_\mu|0\rangle = 0$ to obtain

$$\begin{aligned} P_\mu A_\alpha(x)|0\rangle &= \left[P_\mu, A_\alpha(x)\right]|0\rangle + A_\alpha(x)P_\mu|0\rangle \\ &= \mathrm{i}\partial_\mu A_\alpha(x)|0\rangle \, . \end{aligned}$$

Therefore, repeating the transformation above,

$$HA_\alpha(x)|0\rangle = (-P_\mu^2)A_\alpha(x)|0\rangle = \Box A_\alpha(x)|0\rangle \, . \tag{5.134}$$

As is clear from the fact that the vacuum expectation value of (5.131) is nonzero, this state does not vanish. However, the norm of this state is zero, as we will show below. From (5.132),

$$\begin{aligned} \langle 0|\Box A_\alpha(x)\Box A_\beta(y)|0\rangle &= \langle 0|A_\alpha(x)(-P_\mu^2)(-P_\nu^2)A_\beta(y)|0\rangle \\ &= \langle 0|A_\alpha(x)\Box^2 A_\beta(y)|0\rangle \\ &= 0 \, . \end{aligned} \tag{5.135}$$

Consequently,

$$HA_\alpha(x)|0\rangle \neq 0 \, , \quad H^2 A_\alpha(x)|0\rangle = 0 \, . \tag{5.136}$$

It is obvious from the definition (5.42) that this implies the existence of the dipole ghost for the eigenvalue $\lambda = 0$ of H.

Lorenz Condition

When a dipole ghost exists, we cannot carry out the Fourier expansion. In fact, the Fourier expansion cannot be carried out for the solution of (5.132). This makes it harder to obtain an intuitive understanding of the dipole ghost. In order to investigate its structure intuitively, we consider

$$D_{\mu\nu}^{-1} = \frac{C_{\mu\nu}}{\Box} = \frac{\delta_{\mu\nu}}{\Box} - (1-\alpha)\frac{\partial_\mu\partial_\nu}{\Box^2} \, . \tag{5.137}$$

The main issue is the treatment of the second term. We first write the above equation in the form

$$D_{\mu\nu}^{-1} = \frac{\delta_{\mu\nu}}{\Box} - (1-\alpha)\partial_\mu\partial_\nu \lim_{\epsilon\to 0}\frac{1}{\Box(\Box-\epsilon^2)} \, . \tag{5.138}$$

In later discussions, we will treat ϵ as nonzero, but small, i.e.,

$$D^{-1}_{\mu\nu} = \frac{\delta_{\mu\nu}}{\Box} - (1-\alpha)\frac{1}{\epsilon^2}\left(\frac{1}{\Box - \epsilon^2} - \frac{1}{\Box}\right)\partial_\mu \partial_\nu \ . \tag{5.139}$$

For the inverse $D(\partial, \epsilon)$ of this expression, we have

$$D_{\mu\nu}(\partial, \epsilon) = \delta_{\mu\nu}\Box + \frac{\Box}{\alpha\Box - \epsilon^2}(1-\alpha)\partial_\mu \partial_\nu \ . \tag{5.140}$$

Then the commutation relation corresponding to (5.137) is

$$\left[A_\mu(x), A_\nu(y)\right] = \mathrm{i}\left(\delta_{\mu\nu} + \frac{1-\alpha}{\epsilon^2}\partial_\mu \partial_\nu\right)D(x-y) - \mathrm{i}\frac{1-\alpha}{\epsilon^2}\partial_\mu \partial_\nu \nabla(x-y, \epsilon^2) \ . \tag{5.141}$$

We now write A_μ in such a way as to reproduce this commutation relation:

$$A_\mu(x) = a_\mu(x) + \frac{1-\alpha}{2\epsilon^2}\partial_\mu \partial_\nu a_\nu(x) + \frac{\sqrt{1-\alpha}}{\epsilon}\partial_\mu b(x) \ , \tag{5.142}$$

where

$$\left[a_\mu(x), a_\nu(y)\right] = \mathrm{i}\delta_{\mu\nu} D(x-y) \ , \quad [b(x), b(y)] = \mathrm{i}\Delta(x-y, \epsilon) \ . \tag{5.143}$$

Hence, a_μ is the electromagnetic field in the Fermi gauge and b is a real scalar field with mass ϵ. Of course, these two fields are independent:

$$\left[a_\mu(x), b(y)\right] = 0 \ . \tag{5.144}$$

The field equations are

$$\Box a_\mu = 0 \ , \quad (\Box - \epsilon^2) b = 0 \ . \tag{5.145}$$

From this, we can derive the following equation of motion:

$$D_{\mu\nu}(\partial, \epsilon) A_\nu = 0 \ . \tag{5.146}$$

In the discussion above, we have assumed $1-\alpha > 0$, but if $1-\alpha < 0$, the coefficient of $\partial_\mu b$ in (5.142) becomes $\sqrt{\alpha - 1}/\epsilon$. It turns out that b obeys the indefinite metric. In the following, we discuss the case where $1-\alpha > 0$. Now,

$$\partial_\mu A_\mu = \partial_\mu a_\mu + \epsilon\sqrt{1-\alpha}\, b \ . \tag{5.147}$$

As long as $\epsilon \neq 0$ in the Fourier expansion, a_μ and b are fields of different masses and the Lorenz condition

$$\partial_\mu A_\mu^{(+)}(x)|\Phi\rangle = 0 \tag{5.148}$$

can be decomposed into the following two equations:

$$\partial_\mu a_\mu^{(+)}(x)|\Phi\rangle = 0\,, \quad b^{(+)}(x)|\Phi\rangle = 0\,. \tag{5.149}$$

As can be seen, the new degree of freedom b appears in the Fermi gauge, but the second condition above removes it. Thus, in the end, the condition is reduced to the Lorenz condition in the Fermi gauge expressed by the first relation in (5.149). Due to the assumption that $\epsilon \neq 0$, we can apply the Fourier expansion in the analysis above.

From the discussion above, the Fermi gauge with $\alpha = 1$ is obviously rather special. The Landau gauge with $\alpha = 0$ is also special since C then consists only of the transverse part.

It was Dirac who introduced the indefinite metric for the first time [84], while Gupta and Bleuler did this when quantizing the electromagnetic field [85, 86].

Chapter 6
Quantization of Interacting Systems

So far, we have considered the quantization of free fields, i.e., fields without interactions. In this case, the field equation is linear and can easily be solved. Thanks to this, it can be shown that canonical quantization is a relativistically-invariant quantization method which does not depend on the way we choose our coordinates. However, when quantizing interacting fields, we begin with the equal-time commutation relations and the relativistic invariance is not apparent. In this case, there is a choice between the Schrödinger picture and the Heisenberg picture. But whichever is adopted, it is hard to reach a clear conclusion regarding the issue of relativistic invariance when the canonical quantization method is used. It is the interaction-picture method introduced by Tomonaga [87] and Schwinger [88] that provides a precise solution to this problem.

In this picture, the invariance of both the commutation relations and the Schrödinger equation become clear. Using this manifestly Lorentz-covariant picture, we can then move on to the field theory in terms of Heisenberg's picture. Here, we will explain the supermultiple-time formulation in the interaction picture.

6.1 Tomonaga–Schwinger Equation

We begin with Schrödinger's equation for interacting fields:

$$\mathrm{i}\frac{\partial}{\partial t}\Psi(t) = H\Psi(t)\ . \tag{6.1}$$

Here H is the Hamiltonian of the total system, consisting of a free part and an interaction part:

$$H = \int \mathrm{d}^3x\,\mathscr{H}(\boldsymbol{x}) = \int \mathrm{d}^3x\,\mathscr{H}_{\mathrm{f}}(\boldsymbol{x}) + \int \mathrm{d}^3x\,\mathscr{H}_{\mathrm{int}}(\boldsymbol{x}) = H_{\mathrm{f}} + H_{\mathrm{int}}\ , \tag{6.2}$$

K. Nishijima, *Quantum Field Theory*,
https://doi.org/10.1007/978-94-024-2190-3_6

where f and int indicate the free and interaction parts, respectively. In fact, this separation is not trivial, as will be explained later when we discuss renormalization. However, we shall not go into this yet.

Next, we introduce the unitary transformation

$$U(t) = \exp(\mathrm{i}\, t H_{\mathrm{f}}) , \tag{6.3}$$

and define the field operator

$$\varphi_\alpha(x) = \varphi_\alpha(\boldsymbol{x}, t) = U(t)\varphi_\alpha(\boldsymbol{x})U(t)^{-1} . \tag{6.4}$$

Then, $\varphi_\alpha(x)$ satisfies the free field equation and the commutation relations of the free field:

$$D_{\alpha\beta}(\partial)\varphi_\beta(x) = 0 , \tag{6.5}$$

$$\left[\varphi_\alpha(x), \varphi_\beta^\dagger(y)\right]_\pm = \mathrm{i} C_{\alpha\beta}(\partial)\Delta(x-y) . \tag{6.6}$$

It turns out that the free field obeys the methods of relativistically invariant quantization.

We introduce a new state vector $\Phi(t)$ by

$$\Phi(t) = U(t)\Psi(t) , \tag{6.7}$$

which satisfies

$$\mathrm{i}\frac{\partial}{\partial t}\Phi(t) = \left[U(t)\int \mathrm{d}^3x\, \mathscr{H}_{\mathrm{int}}(\boldsymbol{x})U(t)^{-1}\right]\Phi(t) . \tag{6.8}$$

We consider $\mathscr{H}_{\mathrm{int}}(x)$ as a polynomial in the field operators $\varphi_\alpha(x)$:

$$\begin{aligned} U(t)\mathscr{H}_{\mathrm{int}}\left[\varphi_\alpha(\boldsymbol{x}), \ldots\right]U(t)^{-1} &= \mathscr{H}_{\mathrm{int}}\left[U(t)\varphi_\alpha(\boldsymbol{x})U(t)^{-1}, \ldots\right] \\ &= \mathscr{H}_{\mathrm{int}}\left[\varphi_\alpha(x), \ldots\right] . \end{aligned} \tag{6.9}$$

Then we can rewrite (6.8) in the form

$$\mathrm{i}\frac{\partial}{\partial t}\Phi(t) = \int \mathrm{d}^3x\, \mathscr{H}_{\mathrm{int}}\left[\varphi_\alpha(x), \ldots\right]\Phi(t) . \tag{6.10}$$

In this picture, the field operator obeys the free field equation, but the state vector obeys Schrödinger's equation with a Hamiltonian which is the interaction part of the whole Hamiltonian. This is called the *interaction picture*. Clearly, for free fields, (6.10) shows that Φ is independent of time, and the interaction picture coincides with the Heisenberg picture.

The relativistic invariance of (6.10) is not apparent. This is because we have fixed the time axis. Tomonaga showed how to rewrite this in a Lorentz invariant form using the analogy with Dirac's multi-time formulation [89]. Here we will derive it directly without explaining Dirac's multi-time theory.

First, the way the state vector $\Phi(t)$ has been described means that all observations must be conducted at the same time for the given time axis. This restriction seems much too strong. It would be good if observations at various points in space did not have to be carried out at the same time. For instance, it would be good if observations could be carried out at t_1 for the point (x_1, y_1, z_1) and at t_2 for the point (x_2, y_2, z_2). In this case, if the distance between the two points is space-like, these observations will not interfere with each other, according to Einstein's causality, i.e.,

$$(x_1 - x_2)^2 + (y_1 - y_2)^2 + (z_1 - z_2)^2 - (t_1 - t_2)^2 > 0\,. \tag{6.11}$$

We then consider the time which is not necessarily the same at each point in space, i.e.,

$$t = f(x, y, z)\,. \tag{6.12}$$

This relation specifies a hypersurface in four-dimensional space-time. In addition, we assume that two points on this hypersurface are spatially separated. We call this a *space-like hypersurface*, and denote it by σ. When we introduce the state vector $\Phi[\sigma]$, what type of equation does it obey? This will be a generalization of $\Phi(t)$ corresponding to $t = \text{const}$. To study this, we write (6.10) in the form

$$\Phi(t + \mathrm{d}t) = \Phi(t) - \mathrm{i}\mathrm{d}t \int \mathrm{d}^3x \mathscr{H}_{\text{int}}[\varphi_\alpha(x), \ldots]\Phi(t)\,. \tag{6.13}$$

Then in going from the hypersurface at t to the one at $t+\mathrm{d}t$, we consider that at each point (x, y, z) in space we only shift the volume element with infinitesimal volume d^3x by $\mathrm{d}t$. However, we change it smoothly enough to maintain the condition of being space-like. We do this at all points (see Fig. 6.1). We assume that, for a change $\mathrm{d}\omega = \mathrm{d}^3x\,\mathrm{d}t$ in the infinitesimal volume, the state vector is multiplied by the factor

$$1 - \mathrm{i}\mathrm{d}\omega \mathscr{H}_{\text{int}}(x)\,. \tag{6.14}$$

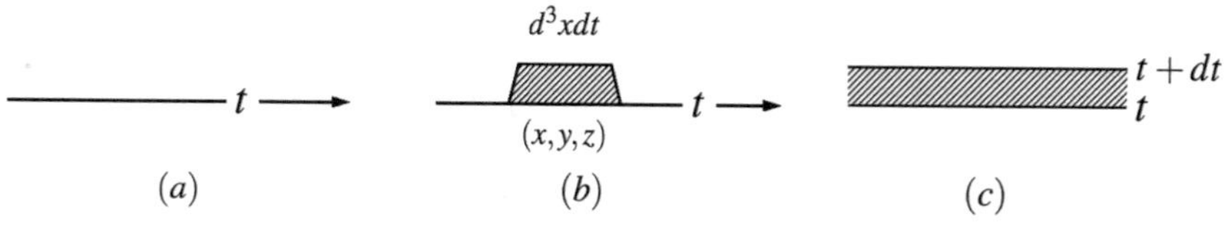

Fig. 6.1 Variation of the hypersurface

Therefore, multiplying by these factors in such a way as to cover all points in space and neglecting infinitesimal terms of order greater than or equal to $d\omega^2$, we obtain (6.13). Generalization to the hypersurface is clear. If the hypersurface σ is altered to σ' solely through a variation of the infinitesimal volume $d\omega$ near the space-time point x, and if both σ and σ' are space-like hypersurfaces, then

$$\Phi[\sigma'] = \left[1 - \mathrm{i}d\omega \mathscr{H}_{\mathrm{int}}(x)\right]\Phi[\sigma] , \tag{6.15}$$

where σ' has been taken to be a future hypersurface of σ. Here we have used the abbreviation

$$\mathscr{H}_{\mathrm{int}}(x) = \mathscr{H}_{\mathrm{int}}\left[\varphi_\alpha(x), \ldots\right] . \tag{6.16}$$

Note that the notion of the space-like hypersurface is independent of the choice of coordinate system. The four-dimensional infinitesimal volume is also a Lorentz-invariant quantity and the Hamiltonian density $\mathscr{H}_{\mathrm{int}}(x)$ is a scalar for simple cases. Therefore, (6.15) is independent of the way we choose the coordinate system. In order to write this equation in a differential form, we define the following functional derivative:

$$\lim_{d\omega \to 0} \frac{\Phi[\sigma'] - \Phi[\sigma]}{d\omega} = \frac{\delta\Phi[\sigma]}{\delta\sigma(x)} . \tag{6.17}$$

Using this symbol, (6.15) can be written in the form

$$\mathrm{i}\frac{\delta\Phi[\sigma]}{\delta\sigma(x)} = \mathscr{H}_{\mathrm{int}}(x)\Phi[\sigma] . \tag{6.18}$$

This is the *Tomonaga–Schwinger equation*. Because we choose a different time at each point in space, this is called the *super multiple-time formulation*.

This theory was first formulated by Tomonaga [87]. It worked well when applied to QED, especially for the renormalized theory. In the United States, its usefulness was also recognized by Schwinger, and it became a central topic in the postwar period.

In simple cases, the Hamiltonian density depends only on the point x and is independent of σ. Let us consider the integrability condition in this case. We thus rewrite the condition

$$\frac{\delta}{\delta\sigma(x)}\frac{\delta}{\delta\sigma(y)}\Phi[\sigma] = \frac{\delta}{\delta\sigma(y)}\frac{\delta}{\delta\sigma(x)}\Phi[\sigma] , \tag{6.19}$$

using (6.18), and find

$$\mathscr{H}_{\mathrm{int}}(y)\mathscr{H}_{\mathrm{int}}(x) - \mathscr{H}_{\mathrm{int}}(x)\mathscr{H}_{\mathrm{int}}(y) = 0 , \tag{6.20}$$

where the points x and y are on the space-like hypersurface σ. This says that the Hamiltonian densities at two space-like separated points commute, and it is called an *integrability condition*. It is closely related to the notion of microscopic causality.

Next, let us consider an initial-value problem. The state vector is given on the space-like hypersurface σ_0:

$$\Phi[\sigma_0] = \Phi \ . \tag{6.21}$$

With this initial condition, the state vector on a general space-like hypersurface is given by

$$\Phi[\sigma] = U[\sigma, \sigma_0]\Phi \ , \tag{6.22}$$

and we choose U in such a way as to satisfy the initial condition and the Tomonaga–Schwinger (TS) equation. First, since (6.22) satisfies the TS equation,

$$\mathrm{i}\frac{\delta}{\delta\sigma(x)}U[\sigma, \sigma_0] = \mathscr{H}_{\text{int}}(x)U[\sigma, \sigma_0] \ . \tag{6.23}$$

Additionally, corresponding to the initial condition (6.21),

$$U[\sigma_0, \sigma_0] = 1 \ . \tag{6.24}$$

When it has this form, U is called a *generalized transformation functional*. This becomes unitary if the Hamiltonian density is Hermitian. To show this, we write the Hermitian conjugate equation of the equation (6.23) as follows:

$$-\mathrm{i}\frac{\delta}{\delta\sigma(x)}U^\dagger[\sigma, \sigma_0] = U^\dagger[\sigma, \sigma_0]\mathscr{H}_{\text{int}}(x) \ . \tag{6.25}$$

Combining (6.23) and (6.25),

$$\frac{\delta}{\delta\sigma(x)}U^\dagger[\sigma, \sigma_0]U[\sigma, \sigma_0] = 0 \ .$$

Thus, the product $U^\dagger U$ does not depend on σ, and we can substitute σ_0 for σ and use (6.24) to obtain

$$U^\dagger[\sigma, \sigma_0]U[\sigma, \sigma_0] = 1. \tag{6.26}$$

Additionally, as is clear from the definition (6.22), the following composition law holds:

$$U[\sigma_2, \sigma_0] = U[\sigma_2, \sigma_1]U[\sigma_1, \sigma_0] \ . \tag{6.27}$$

Since this equation implies that the time evolution of the system is independent of its history, this expresses the causality condition. Moreover, if we set $\sigma_2 = \sigma_0$ in (6.27), then taking into account (6.26),

$$U[\sigma_0, \sigma_1] = (U[\sigma_1, \sigma_0])^{-1} = U^\dagger[\sigma_1, \sigma_0] \,. \tag{6.28}$$

Hence, U is a unitary operator.

In particular, if we choose a space-like hypersurface σ and then introduce a time axis, we can write U as

$$U(t_2, t_1) \,.$$

In this case, we can write the expectation value of a given operator in the interaction picture (but not using the bra-ket symbol here):

$$\big(\Phi(t), \varphi_\alpha(x)\Phi(t)\big) = \big(\Phi, U(t, t_0)^{-1}\varphi_\alpha(x)U(t, t_0)\Phi\big) \,. \tag{6.29}$$

Due to the time-independence of the state vector Φ, this tells us that

$$U(t, t_0)^{-1}\varphi_\alpha(x)U(t, t_0) = \varphi_\alpha^{(\mathrm{H})}(x) \tag{6.30}$$

specifies an operator in the Heisenberg picture. Given the Heisenberg operator $\varphi_\alpha^{(\mathrm{H})}(x)$, it turns out that we can define various interaction pictures depending on the choice of t_0. We write the operators in the interaction picture corresponding to $t_0 \to -\infty$ and $t_0 \to \infty$ as $\varphi_\alpha^{\mathrm{in}}$ and $\varphi_\alpha^{\mathrm{out}}$, respectively. Then,

$$\begin{aligned} \varphi_\alpha^{(\mathrm{H})}(x) &= U(t, -\infty)^{-1}\varphi_\alpha^{\mathrm{in}}(x)U(t, -\infty) \\ &= U(t, \infty)^{-1}\varphi_\alpha^{\mathrm{out}}(x)U(t, \infty) \,. \end{aligned} \tag{6.31}$$

Although it is not strictly rigorous, taking $t \to \infty$ and $U(\infty, \infty) = 1$ in the equation above, we obtain

$$\varphi_\alpha^{\mathrm{out}}(x) = S^{-1}\varphi_\alpha^{\mathrm{in}}(x)S \,, \tag{6.32}$$

$$S = U(\infty, -\infty) \,. \tag{6.33}$$

The operator S is called the *S-matrix*.

6.2 Retarded Product Expansion of the Heisenberg Operators

As mentioned in the last section, field operators in the interaction picture can be quantized in just the same manner as in the free-field case. So, how can we carry out this quantization in the Heisenberg picture? To do this, we usually express the Heisenberg operator in terms of operators quantized in the interaction picture. Since this problem cannot generally be solved, a perturbation method is often used. There are two such methods: the first is a retarded product expansion and the second is the Yang–Feldman method [90]. Here, we will discuss the first of these methods.

In this section, the Heisenberg operator is denoted by φ_α. We choose $\varphi_\alpha^{\text{in}}$ as an operator in the interaction picture. Since the relation between them is given by (6.31), we have

$$\varphi_\alpha(x) = U(t, -\infty)^{-1}\varphi_\alpha^{\text{in}}(x)U(t, -\infty) , \quad t = x_0 . \tag{6.34}$$

We expand this representation perturbatively as a power series in the coupling constants. The interaction Hamiltonian in the interaction picture is

$$H(t) = \int_{x_0=t} \mathrm{d}^3x \mathscr{H}_{\text{int}}(x) . \tag{6.35}$$

We introduce the quantity

$$\mathscr{O}(t_0) = U(t, t_0)^{-1}\mathscr{O}U(t, t_0) . \tag{6.36}$$

Therefore, from the TS equation in the single-time theory, we obtain

$$\mathrm{i}\frac{\partial}{\partial t_0}\mathscr{O}(t_0) = \left[H(t_0), \mathscr{O}(t_0)\right] . \tag{6.37}$$

We convert this equation into the integral form

$$\mathscr{O}(t_0) = \mathscr{O} - \mathrm{i}\int_{t_0}^{t} \mathrm{d}t_0'\left[\mathscr{O}(t_0'), H(t_0')\right] . \tag{6.38}$$

Then iterating by inserting (6.38) back into $\mathscr{O}(t_0')$ in the integrand, we have

$$\begin{aligned}\mathscr{O}(t_0) &= \mathscr{O} - \mathrm{i}\int_{t_0}^{t} \mathrm{d}t_0'\left[\mathscr{O}, H(t_0')\right] + (-\mathrm{i})^2\int_{t_0}^{t} \mathrm{d}t_0' \int_{t_0'}^{t} \mathrm{d}t_0''\left[[\mathscr{O}, H(t'')], H(t_0')\right] + \cdots \\ &= \mathscr{O} - \mathrm{i}\int_{t_0}^{t} \mathrm{d}t'[\mathscr{O}, H(t')] + (-\mathrm{i})^2\int_{t_0}^{t} \mathrm{d}t' \int_{t_0}^{t'} \mathrm{d}t''\left[[\mathscr{O}, H(t')], H(t'')\right] + \cdots .\end{aligned} \tag{6.39}$$

We now make the replacement

$$\mathcal{O}(t_0) \to \varphi_\alpha(x)\,, \quad \mathcal{O} \to \varphi_\alpha^{\text{in}}(x)\,, \quad t_0 \to -\infty\,.$$

This yields the expansion equation

$$\begin{aligned}\varphi_\alpha(x) &= U(t,-\infty)^{-1}\varphi_\alpha^{\text{in}}(\boldsymbol{x},t)U(t,-\infty)\\ &= \varphi_\alpha^{\text{in}}(\boldsymbol{x},t) - \mathrm{i}\int_{-\infty}^{t}\mathrm{d}t'[\varphi_\alpha^{\text{in}}(\boldsymbol{x},t), H(t')]\\ &\quad +(-\mathrm{i})^2\int_{-\infty}^{t}\mathrm{d}t'\int_{-\infty}^{t'}\mathrm{d}t''\big[[\varphi_\alpha^{\text{in}}(\boldsymbol{x},t), H(t')], H(t'')\big] + \cdots\,. \end{aligned} \tag{6.40}$$

Next, we define the retarded product of operators:

$$R\big[\mathcal{O}(t) : A(t_1)\,, \ldots\,, A(t_n)\big] = (-\mathrm{i})^n\big[\ldots[\mathcal{O}(t), A(t_1')]\ldots A(t_n')\big]\,,$$

where $t > t_1' > \ldots > t_n'$ and $(t_1', \ldots, t_n')$ is a permutation of $(t_1, \ldots, t_n)$. If t is smaller than any of the t_j, we define $R = 0$. From the definition, R is a symmetric function of $t_1, \ldots, t_n$. Using R, the general term in (6.40) can be written as

$$\begin{aligned}&(-\mathrm{i})^n\int_{-\infty}^{t}\mathrm{d}t_1\int_{-\infty}^{t_1}\mathrm{d}t_2\ldots\int_{-\infty}^{t_{n-1}}\mathrm{d}t_n\big[\ldots[\varphi_\alpha^{\text{in}}(\boldsymbol{x},t), H(t_1)]\ldots H(t_n)\big]\\ &\quad = \int_{-\infty}^{t}\mathrm{d}t_1\int_{-\infty}^{t_1}\mathrm{d}t_2\ldots\int_{-\infty}^{n-1}\mathrm{d}t_{n-1}\mathrm{d}t_n\,R\big[\varphi_\alpha^{\text{in}}(\boldsymbol{x},t) : H(t_1)\ldots H(t_n)\big]\\ &\quad = \frac{1}{n!}\int_{-\infty}^{t}\mathrm{d}t_1\int_{-\infty}^{t}\mathrm{d}t_2\ldots\int_{-\infty}^{t}\mathrm{d}t_n\,R\big[\varphi_\alpha^{\text{in}}(\boldsymbol{x},t) : H(t_1)\ldots H(t_n)\big]\\ &\quad = \frac{1}{n!}\int_{-\infty}^{\infty}\mathrm{d}t_1\infty_{-\infty}^{\infty}\mathrm{d}t_2\ldots\int_{-\infty}^{\infty}\mathrm{d}t_n\,R\big[\varphi_\alpha^{\text{in}}(\boldsymbol{x},t) : H(t_1)\ldots H(t_n)\big]\,.\end{aligned}$$

Combining this with (6.35), we rewrite the expansion of (6.40) as

$$\varphi_\alpha(x) = \varphi_\alpha^{\text{in}}(x) + \sum_{n=1}^{\infty}\frac{1}{n!}\int_{-\infty}^{\infty}\mathrm{d}^4x_1\ldots\int_{-\infty}^{\infty}\mathrm{d}^4x_n R\ldots[\varphi_\alpha^{\text{in}}(x) : \mathcal{H}_{\text{int}}(x_1)\,, \ldots\,, \mathcal{H}_{\text{int}}(x_n)\ldots]\,. \tag{6.41}$$

This is in fact a generalization of the result in Sect. 4.4.

6.3 Yang–Feldman Expansion of the Heisenberg Operators

Next, we consider a method for expanding the Heisenberg operator using the field equation. This method was originally introduced by Yang and Feldman [90]. The approach used in Sect. 4.4 is in fact based on this.

First, we separate the Lagrangian density into the free part and the interaction part. As mentioned in the previous section, this separation is not trivial:

$$\mathscr{L} = \mathscr{L}_{\mathrm{f}} + \mathscr{L}_{\mathrm{int}} . \tag{6.42}$$

Next, we write down the Euler–Lagrange equation, where we keep the separation above, i.e.,

$$\begin{aligned} [\mathscr{L}]_{\varphi_\alpha^\dagger} &= [\mathscr{L}_{\mathrm{f}}]_{\varphi_\alpha^\dagger} + [\mathscr{L}_{\mathrm{int}}]_{\varphi_\alpha^\dagger} \\ &= D_{\alpha\beta}(\partial)\varphi_\beta + [\mathscr{L}_{\mathrm{int}}]_{\varphi_\alpha^\dagger} = 0 . \end{aligned} \tag{6.43}$$

In order to solve this equation, recall Fig. 4.2 in Sect. 4.4. We assume that interactions exist only in the finite space-time region Ω, i.e., we assume that the coupling constant g is a function of x, with $g \neq 0$ only in Ω and $g = 0$ outside Ω. Therefore, in Fig. 4.2 we assume that $g = 0$ for $t > T_2$ or $t < T_1$. Thus, in these two regions the solution of (6.43) becomes

$$\varphi_\alpha = \begin{cases} \varphi_\alpha^{\mathrm{in}} , & \text{for } t < T_1 \\ \varphi_\alpha^{\mathrm{out}} , & \text{for } t > T_2 . \end{cases} \tag{6.44}$$

At the end, we take the limits $T_1 \to -\infty$ and $T_2 \to \infty$. In addition, the two free fields are defined to be connected throughout space-time via the equation

$$D_{\alpha\beta}(\partial)\varphi_\beta^{\mathrm{in}}(x) = D_{\alpha\beta}(\partial)\varphi_\beta^{\mathrm{out}}(x) = 0 . \tag{6.45}$$

In general, we take

$$\varphi_\alpha(x) = \varphi_\alpha^{\mathrm{in}}(x) + \varphi'_\alpha(x) . \tag{6.46}$$

Inserting this in (6.43) and using (6.45),

$$D_{\alpha\beta}(\partial)\varphi'_\beta(x) = -[\mathscr{L}_{\mathrm{int}}]_{\varphi_\alpha^\dagger}(x) . \tag{6.47}$$

In order to integrate this equation, we introduce the retarded Δ-function Δ^{R} by (4.53):

$$\varphi'_\alpha(x) = \int_\Omega \mathrm{d}^4x' \Delta^{\mathrm{R}}_{\alpha\beta}(x - x')[\mathscr{L}_{\mathrm{int}}]_{\varphi_\beta^\dagger}(x') . \tag{6.48}$$

Extending Ω into the whole space-time and taking $g(x) \to g$,

$$\varphi_\alpha(x) = \varphi_\alpha^{\text{in}}(x) + \int \mathrm{d}^4x' \Delta^{\text{R}}_{\alpha\beta}(x - x')[\mathscr{L}_{\text{int}}]_{\varphi^\dagger_\beta}(x') \,. \tag{6.49}$$

This is a non-linear integral equation for $\varphi_\alpha(x)$. Similarly, using the advanced Δ-function,

$$\varphi_\alpha(x) = \varphi_\alpha^{\text{out}}(x) + \int \mathrm{d}^4x' \Delta^{\text{A}}_{\alpha\beta}(x - x')[\mathscr{L}_{\text{int}}]_{\varphi^\dagger_\beta}(x') \,. \tag{6.50}$$

The quantized field operator can be obtained by solving this integral equation. Taking the difference between (6.49) and (6.50),

$$\varphi_\alpha^{\text{out}}(x) = \varphi_\alpha^{\text{in}}(x) - \int \mathrm{d}^4x' \Delta_{\alpha\beta}(x - x')[\mathscr{L}_{\text{int}}]_{\varphi^\dagger_\beta}(x') \,. \tag{6.51}$$

Alternatively, from (6.32), the S-matrix is determined by

$$S^{-1}\varphi_\alpha^{\text{in}}(x)S = \varphi_\alpha^{\text{in}}(x) - \int \mathrm{d}^4x' \Delta_{\alpha\beta}(x - x')[\mathscr{L}_{\text{int}}]_{\varphi^\dagger_\beta}(x') \,. \tag{6.52}$$

The equation above is called the *Yang–Feldman equation*. The commutation relations are

$$\left[\varphi_\alpha^{\text{in}}(x), \varphi_\beta^{\dagger\text{in}}\right]_\pm = \left[\varphi_\alpha^{\text{out}}(x), \varphi_\beta^{\dagger\text{out}}(y)\right]_\pm = \mathrm{i}C_{\alpha\beta}(\partial)\Delta(x - y) \,. \tag{6.53}$$

The sign chosen above depends on the statistics of the field.

6.4 Examples of Interactions

So far we have discussed the general situation without specifying the form of interaction. We shall now consider some well known examples.

First, let us consider quantum electrodynamics (QED), which describes the interaction between the electromagnetic field and the electron field. A general theory of electromagnetic interactions was given by Dirac. When the field ψ corresponds to a particle with an electric charge e, we only have to make the following replacements in its free-field Lagrangian density:

$$\partial_\mu \psi \to (\partial_\mu - \mathrm{i}eA_\mu) \,, \quad \partial_\mu \bar{\psi} \to (\partial_\mu + \mathrm{i}eA_\mu)\bar{\psi} \,, \tag{6.54}$$

where e is the electric charge of the quantum of the field ψ and $-e$ is the electric charge of its antiparticle. We call the interaction induced by such a replacement a *minimal electromagnetic coupling*. Applying the replacement above for the Lagrangian density of the Dirac field, the interaction part is given by

$$\mathscr{L}_{\text{int}} = j_\mu A_\mu \ , \tag{6.55}$$

where j_μ is a four-current density given by

$$j_\mu = \mathrm{i}e\bar{\psi}\gamma_\mu\psi \ , \quad \text{or} \quad \frac{1}{2}\mathrm{i}e[\bar{\psi}, \gamma_\mu\psi] \ . \tag{6.56}$$

The meaning of the latter case is discussed in Sect. 3.9. We then write the Lagrangian density of the whole system as

$$\mathscr{L} = -\bar{\psi}\big[\gamma_\mu(\partial_\mu - \mathrm{i}eA_\mu) + m\big]\psi - \frac{1}{4}F_{\mu\nu}F_{\mu\nu} - \frac{1}{2\alpha}(\partial_\mu A_\mu)^2 \ . \tag{6.57}$$

QED is characterized by what is known as *gauge invariance*, i.e., making the following changes:

$$\begin{aligned} &\text{gauge transformation of the first kind} \quad \psi \to \mathrm{e}^{\mathrm{i}e\Lambda}\psi \ , \quad \bar{\psi} \to \mathrm{e}^{-\mathrm{i}e\Lambda}\bar{\psi} \ , \\ &\text{gauge transformation of the second kind} \ A_\mu \to A_\mu + \partial_\mu \Lambda \ , \end{aligned} \tag{6.58}$$

the Lagrangian (6.57) is invariant, except for the last term. Note that Λ is an arbitrary function of the space-time coordinate. $D_{\mu\nu}(\partial)$ derived from the free part of the Lagrangian density, i.e., the part that remains when $e = 0$, has no inverse, so quantization does not go through. For this reason, we add the last term in (6.57), which is called, a *gauge-fixing term*. With this procedure, although it is possible to quantize, the general gauge invariance is lost. There is only a restricted gauge invariance, i.e., the Lagrangian density (6.57) is invariant only under the condition

$$\Box\Lambda = 0 \ . \tag{6.59}$$

In Chap. 3, we discussed the free Dirac particle. Here we mention a special feature of the electromagnetic field and its interactions. This concerns the magnetic moment of the electron. From the Lagrangian density (6.57), we obtain the following field equations:

$$\big[\gamma_\mu(\partial_\mu - \mathrm{i}eA_\mu) + m\big]\psi = 0 \ , \quad \bar{\psi}\big[\gamma_\mu(\overleftarrow{\partial}_\mu + \mathrm{i}eA_\mu) - m\big] = 0 \ . \tag{6.60}$$

$$\Box A_\mu + \frac{1-\alpha}{\alpha}\partial_\mu\partial_\nu A_\nu = -j_\mu \ , \tag{6.61}$$

where $\overleftarrow{\partial}_\mu$ in the second equation of (6.60) is an operator which differentiates $\bar{\psi}$ on the left. From this, it is easy to derive the conservation of the current density:

$$\partial_\mu j_\mu = 0 . \tag{6.62}$$

Moreover, taking the divergence of (6.61), we have

$$\Box \partial_\mu A_\mu = 0 . \tag{6.63}$$

Combining this with the Yang–Feldman formalism,

$$\partial_\mu A_\mu = \partial_\mu A_\mu^{\text{in}} = \partial_\mu A_\mu^{\text{out}} . \tag{6.64}$$

Since

$$S^{-1} \partial_\mu A_\mu^{\text{in}} S = \partial_\mu A_\mu^{\text{out}} , \tag{6.65}$$

equation (6.64) shows that the S-matrix and $\partial_\mu A_\mu$ commute:

$$\left[S, \partial_\mu A_\mu \right] = 0 . \tag{6.66}$$

This is (5.120). Therefore, the S-matrix is an observable quantity.

Multiplying the first equation in (6.60) by the operator

$$\gamma_\lambda (\partial_\lambda - \mathrm{i} e A_\lambda) - m ,$$

we obtain

$$\left[\gamma_\lambda \gamma_\mu (\partial_\lambda - \mathrm{i} e A_\lambda)(\partial_\mu - \mathrm{i} e A_\mu) - m^2 \right] \psi = 0 . \tag{6.67}$$

Then we rewrite the product of γ-matrices in the form

$$\gamma_\lambda \gamma_\mu = \delta_{\lambda\mu} + \mathrm{i} \sigma_{\lambda\mu} .$$

Thus, (6.67) can be written in the form

$$\left[(\partial_\lambda - \mathrm{i} e A_\lambda)^2 - m^2 + \frac{e}{2} \sigma_{\lambda\mu} F_{\lambda\mu} \right] \psi = 0 . \tag{6.68}$$

In the non-relativistic approximation, we can derive the Schrödinger equation. Assuming that the electric field is not a quantized field but a static external field, we replace the differential operators by E and $\boldsymbol{p}$:

$$(E - e\varphi)^2 \psi = \left[(\boldsymbol{p} - e\boldsymbol{A})^2 + m^2 - \frac{e}{2} \sigma_{\lambda\mu} F_{\lambda\mu} \right] \psi . \tag{6.69}$$

Writing the energy eigenvalue of this system as $E = m + W$ and introducing the following non-relativistic approximation

$$\left|\frac{W - e\varphi}{2m}\right| \ll 1\,, \tag{6.70}$$

equation (6.69) can be approximated by

$$W\psi = \left[\frac{(\boldsymbol{p} - e\boldsymbol{A})^2}{2m} + e\varphi - \frac{e}{4m}\sigma_{\lambda\mu}F_{\lambda\mu}\right]\psi\,. \tag{6.71}$$

If W stands for the energy in the non-relativistic Schrödinger equation, we pick up only the two large components from the four components of ψ under this approximation, and the last term in (6.71) becomes

$$-\frac{e}{2m}\boldsymbol{\sigma}\boldsymbol{H}\,. \tag{6.72}$$

This means that the magnetic moment of the electron is $-e/2m$. This was an empirical fact in non-relativistic quantum mechanics, but in the Dirac theory it is proven from first principles, and this is one of the experimental reasons in favour of the theory.

For the next example, we consider a system involving nucleons and π-mesons. This interaction was first introduced by Yukawa [17]. In the light of the current quark model, it is viewed as a phenomenological theory rather than a fundamental one. According to Yukawa's theory, the force among nucleons, i.e., the nuclear force, is produced by the exchange of π-mesons between nucleons. Empirically, we know that the nuclear force does not depend on the electric charge of the nucleon in a given angular-momentum state. This is referred to as *charge independence*. This property yields a restriction on the interaction between nucleons and π-mesons. For instance, assume that two nucleons are in the ^1S-state. Here the 1 on the upper left means that the total spin derived from the composition of the two nucleon spins is 0, so it belongs to the spin singlet. S indicates that the relativistic angular momentum is 0. In this state, the potential between two protons, or two neutrons, or a proton and a neutron satisfies the approximate equality

$$V_{\mathrm{pp}}(^1S) = V_{\mathrm{nn}}(^1S) = V_{\mathrm{np}}(^1S)\,. \tag{6.73}$$

On the other hand, due to Pauli's principle, the spin triplet state ^3S exists only in the neutron–proton system. The charge independence reflects a kind of symmetry, and some kind of conserved quantity corresponds to it. This conserved quantity is called *isospin*, because its mathematical properties are similar to those of spin. In the following, we begin by introducing the isospin.

We consider the spin in a two-electron system. Denoting the states with spin-up and spin-down by α and β, respectively, the wave functions of the spin triplet and the spin singlet are:

$$\text{triplet}\begin{cases}\alpha(1)\alpha(2)\,,\\[2ex] \dfrac{\alpha(1)\beta(2)+\beta(1)\alpha(2)}{\sqrt{2}}\,,\\[2ex] \beta(1)\beta(2)\,,\end{cases}\qquad \text{singlet}\ \ \frac{\alpha(1)\beta(2)-\beta(1)\alpha(2)}{\sqrt{2}}\,. \tag{6.74}$$

States in the same multiplet can be mapped to each other by spin rotations.

We consider the proton state p and the neutron state n to be two different states of the same nucleon. Then, by analogy with the spin case, isospin states in the two-nucleon system will belong to either the isospin triplet or the isospin singlet. To be more precise, we rewrite α and β as p and n, respectively. The charge independence of the nuclear force means that members of the same multiplet have the same potential. For example, the spin singlet ^1S-state corresponds to the isospin triplet, while ^3S-state corresponds to the isospin singlet:

$$\text{triplet}\begin{cases}p(1)p(2)\,,\\[2ex] \dfrac{p(1)n(2)+n(1)p(2)}{\sqrt{2}}\,,\\[2ex] n(1)n(2)\,,\end{cases}\qquad \text{singlet}\ \ \frac{p(1)n(2)-n(1)p(2)}{\sqrt{2}}\,. \tag{6.75}$$

For spin, the Pauli matrices σ_x, σ_y, and σ_z are defined by

$$\begin{aligned}&\sigma_x\alpha=\beta\,,\quad \sigma_y\alpha=\mathrm{i}\beta\,,\quad \sigma_z\alpha=\alpha\,,\\ &\sigma_x\beta=\alpha\,,\quad \sigma_y\beta=-\mathrm{i}\alpha\,,\quad \sigma_z\beta=-\beta\,.\end{aligned} \tag{6.76}$$

For isospin, we also introduce three matrices τ_1, τ_2, and τ_3, defining them by the following equations:

$$\begin{aligned}&\tau_1 p=n\,,\quad \tau_2 p=\mathrm{i}n\,,\quad \tau_3 p=p\,,\\ &\tau_1 n=p\,,\quad \tau_2 n=-\mathrm{i}p\,,\quad \tau_3 n=-n\,.\end{aligned} \tag{6.77}$$

As in the case of ordinary spin, we define the isospin of a system with A nucleons as

$$I_a=\frac{1}{2}\sum_{k=1}^{A}\tau_a^{(k)}\,,\quad \text{where } k \text{ is the number of nucleons.} \tag{6.78}$$

Therefore, as in the case of angular momentum the commutation relation among the three components is

$$[I_a, I_b] = \mathrm{i}\epsilon_{abc} I_c \,. \tag{6.79}$$

Then the requirement that the potential V between two nucleons be independent of the charge can be expressed by

$$[I_a, V] = 0 \,. \tag{6.80}$$

In fact, this holds only approximately.

We can consider I_1, I_2, and I_3 as the three components of an abstract vector in isospin space, and we denote it by $\hat{I}$. Although this system is invariant under rotations in the isospin space, due to the Coulomb interaction and the small mass difference between the proton and the neutron, this invariance is approximate. The isospin of the nucleon is $1/2$, while that of the π-meson with three isospin states π^+, π^0, and π^- is 1. Hence, if we denote the eigenvalue of $\hat{I}^2$ by $I'(I'+1)$, it turns out that this multiplet is degenerate with degree $2I'+1$ as in the case of the angular momentum. However, as in the case including π-mesons, we have to generalize (6.78).

The spin of the π-meson will be shown later to be zero. According to (3.70), for a three-component real scalar field, the Lagrangian density is

$$\mathscr{L}_\pi = -\sum_{\alpha=1}^{3} \frac{1}{2}\left[(\partial_\lambda \varphi_\alpha)^2 + \mu^2 \varphi_\alpha^2\right] , \tag{6.81}$$

where φ_1, φ_2, and φ_3 are the three components of the vector $\hat{\varphi}$ in the isospin space. By (3.69), φ_1 and φ_2 correspond to the real and imaginary parts of φ and $\varphi^\dagger$, respectively. φ is the operator which annihilates π^+ and creates π^-, while $\varphi^\dagger$ is its conjugate operator. φ_3 is the operator which creates and annihilates π^0. Since the nucleon belongs to the two-dimensional irreducible representation of the rotation group in the isospin space, it becomes mathematically equivalent to the spinor for ordinary spatial rotations, and is called an *isospinor*. The π-meson is an isovector. The Lagrangian density of the nucleon is

$$\mathscr{L}_\mathrm{N} = -\sum_{\alpha=1}^{2} \bar{\psi}_\alpha (\gamma_\mu \partial_\mu + M)\psi_\alpha \,, \tag{6.82}$$

where ψ_1 stands for the proton component ψ_p and ψ_2 for the neutron component ψ_n. Therefore, referring to (6.77), the τ-matrices have the same representation as the Pauli matrices:

$$\tau_1 = \begin{pmatrix} 0 & 1 \\ 1 & 0 \end{pmatrix} , \quad \tau_2 = \begin{pmatrix} 0 & -\mathrm{i} \\ \mathrm{i} & 0 \end{pmatrix} , \quad \tau_3 = \begin{pmatrix} 1 & 0 \\ 0 & -1 \end{pmatrix} . \tag{6.83}$$

Since ψ is an isospinor, $\bar{\psi}\tau_\alpha\psi$ becomes an isovector in the isospin space. If we take the scalar product of the isovector and the meson field $\hat{\varphi}$, the result is an isoscalar which is invariant under rotations in the isospin space. In order to ensure the charge independence of the nuclear force, the interaction in the nucleon–meson system must be an isoscalar. A similar idea was first introduced by Kemmer [27]. For the interaction, we thus adopt a term proportional to

$$\bar{\psi}\mathscr{O}\hat{\tau}\psi \cdot \hat{\varphi} \,. \tag{6.84}$$

Here $\hat{\tau} \cdot \hat{\varphi}$ stands for the scalar product of two isovectors. $\mathscr{O}$ is a Dirac matrix, and since the π-meson is in fact a pseudo-scalar, we choose it to be $\mathscr{O} = \mathrm{i}\gamma_5$. Considering $\hat{\omega}$ to be a vector generating an infinitesimal rotation, (6.84) is invariant under the following infinitesimal transformation in the isospin space:

$$\psi \to \psi + \frac{\mathrm{i}}{2}(\hat{\omega} \cdot \hat{\tau})\psi \,, \quad \hat{\varphi} \to \hat{\varphi} - \hat{\omega} \times \hat{\varphi} \,, \tag{6.85}$$

where $\times$ denotes the vector product in the isospin space.

The property of charge independence has been observed empirically in many situations. Here we give an example. From the fact that the deuteron d has isospin 0, we can derive the following relation between two differential cross-sections:

$$\frac{\mathrm{d}\sigma(\mathrm{p} + \mathrm{p} \to \mathrm{d} + \pi^+)}{\mathrm{d}\Omega} = 2\frac{\mathrm{d}\sigma(\mathrm{n} + \mathrm{p} \to \mathrm{d} + \pi^0)}{\mathrm{d}\Omega} \,. \tag{6.86}$$

This equation is consistent with experimental results and thus supports the hypothesis of charge independence.

The spin and the parity of the π-meson or pion has been determined as follows. According to the principle of detailed balance, assuming that the mechanical law that governs the nucleon–meson system is time-reversible and comparing with the reverse process,

$$\pi^+ + \mathrm{d} \rightleftarrows \mathrm{p} + \mathrm{p} \,, \tag{6.87}$$

the following relation holds:

$$\frac{\mathrm{d}\sigma(\to)}{\mathrm{d}\Omega} \bigg/ \frac{\mathrm{d}\sigma(\leftarrow)}{\mathrm{d}\Omega} = \frac{4}{3(2S+1)}\frac{p^2}{q^2} \,, \tag{6.88}$$

where p and q are the momenta of the proton and the pion in the centre-of-mass system and S is the spin of the pion. We assume that in any process the initial particle is not polarized. In this equation, it has been taken into account that the spins of the proton and deuteron are 1/2 and 1, respectively. Comparing (6.88) with experiment, we find

$$S = 0 \,, \tag{6.89}$$

i.e., the spin of the pion is zero. Moreover, rewriting (6.88) for the total cross-section, the Pauli principle implies a difference of a factor of two:

$$\frac{\sigma(\rightarrow)}{\sigma(\leftarrow)} = \frac{2}{3(2S+1)} \frac{p^2}{q^2} . \tag{6.90}$$

The spin of π^- is also zero because it is the antiparticle of π^+. From the charge invariance, it is easy to understand that the spin of π^0 must be zero. π^0 decays mainly into two photons:

$$\pi^0 \rightarrow 2\gamma . \tag{6.91}$$

Later, we will use this to prove that the spin of π^0 cannot be 1.

The low-energy π^- loses its energy via electromagnetic interactions when traversing a material, and falls to the $l = 0$ Bohr orbit of the nucleon in a time much shorter than its lifetime. It is subsequently absorbed by the nucleon. In particular, the following absorption process occurs for the deuteron:

$$\pi^- + \mathrm{d} \rightarrow \mathrm{n} + \mathrm{n} . \tag{6.92}$$

Let us compare the angular momenta on each side. Since the spin of the pion is zero and the orbital angular momentum between the pion and the deuteron is zero, the total angular momentum on the left-hand side can be determined by the angular momentum state of the deuteron. The deuteron is a bound state of the proton and the neutron with angular momentum state

$$^3S_1 + {}^3D_1 . \tag{6.93}$$

As mentioned before, the number 3 on the top left indicates that the two spins comprise a parallel triplet term, while S and D indicate that the orbital angular momenta are zero and two, respectively. Moreover, the number 1 on the bottom right indicates that the total angular momentum is unity. Since there are two neutrons, which have Fermi statistics, on the right-hand side of (6.92), their wave function must be anti-symmetric. Consequently, the total wave function of the deuteron must be anti-symmetric as well. Thus, when the spin wave functions are symmetric triplets, the angular momentum must be odd, and when the spin wave function is an anti-symmetric singlet, the angular momentum must be even. Writing down the compositions of spins and orbital angular momenta on the lower right, all the possible states are

$$^1S_0 , \quad {}^3P_{0,1,2} , \quad {}^1D_2 , \quad {}^3F_{2,3,4} , \quad \ldots . \tag{6.94}$$

Choosing the one whose orbital angular momentum is unity, the only angular-momentum state which corresponds to (6.92) is

$$^3S_1 + {}^3D_1 \ \rightarrow \ {}^3P_1 \,. \tag{6.95}$$

Since the parity of the orbital part is altered from even to odd, it turns out that this change originates in the intrinsic parity of the pion. That is, the (intrinsic) parity of the pion is odd. Here we have assumed that the intrinsic parity of the proton and the neutron is even. Therefore, the pion is a spin-0 and parity-odd particle. In general, this is expressed as 0^-. Equivalently, we say that the pion is a pseudo-scalar.

The interactions in the nucleon–meson system which preserve both parity and isospin can be written as follows:

$$\mathscr{L}_{\mathrm{int}} = -\mathrm{i}G\bar{\psi}\gamma_5\hat{\tau}\psi \cdot \hat{\varphi} - \frac{\lambda}{4}(\hat{\varphi}^2)^2 \,, \tag{6.96}$$

where the first term is the so-called *Yukawa interaction*, and the second term expresses the interaction among mesons, a necessary term for the renormalization discussed later. The real (pseudo) scalar representation has been used here. If the complex representation is used,

$$\bar{\psi}\gamma_5\hat{\tau}\psi \cdot \hat{\varphi} = \sqrt{2}(\bar{\psi}_{\mathrm{n}}\gamma_5\psi_{\mathrm{p}}\varphi^\dagger + \bar{\psi}_{\mathrm{p}}\gamma_5\psi_{\mathrm{n}}\varphi) + (\bar{\psi}_{\mathrm{p}}\gamma_5\psi_{\mathrm{p}} - \bar{\psi}_{\mathrm{n}}\gamma_5\psi_{\mathrm{n}})\varphi_3 \,.$$

The full Lagrangian density is therefore

$$\mathscr{L} = -\bar{\psi}(\gamma_\mu\partial_\mu + M)\psi - \frac{1}{2}\left[(\partial_\lambda\hat{\varphi})^2 + \mu^2\hat{\varphi}^2\right] - \mathrm{i}G\bar{\psi}\gamma_5\hat{\tau}\psi \cdot \hat{\varphi} - \frac{\lambda}{4}(\hat{\varphi}^2)^2 \,. \tag{6.97}$$

We end with a remark about the nuclear force, which was the key factor for the development of Yukawa's theory. Since it is difficult to discuss this issue from a completely general point of view, we consider the neutral scalar meson. Assuming that the nucleon is much heavier than the meson, the nucleon can be considered to be at rest, so it can be replaced by an external field. This is analogous to replacing the effects of the proton by a Coulomb potential when we treat the proton as being at rest in the problem of the hydrogen atom. This is called a *static approximation*. In this approximation, for the nucleon–meson interaction, i.e.,

$$H_{\mathrm{int}} = \int \mathrm{d}^3x \; g\bar{\psi}(\boldsymbol{x})\psi(\boldsymbol{x})\varphi(\boldsymbol{x}) \,, \tag{6.98}$$

we consider $\bar{\psi}(\boldsymbol{x})\psi(\boldsymbol{x})$ to be the probability density of the nucleon, and replace it as follows:

$$\bar{\psi}(\boldsymbol{x})\psi(\boldsymbol{x}) \to \sum_{j=1}^{N} \delta^3(\boldsymbol{x} - \boldsymbol{x}_j) , \tag{6.99}$$

where $\boldsymbol{x}_j$ is the position coordinate of the jth nucleon. Hence, the nucleon is not quantized here. Considering a two-body problem, choosing $N = 2$, and substituting (6.99) into (6.98), we have

$$H = H_{\text{f}} + g\big[\varphi(\boldsymbol{x}_1) + \varphi(\boldsymbol{x}_2)\big] \equiv H_{\text{f}} + H' , \tag{6.100}$$

where H_{f} is the Hamiltonian of the free meson and we are working in the Schrödinger picture. We now determine the eigenvalues of this Hamiltonian. In particular, we would like to obtain the energy eigenvalue of the ground state. The equation is

$$H\Psi = E\Psi . \tag{6.101}$$

To solve this problem, we use the Fourier representation

$$H_{\text{f}} = \sum_{\boldsymbol{q}} q_0 c^\dagger(\boldsymbol{q})c(\boldsymbol{q}) , \tag{6.102}$$

$$\varphi(\boldsymbol{x}) = \sum_{\boldsymbol{q}} \frac{1}{\sqrt{2q_0 V}} \big[c(\boldsymbol{q})\mathrm{e}^{\mathrm{i}\boldsymbol{q}\boldsymbol{x}} + c^\dagger(\boldsymbol{q})\mathrm{e}^{-\mathrm{i}\boldsymbol{q}\boldsymbol{x}}\big] . \tag{6.103}$$

H_{f} is already diagonal with respect to the particle number, but the interaction part is not. We thus diagonalize H by a suitable unitary transformation. To do this, we take

$$\Psi = U\Phi , \quad U \equiv \exp(\mathrm{i}S) , \tag{6.104}$$

where U is unitary and S is Hermitian. Therefore,

$$U^{-1}HU\Phi = E\Phi , \tag{6.105}$$

$$\begin{aligned} U^{-1}HU &= U^{-1}(H_{\text{f}} + H')U \\ &= H_{\text{f}} + \mathrm{i}[H_{\text{f}}, S] + \frac{\mathrm{i}^2}{2}\big[[H_{\text{f}}, S], S\big] + \cdots \\ &\quad + \quad H' \quad + \quad \mathrm{i}[H', S] \quad + \cdots \\ &\quad\ \vdots \qquad\ \ \vdots \qquad\qquad \vdots \\ &\quad\ g^0 \qquad g^1 \qquad\quad\ g^2 \end{aligned} \tag{6.106}$$

where, assuming that S is of order of g, we have aligned terms with the same power of g vertically. We then define S by

$$H' + \mathrm{i}[H_\mathrm{f}, S] = 0\,. \tag{6.107}$$

Therefore, (6.106) can be assembled into the following form:

$$U^{-1}HU = H_\mathrm{f} + \sum_{n=1}^{\infty} \mathrm{i}^n \left[\frac{1}{n!} - \frac{1}{(n+1)!}\right] \overbrace{\left[\ldots [H', S] \ldots S\right]}^{n} \text{commutators}\,. \tag{6.108}$$

Since the second term begins with terms of order g^2, it turns out that H has been diagonalized up to order g^2. In the case of the neutral meson, since the term with $n = 1$ is a c-number, any term with more than two commutators vanishes. To solve (6.107), we use

$$\left[H_\mathrm{f}, c(\boldsymbol{q})\right] = -q_0 c(\boldsymbol{q})\,, \quad \left[H_\mathrm{f}, c^\dagger(\boldsymbol{q})\right] = q_0 c^\dagger(\boldsymbol{q})\,, \tag{6.109}$$

to show that S is given by

$$S = -\mathrm{i}g \sum_{\boldsymbol{q}} \frac{1}{\sqrt{2q_0 V}} \frac{1}{q_0} \left[c(\boldsymbol{q})(\mathrm{e}^{\mathrm{i}\boldsymbol{q}\boldsymbol{x}_1} + \mathrm{e}^{\mathrm{i}\boldsymbol{q}\boldsymbol{x}_2}) - c^\dagger(\boldsymbol{q})(\mathrm{e}^{-\mathrm{i}\boldsymbol{q}\boldsymbol{x}_1} + \mathrm{e}^{-\mathrm{i}\boldsymbol{q}\boldsymbol{x}_2})\right]. \tag{6.110}$$

Thus,

$$\begin{aligned} U^{-1}HU &= H_\mathrm{f} + \frac{\mathrm{i}}{2}[H', S] \\ &= H_\mathrm{f} - g^2 \sum_{\boldsymbol{q}} \frac{1}{2q_0^2 V} (\mathrm{e}^{\mathrm{i}\boldsymbol{q}\boldsymbol{x}_1} + \mathrm{e}^{\mathrm{i}\boldsymbol{q}\boldsymbol{x}_2})(\mathrm{e}^{-\mathrm{i}\boldsymbol{q}\boldsymbol{x}_1} + \mathrm{e}^{-\mathrm{i}\boldsymbol{q}\boldsymbol{x}_2})\,. \end{aligned}$$

Taking the limit $V \to \infty$ and making the replacement (4.2),

$$U^{-1}HU = H_\mathrm{f} - \frac{g^2}{(2\pi)^3} \int \frac{\mathrm{d}^3 q}{q_0^2} \left[1 + \cos \boldsymbol{q}(\boldsymbol{x}_1 - \boldsymbol{x}_2)\right]. \tag{6.111}$$

The integral splits into the divergent self-energy δM and the potential $\mathscr{V}(r)$:

$$\delta M = -\frac{g^2}{(2\pi)^3}\int \frac{\mathrm{d}^3 q}{2q_0^2}\,, \tag{6.112}$$

$$\mathscr{V}(r) = -\frac{g^2}{(2\pi)^3}\int \frac{\mathrm{d}^3 q}{q_0^2}\cos \boldsymbol{q}(\boldsymbol{x}_1 - \boldsymbol{x}_2) = -\frac{g^2}{4\pi}\frac{\mathrm{e}^{-mr}}{r}\,, \qquad r = |\boldsymbol{x}_1 - \boldsymbol{x}_2|\,. \tag{6.113}$$

Therefore, the lowest eigenvalue E of H can be written

$$E = 2\delta M + \mathscr{V}(r)\,. \tag{6.114}$$

The function $\mathscr{V}(r)$ is called the *Yukawa potential.*

Chapter 7
Symmetries and Conservation Laws

In this chapter, we study invariance under different types of continuous transformation in a given system, showing that such symmetries are closely related to the existence of conservation laws. The key to this is Noether's theorem. We begin by formulating this idea in a point-particle system, and then extend it to the field case.

7.1 Noether's Theorem for Point-Particle Systems

We define an action integral I starting from the Lagrangian $L(q_j(t), \dot{q}_j(t))$ for a given system:

$$I = \int_{t_1}^{t_2} \mathrm{d}t \, L\big(q_j(t), \dot{q}_j(t)\big) \,. \tag{7.1}$$

We consider the infinitesimal variations

$$t \to t' = t + \delta t \,, \quad q_j(t) \to q'_j(t') = q_j(t) + \delta q_j \,. \tag{7.2}$$

Regarding the dynamical variables q_j, we consider two kinds of transformation. The first does not involve any change in the time variable, while the second does:

$$\delta^* q_j(t) = q'_j(t) - q_j(t) \,, \quad \delta q_j(t) = q'_j(t') - q_j(t) \,. \tag{7.3}$$

The variation δ^* thus commutes with the time derivative and we have the following relation between these two kinds of variation:

$$\delta^* q_j(t) = \delta q_j(t) - \delta t \dot{q}_j(t) \,. \tag{7.4}$$

K. Nishijima, *Quantum Field Theory*,
https://doi.org/10.1007/978-94-024-2190-3_7

We now consider a variation of the action integral corresponding to (7.2):

$$I' = \int_{t_1'}^{t_2'} \mathrm{d}t' L\big(q_j'(t'), \dot{q}_j'(t')\big) . \tag{7.5}$$

The change in the t-derivative can be found from

$$\frac{\mathrm{d}}{\mathrm{d}t'} = \frac{\mathrm{d}t}{\mathrm{d}t'}\frac{\mathrm{d}}{\mathrm{d}t} = \left(1 - \frac{\mathrm{d}}{\mathrm{d}t}\delta t\right)\frac{\mathrm{d}}{\mathrm{d}t} ,$$

and

$$\mathrm{d}t' = \frac{\mathrm{d}t'}{\mathrm{d}t}\mathrm{d}t = \left(1 + \frac{\mathrm{d}}{\mathrm{d}t}\delta t\right)\mathrm{d}t . \tag{7.6}$$

Transforming the integral variable t' of I' back to t, we find the difference from I to be

$$\begin{aligned} I' - I &= \int_{t_1}^{t_2} \mathrm{d}t \left(1 + \frac{\mathrm{d}}{\mathrm{d}t}\delta t\right) L' - \int_{t_1}^{t_2} \mathrm{d}t L \\ &= \int_{t_1}^{t_2} \mathrm{d}t \left(L' - L + \frac{\mathrm{d}}{\mathrm{d}t}\delta t \cdot L\right) . \end{aligned} \tag{7.7}$$

The integrand can be rewritten

$$L' - L + \frac{\mathrm{d}}{\mathrm{d}t}\delta t \cdot L = \sum_j \left(\frac{\partial L}{\partial q_j}\delta q_j + \frac{\partial L}{\partial \dot{q}_j}\delta \dot{q}_j\right) + \frac{\mathrm{d}}{\mathrm{d}t}(\delta t L) - \delta t \frac{\mathrm{d}}{\mathrm{d}t} L .$$

Using (7.4) and taking into account the commutativity of δ^* and the time derivative, the left-hand side becomes

$$\sum_j \left[\frac{\partial L}{\partial q_j} - \frac{\mathrm{d}}{\mathrm{d}t}\left(\frac{\partial L}{\partial \dot{q}_j}\right)\right]\delta^* q_j + \frac{\mathrm{d}}{\mathrm{d}t}\left(\sum_j \frac{\partial L}{\partial \dot{q}_j}\delta^* q_j + \delta t L\right) . \tag{7.8}$$

Although the first term must vanish, according to the Euler–Lagrange equation, we keep it for the moment:

$$I' - I = \left[\sum_j \frac{\partial L}{\partial \dot{q}_j}\delta^* q_j + \delta t L\right]_{t_1}^{t_2} + \int_{t_1}^{t_2} \mathrm{d}t \sum_j \left[\frac{\partial L}{\partial q_j} - \frac{\mathrm{d}}{\mathrm{d}t}\left(\frac{\partial L}{\partial \dot{q}_j}\right)\right]\delta^* q_j . \tag{7.9}$$

If we now use the Euler–Lagrange equation, we obtain

$$I' - I = Q(t_2) - Q(t_1) , \tag{7.10}$$

where

$$Q = \sum_j \frac{\partial L}{\partial \dot{q}_j} \delta^* q_j + \delta t L = \sum_j p_j \delta q_j - H \delta t . \tag{7.11}$$

Therefore, if I is invariant under the infinitesimal transformation (7.2), the equation of motion implies the following conservation law:

$$\frac{\mathrm{d}}{\mathrm{d}t} Q(t) = 0 . \tag{7.12}$$

This correspondence between the invariance of the action integral under the infinitesimal transformation and the conservation law is called *Noether's theorem.* For this theorem within a point-particle system, the reader is referred to *Introduction to Analytical Mechanics for Learning Quantum Mechanics*, written in Japanese by Yasushi Takahashi [91].

Time-like Parallel Translation
If we consider

$$\delta t = \epsilon , \quad \delta q_j(t) = 0 , \tag{7.13}$$

where ϵ is an infinitesimal parameter independent of t, the action integral is invariant in the case where L has no explicit time dependence. Then,

$$Q = -\epsilon H , \tag{7.14}$$

whence H is conserved, i.e., we obtain the conservation law of energy.

Spatial Parallel Translation
If we consider

$$\delta t = 0 , \quad \delta \boldsymbol{x}_j(t) = \boldsymbol{\epsilon} , \tag{7.15}$$

the action integral defined by the Lagrangian for a point-particle system interacting with a two-body central force, viz.,

$$L = \sum_j \frac{1}{2} m_j \dot{\boldsymbol{x}}_j^2 - \sum_{i>j} V(|\boldsymbol{x}_i - \boldsymbol{x}_j|) , \tag{7.16}$$

is invariant. The corresponding conserved quantity is

$$Q = \epsilon \sum_j \boldsymbol{p}_j = \boldsymbol{\epsilon P} , \quad \boldsymbol{p}_j = m_j \dot{\boldsymbol{x}}_j . \tag{7.17}$$

We thus obtain the conservation of momentum $\boldsymbol{P}$.

7.2 Noether's Theorem in Field Theory

It is easy to extend Noether's theorem from point-particle systems to fields. We begin with the action integral

$$I = \int_\Omega \mathrm{d}^4 x \mathscr{L}\big[\varphi_\alpha(x), \varphi_{\alpha,\mu}(x)\big] , \tag{7.18}$$

where $\varphi(x)_\alpha$ is a generic notation for fields in various representations of the Poincaré algebra, and the subscript α distinguishes between different fields, but also between the fields and their conjugates. We introduce the infinitesimal variations

$$x_\mu \to x'_\mu + \delta x_\mu , \quad \varphi_\alpha(x) \to \varphi'_\alpha(x') = \varphi_\alpha(x) + \delta\varphi_\alpha(x) , \tag{7.19}$$

where $\varphi \to \varphi'$ stands for the x-dependence of φ or the variation of the functional form. As in the previous section, we introduce two types of variation:

$$\delta^*\varphi_\alpha(x) = \varphi'_\alpha(x) - \varphi_\alpha(x) , \quad \delta\varphi_\alpha(x) = \varphi'_\alpha(x') - \varphi_\alpha(x) , \tag{7.20}$$

where δ^* is the variation used to derive the Euler–Lagrange equation, which was simply written as δ in Chap. 3. δ^* commutes with the derivative. Therefore, we can consider the variation in Chap. 3 to be δ under the restriction $\delta x_\mu = 0$. The relation between the two variations is

$$\delta^*\varphi_\alpha(x) = \delta\varphi_\alpha(x) - \varphi_{\alpha,\mu}(x)\delta x_\mu . \tag{7.21}$$

Due to the commutativity of δ^* with the derivative,

$$\delta\varphi_{\alpha,\mu}(x) = \partial_\mu \delta\varphi_\alpha(x) - \varphi_{\alpha,\nu}(x)\partial_\mu \delta x_\nu . \tag{7.22}$$

We can now calculate δI:

$$\delta I = \int_{\Omega'} \mathrm{d}^4 x' \mathscr{L}\big[\varphi'_\alpha(x'), \varphi'_{\alpha,\mu}(x')\big] - \int_\Omega \mathrm{d}^4 x \mathscr{L}\big[\varphi_\alpha(x), \varphi_{\alpha,\mu}(x)\big] . \tag{7.23}$$

In order to make the integration regions coincide, we change the variable x' back to x in the first integral. Using

$$\mathrm{d}^4x' = \frac{\partial(x'_0, x'_1, x'_2, x'_3)}{\partial(x_0, x_1, x_2, x_3)}\mathrm{d}^4x \ , \tag{7.24}$$

we obtain

$$\begin{aligned}\delta I = \int_\Omega \mathrm{d}^4x \bigg\{ &\mathscr{L}\Big[\varphi_\alpha(x) + \delta\varphi_\alpha(x), \varphi_{\alpha,\mu}(x) + \delta\varphi_{\alpha,\mu}(x)\Big]\frac{\partial(x'_0, x'_1, x'_2, x'_3)}{\partial(x_0, x_1, x_2, x_3)} \\ &- \mathscr{L}\big[\varphi_\alpha(x), \varphi_{\alpha,\mu}(x)\big]\bigg\} \ .\end{aligned}$$

For infinitesimal δx_μ, we have

$$\frac{\partial(x'_0, x'_1, x'_2, x'_3)}{\partial(x_0, x_1, x_2, x_3)} = 1 + \partial_\nu(\delta x_\nu) \ . \tag{7.25}$$

Using this and the summation convention for repeated indices,

$$\delta I = \int_\Omega \mathrm{d}^4x \left[\frac{\partial\mathscr{L}}{\partial\varphi_\alpha}\delta\varphi_\alpha + \frac{\partial\mathscr{L}}{\partial\varphi_{\alpha,\mu}}\delta\varphi_{\alpha,\mu} + \mathscr{L}\partial_\nu(\delta x_\nu)\right] \ . \tag{7.26}$$

Therefore, if the action integral I is invariant under any choice of Ω,

$$\delta\mathscr{L} + \mathscr{L}\partial_\nu(\delta x_\nu) = 0 \ , \tag{7.27}$$

where

$$\delta\mathscr{L} = \frac{\partial\mathscr{L}}{\partial\varphi_\alpha}\delta\varphi_\alpha + \frac{\partial\mathscr{L}}{\partial\varphi_{\alpha,\mu}}\delta\varphi_{\alpha,\mu} \ . \tag{7.28}$$

Rewriting this in terms of δ^* rather than δ,

$$[\mathscr{L}]_{\varphi_\alpha}\delta^*\varphi_\alpha + \frac{\partial}{\partial x_\mu}\left(\frac{\partial\mathscr{L}}{\partial\varphi_{\alpha,\mu}}\delta^*\varphi_\alpha + \mathscr{L}\delta x_\mu\right) = 0 \ . \tag{7.29}$$

Next, we introduce the canonical energy–momentum tensor

$$\mathscr{T}_{\mu\nu} = -\frac{\partial\mathscr{L}}{\partial\varphi_{\alpha,\mu}}\varphi_{\alpha,\nu} + \delta_{\mu\nu}\mathscr{L} \ . \tag{7.30}$$

We write the second term in (7.29) in terms of δ once again:

$$[\mathscr{L}]_{\varphi_\alpha}\delta^*\varphi_\alpha + \frac{\partial}{\partial x_\mu}\left(\frac{\partial \mathscr{L}}{\partial \varphi_{\alpha,\mu}}\delta\varphi_\alpha + \mathscr{L}_{\mu\nu}\delta x_\nu\right) = 0\,. \tag{7.31}$$

This expresses the invariance of the action integral. Using the field equation, we can drop the first term, and this yields the conservation law

$$\frac{\partial}{\partial x_\mu}\left(\frac{\partial \mathscr{L}}{\partial \varphi_{\alpha,\mu}}\delta\varphi_\alpha + \mathscr{T}_{\mu\nu}\delta x_\nu\right) = 0\,. \tag{7.32}$$

The derivation above is basically the same as the one for a point-particle system. In field theory, Noether's theorem expresses the relationship between invariance and the above conservation law. The conserved four-dimensional current in brackets is called the *Noether current.*

7.3 Applications of Noether's Theorem

Space-Time Translations

If there is no external field, the Lagrangian density will not depend explicitly on any coordinate. This means that, if the Lagrangian density depends on coordinates only through fields, then the action integral will be invariant under the following transformations:

$$\delta x_\mu = \epsilon_\mu\,, \qquad \delta\varphi_\alpha = 0\,. \tag{7.33}$$

Inserting (7.33) into (7.32), the conservation law is in this case

$$\partial_\mu(\mathscr{T}_{\mu\nu}\epsilon_\nu) = 0\,, \quad \text{or} \quad \partial_\mu\mathscr{T}_{\mu\nu} = 0\,. \tag{7.34}$$

This is because ϵ_ν is an arbitrary infinitesimal parameter. From this, we can obtain the energy–momentum vector as a conserved quantity:

$$P_\nu = \int \mathrm{d}^3x\,\mathscr{T}_{0\nu}(x)\,. \tag{7.35}$$

It is easy to check that this definition coincides with the one given in Sect. 3.2. If the Lagrangian density includes an external field,

$$\partial_\mu\mathscr{T}_{\mu\nu} = \frac{\partial\mathscr{L}}{\partial(x_\nu)}\,, \tag{7.36}$$

where the derivative with respect to (x_ν) picks up only the coordinate dependence introduced through the external field, and involves no derivative of the dynamical fields under investigation. We can in fact generalize (3.29). If $\mathscr{O}(x)$ is written in terms of fields and does not depend explicitly on any coordinate,

$$\frac{\partial}{\partial x_\mu}\mathscr{O}[\varphi_\alpha(x)] = \mathrm{i}\big[\mathscr{O}[\varphi_\alpha(x)], P_\mu\big] . \tag{7.37}$$

On the other hand, if $\mathscr{O}(x)$ depends not only on the field $\varphi_\alpha(x)$, but also explicitly on some coordinate, i.e., if $\mathscr{O}$ can be written as $\mathscr{O}[\varphi_\alpha(x), (x_\mu)]$, then (7.37) changes to

$$\frac{\partial}{\partial x_\mu}\mathscr{O}(x) = \mathrm{i}[\mathscr{O}(x), P_\mu] + \frac{\partial}{\partial (x_\mu)}\mathscr{O}(x) . \tag{7.38}$$

This equation will be used later.

Phase Transformations

Next, we consider a phase transformation. Assuming that φ_α is a complex field, the Lagrangian density also includes the complex conjugate φ_α^* of φ_α. Then writing the complex conjugate field explicitly, we consider the Lagrangian density

$$\mathscr{L} = \mathscr{L}\big[\varphi_\alpha(x), \varphi_\alpha^*(x), \varphi_{\alpha,\nu}(x), \varphi_{\alpha,\mu}^*(x)\big] . \tag{7.39}$$

We assume that this $\mathscr{L}$ is invariant under the infinitesimal phase transformation

$$\begin{aligned} \varphi_\alpha &\to \mathrm{e}^{\mathrm{i}q_\alpha\lambda}\varphi_\alpha = \varphi_\alpha + \mathrm{i}q_\alpha\lambda\varphi_\alpha , \\ \varphi_\alpha^* &\to \mathrm{e}^{-\mathrm{i}q_\alpha\lambda}\varphi_\alpha^* = \varphi_\alpha^* - \mathrm{i}q_\alpha\lambda\varphi_\alpha^* , \end{aligned} \tag{7.40}$$

that is,

$$\delta\varphi_\alpha = \mathrm{i}q_\alpha\lambda\varphi_\alpha , \quad \delta\varphi_\alpha^* = -\mathrm{i}q_\alpha\lambda\varphi_\alpha^* . \tag{7.41}$$

Therefore, the Noether current is

$$\lambda\sum_\alpha \mathrm{i}q_\alpha\left(\frac{\partial\mathscr{L}}{\partial\varphi_{\alpha,\mu}}\varphi_\alpha - \frac{\partial\mathscr{L}}{\partial\varphi_{\alpha,\mu}^*}\varphi_\alpha^*\right) . \tag{7.42}$$

When q_α is an electric charge of the quantum of the field φ_α, the conserved current is in fact

$$j_\mu = \sum_\alpha(-\mathrm{i}q_\alpha)\left(\frac{\partial\mathscr{L}}{\partial\varphi_{\alpha,\mu}}\varphi_\alpha - \frac{\partial\mathscr{L}}{\partial\varphi_{\alpha,\mu}^*}\varphi_\alpha^*\right) . \tag{7.43}$$

This yields the conservation law.

Isospin Transformations
Next, we consider an infinitesimal transformation in isospin space. If when we introduce an infinitesimal rotation vector $\hat{\omega}$ the action integral is invariant under the infinitesimal rotation of the field quantity given by (6.85), then, in terms of the nucleon field and the meson field, Noether's current can be written in the form

$$-\hat{\omega}\cdot\left[\frac{\mathrm{i}}{2}\bar{\psi}\gamma_\mu\hat{\tau}\psi - \hat{\varphi}\times\hat{\varphi}_{,\mu}\right]. \tag{7.44}$$

Therefore, the conserved isospin current is

$$\hat{j}_\mu = \frac{\mathrm{i}}{2}\bar{\phi}\gamma_\mu\hat{\tau}\psi - \hat{\varphi}\times\partial_\mu\hat{\varphi}\,. \tag{7.45}$$

The isospin $\hat{L}$ is then given by

$$\hat{I} = \int \mathrm{d}^3x\left[\frac{1}{2}\psi^\dagger\hat{\tau}\psi + \hat{\varphi}\times\hat{\pi}\right], \quad \hat{\pi} = \dot{\hat{\varphi}}\,. \tag{7.46}$$

Various conserved quantities can be derived from the Noether current in this way.

7.4 Poincaré Invariance

Combining the previously mentioned parallel translations in space-time and the Lorentz transformations, we obtain the group of Poincaré transformations. Here we first consider the Lorentz transformation. Using an infinitesimal parameter $\epsilon_{\mu\nu} = -\epsilon_{\nu\mu}$, the infinitesimal Lorentz transformation can be expressed as

$$\delta x_\mu = \epsilon_{\mu\nu}x_\nu\,, \quad \delta\varphi_\alpha(x) = \frac{1}{2}\epsilon_{\mu\nu}(T_{\mu\nu})_{\alpha\beta}\varphi_\beta(x)\,. \tag{7.47}$$

This is a generalization of (3.119). $T_{\mu\nu}$ is a representation of the generator of the Lorentz transformation. We have already encountered the following examples:

$$\text{Spin } 0 \qquad T_{\mu\nu} = 0\,, \tag{7.48a}$$

$$\text{Spin } 1/2 \qquad T_{\mu\nu} = \frac{1}{4}(\gamma_\mu\gamma_\nu - \gamma_\nu\gamma_\mu)\,, \tag{7.48b}$$

$$\text{Spin } 1 \qquad T_{\mu\nu} = a_{\mu\nu}\,. \tag{7.48c}$$

However, since the matrix $a_{\mu\nu}$ in the spin-1 case can be chosen to be the same as the matrix of the coordinate transformation, its components can be determined from the identity

$$\delta x_\mu = \epsilon_{\mu\nu} x_\nu = \frac{1}{2}\epsilon_{\rho\sigma}(a_{\rho\sigma})_{\mu\nu} x_\nu \ .$$

Its specific form is then

$$(a_{\rho\sigma})_{\mu\nu} = \delta_{\mu\rho}\delta_{\nu\sigma} - \delta_{\mu\sigma}\delta_{\nu\rho} \ . \tag{7.49}$$

$T_{\mu\nu}$ given by (7.48a)–(7.48c) satisfies the commutation relation

$$[T_{\rho\sigma}, T_{\lambda\tau}] = \delta_{\sigma\lambda}T_{\rho\tau} - \delta_{\sigma\tau}T_{\rho\lambda} + \delta_{\rho\lambda}T_{\sigma\tau} - \delta_{\rho\tau}T_{\sigma\lambda} \ . \tag{7.50}$$

The Noether current corresponding to the above transformation is

$$\mathscr{M}_{\mu\rho\sigma} = x_\rho \mathscr{T}_{\mu\sigma} - x_\sigma \mathscr{T}_{\mu\rho} - \frac{\partial \mathscr{L}}{\partial \varphi_{\alpha,\mu}}(T_{\rho\sigma})_\beta \ . \tag{7.51}$$

The conservation law is

$$\partial_\mu \mathscr{M}_{\mu\rho\sigma} = 0 \ , \tag{7.52}$$

where the generator $M_{\rho\sigma}$ of the Lorentz transformation is defined by

$$M_{\rho\sigma} = \int \mathrm{d}^3x \mathscr{M}_{0\rho\sigma} \ . \tag{7.53}$$

The same commutation relations hold among the $M_{\rho\sigma}$ as among the $-\mathrm{i}T_{\rho\sigma}$, i.e.,

$$[M_{\rho\sigma}, M_{\lambda\tau}] = -\mathrm{i}(\delta_{\sigma\lambda}M_{\rho\tau} - \delta_{\sigma\tau}M_{\rho\lambda} + \delta_{\rho\lambda}M_{\sigma\tau} - \delta_{\rho\tau}M_{\sigma\lambda}) \ . \tag{7.54}$$

So far we have found the operators generating the Lorentz group. Since the generator of the parallel translation is P_μ, we have to give the commutation relations involving P_μ, i.e.,

$$[M_{\mu\nu}, P_\sigma] = \mathrm{i}(\delta_{\mu\sigma}P_\nu - \delta_{\nu\mu}P_\mu) \ , \tag{7.55}$$

$$[P_\mu, P_\nu] = 0 \ . \tag{7.56}$$

The generators of the Poincaré group form a closed algebra under the commutation relations. Note that even if the conservation law (7.52) holds, not all the $M_{\rho\sigma}$ commute with P_μ, e.g., with the Hamiltonian. This is because a coordinate is included explicitly in (7.51), so we need to use (7.38).

Next, we introduce the symmetric energy–momentum tensor. We write $\mathscr{M}$ as

$$\mathscr{M}_{\mu\nu\sigma} = x_\rho \mathscr{T}_{\mu\sigma} - x_\sigma \mathscr{T}_{\mu\rho} + \mathscr{S}_{\mu\rho\sigma} \tag{7.57}$$

and apply ∂_μ to both sides to obtain

$$\mathscr{J}_{\sigma\rho} - \mathscr{J}_{\rho\sigma} = -\partial_\mu \mathscr{S}_{\mu\sigma\rho} \,. \tag{7.58}$$

This implies that the canonical energy–momentum tensor is not generally symmetric. We thus define a new tensor $\Theta_{\mu\nu}$ by

$$\Theta_{\rho\sigma} = \mathscr{T}_{\rho\sigma} + \partial_\mu \mathscr{G}_{\mu\rho\sigma} \,, \tag{7.59}$$

where

$$\mathscr{G}_{\mu\rho\sigma} = \frac{1}{2}(\mathscr{S}_{\mu\rho\sigma} + \mathscr{S}_{\rho\sigma\mu} + \mathscr{S}_{\sigma\rho\mu}) \,. \tag{7.60}$$

Since $\mathscr{S}_{\mu\rho\sigma}$ is anti-symmetric with respect to ρ and σ,

$$\begin{aligned} \mathscr{G}_{\rho\mu\sigma} &= \frac{1}{2}(\mathscr{S}_{\rho\mu\sigma} + \mathscr{S}_{\mu\sigma\rho} + \mathscr{S}_{\sigma\mu\rho}) \\ &= -\frac{1}{2}(\mathscr{S}_{\rho\sigma\mu} + \mathscr{S}_{\mu\rho\sigma} + \mathscr{S}_{\sigma\rho\mu}) - \mathscr{G}_{\mu\rho\sigma} \,. \end{aligned} \tag{7.61}$$

Therefore, $\mathscr{G}$ is anti-symmetric with respect to the first two indices, i.e.,

$$\partial_\mu \partial_\rho \mathscr{G}_{\mu\rho\sigma} = 0 \,, \tag{7.62}$$

$$\partial_\rho \Theta_{\rho\sigma} = \partial_\rho \mathscr{T}_{\rho\sigma} = 0 \,. \tag{7.63}$$

Moreover,

$$\begin{aligned} \partial_\mu(\mathscr{G}_{\mu\rho\sigma} - \mathscr{G}_{\mu\sigma\rho}) &= \frac{1}{2}\partial_\mu(\mathscr{S}_{\mu\rho\sigma} + \mathscr{S}_{\rho\sigma\mu} + \mathscr{S}_{\sigma\rho\mu} - \mathscr{S}_{\mu\sigma\rho} - \mathscr{S}_{\sigma\rho\mu} - \mathscr{S}_{\rho\sigma\mu}) \\ &= \partial_\mu \mathscr{S}_{\mu\rho\sigma} \,. \end{aligned} \tag{7.64}$$

Combining this with (7.58) and (7.59), it is straightforward to see that the tensor $\Theta_{\rho\sigma}$ is symmetric in ρ and σ, i.e.,

$$\Theta_{\rho\sigma} = \Theta_{\sigma\rho} \,. \tag{7.65}$$

It is thus referred to as the *symmetric energy–momentum tensor*. Since this tensor was first introduced by Belinfante and Rosenfeld [92, 93], it is also called the

Belinfante–Rosenfeld tensor.[1] In addition, taking into account the fact that $\mathscr{G}_{00\nu} = 0$,

$$P_\nu = \int \mathrm{d}^3x\, \mathscr{J}_{0\nu}(x) = \int \mathrm{d}^3x\, \Theta_{0\nu}(x)\ . \tag{7.66}$$

Similarly,

$$M_{\rho\sigma} = \int \mathrm{d}^3x \mathscr{M}_{0\rho\sigma} = \int \mathrm{d}^3x (x_\rho \Theta_{0\sigma} - x_\sigma \Theta_{0\rho})\ . \tag{7.67}$$

To prove this, we have to show that

$$\int \mathrm{d}^3x \left(x_\rho \partial_\mu \mathscr{G}_{\mu 0\sigma} - x_\sigma \partial_\mu \mathscr{G}_{\mu 0\rho}\right) = \int \mathrm{d}^3x \mathscr{S}_{0\rho\sigma}\ .$$

Since $\mu = 4$ does not produce any effects on the left-hand side,

$$\begin{aligned}
\text{(left-hand side)} &= \int \mathrm{d}^3x \partial_\mu (x_\rho \mathscr{G}_{\mu 0\sigma} - x_\sigma \mathscr{G}_{\mu 0\rho}) + \int \mathrm{d}^3x (\mathscr{G}_{\sigma 0\rho} - \mathscr{G}_{\rho 0\sigma}) \\
&= \frac{1}{2} \int \mathrm{d}^3x \left(\mathscr{S}_{\sigma 0\rho} + \mathscr{S}_{0\rho\sigma} + \mathscr{S}_{\rho 0\sigma} - \mathscr{S}_{\rho 0\sigma} - \mathscr{S}_{0\sigma\rho} - \mathscr{S}_{\sigma 0\rho}\right) \\
&= \int \mathrm{d}^3x \mathscr{S}_{0\rho\sigma} = \text{(right-hand side)}\ ,
\end{aligned}$$

which proves (7.67). However, in the above proof we dropped the three-dimensional integral of the three-dimensional divergence. Moreover, it is not the canonical energy–momentum tensor but the symmetric energy–momentum tensor that becomes the source for the gravitational field. Using standard notation, the equation for the gravitational field is

$$R_{\mu\nu} - \frac{1}{2} g_{\mu\nu} R = \kappa \Theta_{\mu\nu}\ . \tag{7.68}$$

Next, we consider the spin of a particle. To do so, we introduce the Pauli–Liubanski operator

$$W_\alpha = \frac{1}{2\mathrm{i}} \epsilon_{\alpha\beta\gamma\delta} M_{\beta\gamma} P_\delta\ , \tag{7.69}$$

where $\epsilon_{\alpha\beta\gamma\delta}$ takes the values $+1$ or -1 when $(\alpha, \beta, \gamma, \delta)$ is an even or odd permutation of $(1, 2, 3, 4)$, respectively, and zero otherwise. Although this axial

[1] In the original Japanese edition, the author referred to this as the Belinfante tensor instead of the Belinfante–Rosenfeld tensor.

vector commutes with P_μ, it transforms as a vector under a proper Lorentz transformation:

$$[P_\mu, W_\alpha] = 0\,. \tag{7.70}$$

The Poincaré algebra has two Casimir operators, i.e., operators which commute with all components of P and M :

$$P^2 = P_\mu P_\mu\,, \quad W^2 = W_\alpha W_\alpha\,. \tag{7.71}$$

We use these two quantities to define the mass and spin of a particle. First, note that W_α has the properties

$$W_\alpha P_\alpha = 0 \tag{7.72}$$

and

$$[W_\alpha, W_\beta] = \epsilon_{\alpha\beta\gamma\delta} W_\gamma P_\delta\,. \tag{7.73}$$

We can diagonalize any set of operators which commute with each other, such as

$$P_\mu\,, \quad W_3\,, \quad P^2\,, \quad W^2\,. \tag{7.74}$$

Assuming that the eigenvalue of $-P^2$ is m^2, m becomes the mass of this particle. When $m \neq 0$, we can pick out the rest frame of this particle, choosing the eigenvalues p_μ of P_μ to be

$$\boldsymbol{p} = 0\,, \quad p_0 = m\,. \tag{7.75}$$

In this case, if we write $\boldsymbol{W} = m\boldsymbol{S}$, the three components of $\boldsymbol{S}$ satisfy the same commutation relations as those of the angular momentum. This is clear from (7.73) and (7.75). Hence, the eigenvalue of W^2 has the form, $m^2 s(s+1)$, where s stands for the spin of this particle. In fact, for the one-particle state in the rest frame, $\boldsymbol{W}$ has the representation

$$W_i \;\rightarrow\; \frac{m}{2\mathrm{i}} \epsilon_{ijk} T_{jk}\,. \tag{7.76}$$

When $m = 0$, things become more difficult. We will discuss this later.

7.5 Representations of the Lorentz Group

The commutation relations for the six creation operators of the Lorentz group are given by (7.54). In this section, we will discuss the issue of representations. To do so, we introduce the notation

$$\boldsymbol{M} = (M_{23}, M_{31}, M_{12}) , \quad \boldsymbol{N} = (M_{14}, M_{24}, M_{34}) . \tag{7.77}$$

Then (7.54) can be decomposed into the following three equations:

$$[M_i, M_j] = \mathrm{i}\epsilon_{ijk} M_k , \quad [M_i, N_j] = \mathrm{i}\epsilon_{ijk} N_k , \quad [N_i, N_j] = \mathrm{i}\epsilon_{ijk} M_k . \tag{7.78}$$

Next, we introduce two linear combinations of $\boldsymbol{M}$ and $\boldsymbol{N}$:

$$\boldsymbol{J} = \frac{1}{2}(\boldsymbol{M} + \boldsymbol{N}) , \quad \boldsymbol{K} = \frac{1}{2}(\boldsymbol{M} - \boldsymbol{N}) . \tag{7.79}$$

Note first that $\boldsymbol{J}$ and $\boldsymbol{K}$ commute, i.e.,

$$[J_i, K_j] = 0 . \tag{7.80}$$

In addition, the commutation relations among the $\boldsymbol{J}$ and those among the $\boldsymbol{K}$ are isomorphic to those of the three components of the angular momentum:

$$[J_i, J_j] = \mathrm{i}\epsilon_{ijk} J_k , \quad [K_i, K_j] = \mathrm{i}\epsilon_{ijk} K_k . \tag{7.81}$$

Therefore, the proper Lorentz group can be decomposed into a direct product of two three-dimensional rotational groups:

$$SO(3) \times SO(3) \sim SU(2) \times SU(2) . \tag{7.82}$$

The representation of the proper Lorentz group is characterized by the eigenvalues of the operators

$$\boldsymbol{J}^2 , \quad J_3 , \quad \boldsymbol{K}^2 , \quad K_3 , \tag{7.83}$$

i.e., $j_1(j_1 + 1)$, m_1, $j_2(j_2 + 1)$, and m_2, respectively. We denote the corresponding eigenvector by

$$|j_1, m_1; j_2, m_2\rangle . \tag{7.84}$$

Here, as in the case of the angular momentum, the possible values of m are

$$m_1 = -j_1, -j_1 + 1, \ldots, j_1 , \quad m_2 = -j_2, -j_2 + 1, \ldots, j_2 . \tag{7.85}$$

When we use the basis in (7.84) to represent the proper Lorentz transformation Λ, we denote this representation by

$$\mathscr{D}^{(j_1,j_2)}(\Lambda) . \tag{7.86}$$

The relationship between this representation and the three-dimensional rotation group is given by

$$\boldsymbol{M} = \boldsymbol{J} + \boldsymbol{K} . \tag{7.87}$$

It turns out that, for three-dimensional rotations, j is given by the composition of j_1 and j_2. The proper Lorentz transformation Λ is expressed by

$$\mathscr{D}(\Lambda) = \exp\left(\frac{1}{2}\mathrm{i}\epsilon_{\mu\nu}M_{\mu\nu}\right) . \tag{7.88}$$

We use the following notation with six parameters $\epsilon_{\mu\nu}$:

$$\boldsymbol{\theta} = (\epsilon_{23}, \epsilon_{31}, \epsilon_{12}) , \quad \mathrm{i}\boldsymbol{\omega} = (\epsilon_{14}, \epsilon_{24}, \epsilon_{34}) , \tag{7.89}$$

where $\boldsymbol{\theta}$ and $\boldsymbol{\omega}$ are real three-dimensional vectors. We can now rewrite (7.88) as

$$\mathscr{D}(\Lambda) = \exp(\mathrm{i}\boldsymbol{\theta}\cdot\boldsymbol{M} - \boldsymbol{\omega}\cdot\boldsymbol{N}) . \tag{7.90}$$

In order to analyze the meaning of the two j's, we compare $\mathscr{D}^{(j,0)}$ with $\mathscr{D}^{(0,j)}$:

$$\mathscr{D}^{(j,0)} : \quad \boldsymbol{M} = \boldsymbol{N} = \boldsymbol{J} , \quad \boldsymbol{K} = 0 , \tag{7.91a}$$

$$\mathscr{D}^{(0,j)} : \quad \boldsymbol{M} = -\boldsymbol{N} = \boldsymbol{K} , \quad \boldsymbol{J} = 0 . \tag{7.91b}$$

Both representations can be written in terms of $\boldsymbol{M}$ as follows:

$$\mathscr{D}^{(j,0)}(\Lambda) = \exp\left[(\mathrm{i}\boldsymbol{\theta} - \boldsymbol{\omega})\cdot\boldsymbol{M}\right] , \tag{7.92a}$$

$$\mathscr{D}^{(0,j)}(\Lambda) = \exp\left[(\mathrm{i}\boldsymbol{\theta} + \boldsymbol{\omega})\cdot\boldsymbol{M}\right] . \tag{7.92b}$$

These imply

$$\mathscr{D}^{(0,j)}(\Lambda) = \mathscr{D}^{(j,0)}(\Lambda^{-1})^{\dagger} . \tag{7.93}$$

To investigate the physical meaning of the two kinds of indices, we recall Weyl's equation. As already mentioned in Sect. 7.3, $M_{\mu\nu}$ can be represented by $-\mathrm{i}T_{\mu\nu}$. In the case of Weyl's equation, from (3.135), depending on the sign of helicity, we have

$$\boldsymbol{M} = \pm\boldsymbol{N} . \tag{7.94}$$

Therefore, j_1 and j_2 correspond to the positive and negative helicity states, respectively.

For space inversion, the transformation properties of these operators are

$$\boldsymbol{M} \to \boldsymbol{M}\,, \quad \boldsymbol{N} \to -\boldsymbol{N}\,, \quad \boldsymbol{J} \rightleftarrows \boldsymbol{K}\,. \tag{7.95}$$

Space inversion does not belong to the proper Lorentz group, so $\mathscr{D}^{(j_1,j_2)}$ is not enough here. It turns out that

$$\mathscr{D}^{(j_1,j_2)} \rightleftarrows \mathscr{D}^{(j_2,j_1)}\,. \tag{7.96}$$

This is the reason why Weyl's equation is not invariant under space inversion. The solution of Dirac's equation preserving parity is

$$\mathscr{D}^{(1/2,0)} \oplus \mathscr{D}^{(0,1/2)}\,. \tag{7.97}$$

7.6 Spin of a Massless Particle

In Sect. 7.3, we discussed the spin of a particle with nonzero mass. In this section, we will study the spin of a massless particle and discuss its properties.

First, we express W_μ using the notation in (7.77):

$$W_0 = \boldsymbol{M}\cdot\boldsymbol{P}\,, \quad \boldsymbol{W} = \mathrm{i}\boldsymbol{N}\times\boldsymbol{P} + \boldsymbol{M}P_0\,. \tag{7.98}$$

Moreover, we know that W_μ is orthogonal to P_μ. Considering the irreducible representation $\mathscr{D}^{(j,0)}$ and noting that (7.91a) implies $\boldsymbol{M} = \boldsymbol{N} = \boldsymbol{J}$ and $\boldsymbol{K} = 0$, (7.98) can be written

$$W_0 = \boldsymbol{J}\cdot\boldsymbol{P}\,, \quad \boldsymbol{W} = \mathrm{i}\boldsymbol{J}\times\boldsymbol{P} + \boldsymbol{J}P_0\,. \tag{7.99}$$

As mentioned before, these imply

$$W^2 = m^2 j(j+1)\,. \tag{7.100}$$

Considering a massless particle moving in the z-direction, we set

$$P_1 = P_2 = 0\,, \quad P_3 = P_0 = P\,, \tag{7.101}$$

where P is a c-number eigenvalue. Therefore,

$$W_0 = PJ_3\,, \quad W_1 = P(J_1 + \mathrm{i}J_2)\,, \quad W_2 = -\mathrm{i}W_1\,, \quad W_3 = W_0\,. \tag{7.102}$$

Since the mass is zero,

$$P_\mu^2 = W_\mu^2 = W_\mu P_\mu = 0 \,. \tag{7.103}$$

If W_μ is a real vector, the equation above means that P_μ is parallel to W_μ. We call this a *real condition*. Denoting the simultaneous eigenstate of $\boldsymbol{J}^2$ and j_3 by $|j, m\rangle$, this condition implies

$$W_\mu |j, m\rangle = \lambda P_\mu |j, m\rangle \,, \tag{7.104}$$

where j is the value of the spin. However, since $P_1 = P_2 = 0$, the equation above implies

$$W_1 |j, m\rangle = W_2 |j, m\rangle = 0 \,. \tag{7.105}$$

Referring to (7.102),

$$(J_1 + i J_2)|j, m\rangle = 0 \,. \tag{7.106}$$

This holds only for $m = j$. If parity is conserved, $\mathscr{D}^{(0,j)}$ is allowed, and in this case $m = -j$ is also possible. It is a property of the massless case that the only possible value of $|m|$ is j. If a different value appears, it means that the state does not belong to an irreducible spin representation. Thus, the form of the equation itself must ensure $|m| = j$. Examples of such particles are the neutrino, the photon, and the graviton. For the neutrino, m is either $1/2$ or $-1/2$, and it breaks parity. On the other hand, since the photon has interactions which preserve parity, $m = \pm 1$ are both possible, while $m = 0$ is not. For the photon, this is well-known: the photon is a transverse wave. We shall now check that both the Weyl equation and Maxwell equations ensure $|m| = j$.

Weyl Equation

Weyl's equation corresponds to representations of the type $\mathscr{D}^{(1/2,0)}$, and (3.135) implies that

$$H = \boldsymbol{\sigma} \cdot \boldsymbol{P} = 2\boldsymbol{J} \cdot \boldsymbol{P} \,. \tag{7.107}$$

Therefore, for a neutrino going in the z-direction,

$$P = 2J_3 P \,, \quad \text{or} \quad J_3 = \frac{1}{2} \,. \tag{7.108}$$

Hence, we confirm that $m = j$.

Maxwell Equations

These equations describe a spin-1 photon. The equations for a free electromagnetic field are

$$\mathrm{rot}\,\boldsymbol{E} + \frac{\partial \boldsymbol{H}}{\partial t} = 0\,, \quad \mathrm{rot}\,\boldsymbol{H} - \frac{\partial \boldsymbol{E}}{\partial t} = 0\,. \tag{7.109}$$

First, let us check the solution corresponding to $\mathscr{D}^{(1,0)}$. In order to obtain the right-handed circularly-polarized solution corresponding to this representation, we choose the following linear combinations of the components of the field strength $F_{\mu\nu}$:

$$F_1^{\mathrm{R}} = F_{23} + F_{14}\,, \quad F_2^{\mathrm{R}} = F_{31} + F_{24}\,, \quad F_3^{\mathrm{R}} = F_{12} + F_{34}\,, \tag{7.110}$$

or in vector form,

$$\boldsymbol{F}^{\mathrm{R}} = \boldsymbol{H} - \mathrm{i}\boldsymbol{E}\,. \tag{7.111}$$

In the following, we shall drop the superscript R. The Maxwell equation for $\boldsymbol{F}$ is

$$\mathrm{i}\frac{\partial}{\partial t}\boldsymbol{F} = \mathrm{rot}\,\boldsymbol{F}\,. \tag{7.112}$$

We introduce the following three components:

$$F_+ = -\frac{1}{\sqrt{2}}(F_1 - \mathrm{i}F_2)\,, \quad F_0 = F_3\,, \quad F_- = \frac{1}{\sqrt{2}}(F_1 + \mathrm{i}F_2)\,. \tag{7.113}$$

Denoting the wave function with these three components by $\mathscr{F}$, (7.112) transforms to

$$\frac{\partial}{\partial t}\mathscr{F} + \boldsymbol{S}\cdot\nabla\mathscr{F} = 0\,, \tag{7.114}$$

where $\boldsymbol{S}$ has the three components

$$S_1 = \begin{pmatrix} 0 & 1 & 0 \\ 1 & 0 & 1 \\ 0 & 1 & 0 \end{pmatrix}\,, \quad S_2 = \frac{1}{\sqrt{2}}\begin{pmatrix} 0 & -\mathrm{i} & 0 \\ \mathrm{i} & 0 & -\mathrm{i} \\ 0 & \mathrm{i} & 0 \end{pmatrix}\,, \quad S_3 = \begin{pmatrix} 1 & 0 & 0 \\ 0 & 0 & 0 \\ 0 & 0 & -1 \end{pmatrix}\,. \tag{7.115}$$

These three components satisfy the same commutation relations as the angular momentum. For $\mathscr{D}^{(1,0)}$, since we have (7.91a),

$$\boldsymbol{M} = \boldsymbol{N} = \boldsymbol{J} = \boldsymbol{S}\,, \quad \boldsymbol{K} = 0\,. \tag{7.116}$$

Here we choose the x-dependence of $\mathscr{F}$ corresponding to a traveling wave in the z-direction:

$$\mathscr{F}(x) = \mathscr{F} \exp\left[\mathrm{i}P(x_3 - x_0)\right] . \tag{7.117}$$

Inserting this into (7.114), we obtain

$$P(1 - S_3)\mathscr{F} = 0 . \tag{7.118}$$

What we can understand from this is that, as expected, only $S_3 = 1$ can be realized.

To obtain the left-handed solution corresponding to $\mathscr{D}^{(0,1)}$, instead of (7.111), we start with

$$\boldsymbol{F}^{\mathrm{L}} = \boldsymbol{H} + \mathrm{i}\boldsymbol{E} . \tag{7.119}$$

In this case, only $S_3 = -1$ can be realized. There is no solution corresponding to $S_3 = 0$.

What we see from the above is that, if we start with the correct equation, only the solution corresponding to $|m| = j$ can exist.

7.7 Pauli–Gürsey Group

In quantum mechanics, an absolute phase of the wave function cannot be observed. Two wave functions ψ_1 and ψ_2 related by

$$\psi_2 = \mathrm{e}^{\mathrm{i}\lambda}\psi_1 \tag{7.120}$$

are completely equivalent, in the sense that there can be no difference in their physical interpretation. Therefore, we can say that a set of wave functions differing by constant phases is an equivalence class as far as physics is concerned. Just the same thing happens in field theory.

For an interacting system including complex fields, the Lagrangian density is

$$\mathscr{L} = \mathscr{L}\left[\varphi_\alpha, \varphi_\alpha^\dagger; \varphi_{\alpha,\mu}, \varphi_{\alpha,\mu}^\dagger\right] . \tag{7.121}$$

This includes physical constants such as a mass m and a coupling constant g. If we compute observable quantities such as differential cross-sections for various reactions and the lifetimes of unstable particles, the results will be given as functions of the physical constants, momenta, and so on. Needless to say, in the final results, the fields themselves do not appear. Metaphorically speaking, the field is like a

Japanese *nabe*[2] used for cooking, and the physical constants are like the ingredients of the dish. Clearly, just as we could use a different *nabe* to cook, we can use a different symbol instead of φ_α in (7.121), say φ'_α, and the final observable result should be the same. Hence, for the complex field in (7.121), we introduce the constant phase transformation

$$\varphi_\alpha \;\rightarrow\; \varphi'_\alpha = \mathrm{e}^{\mathrm{i}\lambda_\alpha}\varphi_\alpha \,. \tag{7.122}$$

Denoting the group formed by such phase transformations by $\mathscr{G}$, we assume that the free part $\mathscr{L}_\mathrm{f}$ of the Lagrangian density is invariant under $\mathscr{G}$, i.e.,

$$\mathscr{L}_\mathrm{f}\big[\mathrm{e}^{\mathrm{i}\lambda_\alpha}\varphi_\alpha, \mathrm{e}^{-\mathrm{i}\lambda_\alpha}\varphi_\alpha^\dagger\,;\, \mathrm{e}^{\mathrm{i}\lambda_\alpha}\varphi_{\alpha,\mu}, \mathrm{e}^{-\mathrm{i}\lambda_\alpha}\varphi_{\alpha,\mu}^\dagger\big] = \mathscr{L}_\mathrm{f}\big[\varphi_\alpha, \varphi_\alpha^\dagger\,;\, \varphi_{\alpha,\mu}, \varphi_{\alpha,\mu}^\dagger\big] \,. \tag{7.123}$$

The problem is to see how the interaction part $\mathscr{L}_\mathrm{int}$ of the Lagrangian density behaves under $\mathscr{G}$. If $\mathscr{L}_\mathrm{int}$ can be kept invariant under $\mathscr{G}$, by choosing $\mathscr{G}$ in a suitable way, a conservation law can be obtained as before. For example, denoting the charge associated with the field quantity φ_α by e_α and choosing the phase λ_α to be

$$\lambda_\alpha = \lambda e_\alpha \,, \tag{7.124}$$

the interaction part does indeed remain invariant. We obtain the law of conservation of electric charge in this way.

But what if the interaction part is not invariant? In this case, the coupling constant transforms. As an example, consider the fields ψ_1, ψ_2, and φ :

$$\mathscr{L}_\mathrm{int} = g\bar{\psi}_1\psi_2\varphi + g^*\bar{\psi}_2\psi_1\varphi^\dagger \,, \tag{7.125}$$

and introduce the phase transformation

$$\psi_1 \rightarrow \psi_1 \,, \quad \psi_2 \rightarrow \psi_2 \,, \quad \varphi \rightarrow \mathrm{e}^{\mathrm{i}\lambda}\varphi \,. \tag{7.126}$$

Inserting (7.126) into (7.125), $\mathscr{L}_\mathrm{int}$ transforms to

$$\mathscr{L}_\mathrm{int} \rightarrow g\mathrm{e}^{\mathrm{i}\lambda}\bar{\psi}_1\psi_2\varphi + g^*\mathrm{e}^{-\mathrm{i}\lambda}\bar{\psi}_2\psi_1\varphi^\dagger \,. \tag{7.127}$$

Comparing this with (7.125), it turns out that the coupling constant must transform according to

$$g \;\rightarrow\; g\mathrm{e}^{\mathrm{i}\lambda} \,. \tag{7.128}$$

---

[2] Hot pot.

Since the transformation above does not change the physical content of the theory, this means that any observable quantity depending on g must be invariant under this transformation, i.e.,

$$\mathcal{O}(g\mathrm{e}^{\mathrm{i}\lambda}) = \mathcal{O}(g) \ . \tag{7.129}$$

For a simple transformation like this, $|g|^2 = g^*g$ is an invariant quantity and all observables become functions of $|g|^2$ alone. This holds even if the phase transformation (7.126) is generalized much further.

Indeed, Pauli generalized the above group of phase transformations, which is commutative, to a non-commutative group [94]. His target was the neutrino field. Here we write down the Lagrangian density of the massless neutrino:

$$\mathcal{L}_\mathrm{f} = -\bar{\psi}\gamma_\mu \partial_\mu \psi \ . \tag{7.130}$$

Strictly speaking, we ought to write $\mathcal{L}_\mathrm{f}$ in a symmetric form with respect to ψ and $\bar{\psi}$. Bearing this in mind, Pauli pointed out the invariance of (7.130) under the transformation

$$\psi \ \rightarrow \ \psi' = a\psi + b\gamma_5 C\bar{\psi}^T \ , \quad \bar{\psi} \ \rightarrow \ \bar{\psi}' = a^*\bar{\psi} - b^*C^{-1}\gamma_5\psi \ . \tag{7.131}$$

The inverse transformation is

$$\psi' \ \rightarrow \ \psi = a^*\psi' - b\gamma_5 C\bar{\psi}'^T \ , \quad \bar{\psi}' \ \rightarrow \ \bar{\psi} = a\bar{\psi}' + b^*C^{-1}\gamma_5\psi' \ , \tag{7.132}$$

where the complex numbers a and b satisfy the condition

$$|a|^2 + |b|^2 = 1 \ . \tag{7.133}$$

This transformation group is a generalization of the group of phase transformations, and we denote it by $\mathcal{G}_1$.

On the other hand, Gürsey noted that $\mathcal{L}_\mathrm{f}$ is also invariant under a transformation group $\mathcal{G}_2$ containing transformations of the form [95]

$$\psi \ \rightarrow \ \psi' = \mathrm{e}^{\mathrm{i}\lambda\gamma_5}\psi \ , \quad \bar{\psi} \ \rightarrow \ \bar{\psi}' = \bar{\psi}\mathrm{e}^{\mathrm{i}\lambda\gamma_5} \ . \tag{7.134}$$

These two groups commute with each other. Using Noether's theorem to derive the conservation law for an infinitesimal transformation which leaves $\mathcal{L}_\mathrm{f}$ invariant, we obtain the following conserved quantities. For $\mathscr{G}_1$,

$$M_1 = \frac{1}{2}\int \mathrm{d}^3x(\psi^\dagger\gamma_5\gamma_4C\psi^\dagger + \psi^T C^{-1}\gamma_4\gamma_5\psi) , \tag{7.135a}$$

$$M_2 = \frac{1}{2\mathrm{i}}\int \mathrm{d}^3x(\psi^\dagger\gamma_5\gamma_4C\psi^\dagger - \psi^T C^{-1}\gamma_4\gamma_5\psi) , \tag{7.135b}$$

$$M_3 = \int \mathrm{d}^3x\psi^\dagger\psi , \tag{7.135c}$$

where we have summed over the spinor indices. In addition, for $\mathscr{G}_2$,

$$N = \int \mathrm{d}^3x\psi^\dagger\gamma_5\psi . \tag{7.136}$$

The generators of this group satisfies the commutation relations

$$\left[\frac{M_i}{2}, \frac{M_j}{2}\right] = \mathrm{i}\epsilon_{ijk}\frac{M_k}{2} , \tag{7.137}$$

$$[M_j, N] = 0 . \tag{7.138}$$

This shows that the generalization $\mathscr{G}_1 \times \mathscr{G}_2$ of the group of phase transformations is isomorphic to $SU(2) \times U(1)$. It was Gürsey who discovered this group structure.

It is entirely because the mass is zero that the Lagrangian density (7.130) permits such a large invariance group. If there is a mass term, only the constant phase transformation mentioned at the beginning of this section is allowed, whence M_3 is the only conserved quantity. Conversely, if there are conserved quantities apart from M_3 in an interacting case, this shows that the mass of the neutrino is zero. In the following, we shall seek such quantities.

Generally, this invariance group turns not to be an invariance group when interactions are introduced, since this induces the transformation among coupling constants, as explained above. Observable quantities must be invariant under this transformation among coupling constants. For beta decay, Pauli derived the quantity that remains invariant under the transformation of the coupling constants.

If the neutrino is massless, the symmetry implemented by the invariance group $\mathscr{G}_1 \times \mathscr{G}_2$ is not completely broken. Some subgroup

$$\mathscr{G}_\nu \subset \mathscr{G}_1 \times \mathscr{G}_2 \tag{7.139}$$

may still be an invariance group. We shall now derive the group $\mathscr{G}_\nu$ which leaves the beta decay interaction invariant.

We write the Lagrangian density of the beta decay interaction as $\mathscr{L}_\beta$ and denote the generator of $\mathscr{G}_\nu$ by M:

$$[\mathscr{L}_\beta, M] = 0 \ . \tag{7.140}$$

Here, M has the general form

$$M = cN + c_1 M_1 + c_2 M_2 + c_3 M_3 \ . \tag{7.141}$$

By a suitable phase transformation, i.e., a transformation belonging to the invariance group $\mathscr{G}_1 \times \mathscr{G}_2$, we transform this M to the standard form

$$M = cN + c_3' M_3' \ , \quad c_3'^2 = c_1^2 + c_2^2 + c_3^2 \ . \tag{7.142}$$

From now on, we shall write M_3' simply as M_3. Therefore, from (7.140),

$$\big[[\mathscr{L}_\beta, cN + c_3 M_3], cN - c_3 M_3\big] = 0 \ . \tag{7.143}$$

Since $\mathscr{L}_\beta$ is linear in the neutrino fields ψ and $\psi^\dagger$ and noting that

$$\big[[\psi, cN + c_3 M_3], cN - c_3 M_3\big] = (c^2 - c_3^2)\psi \ ,$$

$$\big[[\psi^\dagger, cN + c_3 M_3], cN - c_3 M_3\big] = (c^2 - c_3^2)\psi^\dagger \ ,$$

the invariance requirement (7.143) holds if we have

$$c^2 - c_3^2 = 0 \ . \tag{7.144}$$

Therefore, we must choose $N - M_3$ or $N + M_3$ as M. The experimental fact mentioned in Sect. 3.7 requires us to choose

$$M = N - M_3 \ . \tag{7.145}$$

The finite transformation generated by this creation operator is a kind of *chiral transformation*, and is given by

$$\psi \to \psi' = \exp\big[\mathrm{i}(\mathbf{1} - \gamma_5)\lambda\big]\psi \ , \quad \bar{\psi} \to \bar{\psi}' = \bar{\psi}\exp\big[\mathrm{i}(\mathbf{1} + \gamma_5)\lambda\big] \ . \tag{7.146}$$

To be invariant under this transformation, ψ and $\bar{\psi}$ must appear in the interaction terms in the combinations

$$(\mathbf{1} + \gamma_5)\psi \ , \quad \bar{\psi}(\mathbf{1} - \gamma_5) \ . \tag{7.147}$$

This is nothing but ψ_L, introduced in Sect. 3.7. Thus, the neutrino can be described by the two-component Weyl spinor. It turns out that this ensures that the mass of the neutrino is also zero in the interacting case.

Chapter 8
S-Matrix

Reactions among elementary particles or atomic nuclei can be described by the S-matrix. In fact, all our knowledge of the various interactions of these particles is included in the S-matrix. It is the most fundamental quantity in elementary particle physics. Although the S-matrix can in principle be determined if the Lagrangian density is known, it is hard to compute in practice. Here we will discuss how to compute the S-matrix using perturbation theory, which is only valid for small coupling constants. This method was developed by Feynman and Dyson. Two different approaches to QED were greatly developed in the postwar era. One was based on the Tomonaga–Schwinger equation [87, 88], exploiting the traditional canonical formalism. The other started out with Feynman's path integral [96]. This was a rather intuitive approach and its relationship with the traditional method was not clear. Nevertheless, the consequences of the two theories always coincided. Dyson developed a way to calculate the S-matrix, starting with the Tomonaga–Schwinger equation, and finally proved that it was equivalent to Feynman's theory, i.e., he established the equivalence of the two theories [97]. The computational method discussed in the following is based on Dyson's approach.

8.1 Definition of the S-Matrix

The S-matrix has already been discussed in Sect. 6.1, where it was given by the two equations (6.32) and (6.33). In this section, we will see how to compute the expression

$$S = U(\infty, -\infty)\ . \tag{8.1}$$

The S-matrix has several important general properties. For example, the property of unitarity which arose as a consequence of (6.26) is related to the law of

K. Nishijima, *Quantum Field Theory*,
https://doi.org/10.1007/978-94-024-2190-3_8

conservation of probability:

$$S^{\dagger} S = S S^{\dagger} = 1 \,. \tag{8.2}$$

This discussion of unitarity was intuitive. A more rigorous discussion must be based on the asymptotic condition to be introduced later, which concerns the unitarity of the renormalized S-matrix.

Another property is the Poincaré invariance of the S-matrix, which is confirmed by computations. This invariance is expressed by

$$[S, P_{\mu}] = 0 \,, \quad [S, M_{\mu\nu}] = 0 \,. \tag{8.3}$$

Of course, if there are external (background) fields, this property is lost.

8.2 Dyson's Formula for the S-Matrix

Although it is possible to start from the Tomonaga–Schwinger equation, we shall use Schrödinger's equation in the interaction representation because the result is the same for both approaches. We rewrite (6.10) in the form

$$\mathrm{i} \frac{\partial}{\partial t} U(t, t_0) = H_{\mathrm{int}}(t) U(t, t_0) \,, \tag{8.4}$$

where

$$H_{\mathrm{int}}(t) = \int_{x_0 = t} \mathrm{d}^3 x \, \mathscr{H}_{\mathrm{int}} \,. \tag{8.5}$$

With the initial condition $U(t_0, t_0) = 1$, we integrate (8.4) to obtain

$$U(t, t_0) = 1 - \mathrm{i} \int_{t_0}^{t} \mathrm{d}t' H_{\mathrm{int}}(t') U(t', t_0) \,. \tag{8.6}$$

Then we insert this expression for U iteratively into the right-hand side of (8.6) to give

$$U(t, t_0) = 1 - \mathrm{i} \int_{t_0}^{t} \mathrm{d}t' H_{\mathrm{int}}(t') + (-\mathrm{i})^2 \int_{t_0}^{t} \mathrm{d}t' H_{\mathrm{int}}(t') \int_{t_0}^{t'} \mathrm{d}t'' H_{\mathrm{int}}(t'') + \cdots \,. \tag{8.7}$$

However, since this form is not very useful as it stands, we introduce Dyson's time-ordering operator T, where

$$T\big[H_{\text{int}}(t_1)H_{\text{int}}(t_2)\ldots H_{\text{int}}(t_n)\big] = H_{\text{int}}(t'_1)H_{\text{int}}(t'_2)\ldots H_{\text{int}}(t'_n)\,, \tag{8.8}$$

and $t'_1, t'_2, \ldots, t'_n$ is a reordering of $t_1, t_2, \ldots, t_n$ chosen such that

$$t'_1 > t'_2 > \ldots > t'_n\,. \tag{8.9}$$

From this definition, it is clear that the left-hand side of (8.8) becomes a symmetric function of $t_1, t_2, \ldots, t_n$.

If $f(x_1, x_2, \ldots, x_n)$ is a symmetric function of $x_1, x_2, \ldots, x_n$, we may write

$$\int_a^b \mathrm{d}x_1 \int_a^{x_1} \mathrm{d}x_2 \ldots \int_a^{x_{n-1}} \mathrm{d}x_n\, f(x_1, x_2, \ldots, x_n) \tag{8.10}$$
$$= \frac{1}{n!}\int_a^b \mathrm{d}x_1 \int_a^b \mathrm{d}x_2 \ldots \int_a^b \mathrm{d}x_n\, f(x_1, x_2, \ldots, x_n)\,.$$

Using this formula, we have

$$\int_{t_0}^t \mathrm{d}t_1 \int_{t_0}^{t_1} \mathrm{d}t_2 \ldots \int_{t_0}^{t_n-1} \mathrm{d}t_n\, H_{\text{int}}(t_1)H_{\text{int}}(t_2)\ldots H_{\text{int}}(t_n)$$
$$= \int_{t_0}^t \mathrm{d}t_1 \int_{t_0}^{t_1} \mathrm{d}t_2 \ldots \int_{t_0}^{t_{n-1}} \mathrm{d}t_n\, T\big[H_{\text{int}}(t_1)H_{\text{int}}(t_2)\ldots H_{\text{int}}(t_n)\big]$$
$$= \frac{1}{n!}\int_{t_0}^t \mathrm{d}t_1 \int_{t_0}^{t} \mathrm{d}t_2 \ldots \int_{t_0}^{t} \mathrm{d}t_n\, T\big[H_{\text{int}}(t_1)H_{\text{int}}(t_2)\ldots H_{\text{int}}(t_n)\big]\,. \tag{8.11}$$

Using the above expression, we can rewrite the general term in the expansion (8.7). Hence, we obtain Dyson's formula for $U(t, t_0)$ [98]:

$$U(t, t_0) = 1 + \sum_{n=1}^{\infty} \frac{(-\mathrm{i})^n}{n!} \int_{t_0}^t \mathrm{d}t_1 \ldots \int_{t_0}^t \mathrm{d}t_n\, T\big[H_{\text{int}}(t_1)\ldots H_{\text{int}}(t_n)\big]\,. \tag{8.12}$$

Taking the limits $t \to \infty$ and $t_0 \to -\infty$, and using the fact that $H_{\text{int}}(t)$ is a spatial integral of $\mathscr{H}_{\text{int}}(x)$, we obtain Dyson's formula for the S-matrix [98]:

$$S = 1 + \sum_{n=1}^{\infty} \frac{(-\mathrm{i})^n}{n!} \int \mathrm{d}^4x_1 \ldots \int \mathrm{d}^4x_n\big[\mathscr{H}_{\text{int}}(x_1)\ldots \mathscr{H}_{\text{int}}(x_n)\big]\,, \tag{8.13}$$

where the integral is taken over the whole of spacetime. It is obvious now that S commutes with P_μ.

The next problem is the Lorentz invariance of the S-matrix. We note that d^4x is Lorentz invariant and $\mathscr{H}_{\text{int}}(x)$ is a Lorentz scalar. Here, a problem is the time-ordering operator T. Apparently, the definition of T depends on how we choose the time axis. So, does the S-matrix itself depend on the choice of time axis?

If the separation between two spacetime points x_1 and x_2 is time-like, i.e.,

$$(x_1 - x_2)^2 = (\boldsymbol{x}_1 - \boldsymbol{x}_2)^2 - (t_1 - t_2)^2 < 0 ,$$

then $t_1 - t_2$ has the same sign for any Lorentz system, and in this case the time order does not depend on the choice of time axis. However, if the separation is space-like, i.e.,

$$(x_1 - x_2)^2 = (\boldsymbol{x}_1 - \boldsymbol{x}_2)^2 - (t_1 - t_2) > 0 ,$$

then we may have $t_1 > t_2$ or $t_1 < t_2$, depending on how we choose the time axis. On the other hand, in the case where $x_1 - x_2$ is space-like, if

$$\mathscr{H}_{\text{int}}(x_1)\mathscr{H}_{\text{int}}(x_2) = \mathscr{H}_{\text{int}}(x_2)\mathscr{H}_{\text{int}}(x_1) , \quad (x_1 - x_2)^2 > 0 , \tag{8.14}$$

then the ordering is not important. But (8.14) is nothing other than the integrability condition (6.20) for the Tomonaga–Schwinger equation. Therefore, as long as this condition is satisfied, S commutes with $M_{\mu\nu}$, and hence is Lorentz invariant. This proves (8.3).

In fact, the Hamiltonian density may depend on a hypersurface σ, in which case the discussion becomes more complicated. This happens, for example, when the interactions involve derivatives of the fields.

8.3 Wick's Theorem

The Feynman–Dyson theory tells us how to compute the S-matrix, starting with Dyson's formula derived in the last section. It is Wick's theorem that provides the foundation for this method.

In the interaction picture, a field operator satisfies a free field equation, i.e., a linear equation, as mentioned in Sect. 5.4, so it can be decomposed into a positive-frequency part and a negative-frequency part:

$$\varphi_\alpha(x) = \varphi_\alpha^{(+)}(x) + \varphi_\alpha^{(-)}(x) , \tag{8.15}$$

where $\varphi_\alpha^{(+)}(x)$ is an annihilation operator and $\varphi_\alpha^{(-)}(x)$ is a creation operator. It is clear that

$$\varphi_\alpha^{(+)}(x)|0\rangle = 0 , \quad \langle 0|\varphi_\alpha^{(-)}(x) = 0 . \tag{8.16}$$

In the following, we will apply the decomposition (8.15) to field operators in the interaction picture, and express the product of operators using (8.16). First, for simplicity, we consider a neutral scalar field:

$$\begin{aligned}
&\varphi(x_1)\varphi(x_2)\ldots\varphi(x_n) \\
&\quad = \left[\varphi^{(+)}(x_1)+\varphi^{(-)}(x_1)\right]\left[\varphi^{(+)}(x_2)+\varphi^{(-)}(x_2)\right]\ldots\left[\varphi^{(+)}(x_n)+\varphi^{(-)}(x_n)\right].
\end{aligned}$$

Expanding this product, we would like to change the order of the terms, shifting $(-)$ to the left and $(+)$ to the right, to give something of the form

$$\varphi^{(-)}\varphi^{(-)}\ldots\varphi^{(+)}\varphi^{(+)}\ldots . \tag{8.17}$$

However, since $\varphi^{(-)}$ and $\varphi^{(+)}$ do not commute, each time we change the order of two operators, a commutator will show up. For example,

$$\begin{aligned}
\varphi(x)\varphi(y) &= \left[\varphi^{(+)}(x)+\varphi^{(-)}(x)\right]\left[\varphi^{(+)}(y)+\varphi^{(-)}(y)\right] \\
&= \varphi^{(+)}(x)\varphi^{(+)}(y)+\varphi^{(-)}(x)\varphi^{(-)}(y)+\varphi^{(-)}(x)\varphi^{(+)}(y)+\varphi^{(+)}(x)\varphi^{(-)}(y) .
\end{aligned}$$

Except for the last term, all operators are in the order indicated in (8.17). Here, using

$$\varphi^{(+)}(x)\varphi^{(-)}(y) = \varphi^{(-)}(y)\varphi^{(+)}(x) + \left[\varphi^{(+)}(x), \varphi^{(-)}(y)\right],$$

all terms can be set in the standard form, i.e.,

$$\begin{aligned}
\varphi(x)\varphi(y) &= \varphi^{(+)}(x)\varphi^{(+)}(y)+\varphi^{(-)}(x)\varphi^{(-)}(y)+\varphi^{(-)}(x)\varphi^{(+)}(y)+\varphi^{(-)}(y)\varphi^{(+)}(x) \\
&\quad +\left[\varphi^{(+)}(x), \varphi^{(-)}(y)\right] .
\end{aligned} \tag{8.18}$$

The sum of the first four terms, which are in the standard form (8.17), is called the *normal product*, expressed briefly as $:\varphi(x)\varphi(y):$. The last commutator is a c-number. It can be calculated as follows (see Sect. 4.1):

$$\begin{aligned}
\left[\varphi^{(+)}(x), \varphi^{(-)}(y)\right] &= \left[\sum_{\boldsymbol{p}} \frac{1}{\sqrt{2p_0V}} \mathrm{e}^{\mathrm{i}p\cdot x} a(\boldsymbol{p}), \sum_{\boldsymbol{q}} \frac{1}{\sqrt{2q_0V}} \mathrm{e}^{-\mathrm{i}q\cdot y} a^\dagger(\boldsymbol{q})\right] \\
&= \sum_{\boldsymbol{p}} \frac{1}{2p_0V} \mathrm{e}^{\mathrm{i}p\cdot(x-y)} \\
&= \mathrm{i}\Delta^{(+)}(x-y) .
\end{aligned} \tag{8.19}$$

Therefore,

$$\varphi(x)\varphi(y) = :\varphi(x)\varphi(y): + \mathrm{i}\Delta^{(+)}(x-y) . \tag{8.20}$$

The normal product is clearly symmetric, i.e.,

$$:\varphi(x)\varphi(y): = :\varphi(y)\varphi(x): \ . \tag{8.21}$$

This symmetry also holds when there are more field operators in the product. However, if the field obeys Fermi statistics, the normal product becomes anti-symmetric. For a Dirac field ψ, when $x_1', x_2', \ldots, x_n'$ is a permutation of $x_1, x_2, \ldots, x_n$,

$$:\psi(x_1')\psi(x_2')\ldots\psi(x_n'): = (-1)^P \ :\psi(x_1)\psi(x_2)\ldots\psi(x_n): \ , \tag{8.22}$$

where $(-1)^P$ is equal to $+1$ or -1 for an even permutation or an odd permutation, respectively. For a field obeying Bose statistics, there is no such change of sign.

From the definition of the normal product and (8.16), any vacuum expectation value of a normal product vanishes. For example,

$$\langle 0| :\varphi(x)\varphi(y): |0\rangle = 0 \ . \tag{8.23}$$

Therefore, from (8.20),

$$\langle 0|\varphi(x)\varphi(y)|0\rangle = \mathrm{i}\Delta^{(+)}(x-y) \ . \tag{8.24}$$

Generalizing this result, we obtain

$$\langle 0|\varphi(x_1)\varphi(x_2)\ldots\varphi(x_n)|0\rangle = \sum_{k_1<k_2, k_3<k_4, \ldots} \mathrm{i}\Delta^{(+)}(x_{k_1}-x_{k_2})\ldots\mathrm{i}\Delta^{(+)}(x_{k_{n-1}}-x_{k_n}) \ , \tag{8.25}$$

where $k_1, k_2, \ldots, k_n$ is a permutation of $1, 2, \ldots, n$ and n is an even number. Moreover, under the restriction of the inequalities $k_1 < k_2, k_3 < k_4, \ldots$, we sum over all possible combinations. If n is an odd number, then $\varphi(x_1)\varphi(x_2)\ldots\varphi(x_n)|0\rangle$ is an odd number of quantum states, so it is orthogonal to $|0\rangle$ and the expectation value (8.25) vanishes.

Using the normal product, the current density considered in (6.56) can be written as

$$j_\mu = \frac{1}{2}\mathrm{i}e[\bar{\psi}, \gamma_\mu\psi] = \mathrm{i}e :\bar{\psi}\gamma_\mu\psi: \ . \tag{8.26}$$

Therefore, we can deduce the following equation, which says that the vacuum has no charge:

$$\langle 0|j_\mu(x)|0\rangle = 0 \ . \tag{8.27}$$

We now move on to the problem of evaluating S-matrix elements. From Dyson's formula, the matrix element

$$S_{ba} = \langle b|S|a\rangle \tag{8.28}$$

means

$$S_{ba} = \langle b|a\rangle + \sum_{n=1}^{\infty} \frac{(-\mathrm{i})^n}{n!} \int \mathrm{d}^4 x_1 \ldots \int \mathrm{d}^4 x_n \big\langle b\big|T[\mathscr{H}_{\mathrm{int}}(x_1)\ldots\mathscr{H}_{\mathrm{int}}(x_n)]\big|a\big\rangle \,. \tag{8.29}$$

We shall consider here the simplest possible interaction, viz., the Yukawa interaction between a Fermi particle and a Bose particle:

$$\mathscr{H}_{\mathrm{int}}(x) = j(x)\varphi(x)\,, \quad j(x) = g :\bar{\psi}(x)\psi(x):\,, \tag{8.30}$$

where ψ and φ stand for a Dirac field and a neutral scalar field, respectively. We then have the following decomposition:

$$T\big[\mathscr{H}_{\mathrm{int}}(x_1)\ldots\mathscr{H}_{\mathrm{int}}(x_n)\big] = T\big[j(x_1)\ldots j(x_n)\big]\ldots T\big[\varphi(x_1)\ldots\varphi(x_n)\big]\,. \tag{8.31}$$

The states $|a\rangle$ and $|b\rangle$ can also be decomposed into the direct product of the states for the scalar particle and the Dirac particle:

$$|a\rangle = |a^{\mathrm{S}}\rangle \otimes |a^{\mathrm{D}}\rangle\,, \quad |b\rangle = |b^{\mathrm{S}}\rangle \otimes |b^{\mathrm{D}}\rangle\,. \tag{8.32}$$

Hence,

$$\begin{aligned} &\big\langle b\big|T[\mathscr{H}_{\mathrm{int}}(x_1)\ldots\mathscr{H}_{\mathrm{int}}(x_n)]\big|a\big\rangle \\ &\quad = \big\langle b^{\mathrm{D}}\big|T[j(x_1)\ldots j(x_n)]\big|a^{\mathrm{D}}\big\rangle\big\langle b^{\mathrm{S}}\big|T[\varphi(x_1)\ldots\varphi(x_n)]\big|a^{\mathrm{S}}\big\rangle\,. \end{aligned} \tag{8.33}$$

We begin with the second factor, which is easier to evaluate. The scalar particle states are expressed in the form

$$|a^{\mathrm{S}}\rangle = a^\dagger(k_1)\ldots a^\dagger(k_i)|0^{\mathrm{S}}\rangle\,, \quad |b\rangle = a^\dagger(k_1')\ldots a^\dagger(k_f')|0^{\mathrm{S}}\rangle\,. \tag{8.34}$$

In the following, we simply denote $|0^{\mathrm{S}}\rangle$ by $|0\rangle$. Then, extracting i annihilation operators and f creation operators, we obtain

$$\begin{aligned} &\big\langle b^{\mathrm{S}}\big|T[\varphi(x_1)\ldots\varphi(x_n)]\big|a^{\mathrm{S}}\big\rangle \\ &\quad = \big\langle 0\big|a(k_1')\ldots a(k_f')T[\varphi(x_1)\ldots\varphi(x_n)]a^\dagger(k_1)\ldots a^\dagger(k_i)\big|0\big\rangle\,. \end{aligned}$$

In this evaluation, we assume that none of the momenta $k_1, \ldots, k_i$ is equal to any of the momenta $k'_1, \ldots, k'_f$. If there are equal momenta, this corresponds to a situation where particles go straight through without interaction. We begin by eliminating $a^\dagger(k_i)$. To do this, we use (8.16) and bring in the commutator with $a^\dagger(k_i)$:

$$\begin{aligned} \langle 0|AB\ldots a^\dagger(k_i)|0\rangle &= \langle 0|[AB\ldots, a^\dagger(k_i)]|0\rangle + \langle 0|a^\dagger(k_i)AB\ldots|0\rangle \\ &= \langle 0|[AB\ldots, a^\dagger(k_i)]|0\rangle \,. \end{aligned} \tag{8.35}$$

To evaluate the commutator, we use

$$\left[a(k'), a^\dagger(k)\right] = \delta_{k',k} \tag{8.36}$$

and

$$\left[\varphi(x), a^\dagger(k)\right] = \frac{1}{\sqrt{2k_0V}}\mathrm{e}^{\mathrm{i}k\cdot x} = \langle 0|\varphi(x)|k\rangle \,. \tag{8.37}$$

Given the assumption of no equal momenta, a and $a^\dagger$ commute, whence

$$\begin{aligned} &\left\langle 0\middle|a(k'_1)\ldots a(k'_f)T\left[\varphi(x_1)\ldots\varphi(x_n)\right]a^\dagger(k_1)\ldots a^\dagger(k_i)\middle|0\right\rangle \\ &\quad = \sum_{j=1}^{n}\left\langle 0\middle|a(k'_1)\ldots a(k'_f)T\left[\varphi(x_1)\ldots\varphi(x_{j-1})\varphi(x_{j+1})\ldots\varphi(x_n)\right]\right. \\ &\qquad\qquad \left.\times a^\dagger(k_1)\ldots a^\dagger(k_{i-1})\middle|0\right\rangle\langle 0|\varphi(x_j)|0\rangle \,. \end{aligned}$$

Iterating, we eventually obtain

$$\begin{aligned} &\left\langle 0\middle|a(k'_1)\ldots a(k'_f)T\left[\varphi(x_1)\ldots\varphi(x_n)\right]a^\dagger(k_1)\ldots a^\dagger(k_i)\middle|0\right\rangle \\ &\quad = \sum_{\text{comb}}\langle k'_1|\varphi(x'_1)|0\rangle\ldots\langle k'_f|\varphi(x'_f)|0\rangle\langle 0|\varphi(x'_{f+1})|k_1\rangle\ldots\langle 0|\varphi(x'_{f+i})|k_i\rangle \\ &\qquad\qquad \times\left\langle 0\middle|T[\varphi(x'_{f+i+1})\ldots\varphi(x'_n)]\middle|0\right\rangle \,, \end{aligned} \tag{8.38}$$

where $x'_1, \ldots, x'_n$ is a permutation of $x_1, \ldots, x_n$, and selecting $(f+i)$ variables from n variables, we have summed over all the possible combinations that can be written in the form above. We can also write the above equation in the form

$$\begin{aligned} &\left\langle 0\middle|a(k'_1)\ldots a(k'_f)T\left[\varphi(x_1)\ldots\varphi(x_n)\right]a^\dagger(k_1)\ldots a^\dagger(k_i)\middle|0\right\rangle \\ &\quad = \sum_{\text{comb}}\left\langle b^{\mathrm{S}}\right|:\varphi(x'_1)\ldots\varphi(x'_{f+i}):\left|a^{\mathrm{S}}\right\rangle\left\langle 0\middle|T[\varphi(x'_{f+i+1})\ldots\varphi(x'_n)]\middle|0\right\rangle \,, \end{aligned} \tag{8.39}$$

where we have summed over all combinations which separate the n variables into two groups of $(f+i)$ and $(n-f-i)$ variables. Note also that there is a similar equation for operators, viz.,

$$T[\varphi(x_1)\ldots\varphi(x_n)] = \sum_{\text{comb}} :\varphi(x_1')\ldots\varphi(x_j'): \langle 0|T[\varphi(x_{j+1}')\ldots\varphi(x_n')]|0\rangle\,, \tag{8.40}$$

where the sum is over all combinations which separate the n variables into two groups, with one belonging to the normal product and the other to the expectation value.

The next problem is to evaluate the expectation value. To begin with,

$$\begin{aligned}\langle 0|T[\varphi(x)\varphi(y)]|0\rangle &= \theta(x_0-y_0)\langle 0|\varphi(x)\varphi(y)|0\rangle + \theta(y_0-x_0)\langle 0|\varphi(y)\varphi(x)|0\rangle \\ &= \mathrm{i}\theta(x_0-y_0)\Delta^{(+)}(x-y) + \mathrm{i}\theta(y_0-x_0)\Delta^{(+)}(y-x) \\ &= \Delta_{\mathrm{F}}(x-y)\,,\end{aligned} \tag{8.41}$$

where we have used (4.29). Generalizing the equation above,

$$\langle 0|T[\varphi(x_1)\ldots\varphi(x_n)]|0\rangle = \sum_{\text{comb}} \Delta_{\mathrm{F}}(x_1'-x_2')\ldots\Delta_{\mathrm{F}}(x_{n-1}'-x_n')\,, \tag{8.42}$$

where if n is even we have summed over all combinations which separate n variables into $n/2$ pairs and if n is odd the expression vanishes. To prove this, assuming the time order

$$t_1' > t_2' > \ldots > t_n'\,, \quad (t = x_0)\,, \tag{8.43}$$

we obtain

$$\langle 0|T[\varphi(x_1)\ldots\varphi(x_n)]|0\rangle = \langle 0|\varphi(x_1')\ldots\varphi(x_n')|0\rangle\,. \tag{8.44}$$

Since $k_1 < k_2$ means that $t_{k_1}' > t_{k_2}'$, the definition of Δ_{F} implies in this case that

$$\Delta_{\mathrm{F}}(x_{k_1}'-x_{k_2}') = \mathrm{i}\Delta^{(+)}(x_{k_1}'-x_{k_2}')\,. \tag{8.45}$$

Although the right-hand side of (8.44) can be written in the form (8.25), taking into account (8.45), it becomes equivalent to (8.42). Since the time order (8.43) is arbitrary, it turns out that (8.42) holds for any time order.

To summarize so far,

$$T[\varphi(x_1)\ldots\varphi(x_n)] = \sum_{\text{comb}} :\varphi(x_1')\ldots\varphi(x_j'): \Delta_{\mathrm{F}}(x_{j+1}'-x_{j+2}')\ldots\Delta_{\mathrm{F}}(x_{n-1}'-x_n')\,. \tag{8.46}$$

Wick introduced a contraction symbol for Δ_{F}:

$$\overbrace{\varphi(x)\varphi(y)} = \Delta_{\mathrm{F}}(x-y)\ . \tag{8.47}$$

This specifies a pair which appears in a vacuum expectation value. Using this symbol, (8.46) can be viewed as an expansion in terms of different contractions:

$$\begin{aligned} T[\varphi(x_1)\dots\varphi(x_n)] = {} & :\varphi(x_1)\dots\varphi(x_n): \\ & + \sum_{j\neq k} :\varphi(x_1)\dots\overbrace{\varphi(x_j)\dots\varphi(x_k)}\dots\varphi(x_n): \\ & + \sum_{j\neq k}\sum_{l\neq m} :\varphi(x_1)\dots\overbrace{\varphi(x_1)\dots\varphi(x_k)}\dots\overbrace{\varphi(x_l)\dots\varphi(x_m)}\dots\varphi(x_n): \\ & + \cdots\ . \end{aligned} \tag{8.48}$$

As discussed in Sect. 8.2, the T-product is symmetric for boson fields, i.e.,

$$T[\varphi(x_1')\dots\varphi(x_n')] = T[\varphi(x_1)\dots\varphi(x_n)]\ . \tag{8.49}$$

For fermionic fields, it is anti-symmetric in the sense of (8.22):

$$T[\psi(x_1')\dots\psi(x_n')] = (-1)^P T[\psi(x_1)\dots\psi(x_n)]\ . \tag{8.50}$$

Of course, the above equation also holds if ψ is replaced by $\bar{\psi}$. Moreover, the contraction for fermionic fields is

$$\overbrace{\psi_\alpha(x)\bar{\psi}_\beta(y)} = \langle 0|T[\psi_\alpha(x)\bar{\psi}_\beta(y)]|0\rangle = S_{\mathrm{F}\,\alpha\beta}(x-y)\ , \tag{8.51}$$

where S_{F} is given by

$$S_{\mathrm{F}}(x) = -(\gamma_\mu\partial_\mu - m)\Delta_{\mathrm{F}}(x)\ . \tag{8.52}$$

Generally, the contraction functions, also known as Feynman propagators, satisfy the following equations:

$$D_{\alpha\beta}(\partial)\Delta_{\mathrm{F}\,\beta\gamma}(x) = \mathrm{i}\delta_{\alpha\gamma}\delta^4(x)\ , \tag{8.53}$$

$$\Delta_{\mathrm{F}\,\alpha\beta}(x) = C_{\alpha\beta}(\partial)\Delta_{\mathrm{F}}(x)\ . \tag{8.54}$$

Regarding the contraction for fermionic fields, the following equation is trivial:

$$\overbrace{\psi(x)\psi(y)} = \overbrace{\bar{\psi}(x)\bar{\psi}(y)} = 0\ . \tag{8.55}$$

The general form of Wick's theorem is

$$T[AB\ldots Z] =: AB\ldots Z: + \sum_{\text{one cont}} :\overset{\sqcap}{AB}\ldots Z: + \sum_{\text{two cont}} :\overset{\sqcap}{AB}\,\overset{\sqcap}{CD}\ldots Z: + \cdots\,, \tag{8.56}$$

where 'n cont' instructs to sum over all possible terms with n contractions.

Combining Wick's theorem with the above property of the contraction function, we obtain

$$D_{\alpha\beta}(\partial_x)T[\varphi_\beta(x)\ldots] = \mathrm{i}\frac{\delta}{\delta\varphi_\alpha^\dagger}T[\ldots]\,, \tag{8.57}$$

for a complex field. For instance,

$$D_{\alpha\beta}(\partial_x)T[\varphi_\beta(x), \varphi_\gamma^\dagger(y)] = \mathrm{i}\frac{\delta\varphi_\gamma^\dagger(y)}{\delta\varphi_\alpha^\dagger(x)} = \mathrm{i}\delta_{\alpha\gamma}\delta^4(x-y)\,. \tag{8.58}$$

For a Dirac field,

$$-\left(\gamma_\mu\frac{\partial}{\partial x_\mu} + m\right)T[\psi(x)\ldots] = \mathrm{i}\frac{\delta}{\delta\bar{\psi}(x)}T[\ldots]\,. \tag{8.59}$$

In order to carry out a functional derivative with respect to $\bar{\psi}(x)$, we determine the sign by first moving the $\bar{\psi}$ to be differentiated in the T-product to the extreme left and then carrying out the functional derivative. Since this formula plays an important role, e.g., it can be used to transform the S-matrix, it is referred to as a *reduction formula*. A similar formula in the Heisenberg picture will be mentioned later.

8.4 Feynman Diagrams

To understand the structure of the S-matrix intuitively, it is useful to introduce the Feynman diagrams. As an example, we explain the idea for QED in the Fermi gauge.

When we expand the S-matrix in perturbation theory as a power series in the coupling constant, the n th order term is given by

$$S^{(n)} = \frac{(-\mathrm{i})^n}{n!}\int \mathrm{d}^4x_1\ldots\int \mathrm{d}^4x_n\, T\big[\mathscr{H}_{\text{int}}(x_1)\ldots\mathscr{H}(x_n)\big]\,. \tag{8.60}$$

In QED,

$$\mathscr{H}_{\text{int}}(x) = -j_\mu(x)A_\mu(x) = -\mathrm{i}e :\bar{\psi}(x)\gamma_\mu\psi(x): A_\mu(x)\,. \tag{8.61}$$

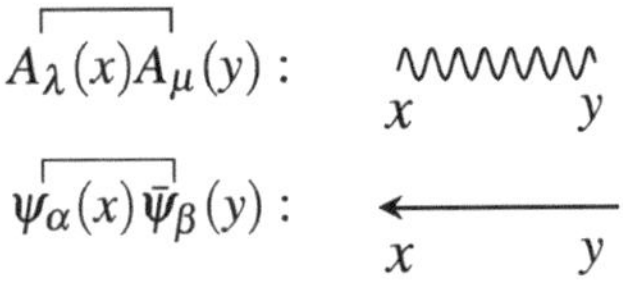

Fig. 8.1 Representation of contractions in a Feynman diagram

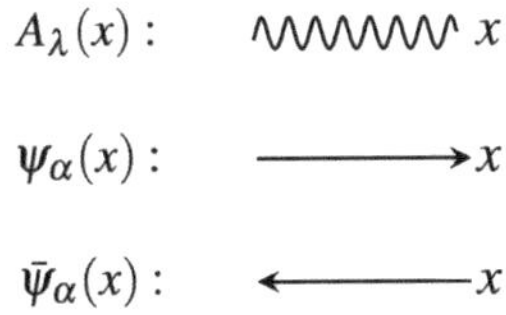

Fig. 8.2 Representation of operators in the normal product

We expand the T-product in (8.60) as a sum of normal products using Wick's theorem. In this case, the operators in the normal product create and annihilate free particles. The contraction functions describe the propagation of virtual particles which cannot be observed. In a Feynman diagram, these operators are represented as lines, providing a way to understand their role visually.

When two operators are contracted, this is represented by a line segment (see Fig. 8.1). In this case, the propagation of a virtual photon is expressed by an undirected wavy line and the propagation of an electron by a directed line, in such a way that the direction coincides with the direction of motion of the electric charge e. If $x_0 > y_0$, this means that a virtual electron created at a point y propagates to a point x and is annihilated there. If $x_0 < y_0$, then a virtual positron created at a point x propagates to a point y and is annihilated there. Regarding the undirected photon, a virtual photon is created at a past point, either x or y, and annihilated at a future point. This interpretation is obvious from (8.41) (Fig. 8.2).

In contrast, operators in the normal product are expressed as undirected and directed half lines for a photon and an electron, respectively. These lines are called *external lines*. In the case of A_λ, the line corresponds to a photon which is either created or absorbed at a point x. In the case of ψ, the line corresponds to either a positron which is created or an electron which is absorbed at a point x. In the case of $\bar{\psi}$, the line corresponds to either an electron which is created or a positron which is absorbed at a point x.

Each term in the expansion of the S-matrix as a sum of normal products corresponds to one Feynman diagram. In the following, we consider the case $n = 2$:

$$S^{(2)} = \frac{(-e)^2}{2}\int \mathrm{d}^4x_1 \int \mathrm{d}^4x_2\, T\big[:\bar{\psi}(x_1)\gamma_\mu\psi(x_1):,\, :\bar{\psi}(x_2)\gamma_\nu\psi(x_2):\big]T\big[A_\mu(x_1),\, A_\nu(x_2)\big]\,. \tag{8.62}$$

We use Wick's theorem to expand this T-product in terms with different numbers of contractions:

1. When there is no contraction,

$$:\bar{\psi}(x_1)\gamma_\mu\psi(x_1)\bar{\psi}(x_2)\gamma_\nu\psi(x_2):\,:A_\mu(x_1)A_\nu(x_2): \;. \tag{8.63}$$

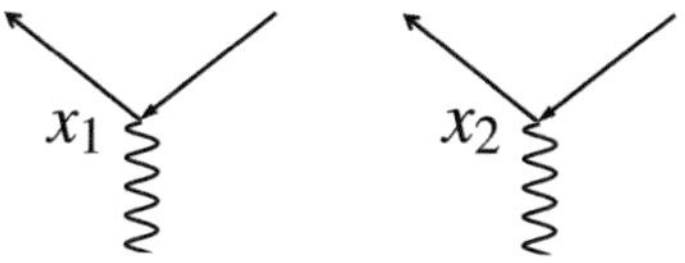

Fig. 8.3 Feynman diagrams for terms in the expansion of the T-product with no contractions

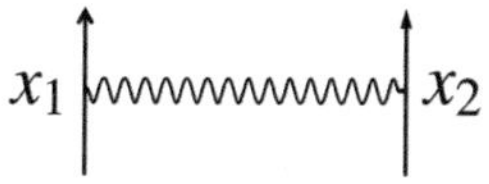

Fig. 8.4 Feynman diagram for the case where only a pair of A's is contracted in the expansion of the T-product

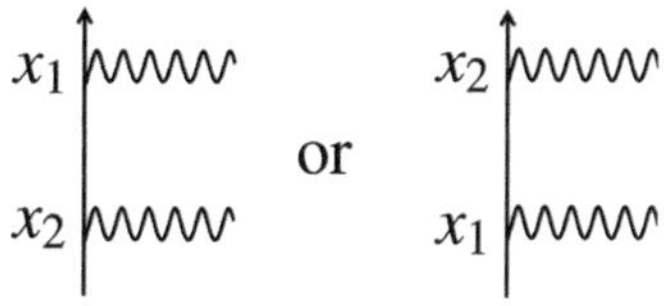

Fig. 8.5 Feynman diagrams for the case where a ψ is contracted with a $\bar{\psi}$ in the expansion of the T-product

Feynman diagrams corresponding to these terms consist of two disconnected parts (Fig. 8.3). They do not correspond to physically realizable processes since they cannot satisfy the energy–momentum conservation law.

2. When only a pair of A's is contracted,

$$:\bar{\psi}(x_1)\gamma_\mu\psi(x_1)\gamma_\nu\bar{\psi}(x_2)\gamma_\nu\psi(x_2):\delta_{\mu\nu}D_F(x_1-x_2)\,. \tag{8.64}$$

This term corresponds to electron–electron scattering, proton–proton scattering, electron–proton scattering, and so on, and the corresponding Feynman diagram is shown in Fig. 8.4.

3. When a ψ is contracted with a $\bar{\psi}$,

$$[:\bar{\psi}(x_1)\gamma_\mu S_F(x_1-x_2)\gamma_\nu\psi(x_2):+:\bar{\psi}(x_2)\gamma_\nu S_F(x_2-x_1)\gamma_\mu\psi(x_1):]:A_\mu(x_1)A_\nu(x_2):\,. \tag{8.65}$$

This term, illustrated in Fig. 8.5, expresses Compton scattering, electron–positron annihilation into two photons or creation from two photons, and so on.

4. When a ψ is contracted with a $\bar{\psi}$ and a pair of A's are contracted,

$$[:\bar{\psi}(x_1)\gamma_\mu S_F(x_1-x_2)\gamma_\nu\psi(x_2):+:\bar{\psi}(x_2)\gamma_\nu S_F(x_2-x_1)\gamma_\mu\psi(x_1):]\delta_{\mu\nu}D_F(x_1-x_2)\,. \tag{8.66}$$

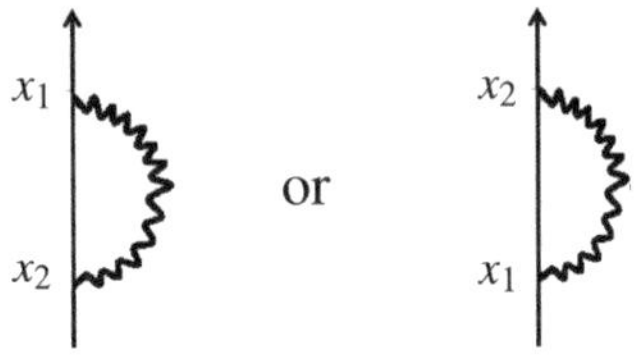

Fig. 8.6 Electron self-energy. Feynman diagrams for the case where a ψ is contracted with a $\bar{\psi}$ and a pair of A's is contracted in the expansion of the T-product

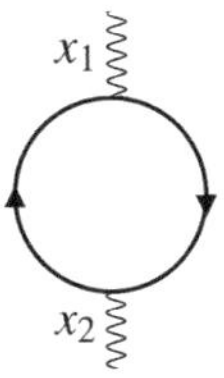

Fig. 8.7 Vacuum polarization. Feynman diagram for the case where two pairs of ψ and $\bar{\psi}$ are contracted in the expansion of the T-product

These terms correspond to the self-energy of the electron. Corresponding Feynman diagrams are shown in Fig. 8.6.

5. When two pairs of ψ and $\bar{\psi}$ are contracted,

$$-\operatorname{Tr}\big[\gamma_\mu S_{\mathrm{F}}(x_1 - x_2)\gamma_\nu S_{\mathrm{F}}(x_2 - x_1)\big] : A_\mu(x_1)A_\nu(x_2): \ . \tag{8.67}$$

This corresponds to the phenomenon called *vacuum polarization*, which gives the photon self-energy. The Feynman diagram is shown in Fig. 8.7. Let us see why the minus sign and the trace have appeared:

$$\begin{aligned}
\bar{\psi}(x_1)\gamma_\mu\psi(x_1)\bar{\psi}(x_2)\gamma_\nu\psi(x_2) &= \bar{\psi}(x_1)\gamma_\mu S_{\mathrm{F}}(x_1 - x_2)\gamma_\nu\psi(x_2) \\
&= \bar{\psi}_\alpha(x_1)\big[\gamma_\mu S_{\mathrm{F}}(x_1 - x_2)\gamma_\nu\big]_{\alpha\beta}\psi_\beta(x_2) \\
&= -\big[\gamma_\mu S_{\mathrm{F}}(x_1 - x_2)\gamma_\nu\big]_{\alpha\beta}\psi_\beta(x_2)\bar{\psi}_\alpha(x_1) \\
&= -\big[\gamma_\mu S_{\mathrm{F}}(x_1 - x_2)\gamma_\nu\big]_{\alpha\beta}\big[S_{\mathrm{F}}(x_2 - x_1)\big]_{\beta\alpha} \\
&= -\operatorname{Tr}\big[\gamma_\mu S_{\mathrm{F}}(x_1 - x_2)\gamma_\nu S_{\mathrm{F}}(x_2 - x_1)\big] \ .
\end{aligned}$$

This shows that the minus sign necessarily comes in when there is a fermion loop, and this will turn out to play an important role in the gauge theory.

6. When all operators are contracted,

$$-\operatorname{Tr}\big[\gamma_\mu S_{\mathrm{F}}(x_1 - x_2)\gamma_\nu S_{\mathrm{F}}(x_1 - x_2)\big]\delta_{\mu\nu} D_{\mathrm{F}}(x_1 - x_2) \ . \tag{8.68}$$

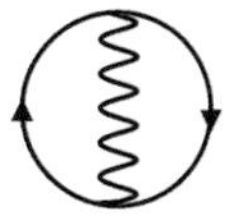

Fig. 8.8 Vacuum bubble. Feynman diagram for the case where all operators are contracted in the expansion of the T-product

This is a process that starts with the vacuum and ends with the vacuum. It is represented by the bubble graph shown in Fig. 8.8. The transition probability amplitude for going from the vacuum to the vacuum, viz.,

$$\langle 0|S|0\rangle \tag{8.69}$$

is obtained using Dyson's formula by summing over all contributions from bubble graphs. By the probability conservation law, the absolute value of this expression should be unity. However, the phase diverges. Adjusting the unobservable phase, we thus redefine the S-matrix by

$$S' = S/\langle 0|S|0\rangle \ . \tag{8.70}$$

To compute S', it turns out that, dropping all bubble graphs without external lines at the outset, we can compute the S-matrix using only the remaining graphs.

8.5 Examples of S-Matrix Elements

Let us compute the S-matrix elements for some examples using the method discussed in the last section. In a real situations, we need to calculate cross-sections and decay amplitudes, but we shall discuss these in the next chapter.

We begin by writing the S-matrix element in the form

$$S_{fi} = \delta_{fi} - \mathrm{i}(2\pi)^4\delta^4(P_f - P_i)T_{fi} \ . \tag{8.71}$$

The first term corresponds to the case where the incoming particles go past each other without interaction. The second term corresponds to the situation where there are interactions. Here we have used the representation which diagonalizes the total energy–momentum tensor, and this gives rise to the four-dimensional δ-function which expresses the conservation law.

8.5.1 Compton Scattering

The normal product corresponding to this process was given in the last section. We consider a transition from the state $|p, k\rangle$ to the state $|p', k'\rangle$, where p and p' express the four-momenta of the electrons and k and k' the four-momenta of the photons. There are two terms in the normal product (8.65), but x_1 and x_2 and μ and ν are all dummy variables, and since the whole expression is symmetric under the replacements $x_1 \leftrightarrow x_2$ and $\mu \leftrightarrow \nu$, we keep only one term and drop the factor of $1/2!$. This yields

$$\begin{aligned}\langle p', k'|S^{(2)}|p, k\rangle &= e^2 \int \mathrm{d}^4x_1 \int \mathrm{d}^4x_2 \langle p'| :\bar{\psi}(x_1)\gamma_\mu S_\mathrm{F}(x_1 - x_2)\gamma_\nu \psi(x_2): |p\rangle \\ &\qquad \times \langle k'| :A_\mu(x_1)A_\nu(x_2): |k\rangle \\ &= e^2 \int \mathrm{d}^4x_1 \int \mathrm{d}^4x_2 \langle p'|\bar{\psi}(x_1)|0\rangle\langle 0|A_\nu(x_2)|k\rangle \\ &\qquad \times \left[\langle k'|A_\mu(x_1)|0\rangle\langle 0|A_\nu(x_2)|k\rangle + \langle k'|A_\nu(x_2)|0\rangle\langle 0|A_\mu(x_1)|k\rangle\right] .\end{aligned} \tag{8.72}$$

Feynman diagrams in the momentum representation are shown in Fig. 8.9.

In order to evaluate this expression, we use the following matrix elements and the Fourier representation:

$$\langle 0|A_\mu(x)|k\rangle = \frac{1}{\sqrt{2k_0 V}} e_\mu^{(\lambda)} \mathrm{e}^{\mathrm{i}k\cdot x} , \quad \langle k'|A_\mu(x)|0\rangle = \frac{1}{\sqrt{2k_0'}} e_\mu^{(\lambda')} \mathrm{e}^{-\mathrm{i}k'\cdot x} ,$$

$$\langle 0|\psi(x)|p\rangle = \frac{1}{\sqrt{V}} u(p)\mathrm{e}^{\mathrm{i}p\cdot x} , \quad \langle p'|\bar{\psi}(x)|0\rangle = \frac{1}{\sqrt{V}} \bar{u}(p')\mathrm{e}^{-\mathrm{i}p'\cdot x} ,$$

$$S_\mathrm{F}(x) = \frac{-\mathrm{i}}{(2\pi)^4} \int \mathrm{d}^4p \frac{-\mathrm{i}p\cdot\gamma + m}{p^2 + m^2 - \mathrm{i}\epsilon} \mathrm{e}^{\mathrm{i}p\cdot x} .$$

Inserting these expressions into (8.72), we obtain T_{fi} as defined in (8.71):

$$\begin{aligned}T_{fi}^{(2)} &= \frac{1}{\sqrt{2k_0 V}} \frac{1}{\sqrt{2k_0' V}} \frac{1}{\sqrt{V}} \frac{1}{\sqrt{V}} \\ &\times e^2 \bar{u}(p')\left[(e'\cdot\gamma')\frac{-\mathrm{i}(p+k)\cdot\gamma + m}{(p+k)^2 + m^2}(e\cdot\gamma) + (e\cdot\gamma)\frac{-\mathrm{i}(p-k')\cdot\gamma + m}{(p-k')^2 + m^2}(e'\cdot\gamma)\right]u(p) ,\end{aligned} \tag{8.73}$$

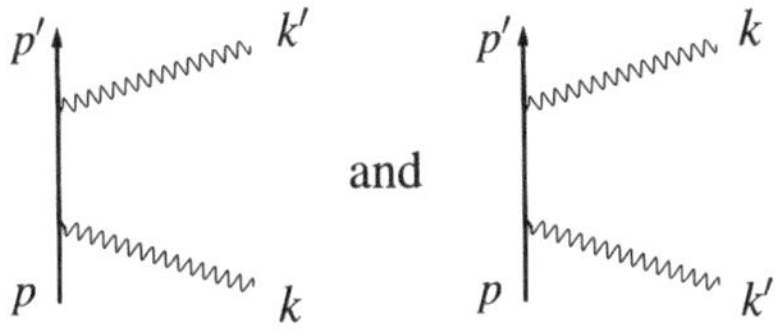

Fig. 8.9 Feynman diagrams for Compton scattering

where e stands for $e^{(\lambda)}$ and e' for $e^{(\lambda')}$. Using the relation for external momenta on the mass shell, i.e., the Einstein relation (3.35), we have

$$(p+k)^2+m^2 = p^2+m^2+k^2+2p\cdot k = 2p\cdot k\ ,$$
$$(p'-k')^2+m^2 = p^2+m^2+k'^2-2p\cdot k' = -2p\cdot k'\ .$$

Using the Dirac equation,

$$(\mathrm{i}p\cdot\gamma+m)u(p) = \bar{u}(p')(\mathrm{i}p'\cdot\gamma+m) = 0\ .$$

We can now define the invariant amplitude $\mathscr{T}_{fi}$ by

$$\mathscr{T}_{fi}^{(2)} = \sqrt{2k_0V}\sqrt{2k_0'V}\sqrt{2p_0V}\sqrt{2p_0'V}\sqrt{2p_0'V}\ T_{fi}^{(2)}\ . \tag{8.74}$$

Equation (8.73) now gives

$$\mathscr{T}_{fi}^{(2)} = e^2\sqrt{2p_0'}\bar{u}(p')\Bigg[(e'\cdot\gamma)\frac{-\mathrm{i}(p+k)\cdot\gamma+m}{2p\cdot k}(e\cdot\gamma) \tag{8.75}$$
$$-(e\cdot\gamma)\frac{-\mathrm{i}(p-k')\cdot\gamma+m}{2p\cdot k'}(e'\cdot\gamma)\Bigg]u(p)\sqrt{2p_0}\ .$$

It is clear from (3.195) that the Dirac spinor multiplied by $\sqrt{2p_0}$ has the Casimir operator

$$\sum_n\sqrt{2p_0}\,u_r(p,h)\bar{u}_s(p,h)\sqrt{2p_0} = (-\mathrm{i}p\cdot\gamma+m)_{rs}\ , \tag{8.76}$$

where we sum over helicities. Therefore, this provides a normalization that does not depend on how we choose the Lorentz system. We will compute the cross-section from (8.75) in the next chapter.

8.5.2 Pion Decay to Muons

The π^+-meson usually decays according to

$$\pi^+ \to \mu^+ + \nu\ . \tag{8.77}$$

We can take the following phenomenological expression for the interaction describing this decay:

$$\mathscr{H}_{\mathrm{int}} = \frac{\mathrm{i}g}{m_\pi}\bar{\psi}_\nu\gamma_\lambda(1+\gamma_5)\psi_\mu\partial_\lambda\varphi + \mathrm{h.c.}\ , \tag{8.78}$$

where m_π is the mass of π^+, ψ_ν and ψ_μ are field operators for the neutrino and the μ-meson, respectively, and h.c. indicates the Hermitian conjugate. If we take $|\mu^+, \nu\rangle = a^\dagger(\mu^+)a^\dagger(\nu)|0\rangle$ as the final state, the lowest order of the decay amplitude has the form

$$T_{fi} = \langle \mu^+, \nu | \mathscr{H}_{\text{int}}(0) | \pi^+ \rangle . \tag{8.79}$$

Hence, denoting the four-momenta of π^+, μ^+, and ν by P, p, and q, respectively, we have $P = p + q$ and

$$\begin{aligned} T_{fi} &= -\frac{\mathrm{i}g}{m_\pi} \langle q | \bar{\psi}_\nu(0) | 0 \rangle \gamma_\lambda (1 + \gamma_5) \langle p | \psi_\mu(0) | 0 \rangle \langle 0 | \partial_\lambda \varphi(0) | \pi^+ \rangle \\ &= \frac{1}{\sqrt{2P_0 V}\sqrt{V}\sqrt{V}} \frac{g}{m_\pi} \bar{u}(q) \gamma_\lambda (1 + \gamma_5) v(p) P_\lambda . \end{aligned} \tag{8.80}$$

In order to simplify this equation, we use

$$\bar{u}(q) \mathrm{i} q \cdot \gamma = (\mathrm{i} p \cdot \gamma - m_\mu) v(p) = 0 .$$

Therefore, the invariant decay amplitude is

$$\begin{aligned} \mathscr{T}_{fi} &= \sqrt{2P_0 V}\sqrt{2q_0 V}\sqrt{2p_0 V}\, T_{fi} \\ &= -\mathrm{i}g \frac{m_\mu}{m_\pi} \sqrt{2q_0} \bar{u}(q)(1 - \gamma_5) v(p) \sqrt{2p_0} . \end{aligned} \tag{8.81}$$

Two-Photon Decay of π^0

Although the two examples mentioned above are rather simple, they are complicated enough to provide a good illustration of this kind of computation. Another good example is the two-photon decay of the π^0-meson, which includes a fermion loop and no divergences, i.e.,

$$\pi^0 \to 2\gamma . \tag{8.82}$$

The interactions involved in this decay process are the Yukawa-type interaction between the π^0 and the proton and the electromagnetic interaction with the proton. Assuming that the former has the form given in (6.97) and using ψ and φ to denote the field operators of the proton and π^0, respectively, we have

$$\mathscr{H}_{\text{int}} = \mathrm{i}G \bar{\psi} \gamma_5 \psi \varphi - \mathrm{i}e \bar{\psi} \gamma_\mu \psi A_\mu \equiv \mathscr{H}_\pi + \mathscr{H}_{\text{em}} . \tag{8.83}$$

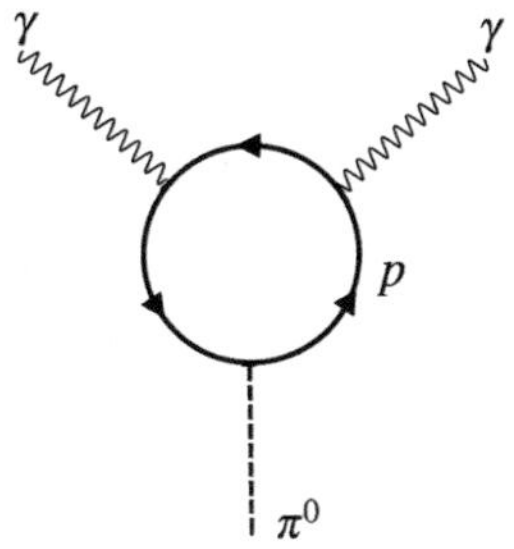

Fig. 8.10 Feynman diagram for the two-photon decay of π^0

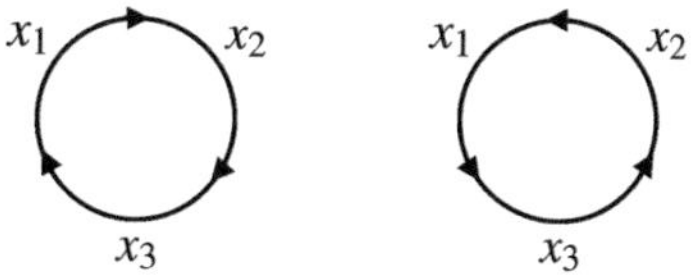

Fig. 8.11 Feynman diagrams for the contraction of fermion operators

The above interactions are sufficient for the lowest order computation. The Feynman diagram is shown in Fig. 8.10.

Denoting the four-momenta of the π^0 and the two photons by q, k', and q'', respectively, the S-matrix element corresponding to the above graph is given by

$$
\begin{aligned}
\langle k', k''|S|q\rangle &= \frac{(-\mathrm{i})^3}{2!}\int \mathrm{d}^4x_1 \int \mathrm{d}^4x_2 \int \mathrm{d}^4x_3 \langle k', k''\big|T'\big[\mathscr{H}_{\mathrm{em}}(x_1)\mathscr{H}_{\mathrm{em}}(x_2)\mathscr{H}_{\pi}(x_3)\big]\big|q\rangle \\
&= \frac{1}{2}e^2G\int \mathrm{d}^4x_1 \int \mathrm{d}^4x_2 \int \mathrm{d}^4x_3 \\
&\quad \times\langle 0\big|T\big[\bar{\psi}(x_1)\gamma_\mu\psi(x_1),\, \bar{\psi}(x_2)\gamma_\nu\psi(x_2),\, \bar{\psi}(x_3)\gamma_5\psi(x_3)\big]\big|0\rangle \\
&\quad \times\langle k', k''\big| :A_\mu(x_1)A_\nu(x_2):\, |0\rangle\langle 0|\varphi(x_3)|q\rangle\ .
\end{aligned}
\tag{8.84}
$$

The reason why the coefficient 1/3! has been replaced by 1/2! is that there are three ways of choosing the coordinate of $\mathscr{H}_\pi$, i.e., x_1, x_2, or x_3, and each of them gives the same contribution, while we have kept only one, i.e., x_3, and multiplied the whole expression by three. Moreover, similarly to the case with Compton scattering, the external photon lines give two terms, but since each gives the same contribution, we keep only one and drop the factor of 1/2. As a consequence,

$$
\begin{aligned}
\langle k', k''|S|q\rangle &= e^2G\int \mathrm{d}^4x_1 \int \mathrm{d}^4x_2 \int \mathrm{d}^4x_3 \\
&\quad \times\langle 0\big|T\big[\bar{\psi}(x_1)\gamma_\mu\psi(x_1),\, \bar{\psi}(x_2)\gamma_\nu\psi(x_2),\, \bar{\psi}(x_3)\gamma_5\psi(x_3)\big]\big|0\rangle \\
&\quad \times\langle k'|A_\mu(x_1)|0\rangle\langle k''|A_\nu(x_2)|0\rangle\langle 0|\varphi(x_3)|q\rangle\ .
\end{aligned}
\tag{8.85}
$$

Note also that there are two ways of contracting fermion operators, as shown in Fig. 8.11. The difference between these two diagrams is that the directions of the fermion loops are opposite:

$$\begin{aligned}\langle 0|T[\ldots]|0\rangle &= -\mathrm{Tr}\big[\gamma_5 S_F(x_3-x_2)\gamma_\nu S_F(x_2-x_1)\gamma_\mu S_F(x_1-x_3)\big] \\ &= -\mathrm{Tr}\big[\gamma_5 S_F(x_3-x_1)\gamma_\mu S_F(x_1-x_2)\gamma_\nu S_F(x_2-x_3)\big]\,. \end{aligned} \tag{8.86}$$

These two terms are in fact equivalent to each other. In order to show this, we make use of the relation

$$C^{-1}S_F(x-y)C = S_F(y-x)^T\,. \tag{8.87}$$

This follows because

$$\begin{aligned} C^{-1}S_F(x-y)C &= C^{-1}\left(-\gamma_\mu\frac{\partial}{\partial x_\mu}+m\right)C\Delta_F(x-y) \\ &= \left(\gamma_\mu^T\frac{\partial}{\partial x_\mu}+m\right)\Delta_F(y-x) \\ &= \left(-\gamma_\mu^T\frac{\partial}{\partial y_\mu}+m\right)\Delta_F(y-x) = S_F(y-x)^T\,. \end{aligned}$$

We also use the following relations:

$$C^{-1}\gamma_\mu C = -\gamma_\mu^T\,,\quad C^{-1}\gamma_5 C = \gamma_5^T\,,\quad \mathrm{Tr}\,A^T = \mathrm{Tr}\,A\,.$$

Therefore,

$$\begin{aligned} &\mathrm{Tr}\big[\gamma_5 S_F(x_3-x_2)\gamma_\nu S_F(x_2-x_1)\gamma_\mu S_F(x_1-x_3)\big]^T \\ &\quad = \mathrm{Tr}\big[S_F(x_1-x_3)^T{\gamma_\mu}^T S_F(x_2-x_1)^T{\gamma_\nu}^T S_F(x_3-x_2)^T{\gamma_5}^T\big] \\ &\quad = \mathrm{Tr}\big[C^{-1}S_F(x_3-x_1)C\cdot C^{-1}(-\gamma_\mu)C\cdot C^{-1}S_F(x_1-x_2)C \\ &\qquad \times C^{-1}(-\gamma_\nu)C\cdot C^{-1}S_F(x_2-x_3)C\cdot C^{-1}\gamma_5 C\big] \\ &\quad = \mathrm{Tr}\big[\gamma_5 S_F(x_3-x_1)\gamma_\mu S_F(x_1-x_2)\gamma_\nu S_F(x_2-x_3)\big]\,. \end{aligned} \tag{8.88}$$

i.e., the first trace is equivalent to the second trace. It is clear from the above example that, if there are an odd number of photons, they will cancel each other by flipping signs. Now,

$$\begin{aligned}\langle k',k''|S|q\rangle = -2e^2G\int \mathrm{d}^4x_1\int \mathrm{d}^4x_2\int \mathrm{d}^4x_3\langle k'|A_\mu(x_1)|0\rangle\langle k''|A_\nu(x_2)|0\rangle\langle 0|\varphi(x_3)|q\rangle \\ \times\mathrm{Tr}\big[\gamma_5 S_F(x_3-x_1)\gamma_\mu S_F(x_1-x_2)\gamma_\nu S_F(x_2-x_3)\big]\,. \end{aligned} \tag{8.89}$$

In the following, we write the polarization vectors of the two photons as e' and e''. Then, inserting the wave function of one particle and the Fourier representation of S_{F}, and integrating over x_1, x_2, and x_3, a factor $(2\pi)^{4\times 3}$ appears. However, this factor cancels the one arising from the Fourier representation of S_{F}. At the same time, three δ-functions of the four-momenta appear, along with three four-dimensional momentum integrals arising from the Fourier representation of S_{F}. Two of the momentum integrals cancel two of the δ-functions, leaving just one four-dimensional momentum integral and one δ-function expressing the conservation of the total energy–momentum. Hence, computing T_{fi}, except for this δ-function, and extracting the invariant decay amplitude $\mathscr{T}_{fi}$, we obtain

$$\mathscr{T}_{fi} = \frac{2e^2G}{(2\pi)^4}e'_\mu e''_\nu \int \mathrm{d}^4 p \tag{8.90}$$
$$\mathrm{Tr}\left[\gamma_5 \frac{-\mathrm{i}(p-k')\cdot\gamma + m}{(p-k')^2+m^2-\mathrm{i}\epsilon}\gamma_\mu \frac{-\mathrm{i}p\cdot\gamma+m}{p^2+m^2-\mathrm{i}\epsilon}\gamma_\nu \frac{-\mathrm{i}(p+k'')\cdot\gamma+m}{(p+k'')^2+m^2-\mathrm{i}\epsilon}\right] .$$

First we compute the trace in the numerator. The details of this calculation can be found in my previous book *Relativistic Quantum Mechanics* [78]. Here we give only the relevant formulas:

$$\mathrm{Tr}(\mathbf{1}) = 4\ , \quad \mathrm{Tr}(\gamma_\mu\gamma_\nu) = 4\delta_{\mu\nu}\ . \tag{8.91}$$

The trace of a product of an odd number of γ-matrices is zero. To find the trace of an even number of γ-matrices, we just do the same calculation as when we contracted all the fermion fields in the Wick contraction. For example,

$$\begin{aligned}\mathrm{Tr}(\gamma_\sigma\gamma_\lambda\gamma_\mu\gamma_\nu) &\longrightarrow (\gamma_\sigma\gamma_\lambda\gamma_\mu\gamma_\nu) + (\gamma_\sigma\gamma_\lambda\gamma_\mu\gamma_\nu) + (\gamma_\sigma\gamma_\lambda\gamma_\mu\gamma_\nu)\\ &\longrightarrow 4(\delta_{\sigma\lambda}\delta_{\mu\nu} + \delta_{\sigma\nu}\delta_{\lambda\mu} - \delta_{\sigma\mu}\delta_{\lambda\nu}).\end{aligned}$$

Clearly, we replace contracted pairs by Kronecker deltas. Signs are determined as though the pairs of γ-matrices to be contracted are next to each other, and when reordering them, treating them as anti-commutative. Since $\gamma_5 = \gamma_1\gamma_2\gamma_3\gamma_4$, we have

$$\mathrm{Tr}(\gamma_5\gamma_\lambda\gamma_\mu\gamma_\nu\gamma_\sigma) = 4\epsilon_{\lambda\mu\nu\sigma}\ , \tag{8.92}$$

which implies

$$\begin{aligned}&\mathrm{Tr}\big[\gamma_5\big(-\mathrm{i}(p-k')\cdot\gamma+m\big)\gamma_\mu(-\mathrm{i}p\cdot\gamma+m)\gamma_\nu\big(-\mathrm{i}(p+k'')\cdot\gamma+m\big)\big]\\ &\quad = 4m\epsilon_{\mu\nu\alpha\beta}\big[p_\alpha(p+k'')_\beta - (p-k')_\alpha(p+k'')_\beta + (p-k')_\alpha p_\beta\big]\\ &\quad = 4m\epsilon_{\mu\nu\alpha\beta}k'_\alpha k''_\beta\ .\end{aligned} \tag{8.93}$$

The Feynman diagram corresponding to this process is shown in Fig. 8.12 for the momentum representation. As is clear from this diagram, the conservation of four-momentum holds at each vertex, just as the conservation of electric current holds in

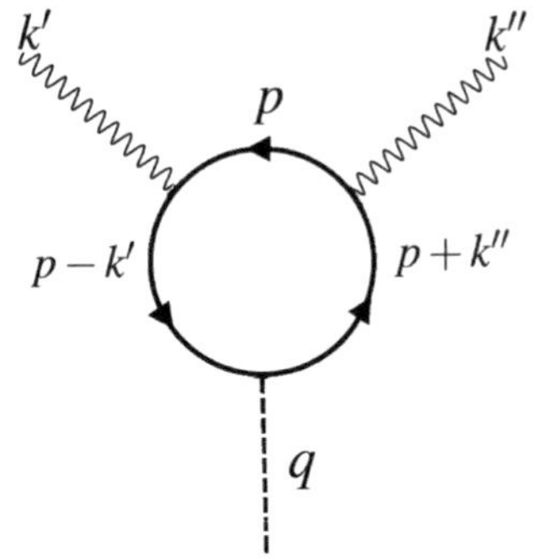

Fig. 8.12 Momentum space Feynman diagram for the two-photon decay of π^0

an electric circuit. This is because, corresponding to each vertex, δ-functions of the momenta appear as a consequence of the four-dimensional coordinate integrals.

Since the trace above does not depend on p, we take it outside the integral:

$$\mathscr{T}_{fi} = \frac{8me^2G}{(2\pi)^4}\epsilon_{\mu\nu\alpha\beta}e'_\mu e''_\nu k'_\alpha k''_\beta \tag{8.94}$$
$$\int \frac{\mathrm{d}^4 p}{\left[(p-k')^2+m^2\right]\left[p^2+m^2-\mathrm{i}\epsilon\right]\left[(p+k'')^2+m^2-\mathrm{i}\epsilon\right)\right]}\,.$$

To carry out the integral over p, we use Feynman's formula:

$$\int \frac{\mathrm{d}^4 p}{(p^2+\Lambda-\mathrm{i}\epsilon)^3} = \frac{\mathrm{i}\pi^2}{2(\Lambda-\mathrm{i}\epsilon)}\,. \tag{8.95}$$

This is a fundamental formula in the Feynman–Dyson theory. We give a proof of this. First, note that

$$\int_0^\infty \mathrm{d}x\, \mathrm{e}^{-\mathrm{i}\alpha x} = \lim_{\epsilon\to+0}\int_0^\infty \mathrm{d}x\, \mathrm{e}^{-\mathrm{i}\alpha x-\epsilon x} = \lim_{\epsilon\to+0}\frac{-\mathrm{i}}{\alpha-\mathrm{i}\epsilon}\,.$$

In the following, we omit reference to the limit. From this,

$$\int_0^\infty \mathrm{d}x\, \exp[-\mathrm{i}(p^2+\Lambda)x] = \frac{-\mathrm{i}}{p^2+\Lambda-\mathrm{i}\epsilon}\,. \tag{8.96}$$

Differentiating this equation twice with respect to Λ yields

$$\int_0^\infty \mathrm{d}x\, x^2 \exp\left[-\mathrm{i}(p^2+\Lambda)x\right] = \frac{2\mathrm{i}}{(p^2+\Lambda-\mathrm{i}\epsilon)^3}\,. \tag{8.97}$$

Moreover, from the formula for the Gaussian integral, if $x > 0$, then

$$\int \mathrm{d}^4 p \exp(-\mathrm{i}p^2 x) = \left[\prod_{j=1}^{3} \int \mathrm{d}p_j \exp(-\mathrm{i}p_j{}^2 x)\right] \int \mathrm{d}p_0 \exp(\mathrm{i}p_0{}^2 x)$$
$$= -\frac{\mathrm{i}\pi^2}{x^2} .$$

Thus,

$$\int \mathrm{d}^4 p \int_0^\infty \mathrm{d}x\, x^2 \exp\big[-\mathrm{i}(p^2 + \Lambda)x\big] = -\mathrm{i}\pi^2 \int_0^\infty \mathrm{d}x\, \exp(-\mathrm{i}\Lambda x)$$
$$= -\frac{\pi^2}{\Lambda - \mathrm{i}\epsilon} .$$

However, since the left-hand side has been obtained by integrating the left-hand side of (8.97) over p, we can use the latter equation to deduce (8.95).

To apply this formula to (8.94), we need another formula, viz.,

$$\frac{1}{A_1 A_2 \ldots A_n} = (n-1)! \int_0^1 \mathrm{d}x_1 \ldots \int_0^1 \mathrm{d}x_n \delta\big(1 - \sum x_i\big) \frac{1}{(x_1 A_1 + x_2 A_2 + \cdots + x_n A_n)^n} . \tag{8.98}$$

A condition for this formula to hold is that the denominator on the right-hand side should not vanish anywhere within the domain of integration. The following are variants of the above formula:

$$\frac{1}{AB} = \int_0^1 \frac{\mathrm{d}x}{\big[xA + (1-x)B\big]^2} , \quad \frac{1}{ABC} = 2\int_0^1 \mathrm{d}x \int_0^1 \mathrm{d}y \frac{x}{\big[A(1-x) + Bxy + Cx(1-y)\big]^3} .$$

We shall use the last of these with

$$A = (p - k')^2 + m^2 - \mathrm{i}\epsilon = p^2 - 2p \cdot k' + m^2 - \mathrm{i}\epsilon ,$$
$$B = p^2 + m^2 - \mathrm{i}\epsilon ,$$
$$C = (p + k'')^2 + m^2 - \mathrm{i}\epsilon = p^2 + 2p \cdot k'' + m^2 - \mathrm{i}\epsilon ,$$

where we have used $k'^2 = k''^2 = 0$. Therefore, the p-integral is

$$I = \int \mathrm{d}^4 p \int_0^1 2x\, \mathrm{d}x \int_0^1 \mathrm{d}y \frac{1}{\big[p^2 + m^2 + (1-x)(-2p \cdot k') + x(1-y)(2p \cdot k'') - \mathrm{i}\epsilon\big]^3}$$
$$= \int \mathrm{d}^4 p' \int_0^1 2x\, \mathrm{d}x \int_0^1 \mathrm{d}y \frac{1}{\big[p'^2 + m^2 + 2x(1-x)(1-y)k' \cdot k'' - \mathrm{i}\epsilon\big]^3} , \tag{8.99}$$

where we have eliminated the linear terms in p by making the replacement

$$p' = p - (1-x)k' + x(1-y)k'' \ . \tag{8.100}$$

We carry out the p'-integral in (8.99) using Feynman's formula (8.101):

$$I = \mathrm{i}\pi^2 \int_0^1 x\,\mathrm{d}x \int_0^1 \mathrm{d}y \frac{1}{m^2 + 2x(1-x)(1-y)k'k''} \ . \tag{8.101}$$

Since $q^2 = (k' + k'')^2 = -\mu^2$, it follows that $2k' \cdot k'' = -\mu^2$, whence

$$I = \mathrm{i}\pi^2 \int_0^1 x\,\mathrm{d}x \int_0^1 \mathrm{d}y \frac{1}{m^2 - \mu^2 x(1-x)(1-y)} \ .$$

Here m is the proton mass and μ is the mass of π^0. Given that the value of $4m^2$ is about 200 times the value of μ^2,

$$\frac{\mu^2}{m^2} x(1-x)(1-y) \leqq \frac{\mu^2}{4m^2} \ll 1 \ .$$

Therefore,

$$I = \frac{\mathrm{i}\pi^2}{m^2} \int_0^1 x\,\mathrm{d}x \int_0^1 \mathrm{d}y \left[1 + \mathscr{O}\left(\frac{\mu^2}{4m^2} \right) \right] \approx \frac{\mathrm{i}\pi^2}{2m^2} \ . \tag{8.102}$$

Inserting this result into (8.94),

$$\mathscr{T}_{fi} = \frac{1}{\pi} \left(\frac{e^2}{4\pi} \right) \frac{G}{m} \left(\mathrm{i}\epsilon_{\mu\nu\alpha\beta} e'_\mu e''_\nu k'_\alpha k''_\beta \right) \ . \tag{8.103}$$

Note that, since the last factor includes i, this is in fact real. In the next chapter, we will use it to compute the lifetime of π^0.

8.6 Furry's Theorem

As shown in the loop calculation for $\pi^0 \to 2\gamma$, if a contribution from a reversed loop has the opposite sign, the two loops cancel each other out, which tells us that the process is forbidden. Let us consider how we could derive this rule without using perturbation theory. QED is invariant under charge conjugation:

$$\mathscr{C}^{-1} \mathscr{H}_{\text{int}}(x) \mathscr{C} = \mathscr{H}_{\text{int}}(x) \ . \tag{8.104}$$

According to the discussion in Sect. 3.9, since the current j_μ changes sign under charge conjugation, (8.104) implies that A_μ also changes its sign:

$$\mathscr{C}^{-1} A_\mu(x)\mathscr{C} = -A_\mu(x) . \tag{8.105}$$

Now, if (8.104) holds, the S-matrix is also invariant under charge conjugation. Dyson's formula for the S-matrix can be written in the form

$$S = T \exp\left[-\mathrm{i}\int \mathrm{d}^4x \mathscr{H}_{\mathrm{int}}(x)\right] . \tag{8.106}$$

This tells us to time-order the Hamiltonian density produced by expanding the exponential in a power series, and then carry out the space-time integral. Thus,

$$\mathscr{C}^{-1} S\mathscr{C} = T \exp\left[-\mathrm{i}\int \mathrm{d}^4x \mathscr{C}^{-1}\mathscr{H}_{\mathrm{int}}(x)\mathscr{C}\right] = S . \tag{8.107}$$

Hence, considering $|a\rangle$ and $|b\rangle$ as eigenstates of $\mathscr{C}$ and setting

$$\mathscr{C}|a\rangle = \epsilon_a|a\rangle , \quad \mathscr{C}|b\rangle = \epsilon_b|b\rangle , \tag{8.108}$$

the transition $a \to b$ will be forbidden whenever $\epsilon_a \neq \epsilon_b$. In the n-photon state,

$$\mathscr{C}|n\gamma\rangle = (-1)^n|n\gamma\rangle . \tag{8.109}$$

Therefore, the transition from odd numbers of photon states to even numbers of photon states is prohibited.

Furthermore, from the charge-conjugation invariance of the Hamiltonian density (8.83), for π^0,

$$\mathscr{C}^{-1}\varphi(x)\mathscr{C} = \varphi(x) . \tag{8.110}$$

Thus, any process whereby π^0 decays into odd numbers of photons is forbidden. This series of selection rules based on charge conjugation is called *Furry's theorem* [98–100].

8.7 Two-Photon Decays of Neutral Mesons

In the last section, we discussed a selection rule based on charge conjugation. Let us consider other grounds for such rules, e.g., a selection rule based on the conservation of angular momentum. In the following, we discuss the process whereby a neutral meson M^0 decays into two photons:

$$M^0 \to 2\gamma . \tag{8.111}$$

It is clear from the discussion in the last section that the invariant decay amplitude $\mathscr{T}$ can be written in the following ways depending on whether M^0 has spin zero or one:

$$\mathscr{T}(M^0 \to 2\gamma) \propto \begin{cases} c_{\mu\nu}(k,k') : A_\mu(k) A_\nu(k') \varphi(k+k') : & \text{spin } 0 \,, \\ c_{\lambda\mu\nu}(k,k') : A_\mu(k) A_\nu(k') \varphi_\lambda(k+k') : & \text{spin } 1 \,. \end{cases} \tag{8.112}$$

Normally, the right-hand side consists of operators, but we have taken suitable matrix elements. Note also that Fourier-transformed field operators are included in the normal product. By (6.66), the normal product above should commute with $\partial_\mu A_\mu(x)$. Looking back to the discussion in Sect. 5.4, this implies that, in the normal product in $\mathscr{T}$, electromagnetic fields only appear in the following combinations:

$$F_{\mu\nu}(k) \,, \quad k_\mu A_\mu(k) \,. \tag{8.113}$$

Taking into account the Lorenz condition, we can set the latter to zero. If $\mathscr{T}$ includes only terms of the form in (8.113), then $\mathscr{T}$ should be invariant under the replacement

$$A_\mu(k) \to A_\mu(k) + \lambda k_\mu \,. \tag{8.114}$$

This is the requirement of gauge invariance. It turns out to be related to the spin property of massless particles discussed in depth in Sect. 8.5.

Case with 0^+

In this case, $c_{\mu\nu}$ must be a tensor built up from k and k', i.e., it must be a linear combination of the following:

$$\delta_{\mu\nu} \,, \quad k_\mu k_\nu \,, \quad k'_\mu k_\nu \,, \quad k_\mu k'_\nu \,, \quad k'_\mu k'_\nu \,.$$

Using the Lorenz condition in (8.112), any term involving k_μ or k'_ν in $c_{\mu\nu}$ has to vanish, so the only possible form is

$$c_{\mu\nu} = a\delta_{\mu\nu} + bk'_\mu k_\nu \,. \tag{8.115}$$

Hence, taking into account the invariance under (8.114),

$$c_{\mu\nu} k_\mu = c_{\mu\nu} k'_\nu = 0 \,, \quad \text{i.e.,} \quad a + bk \cdot k' = 0 \,. \tag{8.116}$$

Therefore, the only possible form of $c_{\mu\nu}$ is

$$c_{\mu\nu} = -b\big[(k \cdot k')\delta_{\mu\nu} - k'_\mu k_\nu\big] \,, \tag{8.117}$$

and the corresponding form of $\mathscr{T}$ is

$$\mathscr{T}(M^0 \to 2\gamma) \propto \left[(k \cdot k')\delta_{\mu\nu} - k'_\mu k_\nu\right] : A_\mu(k) A_\nu(k') \varphi(k+k') :$$
$$\propto : F_{\mu\nu}(k) F_{\mu\nu}(k') \varphi(k+k') : \tag{8.118}$$

This means that the effective interaction of the two-photon decay of the neutral scalar meson is given by

$$F_{\mu\nu} F_{\mu\nu} \varphi \,. \tag{8.119}$$

Case with 0^-

In this case, since $c_{\mu\nu}$ is a pseudo-tensor built up from k and k',

$$c_{\mu\nu} = b\epsilon_{\mu\nu\alpha\beta} k'_\alpha k_\beta \,. \tag{8.120}$$

Therefore, the only possible form of $\mathscr{T}$ is

$$\mathscr{T}(M^0 \to 2\gamma) \propto \; : F_{\mu\nu}(k) \tilde{F}_{\mu\nu}(k') \varphi(k+k') : \,, \tag{8.121}$$

and the effective interaction is given by

$$F_{\mu\nu} \tilde{F}_{\mu\nu} \varphi \,, \tag{8.122}$$

where

$$\tilde{F}_{\mu\nu} = \frac{1}{2} \epsilon_{\mu\nu\alpha\beta} F_{\alpha\beta} \,. \tag{8.123}$$

Case with 1^-

In this case, there is also an irreducibility condition for the spin-1 field, viz.,

$$(k+k')_\lambda \varphi_\lambda (k+k') = 0 \,. \tag{8.124}$$

Thus, k_λ and k'_λ are not independent, and k_μ and k'_ν cannot be used, so we must take $c_{\lambda\mu\nu}$ to be a linear combination of the following:

$$(k_\lambda - k'_\lambda)\delta_{\mu\nu} \,, \quad (k_\lambda - k'_\lambda) k'_\mu k_\nu \,, \quad k'_\mu \delta_{\lambda\nu} \,, \quad k_\nu \delta_{\lambda\mu} \,.$$

However, the first two contradict the boson symmetry, and the last two are excluded by gauge invariance:

$$k_\mu c_{\lambda\mu\nu} = k'_\nu c_{\lambda\mu\nu} = 0 \,. \tag{8.125}$$

Therefore, the 1^--neutral meson cannot decay into two photons. Likewise, the 1^+-neutral meson cannot decay into two photons.

Theorem 8.1 *The two-photon decay of a spin-1 neutral meson is forbidden.*

By a similar argument, we can also prove the following:

Theorem 8.2 *The decay of a spin-0 particle into another spin-0 particle and one photon is forbidden.*

Two-Photon Polarization

For simplicity, if we consider the Coulomb gauge,

$$\boldsymbol{A} \parallel \boldsymbol{e}\,, \quad \boldsymbol{E} \parallel \boldsymbol{e}\,, \quad \boldsymbol{H} \parallel \boldsymbol{k} \times \boldsymbol{e}\,, \tag{8.126}$$

$$\text{decay amplitude of } 0^+ \propto \boldsymbol{e}_1 \cdot \boldsymbol{e}_2\,, \tag{8.127}$$

$$\text{decay amplitude of } 0^- \propto (\boldsymbol{e}_1 \times \boldsymbol{e}_2) \cdot \boldsymbol{k}\,. \tag{8.128}$$

Hence, choosing the direction of $\boldsymbol{k}_1$ to be the z-axis, since $\boldsymbol{k}_2 = -\boldsymbol{k}_1$, the above amplitudes can be expressed in terms of the right polarization R and the left polarization L :

$$\begin{aligned} \boldsymbol{e}_1 \cdot \boldsymbol{e}_2 &= \frac{e_{1x} + \mathrm{i}e_{1y}}{\sqrt{2}} \cdot \frac{e_{2x} - \mathrm{i}e_{2y}}{\sqrt{2}} + \frac{e_{1x} - \mathrm{i}e_{1y}}{\sqrt{2}} \cdot \frac{e_{2x} + \mathrm{i}e_{2y}}{\sqrt{2}} \\ &\propto \Phi^{\mathrm{RR}} + \Phi^{\mathrm{LL}}\,, \end{aligned} \tag{8.129}$$

$$(\boldsymbol{e}_1 \times \boldsymbol{e}_2) \cdot \boldsymbol{k}_1 \propto \Phi^{\mathrm{RR}} - \Phi^{\mathrm{LL}}\,. \tag{8.130}$$

From these, the transformation property under parity $\mathscr{P}$ is

$$\mathscr{P}(\Phi^{\mathrm{RR}} \pm \Phi^{\mathrm{LL}}) = \Phi^{\mathrm{LL}} \pm \Phi^{\mathrm{RR}} = \pm(\Phi^{\mathrm{RR}} \pm \Phi^{\mathrm{LL}})\,. \tag{8.131}$$

Case with Higher Spin

We consider the case with spin greater than 2. We write a field operator whose spin is greater than 2 as follows:

$$\varphi_{\lambda\mu\ldots}\,. \tag{8.132}$$

This should satisfy the following irreducibility conditions:

- Symmetry. The above φ is symmetric under the interchange of any two indices:

$$\varphi_{\ldots\lambda\ldots\mu\ldots} = \varphi_{\ldots\mu\ldots\lambda\ldots}\,. \tag{8.133}$$

- Rule for contractions. The contraction over any pair of indices vanishes:

$$\varphi_{\dots\lambda\dots\lambda\dots} = 0 \,. \tag{8.134}$$

- Rule for divergence. The divergence with respect to an arbitrary index vanishes:

$$\partial_\lambda \varphi_{\dots\lambda\dots} = 0 \,. \tag{8.135}$$

We thus construct scalar effective interactions using two $F_{\mu\nu}$'s and $\varphi_{\mu\nu\dots}$. In this case, we take into account the following properties:

- $F_{\mu\nu}$ and $\tilde{F}_{\mu\nu}$ are anti-symmetric under the interchange of μ and ν.
- $\partial_\mu F_{\mu\nu} = \partial_\mu \tilde{F}_{\mu\nu} = 0$.

Considering first the spin-2 case, there are two possibilities for forming scalars:

$$\text{even parity} \quad (\partial_\alpha \partial_\beta F_{\mu\nu}) F_{\mu\nu} \varphi_{\alpha\beta} \,, \qquad F_{\mu\alpha} F_{\mu\beta} \varphi_{\alpha\beta} \,, \tag{8.136}$$

$$\text{odd parity} \quad (\partial_\alpha \partial_\beta F_{\mu\nu}) \tilde{F}_{\mu\nu} \varphi_{\alpha\beta} \,, \qquad F_{\mu\alpha} \tilde{F}_{\mu\beta} \varphi_{\alpha\beta} \,. \tag{8.137}$$

These can be generalized:

$$(\partial_\alpha \partial_\beta \dots F_{\mu\nu}) F_{\mu\nu} \varphi_{\alpha\beta\dots} \,, \tag{8.138}$$

$$(\partial_\alpha \partial_\beta \dots F_{\mu\rho}) F_{\nu\rho} \varphi_{\mu\nu\alpha\beta\dots} \,. \tag{8.139}$$

Here, both spin and parity are even or odd. However, when both are odd, carrying out the integration by parts to construct the S-matrix, this contradicts the boson symmetry.[1] We thus obtain the following:

Theorem 8.3 *The two-photon decay of a neutral meson whose spin and parity are both odd is forbidden.*

If the spin is even and the parity is odd,

$$(\partial_{\alpha\beta} \dots F_{\mu\nu}) \tilde{F}_{\mu\nu} \varphi_{\alpha\beta\dots} \,. \tag{8.142}$$

[1] Due to (8.118) and (8.121), using the momentum representation, (8.138) and (8.139) can be written in the form

$$\mathscr{T}(M^0 \to 2\gamma) \propto (k-k')_\alpha (k-k')_\beta \dots : F_{\mu\nu}(k) F_{\mu\nu}(k') \varphi_{\alpha\beta\dots}(k+k') : \,, \tag{8.140}$$

$$\mathscr{T}(M^0 \to 2\gamma) \propto (k-k')_\alpha (k-k')_\beta \dots : F_{\mu\rho}(k) F_{\nu\rho}(k') \varphi_{\alpha\beta\dots}(k+k') : \,. \tag{8.141}$$

It is now straightforward to check the boson symmetry.

If the spin is odd and the parity is even,

$$(\partial_\alpha \partial_\beta \ldots F_{\mu\rho}) \tilde{F}_{\nu\rho} \varphi_{\mu\nu\alpha\beta\ldots} \, . \tag{8.143}$$

If both the spin and the parity are even, then general two-photon states are

$$\Phi^{\mathrm{RR}} + \Phi^{\mathrm{LL}} \, , \quad \Phi^{\mathrm{RL}} \, , \quad \Phi^{\mathrm{LR}} \, . \tag{8.144}$$

If the spin is even and the parity is odd, the system decays into the polarization state

$$\Phi^{\mathrm{RR}} - \Phi^{\mathrm{LL}} \, . \tag{8.145}$$

Finally, if the spin is odd and the parity is even, it decays into the polarization states

$$\Phi^{\mathrm{RL}} \, , \quad \Phi^{\mathrm{LR}} \, . \tag{8.146}$$

The above results were derived by Landau and Yang [101, 102] on the basis of angular momentum considerations.[2]

[2] Nowadays, this selection rule is called the Landau–Yang theorem.

Chapter 9
Cross-Sections and Decay Widths

In the last chapter, we discussed ways to evaluate S-matrix elements. However, in order to compare theory with experiment, we must also compute cross-sections and decay widths. We shall now focus on this aspect.

9.1 Møller's Formulas

We start with the transition probability per unit time:

$$w_{fi} = \lim_{\substack{t_1 \to -\infty \\ t_2 \to \infty}} \frac{\left|\langle f|U(t_2, t_1) - 1|i\rangle\right|^2}{t_2 - t_1} . \tag{9.1}$$

Although this ratio is in general a complicated function of t_1 and t_2, it becomes a simple one in the above limit. Since $U(t_2, t_1)$ becomes the S-matrix in this limit, using

$$\langle f|S|i\rangle = \delta_{f_i} - \mathrm{i}(2\pi)^4 \delta^4(P_f - P_i)\langle f|T|i\rangle , \tag{9.2}$$

we have

$$\left|\langle f|S - 1|i\rangle\right|^2 = (2\pi)^4 \delta^4(P_f - P_i)\left|\langle f|T|i\rangle\right|^2 \times (2\pi)^4 \delta^4(0) . \tag{9.3}$$

If we take V and t to be a quantisation volume and a macroscopic time, respectively, then

$$(2\pi)^4 \delta^4(0) = \int \mathrm{d}^4 x \, \mathrm{e}^{\mathrm{i}0\cdot x} = \int \mathrm{d}^4 x \sim Vt ,$$

$$t_2 - t_1 \sim t .$$

K. Nishijima, *Quantum Field Theory*,
https://doi.org/10.1007/978-94-024-2190-3_9

Hence, Eq. (9.1) becomes

$$w_{fi} = (2\pi)^4 \delta^4(P_f - P_i) \left| \langle f|T|i\rangle \right|^2 V \,. \tag{9.4}$$

In addition, the expression for the cross-section is

$$\sigma_{fi} = \frac{V}{v_{\rm rel}} w_{fi} = \frac{(2\pi)^4}{v_{\rm rel}} \delta^4(P_f - P_i) \left| \langle f|T|i\rangle \right|^2 V^2 \,. \tag{9.5}$$

As for a true cross-section, we must sum over the final states. For instance, if there are n particles in the final state, summing over momenta is equivalent to introducing the integral

$$\sum_f \to \frac{V}{(2\pi)^3} \int \mathrm{d}^3 p_1 \ldots \frac{V}{(2\pi)^3} \int \mathrm{d}^3 p_n \,. \tag{9.6}$$

If there is only one particle in the initial state, the width (probability) of its decay to n particles is

$$\begin{aligned} w(1 \to n) &= (2\pi)^4 \frac{V}{(2\pi)^3} \int \mathrm{d}^3 p_1 \ldots \frac{V}{(2\pi)^3} \int \mathrm{d}^3 p_n \left| \langle f|T|i\rangle \right|^2 V \delta^4(P_f - P_i) \\ &= \frac{(2\pi)^{4-3n}}{2P_0} \int \frac{\mathrm{d}^3 p_1}{2p_{10}} \cdots \int \frac{\mathrm{d}^3 p_n}{2p_{n0}} \left| \langle p_1, \ldots, p_n|\mathscr{T}|P\rangle \right|^2 \delta^4(P_f - P_i) \,, \end{aligned} \tag{9.7}$$

where the invariant decay width $\mathscr{T}$ has been defined by

$$\langle p_1, \ldots, p_n|\mathscr{T}|P\rangle = \sqrt{2p_{10}V} \ldots \sqrt{2p_{n0}V} \langle p_1, \ldots, p_n|T|P\rangle \sqrt{2P_0 V} \,. \tag{9.8}$$

This is just the same definition as (8.74).

The integral in (9.7) is Lorentz invariant, as is clear from the discussion in Chap. 4. This can be clearly seen from the relation

$$\int \frac{\mathrm{d}^3 p}{2p_0} \cdots = \int \mathrm{d}^4 p \theta(p_0) \delta(p^2 + m^2) \ldots \,. \tag{9.9}$$

Thus, the decay width in (9.7) is Lorentz invariant except for the appearance of P_0 in the denominator. The average lifetime τ of the decaying particle is given by the reciprocal of the total decay width, i.e.,

$$\frac{1}{\tau} = \sum_n w(1 \to n) \,, \tag{9.10}$$

and τ transforms as P_0. Thus, assuming that the lifetime in the rest frame is τ_0, the lifetime in a moving frame is given by

$$\tau = \frac{\tau_0}{\sqrt{1-\beta^2}} , \tag{9.11}$$

where β is the speed of the particle. Although w is the decay probability per unit time, it is equivalent to the decay width in natural units. In order to make this point clear, it can be denoted by $\Gamma(1 \to n)$.

If there are two particles in the initial state, we can speak of a scattering cross-section. Combining (9.5) with (9.6), the total cross-section for a reaction with n particles in the final state can be calculated from

$$\sigma(2 \to n) = \frac{(2\pi)^{4-3n}}{4v_{\text{rel}} P_{10} P_{20}} \int \frac{\mathrm{d}^3 p_1}{2p_{10}} \cdots \int \frac{\mathrm{d}^3 p_n}{2p_{n0}} |\langle p_1, \ldots, p_n | \mathscr{I} | P_1, P_2 \rangle|^2 \delta^4(P_f - P_i) . \tag{9.12}$$

Then in the center-of-mass frame or the laboratory frame, the relative velocity v_{rel} can be written as

$$v_{\text{rel}} = \left| \frac{\boldsymbol{P}_1}{P_{10}} - \frac{\boldsymbol{P}_2}{P_{20}} \right| . \tag{9.13}$$

We can generalize $v_{\text{rel}} P_{10} P_{20}$ to the Lorentz-invariant form

$$\begin{aligned} v_{\text{rel}} P_{10} P_{20} &= |\boldsymbol{P}_1 P_{20} - \boldsymbol{P}_2 P_{10}| \\ &\to \sqrt{(\boldsymbol{P}_1 P_{20} - \boldsymbol{P}_2 P_{10})^2 - (\boldsymbol{P}_1 \times \boldsymbol{P}_2)^2} \\ &= \sqrt{(P_1 \cdot P_2)^2 - M_1^2 M_2^2} \equiv B , \end{aligned} \tag{9.14}$$

where M_1 and M_2 are the masses of the two incident particles. This B is certainly a Lorentz-invariant quantity, and it coincides with $v_{\text{rel}} P_{10} P_{20}$ in the center-of-mass frame or the laboratory frame. Therefore, the following cross-section is a Lorentz-invariant quantity:

$$\sigma(2 \to n) = \frac{(2\pi)^{4-3n}}{4B} \int \frac{\mathrm{d}^3 p_1}{2p_{10}} \cdots \int \frac{\mathrm{d}^3 p_n}{2p_{n0}} |\langle p_1, \ldots, p_n | \mathscr{I} | P_1, P_2 \rangle|^2 \delta^4(P_f - P_i) . \tag{9.15}$$

We call this *Møller's formula* [103]. If the particles in the final state are of the same kind, then we must of course divide this equation by a suitable factor so as not to count the same state several times.

Moreover, if the initial wave is not polarized and we do not observe spins in the final states, then we have to take an average of the spin states of the initial particles, and sum over the spin states of the particles in the final states. The same thing also holds true for the decay width:

$$|\mathscr{T}_{fi}|^2 \to \frac{1}{2S_i+1}\sum_{S_i}\sum_{S_f}|\mathscr{T}_{fi}|^2 \quad \text{(decay width)} \tag{9.16}$$

$$|\mathscr{T}_{fi}|^2 \to \frac{1}{(2S_1+1)(2S_2+1)}\sum_{S_i}\sum_{S_f}|\mathscr{T}_{fi}|^2 \quad \text{(cross-section)}\,, \tag{9.17}$$

where S_i stands for the spin of the decaying particle in (9.16), S_1 and S_2 stand for the spins of the two incident particles in (9.17), and the summations over S_i and S_f are taken over the spin states in the initial and final states, respectively.

We often have a situation where there are two particles in the final state. Let us consider the integral in that case:

$$\begin{aligned} I &= \int \frac{\mathrm{d}^3 p_1}{2p_{10}} \int \frac{\mathrm{d}^3 p_2}{2p_{20}} \delta^4(P_f - P_i)\ldots \\ &= \int \mathrm{d}^4 p_1 \int \mathrm{d}^4 p_2 \theta(p_{10})\delta(p_1^2+m_1^2)\theta(p_{20})\delta(p_2^2+m_2^2)\delta^4(P_f-P_i)\ldots\,. \end{aligned} \tag{9.18}$$

To carry out this integral, we introduce the change of variables

$$p_1 = \frac{P_f}{2} + \Delta\,, \quad p_2 = \frac{P_f}{2} - \Delta\,, \quad \mathrm{d}^4 p_1 \mathrm{d}^4 p_2 = \mathrm{d}^4 P_f \mathrm{d}^4 \Delta\,. \tag{9.19}$$

Hence,

$$\begin{aligned} I &= \int \mathrm{d}^4\Delta\,\delta\left[\left(\frac{P_i}{2}+\Delta\right)^2 + m_1^2\right]\delta\left[\left(\frac{P_i}{2}-\Delta\right)^2 + m_2^2\right]\ldots \\ &= \frac{1}{2}\int \mathrm{d}^4\Delta\,\delta\left(\Delta^2 + \frac{P_i^2}{4} + \frac{m_1^2+m_2^2}{2}\right)\delta\left(P_i\cdot\Delta + \frac{m_1^2-m_2^2}{2}\right)\ldots\,. \end{aligned}$$

To evaluate this, we first go to the center-of-mass frame, with

$$P_{i0} = W = \sqrt{s} \quad \text{(total energy of the system)}\,. \tag{9.20}$$

Therefore, since $\boldsymbol{P}_i = 0$, Δ_0 can be obtained as

$$\Delta_0 = \frac{m_1^2 - m_2^2}{2W} = \frac{m_1^2-m_2^2}{2\sqrt{s}}\,. \tag{9.21}$$

From the first δ-function, we obtain

$$\vec{\Delta}^2 = \frac{(m_1^2 - m_2^2)^2}{4s} + \frac{s}{4} - \frac{m_1^2 + m_2^2}{2} = \frac{[s - (m_1 - m_2)^2][s - (m_1 + m_2)^2]}{4s} . \tag{9.22}$$

Thus, the integral itself is given by

$$\begin{aligned} I &= \frac{1}{2W} \int \mathrm{d}^3\Delta\, \delta[\Delta^2 + \cdots]\ldots \\ &= \frac{1}{2\sqrt{s}} \int \Delta^2\, \mathrm{d}\Delta\, \mathrm{d}\Omega\, \delta[\Delta^2 + \cdots]\ldots \\ &= \frac{\Delta_f}{4\sqrt{s}} \int \mathrm{d}\Omega \ldots , \end{aligned} \tag{9.23}$$

where Δ_f is the absolute value of the momentum of the final state given by (9.22). Thus,

$$\int \frac{\mathrm{d}^3 p_1}{2p_{10}} \int \frac{\mathrm{d}^3 p_2}{2p_{20}} \delta^4(P_f - P_i) \ldots = \frac{\sqrt{[s - (m_1 - m_2)^2][s - (m_1 + m_2)^2]}}{8s} \int \mathrm{d}\Omega \ldots . \tag{9.24}$$

This yields the differential cross-section in the center-of-mass system:

$$\frac{\mathrm{d}\sigma}{\mathrm{d}\Omega} = \frac{1}{(4\pi)^2 B} \frac{\Delta_f}{4\sqrt{s}} \left| \langle p_1, p_2 | \mathscr{I} | P_1, P_2 \rangle \right|^2 . \tag{9.25}$$

Incidentally, taking Δ_i to be the absolute value of the relativistic momentum in the initial state,

$$\begin{aligned} B^2 &= \frac{1}{4}\left(s - M_1^2 - M_2^2\right)^2 - M_1^2 M_2^2 \\ &= \frac{1}{4}\left[s - (M_1 - M_2)^2\right]\left[s - (M_1 + M_2)^2\right] = s\Delta_i^2 . \end{aligned} \tag{9.26}$$

Therefore, the differential cross-section in the center-of-mass system becomes

$$\frac{\mathrm{d}\sigma}{\mathrm{d}\Omega} = \frac{1}{(8\pi W)^2} \frac{\Delta_f}{\Delta_i} \left| \langle p_1, p_2 | \mathscr{I} | P_1, P_2 \rangle \right|^2 . \tag{9.27}$$

Compton Scattering Cross-Section

Next, we consider Compton scattering in the *laboratory frame*. Letting P_1 and P_2 be the four-momenta of the incident photon and the stationary electron, respectively, and p_1 and p_2 the four-momenta of the scattered photon and electron, we denote the space and time components by

$$P_1 = (\boldsymbol{P}_1, E_1^0)\,, \quad P_2 = (\boldsymbol{0}, m)\,, \quad p_1 = (\boldsymbol{p}_1, E_1)\,, \quad p_2 = (\boldsymbol{p}_2, E_2)\,. \tag{9.28}$$

Thus, the integral in phase space is

$$I = \int \frac{\mathrm{d}^3 p_1}{2E_1} \int \frac{\mathrm{d}^3 p_2}{2E_2} \delta^3(\boldsymbol{p}_1 + \boldsymbol{p}_2 - \boldsymbol{P}_1)\delta(E_1 + E_2 - E_1^0 - m) \ldots\,. \tag{9.29}$$

Integrating over $\boldsymbol{p}_2$ and taking $\boldsymbol{p}_2 = \boldsymbol{P}_1 - \boldsymbol{p}_1$, we have

$$\begin{aligned} I &= \int \frac{\mathrm{d}^3 p_1}{4E_1 E_2} \delta(E_1 + E_2 - E_1^0 - m) \ldots \\ &= \int \frac{p_1^2 \mathrm{d}p_1 \mathrm{d}\Omega_1}{4E_1 E_2} \delta(E_1 + E_2 - E_1^0 - m) \ldots\,, \end{aligned} \tag{9.30}$$

where p_1 is the absolute value of $\boldsymbol{p}_1$ and $\mathrm{d}\Omega_1$ is the differential solid angle of the momentum of the scattered photon. Since

$$E_2^2 = m^2 + (\boldsymbol{P}_1 - \boldsymbol{p}_1)^2 = m^2 + P_1^2 + p_1^2 - 2p_1 P_1 \cos\theta\,, \tag{9.31}$$

we obtain immediately

$$\frac{\partial E_1}{\partial p_1} = \frac{p_1}{E_1}\,, \quad \frac{\partial E_2}{\partial p_1} = \frac{p_1 - P_1 \cos\theta}{E_2}\,. \tag{9.32}$$

Carrying out the integration over p_1,

$$\begin{aligned} I &= \int \mathrm{d}\Omega_1 \frac{p_1^2}{4E_1 E_2 \left(\dfrac{p_1}{E_1} + \dfrac{p_1}{E_2} - \dfrac{P_1}{E_2} \cos\theta \right)} \ldots \\ &= \frac{p_1^2}{4} \int \mathrm{d}\Omega_1 \frac{1}{p_1(E_1 + E_2) - P_1 E_1 \cos\theta} \ldots\,. \end{aligned} \tag{9.33}$$

Introducing the new notation $p_1 = E_1 = \omega'$, $P_1 = E_1^0 = \omega$ and $E_1 + E_2 = \omega + m$, we can rewrite the denominator of the equation above using the Compton formula, to give

$$\omega'(\omega + m) - \omega\omega' \cos\theta = \omega'\big[m + \omega(1 - \cos\theta)\big] = \omega m\,. \tag{9.34}$$

Therefore, the above integral becomes

$$I = \frac{\omega'^2}{4\omega m} \int \mathrm{d}\Omega_1 \ldots . \tag{9.35}$$

Since in this problem $B = \omega m$, the differential cross-section for Compton scattering in the laboratory system is

$$\frac{\mathrm{d}\sigma}{\mathrm{d}\Omega_1} = \frac{1}{(8\pi m)^2} \left(\frac{\omega'}{\omega}\right)^2 \left|\langle f|\mathscr{J}|i\rangle\right|^2 . \tag{9.36}$$

Two-Particle Decay Width
Finally, applying (9.24) to the case where a particle with mass M decays into two particles with masses m_1 and m_2, the decay width is given by

$$\Gamma(1 \to 2) = \frac{1}{16\pi M^3}\sqrt{\left[M^2 - (m_1 - m_2)^2\right]\left[M^2 - (m_1 + m_2)^2\right]}\left|\mathscr{T}(i \to 2)\right|^2 . \tag{9.37}$$

In this case, of course, we must make the replacement indicated in (9.17). If the final state is a two-particle system, the integral in phase space can be carried out as above.

9.2 Examples of Cross-Sections and Decay Widths

Since we now know how to compute S-matrix elements and the formula for the cross-section, we can combine the two and derive some concrete consequences.

Compton Scattering
First, we derive an explicit expression for the cross-section in Compton scattering. We denote the four-momenta of the incident photon and the electron by k and p, respectively, and denote the scattered momenta by k' and p'. In the laboratory system,

$$\boldsymbol{p} = 0 , \quad p_0 = m . \tag{9.38}$$

Writing the polarization vectors of the photons before and after the scattering as e and e', we can assume without loss of generality that

$$p \cdot e = p \cdot e' = 0 , \tag{9.39}$$

because it has been already shown in (5.90) that, for a transverse wave, we can choose e and e' such that they have only spatial components. Moreover, since we can impose the Lorenz condition on the observed photons, we can assume that

$$k \cdot e = k' \cdot e' = 0\,. \tag{9.40}$$

Using these conditions, both $(k \cdot \gamma)$ and $(p \cdot \gamma)$ anticommute with $(e \cdot \gamma)$ and $(e' \cdot \gamma)$. Now we can simplify (8.75) using

$$\begin{aligned}\bar{u}(p')(e' \cdot \gamma)\big[-\mathrm{i}(p+k)\cdot\gamma + m\big](e\cdot\gamma)u(p) \\ &= \bar{u}(p')(e' \cdot \gamma)(e\cdot\gamma)\big[\mathrm{i}(p+k)\cdot\gamma + m\big]u(p) \\ &= \bar{u}(p')(e' \cdot \gamma)(e\cdot\gamma)(\mathrm{i}k\cdot\gamma)u(p)\end{aligned} \tag{9.41}$$

and

$$\bar{u}(p')(e\cdot\gamma)\big[-\mathrm{i}(p-k')\cdot\gamma + m\big](e'\cdot\gamma)U(p) = -\bar{u}(p)(e\cdot\gamma)(e'\cdot\gamma)(\mathrm{i}k'\cdot\gamma)u(p)\,. \tag{9.42}$$

Therefore, the invariant scattering amplitude can be written as

$$\mathscr{T}_{fi} = e^2\sqrt{2p_0'}\bar{u}(p')\left[(e'\cdot\gamma)(e\cdot\gamma)\frac{\mathrm{i}k\cdot\gamma}{2p\cdot k} + (e\cdot\gamma)(e'\cdot\gamma)\frac{\mathrm{i}k'\cdot\gamma}{2p\cdot k'}\right]u(p)\sqrt{2p_0}\,. \tag{9.43}$$

Using (3.195), we take the average over the electron spins in the initial state and sum over the electron spins in the final states. Then, by (9.17), we find

$$\begin{aligned}|\mathscr{T}_{fi}|^2 \to \frac{1}{2}e^4\mathrm{Tr}\Bigg\{&\left[(e'\cdot\gamma)(e\cdot\gamma)\frac{\mathrm{i}k\cdot\gamma}{2p\cdot k} + (e\cdot\gamma)(e'\cdot\gamma)\frac{\mathrm{i}k'\cdot\gamma}{2p\cdot k'}\right](-\mathrm{i}p\cdot\gamma + m) \\ &\times\left[\frac{\mathrm{i}k'\cdot\gamma}{2p\cdot k'}(e'\cdot\gamma)(e\cdot\gamma) + \frac{\mathrm{i}k\cdot\gamma}{2p\cdot k}(e\cdot\gamma)(e'\cdot\gamma)\right](-\mathrm{i}p'\cdot\gamma + m)\Bigg\}\,.\end{aligned} \tag{9.44}$$

Inserting this into (9.36), the differential cross-section in the laboratory system is

$$\left.\frac{\mathrm{d}\sigma}{\mathrm{d}\Omega}\right|_{\mathrm{lab}} = \frac{e^4}{128\pi^2}\frac{\omega'^2}{m^2\omega^2}\mathrm{Tr}[\ldots]\,. \tag{9.45}$$

When computing the trace, it is useful to introduce the vector

$$a = \frac{k'}{2p\cdot k'} - \frac{k}{2p\cdot k}\,, \tag{9.46}$$

which is orthogonal to p. Then,

$$(e' \cdot \gamma)(e \cdot \gamma)\frac{ik \cdot \gamma}{2p \cdot k'} + (e \cdot \gamma)(e' \cdot \gamma)\frac{ik' \cdot \gamma}{2p \cdot k'} = (e \cdot \gamma)(e' \cdot \gamma)(ia \cdot \gamma) + 2(e \cdot e')\frac{ik \cdot \gamma}{2p \cdot k} .$$

The trace now reduces to

$$\mathrm{Tr}[\ldots] = 8(e \cdot e')^2 + 2\frac{(\omega - \omega')^2}{\omega\omega'} . \tag{9.47}$$

Hence,

$$\left.\frac{d\sigma}{d\Omega}\right|_{\text{lab}} = \left(\frac{e^2}{4\pi}\right)^2 \frac{1}{4m^2}\left(\frac{\omega'}{\omega}\right)^2 \left[4(e \cdot e')^2 + \frac{(\omega - \omega')^2}{\omega\omega'}\right]. \tag{9.48}$$

This is valid when the polarizations of the incident photon and the scattered photon are both specified. Then averaging the polarizations of the incident wave and summing over those of the scattered wave, we obtain

$$(e \cdot e')^2 \to \frac{1}{2}\sum_{\lambda}\sum_{\lambda'}(e, e')^2 = \frac{1}{2}(1 + \cos^2\theta) . \tag{9.49}$$

The part without polarization vectors is simply doubled.

To derive (9.49), we argue as follows. If we choose the direction of k to be the z-direction, then we can chose the x- and the y-directions along the polarization vectors:

$$\sum_{\lambda}(\boldsymbol{a} \cdot \boldsymbol{e}^{(\lambda)})^2 = a_x^2 + a_y^2 = \boldsymbol{a}^2 - a_z^2 = \boldsymbol{a}^2 - \frac{(\boldsymbol{a} \cdot \boldsymbol{k})^2}{\boldsymbol{k}^2} .$$

Using this equation twice, we obtain (9.49). Here, θ is the scattering angle of the photon. The resulting cross-section is given finally by the Klein–Nishina formula [104]

$$\left.\frac{d\sigma}{d\Omega}\right|_{\text{lab}} = \frac{1}{2}\left(\frac{e^2}{4\pi}\right)^2 \frac{1}{[m + \omega(1 - \cos\theta)]^2}\left\{\frac{\omega^2(1 - \cos\theta)^2}{m[m + \omega(1 - \cos\theta)]} + 1 + \cos^2\theta\right\} , \tag{9.50}$$

where we have used (9.34) to express ω' in terms of ω and $\cos\theta$. In particular, if $\omega \ll m$, this reduces to the classical Thomson formula

$$\left.\frac{d\sigma}{d\Omega}\right|_{\text{lab}} = \left(\frac{e^2}{4\pi m}\right)^2 \frac{1 + \cos^2\theta}{2} = r_0^2\frac{1 + \cos^2\theta}{2} , \tag{9.51}$$

where r_0 is the classical electron radius given by

$$r_0 = \frac{e^2}{4\pi mc^2} = \frac{e^2}{4\pi\hbar c}\frac{\hbar}{mc} = 2.818 \times 10^{-13}\text{cm} \quad \text{(CGS units)} . \tag{9.52}$$

The total cross-section in this case is called the *Thomson cross-section*, given by

$$\sigma = \frac{8\pi}{3} r_0^2 = \sigma_{\text{Thomson}} = 6.65 \times 10^{-25}\text{cm}^2 . \tag{9.53}$$

Integrating (9.50) over the solid angle, we obtain the total cross-section as

$$\sigma = \frac{3}{4}\left\{\frac{1+\gamma}{\gamma^3}\left[\frac{2\gamma(1+\gamma)}{1+2\gamma} - \log(1+2\gamma)\right] + \frac{1}{2\gamma}\log(1+2\gamma) - \frac{1+3\gamma}{(1+2\gamma)^2}\right\}\sigma_{\text{Thomson}} , \tag{9.54}$$

where $\gamma = \omega/m$. Two approximations can be given:

$$\sigma = \begin{cases} \left(1 - 2\gamma + \dfrac{26}{5}\gamma^2 + \cdots\right)\sigma_{\text{Thomson}} , & \text{for } \gamma \ll 1 , \\ \dfrac{3}{8\gamma}\left[\log(2\gamma) + \dfrac{1}{2}\right]\sigma_{\text{Thomson}} , & \text{for } \gamma \gg 1 . \end{cases} \tag{9.55}$$

Pion Decay

Next, let us apply (9.37) to the decay of pions π into muons μ. We use (8.81) to compute

$$|\mathscr{T}_{fi}|^2 \to \sum_{S_f} |\mathscr{T}_{fi}|^2 , \tag{9.56}$$

where we have summed over spins in the final states:

$$\sum_{S_f} |\mathscr{T}_{fi}|^2 = g^2 \left(\frac{m_\mu}{m_\pi}\right)^2 \text{Tr}\big[(1-\gamma_5)(-\mathrm{i}p\cdot\gamma - m_\mu)(1+\gamma_5)(-\mathrm{i}q\cdot\gamma)\big] . \tag{9.57}$$

The trace here becomes

$$\text{Tr}[\ldots] = -8p\cdot q = 4(m_\pi^2 - m_\mu^2) . \tag{9.58}$$

Therefore, using (9.37), we obtain the decay width

$$\Gamma(\pi \to \mu + \nu) = \left(\frac{g^2}{4\pi}\right) m_\pi \left(\frac{m_\mu}{m_\pi}\right)^2 \left[1 - \left(\frac{m_\mu}{m_\pi}\right)^2\right] . \tag{9.59}$$

Since its reciprocal is the lifetime $\tau = 2.56 \times 10^{-8}$ s of π^+, g^2 must be given by

$$\frac{g^2}{4\pi} = 1.76 \times 10^{-15} . \tag{9.60}$$

We also assume the interaction (8.78) for $\pi^+ \to e^+ + \nu$, with the same value of g^2. This is called the *universality assumption*. We obtain the branching ratio

$$\frac{\Gamma(\pi^+ \to e^+ + \nu)}{\Gamma(\pi \to \mu^+ + \nu)} = \left(\frac{m_e}{m_\mu}\right)^2 \frac{m_\pi^2 - m_e^2}{m_\pi^2 - m_\mu^2} = 1.29 \times 10^{-4} . \tag{9.61}$$

This ratio is confirmed by experiment, and will later provide an important insight when investigating the structure of the interaction.

Lastly, we shall consider the decay $\pi^0 \to 2\gamma$. $\mathscr{T}_{fi}$ has been given in (8.103). To find its absolute value squared, we must sum over the polarization states of the two photons. Regarding $e^{(\lambda')}$ and $e^{(\lambda'')}$, we use (5.91) to sum over λ' and λ'':

$$\begin{aligned} \sum_{\lambda'}\sum_{\lambda''} \left(\mathrm{i}\epsilon_{\mu\nu\alpha\beta} e'_\mu e''_\nu k'_\alpha k''_\beta\right)^2 &= -\epsilon_{\mu\nu\alpha\beta}\epsilon_{\mu\nu\alpha'\beta'} k'_\alpha k'_{\alpha'} k''_\beta k''_{\beta'} \\ &= 2(\delta_{\alpha\beta'}\delta_{\alpha'\beta} - \delta_{\alpha\alpha'}\delta_{\beta\beta'}) k'_\alpha k'_{\alpha'} k''_\beta k''_{\beta'} \\ &= 2\left[(k' \cdot k'')^2 - k'^2 k''^2\right] \\ &= \frac{1}{2}\mu^4 . \end{aligned} \tag{9.62}$$

Here we denote the mass of π^0 and the mass of the proton by μ and m, respectively. We use (9.37) and divide by 2 so as not to count the same state twice, since there are two identical particles in the final state. Then,

$$\begin{aligned} \Gamma(\pi^0 \to 2\gamma) &= \frac{1}{2}\frac{1}{16\pi\mu}\left(\frac{1}{\pi}\right)^2\left(\frac{e^2}{4\pi}\right)^2 \frac{G^2}{m^2}\frac{\mu^4}{2} \\ &= \frac{\alpha^2}{64\pi^3} G^2 \left(\frac{\mu}{m}\right)^2 \mu , \quad \text{where } \alpha = \frac{e^2}{4\pi} \approx \frac{1}{137} . \end{aligned} \tag{9.63}$$

Taking the reciprocal,

$$\tau = 0.7 \times 10^{-15} \left(\frac{G^2}{4\pi}\right)^{-1} \text{s} . \tag{9.64}$$

If we take $G^2/4\pi = 15$, we find $\tau \approx 0.5 \times 10^{-16}$ s. However, experiment yields 2×10^{-16} s, so the agreement with theory is not very good. We will discuss this issue later in relation to the quark model.

Naturally, when there is no good agreement between theory and experiment, this leaves us with a problem, i.e., to find out whether perturbation theory is not a good approximation or the model is not good.

9.3 Inclusive Reactions

The cross-section considered up to now is the one for the exclusive reaction, which corresponds to observing all particles in the final state. On the other hand, we can also consider the case where certain particles in the final state are observed and others are not. We then refer to an *inclusive reaction*. In this section, we shall discuss how to compute the cross-section for an inclusive reaction.

We thus consider the inelastic scattering of an electron and a nucleon. As an example, we study the reaction

$$\mathrm{e}^- + \mathrm{p} \to \mathrm{e}^- + \cdots \,, \tag{9.65}$$

where we only observe the electron in the final state (see Fig. 9.1). We assume that an electron with four-momentum q exchanges a virtual photon with the proton and observe the electron whose four-momentum changes to q'. We assume that the proton absorbing the virtual photon goes into a state comprising n hadrons. We denote the four-momentum of the proton in the initial state by p, and denote the n-hadron state by $|n\rangle$. We take the electromagnetic interaction into account up to second order, splitting the Hamiltonian density as follows:

$$\mathscr{H}_{\mathrm{int}}(x) = -j_\mu(x) A_\mu(x) \,, \tag{9.66}$$

$$j_\mu = j^{\mathrm{e}}_\mu(x) + j^{\mathrm{h}}_\mu(x) \,, \tag{9.67}$$

where j^{e}_μ and j^{h}_μ are the four-current densities of the electron and the hadrons, respectively. Therefore, to lowest order, the relevant S-matrix element becomes

$$S_{fi} = (-\mathrm{i})^2 \int \mathrm{d}^4 x \int \mathrm{d}^4 y \langle q' | j^{\mathrm{e}}_\mu(x) | q \rangle D_{\mathrm{F}}(x-y) \langle n | j^{\mathrm{h}}_\mu(y) | p \rangle \,, \tag{9.68}$$

which describes elastic scattering. However, for the strong interaction, we must include higher order corrections. Denoting the Hamiltonian density of the strong

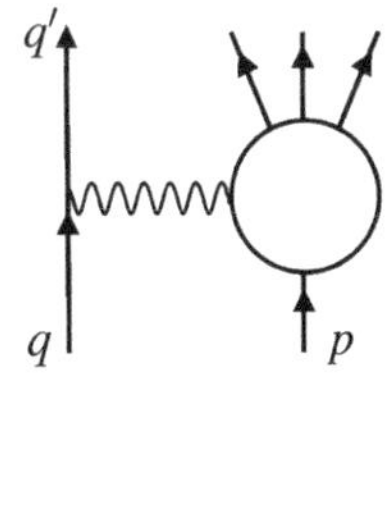

Fig. 9.1 Inclusive reaction. An electron scatters on a proton

interaction by $\mathscr{H}_s$, higher order corrections are taken into account by making the substitution

$$\langle n|j_\mu^{\rm h}(x)|p\rangle \longrightarrow \frac{\langle n|T\left\{j_\mu^{\rm h}(x)\exp\left[-{\rm i}\int {\rm d}^4y\mathscr{H}_s(y)\right]\right\}|p\rangle}{\langle 0|T\exp\left[-{\rm i}\int {\rm d}^4y\mathscr{H}_s(y)\right]|0\rangle}\,,$$

where we have dropped the bubble graphs from (8.70) and we have used the notation in (8.106). The right-hand side of this is equivalent to

$$\langle \boldsymbol{n}|\boldsymbol{j}_\mu^{\rm h}(x)|\boldsymbol{p}\rangle\,, \tag{9.69}$$

as will be shown in Chap. 11. Here the bold face indicates the Heisenberg picture, but for the time being we will write (9.69) using the usual typeface, as there is no risk of confusion. We reinterpret the matrix elements of the hadron parts in (9.68) as standing for the Heisenberg picture. We rewrite (9.68) in the momentum representation and set $V = 1$:

$$\begin{aligned}
S_{fi} &= -{\rm i}e\int {\rm d}^4x\int {\rm d}^4y\ \bar{u}(q')\gamma_\mu u(q){\rm e}^{{\rm i}(q-q')\cdot x}\left[\frac{-{\rm i}}{(2\pi)^4}\int {\rm d}^4k\frac{{\rm e}^{{\rm i}k\cdot(x-y)}}{k^2-{\rm i}\epsilon}\right]\langle n|j_\mu^{\rm h}(0)|p\rangle{\rm e}^{{\rm i}(p-n)\cdot y}\\
&= (-{\rm i})^2(2\pi)^4e\int {\rm d}^4k\ \delta^4(q-q'+k)\delta^4(p-n-k)\frac{1}{k^2-{\rm i}\epsilon}\bar{u}(q')\gamma_\mu u(q)\langle n|j_\mu^{\rm h}(0)|p\rangle\\
&= {\rm i}e(2\pi)^4\delta^4(p-n-q'+q)\frac{1}{(q'-q)^2}\bar{u}(q'){\rm i}\gamma_\mu u(q)\langle n|j_\mu^{\rm h}(0)|p\rangle\,,
\end{aligned} \tag{9.70}$$

where e stands for the electric charge of the electron, including its sign. The matrix T is now given by

$$T_{fi} = -e\frac{1}{(q'-q)^2}\bar{u}(q'){\rm i}\gamma_\mu u(q)\langle n|j_\mu^{\rm h}(0)|p\rangle\,. \tag{9.71}$$

Thus, the total cross-section is

$$\begin{aligned}
\sigma &= \frac{(2\pi)^4}{v_{\rm rel}}\int\frac{{\rm d}^3q'}{(2\pi)^3}\sum_n \delta^4(n+q'-q-p)|T_{fi}|^2\\
&= \frac{(2\pi)^4}{v_{\rm rel}}\int\frac{{\rm d}^3q'}{(2\pi)^3}\sum_n\frac{e^2}{\left[(q'-q)^2\right]^2}\delta^4(n+q'-q-p)|\bar{u}(q'){\rm i}\gamma_\mu u(q)\langle n|j_\mu^{\rm h}(0)|p\rangle|^2\,.
\end{aligned} \tag{9.72}$$

We introduce the four-vector $Q = q - q'$ and decompose the right-hand side of (9.72) into the electron part and the hadronic part:

$$\sigma = \frac{M}{2p_0q_0v_{\text{rel}}}\frac{1}{(2\pi)^3}\int\frac{\mathrm{d}^3q'}{2q'_0}\frac{e^2}{Q^4}4q_0q'_0\bar{u}(q')\mathrm{i}\gamma_\nu u(q)\cdot\bar{u}(q)\mathrm{i}\gamma_\mu u(q')$$
$$\times\sum_n\frac{p_0}{M}\langle p|j^{\text{h}}_\mu(0)|n\rangle\langle n|j^{\text{h}}_\nu(0)|p\rangle(2\pi)^4\delta^4(n-p-Q)\,, \tag{9.73}$$

where M is the proton mass. We also set $B = p_0q_0v_{\text{rel}}$. Averaging the electron spins over the initial states and summing them over the final states, we obtain

$$4q_0q'_0\bar{u}(q')\mathrm{i}\gamma_\nu u(q)\bar{u}(q)\mathrm{i}\gamma_\mu u(q') \to \frac{1}{2}\mathrm{Tr}\big[\mathrm{i}\gamma_\nu(-\mathrm{i}q\cdot\gamma+m)\mathrm{i}\gamma_\mu(-\mathrm{i}q'\cdot\gamma+m)\big]\,. \tag{9.74}$$

The value of the trace is

$$\frac{1}{2}\mathrm{Tr}[\ldots] = -2\big[\delta_{\mu\nu}(q\cdot q'+m^2) - q_\mu q'_\nu - q'_\mu q_\nu\big]\,. \tag{9.75}$$

Furthermore, the hadron part transforms as a second-rank tensor:

$$\sum_n\frac{p_0}{M}\langle p|j^{\text{h}}_\mu(0)|n\rangle\langle n|j^{\text{h}}_\nu(0)|p\rangle(2\pi)^4\delta^4(n-p-Q)$$
$$= \int\mathrm{d}^4x\frac{p_0}{M}\big\langle p|j^{\text{h}}_\mu(x)j^{\text{h}}_\nu(0)|p\big\rangle\mathrm{e}^{-\mathrm{i}Q\cdot x}\,. \tag{9.76}$$

The total cross-section is now

$$\sigma = \frac{M}{2B}\frac{1}{(2\pi)^3}\int\frac{\mathrm{d}^3q'}{2q'_0}\frac{e^2}{Q^4}\left(\frac{1}{2}\mathrm{Tr}\right)_{\mu\nu}\int\mathrm{d}^4x\frac{p_0}{M}\big\langle p|j^{\text{h}}_\mu(x)j^{\text{h}}_\nu(0)|p\big\rangle\mathrm{e}^{-\mathrm{i}Q\cdot x}\,. \tag{9.77}$$

In the following, we consider the laboratory system:

$$B = qM\,,\quad q = |\boldsymbol{q}|\,. \tag{9.78}$$

Then considering a very high energy and setting $q \approx q_0 = E$ and $q' \approx q'_0 = E'$,

$$\sigma = \frac{1}{2E}\frac{1}{(2\pi)^3}\int\frac{E'^2\mathrm{d}E'\mathrm{d}\Omega'}{2E'}\frac{e^2}{Q^4}\left(\frac{1}{2}\mathrm{Tr}\right)_{\mu\nu}\left(\int\right)_{\mu\nu}\,. \tag{9.79}$$

Inserting (9.75),

$$\frac{\mathrm{d}^2\sigma}{\mathrm{d}\Omega'\mathrm{d}E'} = \frac{1}{(2\pi)^3}\frac{E'}{2E}\frac{e^2}{Q^4}\left[q_\mu q'_\nu + q'_\mu q_\nu - \delta_{\mu\nu}(q\cdot q' + m^2)\right]\left(\int\right)_{\mu\nu} . \tag{9.80}$$

The last factor on the right-hand side is a second-rank tensor proportional to e^2, whence we may write

$$\int \mathrm{d}^4x \, \frac{p_0}{M}\langle p|j^{\mathrm{h}}_\mu(x) j^{\mathrm{h}}_\nu(0)|p\rangle \mathrm{e}^{-\mathrm{i}Q\cdot x} = 2\pi e^2 W_{\mu\nu} . \tag{9.81}$$

Here, $W_{\mu\nu}$ is a function of the two four-momenta p and Q with the general form

$$W_{\mu\nu} = W_1\left(\delta_{\mu\nu} - \frac{Q_\mu Q_\nu}{Q^2}\right) + \frac{1}{M^2}W_2\left(p_\mu - \frac{p\cdot Q}{Q^2}Q_\mu\right)\left(p_\nu - \frac{p\cdot Q}{Q^2}Q_\nu\right) . \tag{9.82}$$

Equation (9.82) has been chosen in such a way that the following relation holds, corresponding to the conservation of electric current:

$$Q_\mu W_{\mu\nu} = Q_\nu W_{\mu\nu} = 0 . \tag{9.83}$$

Note also the following relations, which hold at high energies:

$$\left[q_\mu q'_\nu + q'_\mu q_\nu - \delta_{\mu\nu}(q\cdot q' + m^2)\right]\left(\delta_{\mu\nu} - \frac{Q_\mu Q_\nu}{Q^2}\right) \approx Q^2 \approx 4EE'\sin^2\frac{\theta}{2} , \tag{9.84}$$

$$\left[q_\mu q'_\nu + q'_\mu q_\nu - \delta_{\mu\nu}(q\cdot q' + m^2)\right]\left(p_\mu - \frac{p\cdot Q}{Q^2}Q_\mu\right)\left(p_\nu - \frac{p\cdot Q}{Q^2}Q_\nu\right) \approx 2M^2EE'\cos^2\frac{\theta}{2} , \tag{9.85}$$

Substituting this into (9.80),

$$\frac{\mathrm{d}^2\sigma}{\mathrm{d}\Omega'\mathrm{d}E'} = 4\alpha^2\frac{E'^2}{Q^4}\left(2W_1\sin^2\frac{\theta}{2} + W_2\cos^2\frac{\theta}{2}\right) . \tag{9.86}$$

The quantities W_1 and W_2 are called *structure functions*. Both are scalar functions constructed from p and Q. Taking into account the fact that $p^2 = -M^2$,

$$Q^2 , \qquad p\cdot Q \equiv M\nu , \tag{9.87}$$

are independent scalar quantities, where ν is equal to the energy loss of the electron in the laboratory system:

$$\nu = E - E' . \tag{9.88}$$

According to Bjorken, if $Q^2/2M\nu = x$ is finite in the limit $Q^2 \to \infty$ and $M\nu \to \infty$, then both become functions of the ratio x alone, i.e.,

$$MW_1 \to F_1(x) , \qquad \nu W_2 \to F_2(x) . \tag{9.89}$$

This is *Bjorken's scaling law* [105], which would later provide a basis for the gauge theory of strong interactions.

9.4 Optical Theorem

In quantum mechanics, the scattering amplitude for spin-0 particles passing through a central force field can be expressed by the following partial wave expansion:

$$f(\theta) = \frac{1}{k}\sum_{l=0}^{\infty}(2l+1)e^{i\delta_l}\sin\delta_l\, P_l(\cos\theta) . \tag{9.90}$$

Here k is the modulus of the momentum and P_l is the l th Legendre polynomial. If the scattering is purely elastic, the phase shift δ_l is a real number, so the total cross-section σ satisfies

$$\sigma = \int d\Omega |f(\theta)|^2 = \frac{4\pi}{k^2}\sum_{l=0}^{\infty}(2l+1)\sin^2\delta_l = \frac{4\pi}{k}\mathrm{Im} f(0) . \tag{9.91}$$

This is called the *optical theorem*. We shall now consider its generalization to field theory.

Inserting (8.71) into (8.2), we express the unitarity of the S-matrix in terms of T:

$$\mathrm{Im}\mathscr{T}_{fi} = -\frac{(2\pi)^4}{2}\sum_j T^\dagger_{fj}\delta^4(P_f - P_i)T_{ji} . \tag{9.92}$$

Setting $f = i$,

$$\mathrm{Im}\mathscr{T}_{ii} = -\frac{(2\pi)^4}{2}\sum_j \delta^4(P_i - P_j)|T_{ji}|^2 . \tag{9.93}$$

We rewrite this relation in terms of the invariant amplitude $\mathscr{T}$:

$$\begin{aligned}\mathrm{Im}\mathscr{T}_{ii} &= -\sum_n \frac{(2\pi)^{4-3n}}{2}\int\frac{\mathrm{d}^3p_1}{2p_{10}}\cdots\int\frac{\mathrm{d}^3p_n}{2p_{n0}}\delta^4(P_i-P_n)|\mathscr{T}_{ni}|^2 \\ &= -2B\sum_{n=2}^{\infty}\sigma(i\to n)\,,\end{aligned} \tag{9.94}$$

where we have assumed that the initial state $|i\rangle$ is a two-particle state. Denoting the total cross-section by σ,

$$\mathrm{Im}\mathscr{T}_{ii} = -2B\sigma \qquad (|i\rangle \text{ a two-particle state})\,. \tag{9.95}$$

This is the optical theorem in field theory. Similarly, for the total decay width Γ, we obtain

$$\mathrm{Im}\mathscr{T}_{ii} = -P_0\Gamma \qquad (|i\rangle \text{ a one-particle state})\,. \tag{9.96}$$

Note that the optical theorem in field theory holds for inelastic scattering as well as elastic scattering.

As an example of the optical theorem, let us consider the process where an electron–positron pair annihilates to produce hadrons:

$$\mathrm{e}^- + \mathrm{e}^+ \longrightarrow \text{hadrons}\,. \tag{9.97}$$

We consider the propagators, including the lowest order for the electromagnetic interaction and higher order corrections for the strong interaction:

$$D'_{F\mu\nu} = \frac{\langle 0|T\big[A_\mu(x)A_\nu(y)\exp[-\mathrm{i}\int\mathrm{d}^4z\mathscr{H}_\mathrm{s}(z)]\big]|0\rangle}{\langle 0|T\exp\big[-\mathrm{i}\int\mathrm{d}^4z\mathscr{H}_\mathrm{s}(z)\big]|0\rangle}\,. \tag{9.98}$$

Now corresponding to $\mathscr{T}_{ii}$, we consider the process

$$\mathrm{e}^-(p) + \mathrm{e}^+(\bar{p}) \longrightarrow \mathrm{e}^-(p') + \mathrm{e}^+(\bar{p}')\,, \tag{9.99}$$

where the four-momenta are indicated in brackets. If only hadrons exist between the in and out states, the S-matrix is

$$S^{(2)} = -\frac{1}{2}\int\mathrm{d}^4x\int\mathrm{d}^4y\,:j^\mathrm{e}_\mu(x)D'_{F\mu\nu}(x-y)j^\mathrm{e}_\mu(y):\,. \tag{9.100}$$

Setting $V=1$, the T-matrix element for this process can be written as

$$T_{fi} = -e^2\bar{u}(p')\mathrm{i}\gamma_\mu v(\bar{p}')D'_{F\mu\nu}(p+\bar{p})\bar{v}(\bar{p})\mathrm{i}\gamma_\nu u(p)\,. \tag{9.101}$$

Fig. 9.2 Hadrons produced by electron–positron annihilation

Thus, the invariant scattering amplitude for forward scattering is

$$\mathscr{T}_{ii} = -e^2 D'_{F_{\mu\nu}}(p+\bar{p})(2p_0)(2\bar{p}_0)\bar{u}(p)\mathrm{i}\gamma_\mu v(\bar{p})\bar{v}(\bar{p})\mathrm{i}\gamma_\nu u(p) . \tag{9.102}$$

Averaging over the spins in the initial state,

$$\begin{aligned}(2p_0)(2\bar{p}_0)\bar{u}(p)\mathrm{i}\gamma_\mu v(\bar{p})\bar{v}(\bar{p})\mathrm{i}\gamma_\nu u(p) &\to \frac{1}{4}\mathrm{Tr}\big[\mathrm{i}\gamma_\mu(-\mathrm{i}\bar{p}\cdot\gamma - m)\mathrm{i}\gamma_\nu(-\mathrm{i}p\cdot\gamma + m)\big] \\ &= p_\mu\bar{p}_\nu + \bar{p}_\mu p_\nu + \delta_{\mu\nu}(m^2 - p\cdot\bar{p}) . \end{aligned} \tag{9.103}$$

Setting

$$s = -(p+\bar{p})^2 = 2(m^2 - p\cdot\bar{p}) , \tag{9.104}$$

we have

$$B = \sqrt{(p\cdot\bar{p} - m^4)^2} = \frac{1}{2}\sqrt{s(s-4m^2)} . \tag{9.105}$$

To consider the imaginary part of $\mathscr{T}_{ii}$, we introduce only the contribution of $D'_{F_{\mu\nu}}$ for hadrons between the in and out states (see the Feynman diagram in Fig. 9.2):

$$D'_{F_{\mu\nu}} \to \frac{1}{k^4}(k_\mu k_\nu - \delta_{\mu\nu}k^2)\sigma^{\mathrm{h}}(-k^2) = \frac{1}{k^4}\Pi^{\mathrm{h}}_{\mu\nu}(-k^2) , \quad k = p + \bar{p} . \tag{9.106}$$

This can be written in this form for basically the same reasons as were given to justify (9.82). As can be read off from the Feynman diagram, $\Pi^{\mathrm{h}}_{\mu\nu}(k^2)$ can be written in the form

$$\langle 0|T\big[j^{\mathrm{h}}_\mu(x)j^{\mathrm{h}}_\nu(y)\big]|0\rangle = \frac{-\mathrm{i}}{(2\pi)^4}\int \mathrm{d}^4k\, \mathrm{e}^{\mathrm{i}k\cdot(x-y)}\Pi^{\mathrm{h}}_{\mu\nu}(-k^2) . \tag{9.107}$$

In this regard, however, it should be remembered that the left-hand side has been defined in the Heisenberg picture. Writing

$$\sigma^{\mathrm{h}}(-k^2) = \Pi^{\mathrm{h}}(s) , \quad s = -k^2 , \tag{9.108}$$

we obtain

$$-\mathrm{Im}\mathscr{T}_{ii} = e^2 \left(p_\mu \bar{p}_\nu + \bar{p}_\mu p_\nu + \frac{1}{2}\delta_{\mu\nu} s \right) \frac{1}{s^2}(k_\mu k_\nu - \delta_{\mu\nu} k^2)\mathrm{Im}\Pi^{\mathrm{h}}(s)$$
$$= e^2 \frac{s + 2m^2}{s} \mathrm{Im}\Pi^{\mathrm{h}}(s) \, . \tag{9.109}$$

Therefore, from the optical theorem the total cross-section of (9.97) is

$$\sigma(\mathrm{e}^- + \mathrm{e}^+ \to \text{hadrons}) = \frac{1}{2B}(-\mathrm{Im}\mathscr{T}_{ii}) = \frac{e^2}{s} \frac{s + 2m^2}{\sqrt{s(s - 4m^2)}} \mathrm{Im}\Pi^{\mathrm{h}}(s) \, . \tag{9.110}$$

In particular, assuming $s \gg m^2$ at high energies,

$$\sigma(\mathrm{e}^- + \mathrm{e}^+ \to \text{hadrons}) = \frac{e^2}{s} \mathrm{Im}\Pi^{\mathrm{h}}(s) \, . \tag{9.111}$$

We shall see later that measurements of this cross-section can tell us how many types of quarks (flavours) there are.

9.5 Three-Body Decays

In Sect. 9.1, we discussed the integral in phase space when the final state is a two-particle state. In this section, we consider the integral for the three-particle state. Although this sort of integral is complicated in many cases, we will discuss several tractable examples here.

To begin with, we consider the decay of the μ-meson:

$$\mu^+ \to \mathrm{e}^+ + \nu + \bar{\nu} \, . \tag{9.112}$$

We shall not be concerned whether or not $\bar{\nu}$ appearing here is an antiparticle of ν because it is not relevant to the following discussion. In this process, the only quantity observed in the final state is the electron. We will discuss the form of the electron energy spectrum. The reaction (9.112) can be described by what is known as the *Fermi interaction*. The invariant decay width $\mathscr{T}_{fi}$ can be expressed in terms of four Dirac spinors and suitable γ-matrices. If we average or sum spins, then the following can be expressed in terms of four Casimir operators and traces of products of suitable γ-matrices:

$$\sum_{S_i} \sum_{S_f} |\mathscr{T}_{fi}|^2 \, . \tag{9.113}$$

Therefore, the result must be linear scalar functions of p_ν, $p_{\bar{\nu}}$, and p_e, or m_e, p_μ, and m_μ. The possible combinations are:

$$m_\mu m_e(p_\nu \cdot p_{\bar{\nu}}), \quad (p_\mu \cdot p_e)(p_\nu \cdot p_{\bar{\nu}}), \quad (p_\mu \cdot p_\nu)(p_e \cdot p_{\bar{\nu}}), \quad (p_\mu \cdot p_{\bar{\nu}})(p_e \cdot p_\nu) . \tag{9.114}$$

Inserting these into the formula for the decay width, we carry out the integration. Since the μ-meson is about 200 times more massive than the electron, the speed of the emerging electron is close to the speed of light, so m_e can be neglected. Then we do not need to consider the first term in (9.114). The third and fourth terms will yield the same results if we integrate over p_ν and $p_{\bar{\nu}}$ without observing the neutrino, so we cannot tell the difference between them as long as we discuss possible shapes of the energy spectrum. Hence, in order to determine the form of the energy spectrum, it is enough to consider only the two terms

$$(p_\mu \cdot p_e)(p_\nu \cdot p_{\bar{\nu}}) , \quad (p_\mu \cdot p_\nu)(p_e \cdot p_{\bar{\nu}}) . \tag{9.115}$$

The parameter which determines the form of the spectrum is simply a ratio of their coefficients, and chosen in a suitable way, it is known as the *Michel parameter* [106].

Up to a multiplicative constant, the decay rate is given by

$$\begin{aligned} \Gamma &\propto \int \frac{\mathrm{d}^3 p_e}{2p_{e0}} \int \frac{\mathrm{d}^3 p_\nu}{2p_{\nu 0}} \int \frac{\mathrm{d}^3 p_{\bar{\nu}}}{2p_{\bar{\nu}0}} \delta^4(P - p_e - p_\nu - p_{\bar{\nu}}) \langle |\mathscr{T}_{fi}|^2 \rangle \\ &= \int \frac{\mathrm{d}^3 p_\nu}{2p_{e0}} \int \mathrm{d}^4 Q\, \delta^4(P - p_e - Q) \int \frac{\mathrm{d}^3 p_\nu\, \mathrm{d}^3 p_{\bar{\nu}}}{2p_{\nu 0}\, 2p_{\bar{\nu}0}} \delta^4(Q - p_\nu - p_{\bar{\nu}}) \langle |\mathscr{T}_{fi}|^2 \rangle , \end{aligned} \tag{9.116}$$

where $\langle |\mathscr{T}_{fi}|^2 \rangle$ is averaged over the spins. It will be a linear combination of the two terms in (9.115). In order to integrate over the neutrino momenta, we must evaluate

$$\begin{aligned} I_{\alpha\beta} &= \int \frac{\mathrm{d}^3 p_\nu}{2p_{\nu 0}} \int \frac{\mathrm{d}^3 p_{\bar{\nu}}}{2p_{\bar{\nu}0}} \delta^4(Q - p_\nu - p_{\bar{\nu}})(p_\nu)_\alpha (p_{\bar{\nu}})_\beta \\ &= \int \mathrm{d}^4 p_\nu \int \mathrm{d}^4 p_{\bar{\nu}} \delta_+(p_\nu^2) \delta_+(p_{\bar{\nu}}^2) \delta^4(Q - p_\nu - p_{\bar{\nu}})(p_\nu)_\alpha (p_{\bar{\nu}})_\beta , \end{aligned} \tag{9.117}$$

where we have used the notation $\delta_+(p^2) = \theta(p_0)\delta(p^2)$. We make the change of variables

$$p_\nu = \frac{Q'}{2} + \Delta , \quad p_{\bar{\nu}} = \frac{Q'}{2} - \Delta . \tag{9.118}$$

Therefore, the integral is now over $d^4Q'd^4\Delta$. Now integrating over Q', we can take $Q' = Q$ in the remaining integral,

$$I_{\alpha\beta} = \int d^4\Delta\delta\left[\left(\frac{Q}{2}+\Delta\right)^2\right]\delta\left[\left(\frac{Q}{2}-\Delta\right)^2\right]\left(\frac{Q}{2}+\Delta\right)_\alpha\left(\frac{Q}{2}-\Delta\right)_\beta$$

$$= \int d^4\Delta\delta\left(\frac{Q^2}{4}+\Delta^2\right)\delta(2Q\cdot\Delta)\left(\frac{Q}{2}+\Delta\right)_\alpha\left(\frac{Q}{2}-\Delta\right)_\beta , \qquad (9.119)$$

where we have assumed that the integration is carried out only in the region $Q_0 > 2|\Delta_0|$. Since Q is a time-like vector, we choose the direction of Q to be the time axis. Then integrating $\delta(Q\cdot\Delta)$ with respect to Δ_0, we can take

$$\Delta_0 = 0 \qquad (9.120)$$

in the remaining integral and only terms of even order in $\vec{\Delta}$ survive:

$$I_{\alpha\beta} = \frac{1}{2Q_0}\int \frac{\vec{\Delta}^2 d|\vec{\Delta}^2|d\Omega}{2|\vec{\Delta}|}\delta\left(\frac{Q^2}{4}+\vec{\Delta}^2\right)\left(\frac{1}{4}Q_\alpha Q_\beta - \Delta_\alpha\Delta_\beta\right) . \qquad (9.121)$$

By symmetry, the integral in the second term survives only when $\alpha = \beta \neq 4$, so we can make the following replacement:

$$\Delta_\alpha\Delta_\beta \to \frac{1}{3}\delta_{\alpha\beta}(1-\delta_{\alpha 4})\vec{\Delta}^2 = \frac{1}{3}\left(\delta_{\alpha\beta} - \frac{Q_\alpha Q_\beta}{Q^2}\right)\vec{\Delta}^2$$

$$= -\frac{1}{12}(\delta_{\alpha\beta}Q^2 - Q_\alpha Q_\beta). \qquad (9.122)$$

Integrating the first term and the remaining one,

$$I_{\alpha\beta} = \frac{\pi}{24}(\delta_{\alpha\beta}Q^2 + 2Q_\alpha Q_\beta) . \qquad (9.123)$$

Inserting this result into (9.116), we are left with two integrals:

$$\int \frac{d^3p}{2p_0}\int d^4Q\,\delta^4(P-p-Q)(P\cdot p)(4Q^2+2Q^2) , \qquad (9.124)$$

$$\int \frac{d^3p}{2p_0}\int d^4Q\,\delta^4(P-p-Q)\big[(P\cdot p)Q^2 + 2(P\cdot Q)(p\cdot Q)\big] , \qquad (9.125)$$

where we have taken $p = p_e$ and $P = p_\mu$. We consider the rest frame of the μ-meson, i.e.,

$$P_0 = m_\mu , \quad \boldsymbol{P} = 0 . \qquad (9.126)$$

The energy of the electron attains a maximum value of $m_\mu/2$ when the neutrinos are emitted in opposite directions, and since we are assuming $m_e = 0$, we can set

$$p_0 = |\boldsymbol{p}| = \frac{1}{2} m_\mu x \, , \tag{9.127}$$

where the possible values of x lie between 0 and 1, and

$$\frac{d^3 p}{2 p_0} \propto x \, dx \, . \tag{9.128}$$

Then the x-dependences of the scalar products in the integral are

$$P \cdot p = -\frac{1}{2} m_\mu^2 x \, , \quad -Q^2 = -P^2 + 2p \cdot P = m_\mu^2 (1 - x) \, ,$$

$$P \cdot Q = P^2 - p \cdot P = m_\mu^2 \left(-1 + \frac{x}{2} \right) \, , \quad p \cdot Q = p \cdot P = -\frac{1}{2} m_\mu^2 x \, .$$

After a trivial integration over the solid angle, we now obtain

$$\Gamma \propto \int_0^1 dx \, x^2 (a + bx) \, , \tag{9.129}$$

whence the spectrum is a linear combination of x^2 and x^3. Normalizing the integral to unity and leaving the ratio of two suitable linear combinations of x^2 and x^3 as a parameter, the energy spectrum of the electron becomes

$$dN = 4x^2 \left[3(1 - x) + \frac{2}{3} \rho (4x - 3) \right] dx \, . \tag{9.130}$$

This has been chosen so that the first term vanishes when $x = 1$ and the integral over x from 0 to 1 is unity. Carrying out the integral over x, the second term vanishes, and its coefficient ρ is called the *Michel parameter*. The value of ρ can be determined from the form of the spectrum when $x \sim 1$. The value of ρ depends on the choice of the γ-matrices in the Fermi four-fermion interaction. For the vector–axial vector (VA) type, $\rho = 0.75$ and for the scalar–pseudoscalar (SP) type, $\rho = 0$. This will be discussed later.

In association with the three-body decay, another important idea is the *Dalitz plot* [43, 107]. Dalitz proposed this to determine the parity and spin of the K^+-meson from the energy distribution of the π-meson in the decay process

$$K^+ \to \pi^+ + \pi^+ + \pi^- \, . \tag{9.131}$$

The decay width is given by

$$\Gamma(\mathrm{K}^+ \to \pi^+ + \pi^+ + \pi^-) \tag{9.132}$$
$$= \frac{(2\pi)^{-5}}{2M} \int \frac{\mathrm{d}^3 p_1}{2p_{10}} \int \frac{\mathrm{d}^3 p_2}{2p_{20}} \int \frac{\mathrm{d}^3 p_3}{2p_{30}} \delta^4(P_f - P_i) |\mathscr{T}(\mathrm{K}^+ \to \pi^+ + \pi^+ + \pi^-)|^2 ,$$

where P_i is the four-momentum of the K^+ meson, p_1 and p_2 are the four-momenta of the two π^+ mesons, and p_3 is the four-momentum of the π^- meson. These satisfy the relation $P_f = p_1+p_2+p_3$. In order not to count the same state twice, we restrict the integration to the region

$$p_{10} < p_{20} \, . \tag{9.133}$$

$|\mathscr{T}|^2$ is a function of the scalar products $(p_i \cdot p_j)$. Only two of them are independent so we choose

$$p_1 \cdot P_i \, , \qquad p_3 \cdot P_i \, , \tag{9.134}$$

as independent variables. In the rest frame of the K-meson, these scalar products can be written as $-M\omega_1$ and $-M\omega_3$, respectively. Here, M is the mass of the K-meson and ω_1 and ω_3 are the energies of the π-mesons in the rest frame of the K-meson. From the law of conservation of energy,

$$\omega_1 + \omega_2 + \omega_3 = M \, . \tag{9.135}$$

Since $|\mathscr{T}|^2$ can be viewed as a function of ω_1 and ω_3, the integrals in (9.132) can be written in the form

$$I = \int \frac{\mathrm{d}^3 p_1}{\omega_1} \int \frac{\mathrm{d}^3 p_2}{\omega_2} \int \frac{\mathrm{d}^3 p_3}{\omega_3} \delta^3(\boldsymbol{p}_1 + \boldsymbol{p}_2 + \boldsymbol{p}_3) \delta(\omega_1 + \omega_2 + \omega_3 - M) f(\omega_1, \omega_3)$$
$$= \int \mathrm{d}\omega_1' \int \mathrm{d}\omega_3' \, K(\omega_1', \omega_3') f(\omega_1', \omega_3') \, , \tag{9.136}$$

where

$$K(\omega_1', \omega_3') = \int \frac{p_1^2 \mathrm{d}p_1 \mathrm{d}\Omega_1}{\omega_1} \frac{p_3^2 \mathrm{d}p_3 \mathrm{d}\Omega_3}{\omega_3} \frac{\delta(\omega_1 + \omega_2 + \omega_3 - M)}{\omega_2} \delta(\omega_1 - \omega_1') \delta(\omega_3 - \omega_3') \, . \tag{9.137}$$

Here $p_1 = |\boldsymbol{p}_1|$ and $p_3 = |\boldsymbol{p}_3|$. We begin by integrating with respect to ω_3. Choosing the direction of $\boldsymbol{p}_1$ as the z-direction and taking the angle between $\boldsymbol{p}_1$

and $\boldsymbol{p}_3$ as θ,

$$\int \mathrm{d}\Omega_1 \int \mathrm{d}\Omega_3 \ldots = 8\pi^2 \int_{-1}^{1} \mathrm{d}(\cos\theta) \ldots . \tag{9.138}$$

We have left $\cos\theta$ in the above because, from the law of conservation of momentum, ω_2 is a function of $\cos\theta$:

$$\begin{aligned} &\omega_1^2 = p_1^2 + \mu^2 , \quad \omega_3^2 = p_3^2 + \mu^2 , \\ &\omega_2^2 = (\boldsymbol{p}_1 + \boldsymbol{p}_3)^2 + \mu^2 = p_1^2 + p_3^2 + 2p_1p_3\cos\theta + \mu^2 . \end{aligned} \tag{9.139}$$

In order to change the integration variables from p_1, p_3, and $\cos\theta$ to ω_1, ω_2, and ω_3, we use the Jacobian

$$\frac{\partial(\omega_1, \omega_2, \omega_3)}{\partial(p_1, p_3, \cos\theta)} = \frac{p_1^2 p_3^2}{\omega_1\omega_2\omega_1} . \tag{9.140}$$

Hence,

$$\begin{aligned} K(\omega_1', \omega_3') = 8\pi^2 &\int \mathrm{d}p_1\mathrm{d}p_2\mathrm{d}(\cos\theta)\frac{\partial(\omega_1, \omega_2, \omega_3)}{\partial(p_1, p_3, \cos\theta)} \\ &\times \delta(\omega_1 + \omega_2 + \omega_3 - M)\delta(\omega_1 - \omega_1')\delta(\omega_3 - \omega_3') \\ = 8\pi^2 &\int \mathrm{d}\omega_1\mathrm{d}\omega_2\mathrm{d}\omega_3\delta(\omega_1 + \omega_2 + \omega_3 - M)\delta(\omega_1 - \omega_1')\delta(\omega_3 - \omega_3') . \end{aligned} \tag{9.141}$$

If the values of ω_1' and ω_3' are physically realizable, then this integral yields $8\pi^2$. If not, it yields zero. All unnecessary terms can thus be completely dropped. This is an amazing simplification. The equation (9.132) for the decay width becomes

$$\Gamma(\mathrm{K}^+ \to \pi^+ + \pi^+ + \pi^-) = \frac{(4\pi)^{-3}}{M} \int_{\omega_1 < \omega_2} \mathrm{d}\omega_1\mathrm{d}\omega_3 \left|\mathscr{T}(\mathrm{K}' \to \pi^+ + \pi^+ + \pi^-)\right|^2 . \tag{9.142}$$

What is important here is the proportionality relation

$$\frac{\mathrm{d}^2\Gamma}{\mathrm{d}\omega_1\mathrm{d}\omega_3} \propto |\mathscr{T}|^2 . \tag{9.143}$$

Thus, plotting examples of each decay as points in the (ω_1,ω_3) plane, the density of the plot turns out to be proportional to the modulus squared of the decay width, so that $|\mathscr{T}|^2$ can be found from experimental results as a function of ω_1 and ω_3.

Dalitz showed how to produce this plot inside an equilateral triangle as follows. Defining the kinetic energy of each π-meson as $T_i = \omega_i - \mu$, energy conservation implies

$$T_1 + T_2 + T_3 = M - 3\mu \equiv Q\,. \tag{9.144}$$

Then, picking one point inside the triangle and drawing perpendiculars from that point to the three edges, the sum of the perpendiculars will be equal to the height of the triangle. We choose the height of the triangle as Q in suitable units. Then we express an example decay observed by experiment as an internal point so that the lengths of each perpendicular connecting the point with each edge are equal to T_1, T_2, and T_3, respectively. This gives what is known as the Dalitz plot.

Since the value of Q is about 75 MeV, the pions emitted in the decay process can be treated as non-relativistic to a good approximation. Let us investigate what kind of plot we obtain in this case. The laws of conservation of energy and momentum are

$$p_1^2 + p_2^2 + p_3^2 = 2\mu Q\,, \tag{9.145}$$

$$\boldsymbol{p}_1 + \boldsymbol{p}_2 + \boldsymbol{p}_3 = 0\,. \tag{9.146}$$

This means that p_1, p_2, and p_3 must satisfy the triangle inequality

$$(p_1 + p_3)^2 \geq p_2^2 \geq (p_1 - p_3)^2\,, \tag{9.147}$$

or

$$2p_1p_2 \geq \left|p_2^2 - p_1^2 - p_3^2\right|\,.$$

Writing this in terms of T,

$$4T_1T_3 \geq (T_2 - T_1 - T_3)^2 = (Q - 2T_1 - 2T_3)^2\,. \tag{9.148}$$

This describes the interior of an ellipse in an oblique coordinate system. However, since this diagram is symmetric in T_1, T_2, and T_3, it must remain invariant under a rotation of 120 degrees around the center of mass of the regular triangle. This means it has to be a circle. Hence, this is the interior of the circle inscribed inside an equilateral triangle. In Fig. 9.3, the lengths of the perpendiculars connecting the point in the interior of the circle with the three edges of the triangle T_1, T_2 and T_3, respectively. Due to the condition $T_1 < T_2$, points are restricted to lie within the semicircle.

Dalitz's plot was originally proposed to determine the spin and parity of K^+. To do so, we must determine the spin and parity of the final state. We assume that the relative orbital angular momentum of the two π^+ particles is $\boldsymbol{l}$, and that the relative

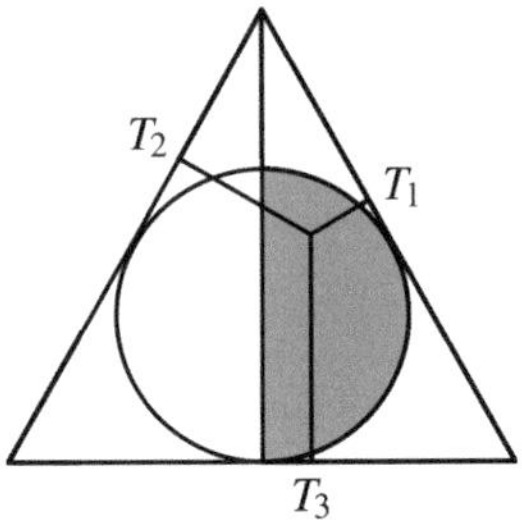

Fig. 9.3 Dalitz plot

Table 9.1 Smallest combinations of (l, l') for combinations of spins and parities in several kinds of 3π system

Spins and parities in 3π-system	Smallest value of (l, l')
0^+	Nothing
0^-	(0,0)
1^+	(0,1)
1^-	(2,2)
2^+	(2,1)
2^-	(0,2), (2,0)
$\vdots$	$\vdots$

orbital angular momentum of the center of the two π^+ particles and π^- is $\boldsymbol{l}'$. Since the spin of π is zero, denoting the spin of K^+ by $\boldsymbol{J}$, we have

$$\boldsymbol{J} = \boldsymbol{l} + \boldsymbol{l}' \,. \tag{9.149}$$

1. **Spinless K Meson.** In this case, since $J = 0$, we obviously have $l = l'$. Then the parity of the system with three π-mesons is $(-1)^{l+l'} = 1$ from the orbital part. The intrinsic parity of π itself is -1, so it will be $(-1)^3 = -1$ for three pions. The product of these two is thus -1. Therefore, if the K meson is spinless, the parity is odd and it can be denoted by 0^-.
2. **General Case.** From the Bose symmetry for particles of the same kind, l is even. Table 9.1 gives the smallest combinations of (l, l') for combinations of spins and parities in several kinds of 3π system. It should be obvious that the parity is given by $(-1)^{l'+1}$.

Since the value of Q for this decay is small, viz., 75 MeV, the large decay width corresponding to (l, l') becomes small due to the centrifugal force. Denoting the relative momenta corresponding to l and l' by p and q,

$$\mathscr{T} \sim (pR)^l (qR)^{l'} \,, \tag{9.150}$$

where R is the range of the force among the π mesons. Since the density of the Dalitz plot is shown experimentally to be homogeneous,

$$|\mathscr{T}|^2 \sim \text{const.} \tag{9.151}$$

Thus, $l = l' = 0$ and we find that the 3π-system is in the 0^- state. If K^+ is spinless, the process

$$K^+ \to \pi^+ + \gamma$$

is forbidden, as discussed in Sect. 8.7. Indeed, this decay pattern has not been observed in experiments.

Similarly, given the density of the Dalitz plot for the process $\eta^0 \to \pi^+ + \pi^- + \pi^0$ and the existence of the process $\eta^0 \to 2\gamma$, the neutral meson η^0 turns out to be 0^-.

We observe that K^+ can also decay into two π mesons, viz.,

$$K^+ \to \pi^+ + \pi^0 \,. \tag{9.152}$$

Since the spin of K^+ is zero, the relative angular momentum between π^+ and π^0 must be zero. The spin and the parity of this final state with two π mesons can be written as 0^+. This contradicts the result for the 3π decay, which gave 0^-. The decay patterns of (9.152) and (9.131) were originally referred to as θ and τ, respectively, and this contradiction was known as the θ–τ puzzle. Although θ and τ were originally considered to be different particles, more accurate detections showed that their mass and lifetime were actually the same. Furthermore, it gradually transpired that the ratio of the frequencies of the two reactions (decay to three vs. decay to two pions) is a constant, independent of the experimental conditions for their creation, so it was finally accepted that they must be same particles. Then Lee and Yang suggested that parity might not be conserved in weak interactions. To check this idea, they proposed several kinds of experiment [45], and in 1956 it was eventually proven that their idea was correct.

Chapter 10
Discrete Symmetries

In Chap. 7, we discussed Noether's theorem, which shows how a conservation law can be derived from the invariance of a theory under a continuous transformation group. These conservation laws hold in both classical and quantum theories. In contrast, discrete symmetries such as space inversion and charge conjugation are specific to quantum theory, where there are several kinds of conservation law that do not arise classically. Since physically observable quantities always vary continuously in classical theory, discrete symmetries are not connected to conservation laws. However, given one classical solution, this kind of symmetry leads us to another classical solution.

10.1 Symmetries and Unitary Transformations

When we discussed the invariance of a theory under a group of continuous transformations in Chap. 7, we worked from the invariance under an infinitesimal transformation group. In quantum theory, for a transformation like this, the state vector transforms under a unitary transformation generated as a product of infinitesimal transformations:

$$\Phi \to U\Phi = \exp(\mathrm{i}\lambda Q)\Phi \ , \tag{10.1}$$

where Q is a conserved quantity derived by Noether's theorem and represented by a Hermitian operator, and λ is an arbitrary real number. Therefore, U is a unitary operator.

A discrete unitary transformation can be expressed by a unitary operator that does not have a generator like this. Charge conjugation, discussed in Sects. 3.9 and 8.6, can be represented by the discrete unitary transformation $\mathscr{C}$. This transformation does not change the Lagrangian density, nor the form of the equation of motion,

K. Nishijima, *Quantum Field Theory*,
https://doi.org/10.1007/978-94-024-2190-3_10

and it commutes with the S-matrix. Thus, $\mathscr{C}$ is a conserved quantity. In QED, for example, the field operators transform according to

$$\mathscr{C}^{-1}\psi\mathscr{C} = C\bar{\psi}\ , \quad \mathscr{C}^{-1}\bar{\psi}\mathscr{C} = C^{-1}\psi\ , \quad \mathscr{C}^{-1}A_\mu\mathscr{C} = -A_\mu\ , \tag{10.2}$$

under this transformation. As already discussed in Sect. 3.9, the Dirac matrix C for charge conjugation is uniquely determined up to a complex phase. Denoting C for the spinor field ψ_a by C_a and using a standard matrix C, we may write

$$C_a = \eta_a C\ , \quad |\eta_a| = 1\ . \tag{10.3}$$

When a system possesses charge conjugation invariance, the phase η_a is chosen in such a way that the Lagrangian density is invariant under the charge conjugation transformation. In a system which is not invariant under any choice of the phase, charge conjugation is not a symmetry. Incidentally, the relations in (10.2) imply

$$\mathscr{C}^{-2}\psi\mathscr{C}^2 = \psi\ , \quad \mathscr{C}^{-2}\bar{\psi}\mathscr{C}^2 = \bar{\psi}\ , \quad \mathscr{C}^{-2}A_\mu\mathscr{C}^2 = A_\mu\ , \tag{10.4}$$

that is, $\mathscr{C}^2$ commutes with all field operators. Thus, we can deduce that $\mathscr{C}^2$ is a c-number. By a suitable choice of phase for $\mathscr{C}$ itself, we can thus arrange for

$$\mathscr{C}^2 = 1\ . \tag{10.5}$$

In general, charge conjugation is conserved under strong and weak interactions.

In a similar way, space inversion $\mathscr{P}$ is also conserved in all the fundamental interactions except the weak interaction, i.e., the form of the Lagrangian density and the equation of motion remain invariant, and $\mathscr{P}$ also commutes with the S-matrix. As already stated in (3.128), for the spinor field ψ_a,

$$\mathscr{P}^{-2}\psi_a(\boldsymbol{x},t)\mathscr{P} = \epsilon_a\gamma_4\psi_a(-\boldsymbol{x},t)\ , \quad |\epsilon_a| = 1\ . \tag{10.6}$$

Implementing the space inversion twice, this just becomes a phase transformation:

$$\mathscr{P}^{-1}\psi(\boldsymbol{x},t)\mathscr{P}^2 = \epsilon_a^2\psi(\boldsymbol{x},t)\ . \tag{10.7}$$

If a theory is invariant under a phase transformation like this, then redefining a suitable combination of the original $\mathscr{P}$ in (10.6) and this phase transformation to give a new $\mathscr{P}$, we have

$$\mathscr{P}^{-2}\psi(\boldsymbol{x},t)\mathscr{P}^2 = \psi(\boldsymbol{x},t)\ . \tag{10.8}$$

In this case, the following restriction is imposed on ϵ_a :

$$\epsilon_a^2 = 1\ , \quad \epsilon_a = \pm 1\ . \tag{10.9}$$

This ϵ_a is called the *intrinsic parity* of the spinor field labeled by a. We chose the intrinsic parity of both the proton and the neutron to be $+1$ in Sect. 6.4. In addition, since $\mathscr{P}^2$ itself commutes with all field operators, we can adjust the overall phase so that

$$\mathscr{P}^2 = 1 . \tag{10.10}$$

If the square of a unitary operator is unity, as in (10.5) and (10.10), this operator is also Hermitian. The intrinsic parity of one spinor field can be fixed freely if the system is invariant under a group of continuous phase transformations, in which case the phase of parity can be absorbed by the continuous transformations. If the system is invariant under a number of phase transformations, then we can correspondingly fix the intrinsic parities of an equal number of spinor fields. A useful convention is

$$\epsilon_{\mathrm{n}} = \epsilon_{\mathrm{p}} = \epsilon_{\Lambda} = 1 . \tag{10.11}$$

This convention is a slight extension of the one mentioned above and is associated with the Sakata model. In the quark model, this convention comes down to the convention that all quarks have intrinsic parity $+1$.

Generalizing the above discussion, we introduce a unitary transformation $\mathscr{U}$ corresponding to one of the discrete symmetries. This must commute with the S-matrix:

$$\mathscr{U}^{-1} S \mathscr{U} = S . \tag{10.12}$$

Now consider two eigenstates of $\mathscr{U}$:

$$\mathscr{U}|a\rangle = \epsilon_a |a\rangle , \quad \mathscr{U}|b\rangle = \epsilon_b |b\rangle . \tag{10.13}$$

If $\epsilon_a \neq \epsilon_b$, the transition $a \to b$ is prohibited, i.e.,

$$\langle b|S|a\rangle = 0 , \quad \text{if } \epsilon_a \neq \epsilon_b . \tag{10.14}$$

This is a generalization of Furry's theorem. As a property of the unitary transformation, taking

$$|a'\rangle = \mathscr{U}|a\rangle , \quad |b'\rangle = \mathscr{U}|b\rangle , \tag{10.15}$$

we obtain

$$\langle b'|a'\rangle = \langle b|a\rangle . \tag{10.16}$$

10.2 Parity of Antiparticles

When discussing charge conjugation and space inversion, an important issue is the commutation relation. We assume that the vacuum is simultaneously an eigenstate of both operators with eigenvalue 1. This is a consequence of the assumption that the vacuum is not degenerate:

$$\mathscr{P}|0\rangle = \mathscr{C}|0\rangle = |0\rangle \, . \tag{10.17}$$

In the following, we adopt the convention

$$\mathscr{P}^2 = \mathscr{C}^2 = 1 \, . \tag{10.18}$$

We now carry out these two transformations in two different orders on a Dirac field, and compare the results:

$$\begin{aligned}\mathscr{P}^{-1}\mathscr{C}^{-1}\psi(\boldsymbol{x},t)\mathscr{C}\mathscr{P} &= \mathscr{P}^{-1}\big[\eta C\bar{\psi}^{\mathrm{T}}(\boldsymbol{x},t)\big]\mathscr{P} \\ &= \eta\epsilon C\big[\bar{\psi}(-\boldsymbol{x},t)\gamma_4\big]^{\mathrm{T}} \\ &= \eta\epsilon C\gamma_4^{\mathrm{T}}\bar{\psi}^{\mathrm{T}}(-\boldsymbol{x},t) \\ &= -\eta\epsilon\gamma_4 C\bar{\psi}^{\mathrm{T}}(-\boldsymbol{x},t)\end{aligned}$$

and

$$\begin{aligned}\mathscr{C}^{-1}\mathscr{P}^{-1}\psi(\boldsymbol{x},t)\mathscr{P}\mathscr{C} &= \mathscr{C}^{-1}\big[\epsilon\gamma_4\psi(-\boldsymbol{x},t)\big]\mathscr{C} \\ &= \eta\epsilon\gamma_4 C\bar{\psi}^{\mathrm{T}}(-\boldsymbol{x},t) \\ &= -\mathscr{P}^{-1}\mathscr{C}^{-1}\psi(\boldsymbol{x},t)\mathscr{C}\mathscr{P} \, .\end{aligned} \tag{10.19}$$

Note that, since $\bar{\psi}$ is a 1×4 matrix, the transpose $\bar{\psi}^{\mathrm{T}}$ is a 4×1 matrix. Generalizing (10.19),

$$\mathscr{C}^{-1}\mathscr{P}^{-1}\psi(x_1)\ldots\psi(x_n)\mathscr{P}\mathscr{C} = (-1)^n\mathscr{P}^{-1}\mathscr{C}^{-1}\psi(x_1)\ldots\psi(x_n)\mathscr{C}\mathscr{P} \, . \tag{10.20}$$

It is possible for $\bar{\psi}$ to appear in the product of these n operators. Multiplying the vacuum state by (10.20) and denoting the state containing n Dirac particles by $|n\rangle$,

$$\mathscr{C}\mathscr{P}|n\rangle = (-1)^n\mathscr{P}\mathscr{C}|n\rangle \, . \tag{10.21}$$

We assume that the eigenstate of $\mathscr{P}$ and its eigenvalue are $|n\rangle$ and $\epsilon(n)$, respectively. Moreover, taking $\mathscr{C}|n\rangle = |\bar{n}\rangle$, we have

$$\mathscr{P}|\bar{n}\rangle = (-1)^n \epsilon(n)|\bar{n}\rangle \ . \tag{10.22}$$

Thus, considering the state with one Dirac particle ($n = 1$), we observe that the intrinsic parities of $|1\rangle$ and $|\bar{1}\rangle$ have opposite signs. Assuming that a Bose particle can transit to a state with an even number of Fermi particles, it must have the same intrinsic parity as its antiparticle. Therefore, we obtain the following theorem:

Theorem 10.1 *A Bose particle and its antiparticle have the same intrinsic parity, while a Fermi particle and its antiparticle have opposite intrinsic parities.*

If N is the number of Fermi particles, (10.21) implies the commutation relation

$$\mathscr{C}\mathscr{P} = \mathscr{P}\mathscr{C}(-1)^N = (-1)^N \mathscr{P}\mathscr{C} \ . \tag{10.23}$$

Since $(-1)^N$ commutes with both $\mathscr{C}$ and $\mathscr{P}$, considering the group generated by $\mathscr{C}$, $\mathscr{P}$, and $(-1)^N$, we note that all raised to the second power are equal to 1 and that $(-1)^N$ thus belongs to the center of the group.

π^+–π^- System

Next, we consider a two-particle system in the center-of-mass frame so that it is in an eigenstate of $\mathscr{P}$. We express this state in terms of the creation operator and act on it with $\mathscr{C}$ and $\mathscr{P}$:

$$\Phi = \varphi(\boldsymbol{p})\varphi^\dagger(-\boldsymbol{p})|0\rangle \ , \tag{10.24}$$

$$\mathscr{C}\Phi = \varphi^\dagger(\boldsymbol{p})\varphi(-\boldsymbol{p})|0\rangle \ , \tag{10.25}$$

$$\mathscr{P}\Phi = \varphi(-\boldsymbol{p})\varphi^\dagger(\boldsymbol{p})|0\rangle = \varphi^\dagger(\boldsymbol{p})\varphi(-\boldsymbol{p})|0\rangle \ , \tag{10.26}$$

where $\mathscr{C}^{-1} = \mathscr{C}$ and we have used the following:

$$\mathscr{C}^{-1}\varphi(x)\mathscr{C} = \varphi^\dagger(x) \ , \quad \mathscr{C}^{-1}\varphi^\dagger(x)\mathscr{C} = \varphi(x) \ . \tag{10.27}$$

From these, we immediately obtain

$$\mathscr{C}^{-1}\Phi = \mathscr{P}\Phi \ , \quad \mathscr{C}\mathscr{P}\Phi = \Phi \ , \tag{10.28}$$

i.e., the π^+–π^- system is in an eigenstate of $\mathscr{C}\mathscr{P}$ with eigenvalue 1. Assuming that a neutral K meson is an eigenstate of $\mathscr{C}\mathscr{P}$, it has two components, one corresponding to the eigenvalue $+1$ and one corresponding to the eigenvalue -1. These are denoted by K^0_{S} or K^0_{L}, respectively. The former has a short lifetime ($\sim 10^{-10}$ s), and the latter has a long lifetime ($\sim 5 \times 10^{-8}$ s), as indicated by the subscripts S and L, respectively. Thus, if $\mathscr{C}\mathscr{P}$ is conserved, the component with a long lifetime cannot decay into two π mesons. However, in 1964 Christenson, Cronin, Fitch, and Turlay observed this process, and thereby demonstrated that $\mathscr{C}\mathscr{P}$ is not conserved [108]:

$$K_L^0 \longrightarrow \pi^+ + \pi^- . \tag{10.29}$$

^1S and ^3S Positronium States

An electron and positron can form a bound state called *positronium*, which is unstable and decays into a number of photons. Writing this state as Φ, the parity is determined by the quantum L of their relative orbital angular momentum, i.e.,

$$\mathscr{P}\Phi = (-1)^L \epsilon(e^-)\epsilon(e^+)\Phi = (-1)^{L+1}\Phi , \tag{10.30}$$

where we have used Theorem 10.1. Therefore, in the S-state ($L = 0$), the parity is odd (-1). In order to investigate the eigenvalues of the charge conjugation, we write Φ in the form

$$\Phi = a^\dagger(\boldsymbol{p}, \sigma_z, e)a^\dagger(-\boldsymbol{p}, \sigma_z', -e)|0\rangle , \tag{10.31}$$

where $a^\dagger$ is the creation operator for a particle whose momentum, spin, and electric charge are specified. Therefore,

$$\begin{aligned}\mathscr{C}a^\dagger(\boldsymbol{p}, \sigma_z, e)a^\dagger(-\boldsymbol{p}, \sigma_z', -e)|0\rangle &= a^\dagger(\boldsymbol{p}, \sigma_z, -e)a^\dagger(-\boldsymbol{p}, \sigma_z', e)|0\rangle \\ &= -a^\dagger(-\boldsymbol{p}, \sigma_z', e)a^\dagger(\boldsymbol{p}, \sigma_z, -e)|0\rangle \\ &= -P_x P_\sigma \Phi , \end{aligned}\tag{10.32}$$

where P_x and P_σ are operators that swap the coordinates and spins, respectively, of the two particles. The two-particle system can be in a spin triplet state $S = 1$ or a spin singlet state $S = 0$. Since these correspond to a symmetric spin state and an anti-symmetric spin state, respectively, we have

$$P_x\Phi = (-1)^L\Phi , \quad P_\sigma\Phi = (-1)^{S+1}\Phi . \tag{10.33}$$

Therefore, inserting (10.33) into (10.32),

$$\mathscr{C}\Phi = (-1)^{L+S}\Phi . \tag{10.34}$$

Of course, the positronium state is only ever realized as a sum over states like (10.31) with a variety of different values of $\boldsymbol{p}$. Thus, for positronium states ^3S and ^1S,

$$\mathscr{C}\Phi(^3\mathrm{S}) = -\Phi(^3\mathrm{S}) , \quad \mathscr{C}\Phi(^1\mathrm{S}) = \Phi(^1\mathrm{S}) . \tag{10.35}$$

The former is a vector and the latter is a pseudoscalar. The decay pattern predicted from Furry's theorem is

$$^3\mathrm{S} \to 3\gamma\,, \quad ^1\mathrm{S} \to 2\gamma\,. \tag{10.36}$$

10.3 Isospin Parity and G-Conjugation

We define the isospin parity T as a rotation in isospin space:

$$T = \exp(\mathrm{i}\pi I_2)\,. \tag{10.37}$$

Clearly,

$$T^{-1} I_3 T = -I_3\,. \tag{10.38}$$

Since I_3 is half of the difference between the number of protons and the number of neutrons in a nucleus, it is diagonalized. Thus, T can only be simultaneously diagonalized when the eigenvalue of I_3 for a certain state is zero. In this case, the quantum number I is an integer, so all eigenvalues of I_2 become integers. Thus, it turns out that we can make the replacement

$$T^2 = \exp(2\pi\mathrm{i}I_2) \quad \longrightarrow \quad 1\,. \tag{10.39}$$

In this case, the eigenvalues of T are

$$T = \pm 1\,, \tag{10.40}$$

which is the isospin parity of the given nucleus. This quantity can be defined only for a nucleus in which the numbers of protons and neutrons are the same.

For a system with one nucleus, T has the representation

$$T = \exp\left(\frac{1}{2}\mathrm{i}\pi\tau_2\right) = \mathrm{i}\tau_2 = \begin{pmatrix} 0 & 1 \\ -1 & 0 \end{pmatrix}\,. \tag{10.41}$$

Using this matrix, we can swap the proton and the neutron.

Theorem 10.2 *The Kroll–Foldy theorem. The isospin parity of the state with* $I_3 = 0$ *is given by*

$$T = (-1)^I\,. \tag{10.42}$$

Proof Considering the isospin eigenstate $|I', I'_3\rangle$, the theorem states that

$$T|I', 0\rangle = (-1)^{I'}|I', 0\rangle \ . \tag{10.43}$$

To prove this, it is enough to prove the algebraically equivalent equation

$$e^{i\pi L_y} Y_{l,0} = (-1)^l Y_{l,0} \ , \tag{10.44}$$

where $Y_{l,0}$ are spherical harmonics, i.e., common eigenstates of I^2 and I_3. However, the operator corresponding to T is a rotation through π radians about the y axis, so

$$x \to -x \ , \quad y \to y \ , \quad z \to -z \ . \tag{10.45}$$

Thus,

$$\cos\theta \to -\cos\theta \ . \tag{10.46}$$

Replacing $Y_{l,0}$ by the Legendre polynomials $P_l(\cos\theta)$, we obtain the same equation as (10.44):

$$e^{i\pi L_y} P_l(\cos\theta) = P_l(-\cos\theta) = (-1)^l P_l(\cos\theta) \ . \tag{10.47}$$

■

According to (10.45), the three components of the isospin transform under T as

$$T^{-1} I_1 T = -I_1 \ , \quad T^{-1} I_2 T = I_2 \ , \quad T^{-1} I_3 T = -I_3 \ . \tag{10.48}$$

Transforming the field theoretical expression (7.46) for the isospin of nucleons under charge conjugation, we obtain

$$\mathscr{C}^{-1} \hat{I} \mathscr{C} = \mathscr{C}^{-1} \left(\frac{1}{2} \int d^3x \, \psi^\dagger \hat{\tau} \psi \right) \mathscr{C} = -\frac{1}{2} \int d^3x \, \psi^\dagger \hat{\tau}^T \psi \ . \tag{10.49}$$

Now, using the Pauli representation for $\hat{\tau}$,

$$\mathscr{C}^{-1} I_1 \mathscr{C} = -I_1 \ , \quad \mathscr{C}^{-1} I_2 \mathscr{C} = I_2 \ , \quad \mathscr{C}^{-1} I_3 \mathscr{C} = -I_3 \ . \tag{10.50}$$

This result also holds for the isospin of the meson. Comparing this with (10.48), the product $\mathscr{C}T$ commutes with the three isospin components. Combining $\mathscr{C}$ and T into G, we define G-conjugation by

$$G = \mathscr{C}T = \mathscr{C} e^{i\pi I_2} \ . \tag{10.51}$$

Table 10.1 Transformation properties of quantities specifying particle states

	Angular momentum	Momentum	Coordinate	Charge
Charge conjugation	+	−	−	+
Space inversion	+	−	−	+
Time reversal	−	−	+	+

Therefore,

$$G^{-1}\hat{I}G = \hat{I} \ . \tag{10.52}$$

For the meson field,

$$G^{-1}\hat{\varphi}G = -\hat{\varphi} \ . \tag{10.53}$$

As just described, all isospin states of the π meson are eigenstates of G with eigenvalue -1. This is useful, since in contrast with the charge conjugation case it can be used not only for the neutral case but also for the charged states. In this regard, however, the law of conservation of isospin is true only for the nuclear force. It is broken by electromagnetic interactions, and the same goes for G-invariance.

On the other hand, in the case of strong interactions, Furry's theorem holds for G-invariance:

$$\begin{array}{cccccccc} & \rho \rightarrow & \pi + & \pi \ , & \rho \not\rightarrow & \eta + & \pi \ , \\ G & + & - & - & + & + & - \ . \end{array}$$

The former reaction is realized and the latter is forbidden. This is thus another example of Furry's theorem.

Although the transformations discussed above, namely, space inversion, charge conjugation, and G-conjugation, are unitary transformations, time reversal is an *antiunitary transformation*. This is the subject of the next section. For later convenience, Table 10.1 shows how several quantities used to specify elementary particle states are affected by the various transformations. This table will be useful in the next section.

10.4 Antiunitary Transformations

Charge conjugation and space inversion discussed in the previous section are unitary transformations, but time reversal is an antiunitary transformation. Denoting charge conjugation, space inversion, and time reversal by C, P, and T, respectively, several kinds of product transformation can be defined and categorized as follows:

$$\begin{aligned} &\text{unitary transformations} \quad C\,,\ P\,,\ CP\,,\\ &\text{antiunitary transformations}\ T\,,\ CT\,,\ PT\,,\ CPT\,. \end{aligned}$$

We first discuss the general characteristics of unitary and antiunitary transformations:

1. Unitary transformation. This is defined by

$$\Phi \to \Phi' = \mathscr{U}\Phi\ , \tag{10.54}$$

where $\mathscr{U}$ is a unitary operator. The inner product of two state vectors is invariant, i.e., expressing the inner product by a bracket, we have

$$(\Phi', \Psi') = (\Phi, \Psi)\ . \tag{10.55}$$

2. Antiunitary transformation. This is defined by

$$\Phi \to \Phi' = \mathscr{U}^{\mathrm{T}}\Phi^* \equiv \Phi^\star\ , \tag{10.56}$$

which is clearly a generalization of complex conjugation and has thus been indicated by the notation $\star$. In this case,

$$(\Phi', \Psi') = (\Phi^\star, \Psi^\star) = (\Psi, \Phi)\ . \tag{10.57}$$

This equation clearly holds true because $\star$ is a generalization of complex conjugation. This transformation clearly converts a ket vector into a bra vector.

We can also define the transformation of an operator by

$$(\Phi^\star, Q\Psi^\star) = (\Psi, Q^\star\Psi)\ . \tag{10.58}$$

We can thus choose to apply the antiunitary transformation to the state vector or to the operator, which corresponds to the difference between the Schrödinger picture and the Heisenberg picture, where we choose to put the time dependence on the state vector or on the operator. Since in the Heisenberg picture the form of the equation of motion is the same for both the quantum and classical theories, the latter choice is useful in the sense that knowledge of classical theory can be used. Moreover, it is straightforward to show that

$$Q^\star = \mathscr{U}Q^{\mathrm{T}}\mathscr{U}^{-1}\ . \tag{10.59}$$

Hence, for operator transformations, we have

$$(aA + bB)^\star = aA^\star + bB^\star \quad \text{(linearity)} \tag{10.60}$$

and

$$(AB)^\star = B^\star A^\star \; , \tag{10.61}$$

where a and b are c-numbers. This linearity does not hold for transformations of state vectors, where coefficients are replaced by their complex conjugates. This is another advantage of considering operator transformations.

Theorem 10.3 *If an operator Q corresponding to a physical quantity satisfies $Q^\star = \pm Q$ and if Φ is an eigenstate of Q with eigenvalue q, then $\Phi^\star$ is also an eigenstate of Q and its eigenvalues are $\pm q$.*

Proof We have

$$(Q\Phi^\star, \Psi^\star) = (\Phi^\star, Q\Psi^\star) = (\Psi, Q^\star\Phi) = \pm(\Psi, Q\Phi) = \pm q(\Psi, \Phi) = \pm q(\Phi^\star, \Psi^\star) \, .$$

Therefore, since $\Psi^\star$ is an arbitrary state,

$$Q\Phi^\star = \pm q\Phi^\star \; .$$

■

If a system is invariant under an antiunitary transformation $\star$, then we have

$$S^\star = S \; . \tag{10.62}$$

In the Schrödinger picture, we decompose the Hamiltonian into the free part and the interacting part:

$$H = H_\mathrm{f} + H_\mathrm{int} \; . \tag{10.63}$$

The condition for invariance (10.62) is satisfied if

$$H_\mathrm{f}^\star = H_\mathrm{f} \; , \qquad H_\mathrm{int}^\star = H_\mathrm{int} \; . \tag{10.64}$$

In order to prove this, we change to the interaction picture. In this picture, the interaction Hamiltonian is

$$H_\mathrm{int}(t) = \mathrm{e}^{\mathrm{i}tH_\mathrm{f}} H_\mathrm{int} \mathrm{e}^{-\mathrm{i}tH_\mathrm{f}} \; . \tag{10.65}$$

Thus,

$$\begin{aligned} H_{\text{int}}(t)^\star &= \mathrm{e}^{-\mathrm{i}t H_\mathrm{f}^\star} H_{\text{int}}^\star \mathrm{e}^{\mathrm{i}t H_\mathrm{f}^\star} \\ &= \mathrm{e}^{-\mathrm{i}t H_\mathrm{f}} H_{\text{int}} \mathrm{e}^{\mathrm{i}t H_\mathrm{f}} \\ &= H_{\text{int}}(-t) \ . \end{aligned} \tag{10.66}$$

This will be our starting point. We write Dyson's formula in the form

$$S = T_t \exp\left[-\mathrm{i}\int \mathrm{d}t \ H_{\text{int}}(t)\right] , \tag{10.67}$$

where T_t indicates that we take into account the ordering with respect to the parameter t. Since the ordering is reversed when we apply $\star$, we introduce the operator $\tilde{T}_t$:

$$\begin{aligned} S^\star &= \tilde{T}_t \exp\left[-\mathrm{i}\int \mathrm{d}t \ H_{\text{int}}(t)^\star\right] \\ &= \tilde{T}_t \exp\left[-\mathrm{i}\int \mathrm{d}t \ H_{\text{int}}(-t)\right] \\ &= T_{t'} \exp\left[-\mathrm{i}\int \mathrm{d}t' H_{\text{int}}(t')\right] = S \ , \end{aligned} \tag{10.68}$$

where we have used $t' = -t$ and $\tilde{T}_t = T_{-t}$.

Although invariance under the antiunitary transformation does not imply the selection rule, the principle of detailed balance can be derived, as discussed in Sect. 6.4. Equation (10.68) implies that

$$(\Phi^\star, S\Psi^\star) = (\Psi, S\Phi) \ , \tag{10.69}$$

so summing over spin states,

$$\sum_{S_i}\sum_{S_f} \left|\mathscr{T}(i \to f)\right|^2 = \sum_{S_i}\sum_{S_f} \left|\mathscr{T}(f^\star \to i^\star)\right|^2 . \tag{10.70}$$

Combining this with (9.27), we obtain a relation between differential cross-sections for inverse processes:

1. Invariance under T and TP :

$$\frac{\dfrac{\mathrm{d}}{\mathrm{d}\Omega}\sigma(a+b \to c+d)}{\dfrac{\mathrm{d}}{\mathrm{d}\Omega}\sigma(c+d \to a+b)} = \frac{(2S_c+1)(2S_d+1)}{(2S_a+1)(2S_b+1)} \left(\frac{p_{cd}}{p_{ab}}\right)^2 , \tag{10.71}$$

where S_a stands for the spin of particle a, etc., and p_{ab} (p_{cd}) are the moduli of the relative momenta between particles a and b (c and d) in the center-of-mass systems. Equation (6.88) used to determine the spin of the π meson or pion is a special case of (10.71), derived on the assumption of invariance under the time reversal of strong interactions.

2. Invariance under CT and CPT :

$$\frac{\dfrac{\mathrm{d}}{\mathrm{d}\Omega}\sigma(a+b\to c+d)}{\dfrac{\mathrm{d}}{\mathrm{d}\Omega}\sigma(\bar{c}+\bar{d}\to\bar{a}+\bar{b})}=\frac{(2S_c+1)(2S_d+1)}{(2S_a+1)(2S_b+1)}\left(\frac{p_{cd}}{p_{ab}}\right)^2 . \tag{10.72}$$

10.5 CPT Theorem

We now discuss the CPT theorem or Pauli–Lüders theorem, a theorem about antiunitary transformations which always holds true in a local field theory. This theorem was given by Lüders in [109] and by Pauli in [110]. It attracted a lot of attention in 1956, when it was discovered that parity is not always conserved.

In this section, $\star$ stands for the CPT transformation. It is defined by the following equations:

$$\psi(x)^\star = \mathrm{i}\gamma_5\psi(-x) , \quad \bar{\psi}(x)^\star = \mathrm{i}\bar{\psi}(-x)\gamma_5 \qquad \text{(spinor)} , \tag{10.73}$$

$$\varphi_{\lambda_1\ldots\lambda_n}(x)^\star = (-1)^n\varphi_{\lambda_1\ldots\lambda_n}(-x) \qquad \text{(tensor)} . \tag{10.74}$$

Here the field operators are in time-dependent representations, such as the Heisenberg picture or the interaction picture.

Theorem 10.4 *CPT theorem. Any local field theory is invariant under the CPT transformation.*

The content of this theorem can be expressed by

$$\mathscr{L}(x)^\star = \mathscr{L}(-x) , \tag{10.75}$$

or more specifically by

$$\mathscr{L}_\mathrm{f}(x)^\star = \mathscr{L}_\mathrm{int}(-x) , \quad \mathscr{L}_\mathrm{int}(x)^\star = \mathscr{L}_\mathrm{int}(-x) . \tag{10.76}$$

To prove this, we consider a simple Lagrangian density:

$$\mathscr{L}_\mathrm{int}(x) = g\bar{\psi}_{\alpha_1}(x)\ldots\bar{\psi}_{\alpha_n}(x)O^{\lambda_1\ldots\lambda_n}_{\alpha_1\ldots\alpha_n\beta_1\ldots\beta_n}\psi_{\beta_1}(x)\ldots\psi_{\beta_n}(x)\varphi_{\lambda_1\ldots\lambda_m}(x) , \tag{10.77}$$

where we have assumed that there are as many occurrences of $\bar{\psi}$ as of ψ, and O is a product of the Dirac matrices, e.g., if $m = n$,

$$(\gamma_{\lambda_1})_{\alpha_1\beta_1} \cdots (\gamma_{\lambda_n})_{\alpha_n\beta_n}(1+\gamma_5) \,. \tag{10.78}$$

Furthermore, $\varphi_{\lambda_1 \ldots \lambda_m}$ is a bosonic part and if $\lambda_1, \ldots, \lambda_m$ are vector indices, it can be a product of multi-operators, e.g.,

$$\varphi_{\alpha\beta\gamma} = \varphi_\alpha(x)\varphi_\beta(x)\frac{\partial\varphi(x)}{\partial x_\gamma} \,. \tag{10.79}$$

Clearly, however, we always have

$$\varphi_{\lambda_1 \ldots \lambda_m}(x)^\star = (-1)^m \varphi_{\lambda_1 \ldots \lambda_m}(-x) \,. \tag{10.80}$$

The order of spinor operators is reversed under CPT. Assuming that the bosonic part is defined by the normal product, the ordering makes no difference under CPT. Restoring the order of the spinor part introduces a sign factor $(-1)^n$. Thus, in order to prove (10.75), we must prove that

$$(-1)^{m+n}(\mathrm{i}\gamma_5)_{\alpha_1\alpha'_1} \cdots (\mathrm{i}\gamma_5)_{\alpha_n\alpha'_n} O^{\lambda_1 \ldots \lambda_m}_{\alpha'_1 \ldots \alpha'_n \beta'_1 \ldots \beta'_n} (\mathrm{i}\gamma_5)_{\beta'_1\beta_1} \cdots (\mathrm{i}\gamma_5)_{\beta'_n\beta_n} = O^{\lambda_1 \ldots \lambda_m}_{\alpha'_1 \ldots \alpha'_n \beta'_1 \ldots \beta'_n} \,. \tag{10.81}$$

To do this, we use the following:

$$\mathrm{i}\gamma_5(1 \text{ or } \gamma_5)\mathrm{i}\gamma_5 = -(1 \text{ or } \gamma_5) \qquad (n-m) \text{ combinations}\,,$$
$$\mathrm{i}\gamma_5(\gamma_\mu \text{ or } \gamma_\mu\gamma_5)\mathrm{i}\gamma_5 = \gamma_\mu \text{ or } \gamma_\mu\gamma_5 \; m \text{ combinations}\,.$$

Therefore, the change in the overall sign is $(-1)^{n-m} = (-1)^{n+m}$, which proves (10.81).

Next, we consider the case where the number of occurrences of ψ and $\bar{\psi}$ in the interaction Lagrangian are different. For example,

$$\mathscr{L}_{\text{int}}(x) = g\bar{\psi}(x)\mathrm{i}\gamma_5 C\bar{\psi}^{\mathrm{T}}(x)\varphi(x) + \text{h.c.} \tag{10.82}$$

It is clear from the discussion about charge conjugation that $C\bar{\psi}^{\mathrm{T}}$ transforms in the same way as ψ under a proper Lorentz transformation. Hence, (10.82) is invariant under a proper Lorentz transformation. Then, taking

$$\psi'(x) = C\bar{\psi}^{\mathrm{T}}(x) \,, \tag{10.83}$$

the fields ψ' and ψ transform in the same way under the CPT transformation. This can be proven as follows:

$$\begin{aligned}\psi'(x)^\star &= \left[C\bar{\psi}^{\mathrm{T}}(x)\right]^\star = \mathrm{i}C\left[\bar{\psi}(-x)\gamma_5\right]^{\mathrm{T}} = \mathrm{i}C\gamma_5^{\mathrm{T}}\bar{\psi}^{\mathrm{T}}(-x)\\ &= \mathrm{i}\gamma_5 C\bar{\psi}^{\mathrm{T}}(-x) = \mathrm{i}\gamma_5\psi'(-x)\ . \end{aligned} \tag{10.84}$$

Thus, the assumption that the numbers of ψ and $\bar{\psi}$ are the same is not necessary if one rewrites $\bar{\psi}$ as ψ'. Let us examine what we have said above regarding the transformation properties of $C\bar{\psi}^{\mathrm{T}}$.

Transformation Properties of $C\bar{\psi}^{\mathrm{T}}$

In Sect. 3.6, we looked carefully at the way ψ transforms under proper Lorentz transformations. Now we investigate the transformation of $C\bar{\psi}^{\mathrm{T}}$. From (3.116),

$$C\bar{\psi}^{\mathrm{T}} \to C\left(\bar{\psi}S^{-1}\right)^{\mathrm{T}} = C(S^{\mathrm{T}})^{-1}\bar{\psi}^{\mathrm{T}}\ . \tag{10.85}$$

We consider first an infinitesimal transformation (3.119), whence

$$S^{\mathrm{T}} = 1 + \frac{1}{2}\epsilon_{\mu\nu}T_{\mu\nu}^{\mathrm{T}}\ . \tag{10.86}$$

Using (3.121),

$$CS^{\mathrm{T}}C^{-1} = 1 + \frac{1}{2}\epsilon_{\mu\nu}CT_{\mu\nu}^{\mathrm{T}}C^{-1} = 1 - \frac{1}{2}\epsilon_{\mu\nu}T_{\mu\nu} = S^{-1}\ . \tag{10.87}$$

Thus, since this relation should also hold true for a finite proper Lorentz transformation, we have

$$C(S^{\mathrm{T}})^{-1}\bar{\psi}^{\mathrm{T}} = C(S^{\mathrm{T}})^{-1}C^{-1}C\bar{\psi}^{\mathrm{T}} = S\left(C\bar{\psi}^{\mathrm{T}}\right)\,, \tag{10.88}$$

i.e., for a proper Lorentz transformation,

$$C\bar{\psi}^{\mathrm{T}} \to\ S\left(C\bar{\psi}^{\mathrm{T}}\right)\,. \tag{10.89}$$

This shows that it obeys the same transformation law as ψ.

In contrast, for space inversion, (10.6) implies

$$C\bar{\psi}^{\mathrm{T}} \to\ \epsilon C(\bar{\psi}\gamma_4)^{\mathrm{T}} = \epsilon C\gamma_4^{\mathrm{T}}\bar{\psi}^{\mathrm{T}} = -\epsilon\gamma_4\left(C\bar{\psi}^{\mathrm{T}}\right)\,. \tag{10.90}$$

The sign is different compared with the transformation of ψ under space inversion. This confirms Theorem 10.1 in Sect. 10.2.

Origin of the Name 'CPT Transformation'

We now prove that the combination of the three transformations C, P, and T yields $\star$. Since for each transformation the phase is not necessarily determined, we apply the simplest one. Since time reversal swaps the creation and annihilation of a particle, it turns out that $\psi \rightleftarrows \bar{\psi}$ and $t \to -t$. Furthermore, the sign of the spatial components of the four-current $\mathrm{i}\bar{\psi}\gamma_\mu\psi$ is reversed. Thus, the time reversal of the spinor becomes

$$T: \qquad \psi(\boldsymbol{x}, t) \to \bar{\psi}(\boldsymbol{x}, -t)\gamma_4\gamma_5 C \;, \quad \bar{\psi}(\boldsymbol{x}, t) \to C^{-1}\gamma_5\gamma_4\psi(\boldsymbol{x}, -t) \;, \tag{10.91}$$

up to a phase. Next, consecutively carrying out space inversion and then charge conjugation,

$$PT: \qquad \psi(\boldsymbol{x}, t) \to \bar{\psi}(-\boldsymbol{x}, -t)\gamma_5 C \;, \quad \bar{\psi}(\boldsymbol{x}, t) \to C^{-1}\gamma_5\psi(-\boldsymbol{x}, -t) \;, \tag{10.92}$$

and

$$CPT: \qquad \psi(\boldsymbol{x}, t) \to (\gamma_5 C)^{\mathrm{T}} C^{-1}\psi(-\boldsymbol{x}, -t) = -\gamma_5\psi(-\boldsymbol{x}, -t) \;. \tag{10.93}$$

Similarly,

$$CPT: \qquad \bar{\psi}(\boldsymbol{x}, t) \to \bar{\psi}(-\boldsymbol{x}, -t)\gamma_5 \;. \tag{10.94}$$

Apart from the undetermined factor, this coincides with $\star$ in (10.73). This is the reason why $\star$ is called the CPT transformation. Although the CPT transformation in (10.73) is uniquely determined, the phase of each transformation is not necessarily uniquely determined.

From the fact that the product of these three transformations is always invariant, it follows that, if one of them does not make the theory invariant, at least one more must also break the invariance. The non-conservation of parity was discovered in 1956, implying the violation of invariance under space inversion. It then turned out that charge conjugation invariance was also broken. At the time, it was expected that invariance under CP would hold. But as already mentioned when discussing the decay pattern in (10.29), this too turned out to be incorrect.

Transformation Properties of the Hamiltonian

The energy–momentum four-vector is

$$P_\nu = \int \mathrm{d}^3x\, \mathscr{T}_{0\nu}(x) \;. \tag{10.95}$$

Under the CPT transformation,

$$\mathscr{T}_{\mu\nu}(x)^{\star} = \mathscr{T}_{\mu\nu}(-x)\ . \tag{10.96}$$

Thus,

$$P_{\nu}^{\star} = P_{\nu}\ . \tag{10.97}$$

At $t = 0$ or in the Schrödinger picture, this implies

$$H_{\mathrm{f}}^{\star} = H_{\mathrm{f}}\ , \qquad H_{\mathrm{int}}^{\star} = H_{\mathrm{int}}\ . \tag{10.98}$$

Since (10.58) can be derived from (10.64), this equation ensures that the S-matrix is invariant under the CPT transformation.

Equality of Particle and Antiparticle Masses

We use Theorem 10.2 in Sect. 10.3. Assuming that Φ is a state containing a particle of mass m, $\Phi^{\star}$ is a state containing its antiparticle, i.e.,

$$(P_{\nu}^2 + m^2)\Phi = 0\ . \tag{10.99}$$

Therefore, from (10.97),

$$(P_{\nu}^2 + m^2)\Phi^{\star} = 0\ . \tag{10.100}$$

These equations imply that the masses of a particle and its antiparticle are the same.

Equality of Particle and Antiparticle Lifetimes

We have

$$\mathscr{T}(a \to a) = \mathscr{T}(a^{\star} \to a^{\star})\ , \tag{10.101}$$

due to the invariance of S. Taking the imaginary part as an application of the optical theorem to the decay process, it turns out that the total decay widths of a particle and its antiparticle are the same, whence their lifetimes are also the same.

Chapter 11
Green's Functions

So far we have talked a lot about the properties of the S-matrix, but there is an important concept closely related to the S-matrix, namely, the concept of Green's functions. We first define Green's functions in the interaction picture, and then rewrite them in the Heisenberg picture. This leads to the so-called *Gell-Mann–Low relation* [111]. We then discuss Matthews' theorem [112] for the relation between the Hamiltonian formalism and the Lagrangian formalism. Finally, we turn our attention to the reduction formula connecting the S-matrix and Green's functions. This formula relates to fundamental conditions called *asymptotic conditions* in the Heisenberg picture.

11.1 Gell-Mann–Low Relation

As in Chap. 3, writing the free-field energy–momentum four-vector in the interaction picture as $P_\mu^{(0)}$, we find that (3.29) holds true for a field operator $\mathscr{O}(x)$ without external fields, i.e.,

$$\left[P_\mu^{(0)}, \mathscr{O}(x)\right] = \mathrm{i}\frac{\partial}{\partial x_\mu}\mathscr{O}(x) . \tag{11.1}$$

In particular, for $\mu = 0$,

$$\left[P_0^{(0)}, \mathscr{O}(x)\right] = -\mathrm{i}\frac{\partial}{\partial t}\mathscr{O}(x) . \tag{11.2}$$

We now introduce the following transformation function as $\mathscr{O}$:

$$U(t, t_0) = T \exp\left[-\mathrm{i}\int_{t_0}^{t} \mathrm{d}t'\, H_{\mathrm{int}}(t')\right] . \tag{11.3}$$

K. Nishijima, *Quantum Field Theory*,
https://doi.org/10.1007/978-94-024-2190-3_11

We insert this into (11.2), but bearing in mind that

$$\left[P_0^{(0)}, -\mathrm{i}\int_{t_0}^{t} \mathrm{d}t' H_{\mathrm{int}}(t')\right] = -\int_{t_0}^{t} \mathrm{d}t' \frac{\partial}{\partial t'} H_{\mathrm{int}}(t') = -\big[H_{\mathrm{int}}(t) - H_{\mathrm{int}}(t_0)\big] . \tag{11.4}$$

This immediately yields

$$\big[P_0^{(0)}, U(t, t_0)\big] = -T\big[[H_{\mathrm{int}}(t) - H_{\mathrm{int}}(t_0)]U(t, t_0)\big] = -H_{\mathrm{int}}(t)U(t, t_0) + U(t, t_0)H_{\mathrm{int}}(t_0) . \tag{11.5}$$

We now set $t = 0$ and take the limit $t_0 \to -\infty$. In this case, putting

$$P_0^{(0)} + H_{\mathrm{int}}(0) = H_{\mathrm{total}} , \tag{11.6}$$

we obtain

$$H_{\mathrm{total}} U(0, -\infty) = U(0, -\infty)\big[P_0^{(0)} + H_{\mathrm{int}}(-\infty)\big] .$$

Assuming that interactions exist only in a finite space-time region, as in Sects. 4.4 and 7.3, we consider the limit as this space-time region extends to infinity. This corresponds to introducing interactions adiabatically, and a variety of consequences depend on how the limit is actually taken. With some loss of rigour, we assume that

$$H_{\mathrm{int}}(-\infty) = 0 , \tag{11.7}$$

whence

$$H_{\mathrm{total}} U(0, -\infty) = U(0, -\infty) P_0^{(0)} . \tag{11.8}$$

Then writing the vacuum in the interaction picture as Φ_0,

$$P_0^{(0)} \Phi_0 = 0 . \tag{11.9}$$

Hence, writing the vacuum in the Heisenberg picture as Ψ_0, we find

$$H_{\mathrm{total}} \Psi_0 = 0 , \quad \Psi_0 = U(0, -\infty)\Phi_0 . \tag{11.10}$$

As shown in Chap. 6, the Heisenberg picture and the interaction picture are related by (6.30), viz.,

$$U(t, t_0)^{-1} \varphi_\alpha(\boldsymbol{x}, t) U(t, t_0) = \varphi_\alpha^{(\mathrm{H})}(\boldsymbol{x}, t) . \tag{11.11}$$

In this section, we set $t_0 = 0$. Therefore, for $t_1 > t_2$,

$$
\begin{aligned}
&\big(\Phi_0, T\big[U(\infty,-\infty)A(\boldsymbol{x}_1,t_1)B(\boldsymbol{x}_2,t_2)\big]\Phi_0\big)\\
&\quad= \big(\Phi_0, U(\infty,t_1)A(\boldsymbol{x}_1,t_1)U(t_1,t_2)B(\boldsymbol{x}_2,t_2)U(t_2,-\infty)\Phi_0\big)\\
&\quad= \big(\Phi_0, U(\infty,0)A^{(\mathrm{H})}(\boldsymbol{x}_1,t_1)B^{(\mathrm{H})}(\boldsymbol{x}_2,t_2)U(0,-\infty)\Phi_0\big)\\
&\quad= \big(\Phi_0, U(\infty,-\infty)U(0,-\infty)^{-1}A^{(\mathrm{H})}(\boldsymbol{x}_1,t_1)B^{(\mathrm{H})}(\boldsymbol{x}_2,t_2)U(0,-\infty)\Phi_0\big)\\
&\quad= \big(\Phi_0, U(\infty,-\infty)\Phi_0)(\Phi_0, U(0,-\infty)^{-1}A^{(\mathrm{H})}(\boldsymbol{x}_1,t_1)B^{(\mathrm{H})}(\boldsymbol{x}_2,t_2)U(0,-\infty)\Phi_0\big)\\
&\quad= \big(\Phi_0, U(\infty,-\infty)\Phi_0\big)\big(\Psi_0, A^{(\mathrm{H})}(\boldsymbol{x}_1,t_1)B^{(\mathrm{H})}(\boldsymbol{x}_2,t_2)\Psi_0\big)\,,
\end{aligned}
\tag{11.12}
$$

where we have used the composition rule (6.27) for the transformation function and $U(\infty,-\infty)$, and the fact that the vacuum Φ_0 transforms to the vacuum up to a phase. This can be written in the form

$$
\big(\Psi_0, T\big[A^{(\mathrm{H})}(x_1)B^{(\mathrm{H})}(x_2)\big]\Psi_0\big) = \frac{\big(\Phi_0, T\big[U(\infty,-\infty)A(x_1)B(x_2)\big]\Phi_0\Phi_0\big)}{\big(\Phi_0, U(\infty,-\infty)\Phi_0\big)}\,.
\tag{11.13}
$$

The presence of the denominator corresponds to neglecting bubble graphs, as discussed in connection with (8.70). This equation is easily generalized:

$$
\begin{aligned}
&\big(\Psi_0, T\big[A^{(\mathrm{H})}(x_1)B^{(\mathrm{H})}(x_2)\ldots Z^{(\mathrm{H})}(x_n)\big]\Psi_0\big)\\
&\qquad= \frac{\big(\Phi_0, T\big[U(\infty,-\infty)A(x_1)B(x_2)\ldots Z(x_n)\big]\Phi_0\big)}{\big(\Phi_0, U(\infty,-\infty)\Phi_0\big)}\,.
\end{aligned}
\tag{11.14}
$$

This is called the *Gell-Mann–Low relation* [111]. It gives the relationship between the Heisenberg picture and the interaction picture. Additionally, any quantity like the left-hand side, i.e., a vacuum expectation value of a product of time-ordered field operators, is called a *Green's function*.

11.2 Green's Functions and Their Generating Functionals

When we discuss properties of a Green's function, instead of considering them one by one, it is sometimes useful to discuss their generating function, or more precisely their generating functional. In this section, we describe the simplest generating functional, which produces the Green's functions of a neutral scalar field.

We consider the Lagrangian density

$$
\mathscr{L} = -\frac{1}{2}\big[(\partial_\mu\varphi)^2 + m^2\varphi^2\big] - \frac{g}{4!}\varphi^4\,.
\tag{11.15}
$$

We split this into the free part and the interaction part. In fact, such a separation is not trivial, and becomes particularly complicated when we take into account renormalization in the next chapter. However, in this section, we simply choose

$$\mathscr{H}_{\mathrm{int}}(x) = -\mathscr{L}_{\mathrm{int}}(x) = \frac{g}{4!}\varphi^4 . \tag{11.16}$$

In the following, we shall express the vacuum Φ_0 simply as $|0\rangle$. We introduce a c-number external field $J(x)$ and define the functional

$$\mathscr{T}^{(0)}[J] = \frac{\Big\langle 0\Big| T \exp\Big\{ - \mathrm{i} \int \mathrm{d}^4x \big[\mathscr{H}_{\mathrm{int}}(x) + J(x)\varphi(x)\big]\Big\}\Big|0\Big\rangle}{\big\langle 0\big| T \exp\big[- \mathrm{i}\int \mathrm{d}^4x \mathscr{H}_{\mathrm{int}}(x)\big]\big|0\big\rangle} . \tag{11.17}$$

We now derive the equation satisfied by the generating functional of the Green's functions $\mathscr{T}^{(0)}[J]$. To do this, we just have to use the reduction formula (8.57), i.e.,

$$(\Box_x - m^2) T[\varphi(x)AB\ldots] = \mathrm{i}\frac{\delta}{\delta\varphi(x)} T[AB\ldots] . \tag{11.18}$$

Writing the denominator of (11.17) as $\langle 0|U(\infty, -\infty)|0\rangle$,

$$\begin{aligned}
&(\Box_x - m^2)\mathrm{i}\frac{\delta}{\delta J(x)}\mathscr{T}^{(0)}[J] \\
&\quad = \frac{(\Box_x - m^2)\langle 0|T\big(\varphi(x)\exp\{ - \mathrm{i}\int \mathrm{d}^4y[\mathscr{H}_{\mathrm{int}}(y) + J(y)\varphi(y)]\}\big)|0\rangle}{\langle 0|U(\infty, -\infty)|0\rangle} \\
&\quad = \frac{\Big\langle 0\Big|\mathrm{i}\frac{\delta}{\delta\varphi(x)} T\big(\exp\{ - \mathrm{i}\int \mathrm{d}^4y[\mathscr{H}_{\mathrm{int}}(y) + J(y)\varphi(y)]\}\big)\Big|0\Big\rangle}{\langle 0|U(\infty, -\infty)|0\rangle} \\
&\quad = \frac{\Big\langle 0\Big|T\Big(\Big[\frac{g}{3!}\varphi^3(x) + J(x)\Big]\exp\{-\mathrm{i}\int \mathrm{d}^4y[\mathscr{H}_{\mathrm{int}}(y) + J(y)\varphi(y)]\}\Big)\Big|0\Big\rangle}{\langle 0|U(\infty, -\infty)|0\rangle} \\
&\quad = \left\{\frac{g}{3!}\left[\mathrm{i}\frac{\delta}{\delta J(x)}\right]^3 + J(x)\right\}\mathscr{T}^{(0)}[J] .
\end{aligned} \tag{11.19}$$

Integrating this equation, we obtain the functional equation for $\mathscr{T}^{(0)}[J]$:

$$\mathrm{i}\frac{\delta}{\delta J(x)}\mathscr{T}^{(0)}[J] = -\mathrm{i}\int \mathrm{d}^4y \Delta_{\mathrm{F}}(x-y)\left\{\frac{g}{3!}\left[\mathrm{i}\frac{\delta}{\delta J(y)}\right]^3 + J(y)\right\}\mathscr{T}^{(0)}[J] . \tag{11.20}$$

Differentiating $\mathscr{T}^{(0)}[J]$ n times with respect to J and setting $J = 0$, we obtain the n-point Green's function:

$$\mathrm{i}^n \frac{\delta^n}{\delta J(x_1)\dots\delta J(x_n)} \mathscr{T}^{(0)}[J]\bigg|_{J=0} = \frac{\langle 0|T\big[\varphi(x_1)\dots\varphi(x_n)U(\infty,-\infty)\big]|0\rangle}{\langle 0|U(\infty,-\infty)|0\rangle}. \tag{11.21}$$

We now turn to the Heisenberg picture, defining the generating functional by

$$\mathscr{T}[J] = \langle \mathbf{0}|T\exp\Big[-\mathrm{i}\int \mathrm{d}^4x\, J(x)\boldsymbol{\varphi}(x)\Big]|\mathbf{0}\rangle\,, \tag{11.22}$$

where we have written Ψ_0 and $\varphi^{(\mathrm{H})}$ in bold face as $|\mathbf{0}\rangle$ and $\boldsymbol{\varphi}$, respectively. Therefore,

$$(\Box_x - m^2)\mathrm{i}\frac{\delta}{\delta J(x)}\mathscr{T}[J] = (\Box_x - m^2)\langle \mathbf{0}|T\Big\{\boldsymbol{\varphi}(x)\exp\Big[-\mathrm{i}\int \mathrm{d}^4y\, J(y)\boldsymbol{\varphi}(y)\Big]\Big\}|\mathbf{0}\rangle\,. \tag{11.23}$$

In this picture,

$$(\Box_x - m^2)T[\boldsymbol{\varphi}(x)\boldsymbol{AB}\dots] = T\big[(\Box_x - m^2)\boldsymbol{\varphi}(x), \boldsymbol{AB}\dots\big] + \mathrm{i}\frac{\delta}{\delta\boldsymbol{\varphi}(x)}T[\boldsymbol{AB}\dots]\,. \tag{11.24}$$

Inserting this into (11.23) and using the equation of motion for $\boldsymbol{\varphi}$, we recover (11.19). Hence, (11.20) can also be recovered, i.e.,

$$\mathrm{i}\frac{\delta}{\delta J(x)}\mathscr{T}[J] = -\mathrm{i}\int \mathrm{d}^4y\, \Delta_{\mathrm{F}}(x-y)\bigg\{\frac{g}{3!}\Big[\mathrm{i}\frac{\delta}{\delta J(y)}\Big]^3 + J(y)\bigg\}\mathscr{T}[J]\,. \tag{11.25}$$

This implies that, even if the boundary conditions are the same, we can expect

$$\mathscr{T}[J] = \mathscr{T}^{(0)}[J]\,. \tag{11.26}$$

In fact, differentiating this n times with respect to J and setting $J = 0$, it becomes

$$\langle \mathbf{0}|T[\boldsymbol{\varphi}(x_1)\dots\boldsymbol{\varphi}(x_n)]|\mathbf{0}\rangle = \frac{\langle 0|T\big[\varphi(x_1)\dots\varphi(x_n)U(\infty,-\infty)\big]|0\rangle}{\langle 0|U(\infty,-\infty)|0\rangle}\,, \tag{11.27}$$

which is nothing other than the Gell-Mann–Low relation. Thus, it turns out that (11.26) always holds true.

When a Green's function is expressed by a Feynman diagram, it will generally give a set of non-connected parts. We focus only on the graphs in which n points are connected to each other by lines. We write the contribution corresponding to such a graph as

$$\langle \mathbf{0} | T[\boldsymbol{\varphi}(x_1) \dots \boldsymbol{\varphi}(x_n)] | \mathbf{0} \rangle_{\text{conn}} , \tag{11.28}$$

where conn stands for 'connected.' So what is the relationship between the connected Green's function and the original Green's function? To find out, we start with some point x and separate into the part connected with x and the other parts. This gives a recursion formula:

$$\begin{aligned} &\langle \mathbf{0} | T[\boldsymbol{\varphi}(x)\boldsymbol{\varphi}(x_1) \dots \boldsymbol{\varphi}(x_n)] | \mathbf{0} \rangle \\ &\quad = \sum_{\text{comb}} \langle \mathbf{0} | T[\boldsymbol{\varphi}(x)\boldsymbol{\varphi}(x_1') \dots \boldsymbol{\varphi}(x_k')] | \mathbf{0} \rangle_{\text{conn}} \langle \mathbf{0} | T[\boldsymbol{\varphi}(x_{k+1}') \dots \boldsymbol{\varphi}(x_n')] | \mathbf{0} \rangle , \end{aligned} \tag{11.29}$$

where the summation has been taken over all combinations separating $x_1, \dots, x_n$ into $x_1', \dots, x_k'$ and $x_{k+1}', \dots, x_n'$. In order to express this relation in a closed form, we introduce a generating functional $\mathscr{R}[J]$ for the connected Green's functions:

$$\mathscr{R}[J] = \sum_{n=1}^{\infty} \frac{(-\mathrm{i})^n}{n!} \int \mathrm{d}^4 x_1 \dots \int \mathrm{d}^4 x_n \langle \mathbf{0} | T[\boldsymbol{\varphi}(x_1) \dots \varphi(x_n)] | \mathbf{0} \rangle_{\text{conn}} J(x_1) \dots J(x_n) , \tag{11.30}$$

$$\mathrm{i}^n \frac{\delta^n}{\delta J(x_1) \dots \delta J(x_n)} \mathscr{R}[J] \bigg|_{J=0} = \langle \mathbf{0} | T[\boldsymbol{\varphi}(x_1) \dots \varphi(x_n)] | \mathbf{0} \rangle_{\text{conn}} . \tag{11.31}$$

The recursion formula can now be expressed in closed form:

$$\mathrm{i} \frac{\delta}{\delta J(x)} \mathscr{T}[J] = \left(\mathrm{i} \frac{\delta}{\delta J(x)} \mathscr{R}[J] \right) \mathscr{T}[J] . \tag{11.32}$$

We solve this functional equation with the boundary conditions

$$\mathscr{R}[0] = 0 , \quad \mathscr{T}[0] = 1 . \tag{11.33}$$

The solution is

$$\mathscr{T}[J] = \exp \mathscr{R}[J] . \tag{11.34}$$

Then writing the expectation value of $\boldsymbol{\varphi}(x)$ when there exists an external field J as $\langle \boldsymbol{\varphi}(x) \rangle$, we have

$$\langle \boldsymbol{\varphi}(x) \rangle = \frac{\mathrm{i} \frac{\delta}{\delta J(x)} \mathscr{T}[J]}{\mathscr{T}[J]} = \mathrm{i} \frac{\delta}{\delta J(x)} \mathscr{R}[J] . \tag{11.35}$$

In some situations, it is more useful to consider $\langle\varphi(x)\rangle$ as an independent external field instead of $J(x)$, so we introduce the following Legendre transformation:

$$\mathscr{F} = \mathscr{R} + \mathrm{i}\int \mathrm{d}^4x\, J(x)\langle\varphi(x)\rangle\,. \tag{11.36}$$

This yields

$$\begin{aligned}\delta\mathscr{F} &= \delta\mathscr{R} + \mathrm{i}\int \mathrm{d}^4x\big[J(x)\delta\langle\varphi(x)\rangle + \delta J(x)\langle\varphi(x)\rangle\big] \\ &= -\mathrm{i}\int \mathrm{d}^4x\,\delta J(x)\delta\langle\varphi(x)\rangle + \mathrm{i}\int \mathrm{d}^4x\,\big[J(x)\delta\langle\varphi(x)\rangle + \delta J(x)\langle\varphi(x)\rangle\big] \\ &= \mathrm{i}\int \mathrm{d}^4x\; J(x)\delta\langle\varphi(x)\rangle\,,\end{aligned} \tag{11.37}$$

i.e., taking $\langle\varphi(x)\rangle$ as an independent variable instead of $J(x)$,

$$\frac{\delta\mathscr{F}}{\delta\langle\varphi(x)\rangle} = \mathrm{i}J(x)\,. \tag{11.38}$$

Variable transformations like this will play an important role in the discussion of spontaneous symmetry breaking later on. Moreover, differentiating (11.35) again,

$$\mathrm{i}\frac{\delta\langle\varphi(x)\rangle}{\delta J(y)} = -\frac{\delta^2\mathscr{R}}{\delta J(x)\delta J(y)} = \Delta_{\mathrm{F}}(x, y)\,. \tag{11.39}$$

This is the two-point Green's function in the case where an external field exists. Differentiating (11.38),

$$\frac{\delta^2\mathscr{F}}{\delta\langle\varphi(x)\rangle\delta\langle\varphi(y)\rangle} = \mathrm{i}\frac{\delta J(y)}{\delta\langle\varphi(x)\rangle} = -\Delta_{\mathrm{F}}^{-1}(x, y)\,, \tag{11.40}$$

which is the inverse of the two-point Green's function.

11.3 Different Time-Orderings in the Lagrangian Formalism

In the previous section, we investigated the relationship between representations of Green's functions in the Heisenberg picture and in the interaction picture. In this section, we shall discuss the difference between time-orderings in the Hamiltonian formalism and the Lagrangian formalism.

For simplicity, we start with particle dynamics. Considering a one-dimensional system, we assume that

$$p(t) = m\dot{q}(t) . \tag{11.41}$$

Therefore, under the time-ordering operator T, $p(t)$ is ordered as an operator corresponding to time t. This is because in the Hamiltonian formalism, the operators $q(t)$ and $p(t)$ are treated as independent variables, so these are considered to be quantities that need no further temporal decomposition. However, in the Lagrangian formalism, only $q(t)$ is an independent quantity, so $\dot{q}(t)$ becomes

$$\dot{q}(t) = \lim_{\epsilon\to 0} \frac{q(t+\epsilon) - q(t)}{\epsilon} . \tag{11.42}$$

Assuming that ϵ is small but finite, it turns out that, in the Lagrangian formalism, $\dot{q}(t)$ is associated with two clock times in time-orderings. We denote the time-ordering operator in this treatment by T^*. It should be noted here that, in the path-integral method, the same decomposition (11.42) is used. Therefore, the method using T^* is closely related to the path-integral method. So what is the difference between using T or T^*?

In the Hamiltonian formalism and hence under T, we assume that $\dot{q}(t)$ can be treated as a one-clock-time quantity, as in (11.41). Now,

$$\begin{aligned} T^*[\dot{q}(t), q(t')] &= T^*\left[\lim_{\epsilon\to 0} \frac{q(t+\epsilon) - q(t)}{\epsilon}, q(t')\right] \\ &= \lim_{\epsilon\to 0} T\left[\frac{q(t+\epsilon) - q(t)}{\epsilon}, q(t')\right] \\ &= \frac{\partial}{\partial t} T\big[q(t), q(t')\big] . \end{aligned} \tag{11.43}$$

However, if q obeys a second order differential equation, then only $\dot{q}$ can enter in T^*, because $\ddot{q}$ is not independent of q. Another example is

$$T^*\big[\dot{q}(t), \dot{q}(t')\big] = \frac{\partial^2}{\partial t \partial t'} T\big[q(t), q(t')\big] . \tag{11.44}$$

The derivation of this equation is perfectly analogous to the one in the previous example. However, using T, we find

$$\begin{aligned} \frac{\partial}{\partial t} T\big[q(t), q(t')\big] &= T[\dot{q}(t), q(t')] + \frac{\partial}{\partial t}\left[\frac{1}{2}\epsilon(t-t')\right]\big[q(t), q(t')\big] \\ &= T\big[\dot{q}(t), q(t')\big] + \delta(t-t')\big[q(t), q(t')\big] \\ &= T\big[\dot{q}(t), q(t')\big] . \end{aligned} \tag{11.45}$$

Therefore, this is the same as T^* in (11.43). However, assuming (11.41),

$$\frac{\partial}{\partial t}T\big[q(t), \dot{q}(t')\big] = T\big[\dot{q}(t), \dot{q}(t')\big] + \delta(t-t')\big[q(t), \dot{q}(t')\big]$$
$$= T\big[\dot{q}(t), \dot{q}(t')\big] + \frac{\mathrm{i}}{m}\delta(t-t') \,. \tag{11.46}$$

The left-hand side is equal to the right-hand side of (11.44). So setting the left-hand side of (11.46) equal to the left-hand side of (11.44),

$$T\big[\dot{q}(t), \dot{q}(t')\big] = T^*\big[\dot{q}(t), \dot{q}(t')\big] - \frac{\mathrm{i}}{m}\delta(t-t') \,. \tag{11.47}$$

The difference between T and T^* becomes clear in this way.

This argument can be extended to field theory. For simplicity, we consider a neutral scalar field in the interaction picture:

$$\frac{\partial}{\partial x_\mu}T\big[\partial_\mu\varphi(x), \varphi(y)\big] = T\big[\partial_\mu\varphi(x), \varphi(y)\big] + \frac{1}{\mathrm{i}}\delta_{\mu 4}\delta(x_0-y_0)\big[\varphi(x), \varphi(y)\big]$$
$$= T\big[\partial_\mu\varphi(x), \varphi(y)\big] \,. \tag{11.48}$$

In order to differentiate this one more time, we introduce a unit time-like vector n_μ. Here we assume that $n_1 = n_2 = n_3 = 0$ and $n_4 = \mathrm{i}$:

$$\frac{\partial}{\partial y_\nu}T\big[\partial_\mu\varphi(x), \varphi(y)\big] = T\big[\partial_\mu\varphi(x), \varphi(y)\big] + \mathrm{i}\delta_{\nu 4}\delta(x_0-y_0)\big[\partial_\mu\varphi(x), \varphi(y)\big] \,.$$

The second term becomes

$$\delta_{\mu 4}\delta_{\nu 4}\delta(x_0-y_0)\big[\dot{\varphi}(x), \varphi(y)\big] = -\mathrm{i}\delta_{\mu 4}\delta_{\nu 4}\delta^4(x-y)$$
$$= \mathrm{i}n_\mu n_\nu \delta^4(x-y) \,.$$

Taking the vacuum expectation value,

$$\big\langle 0\big|T\big[\partial_\mu\varphi(x), \partial_\nu\varphi(y)\big]\big|0\big\rangle = \frac{\partial^2}{\partial x_\mu \partial y_\nu}\Delta_{\mathrm{F}}(x-y) - \mathrm{i}n_\mu n_\nu\delta^4(x-y) \,. \tag{11.49}$$

On the other hand,

$$\big\langle 0\big|T^*\big[\partial_\mu\varphi(x), \partial_\nu\varphi(y)\big]\big|0\big\rangle = \frac{\partial^2}{\partial x_\mu \partial y_\nu}\Delta_{\mathrm{F}}(x-y) \,. \tag{11.50}$$

This difference corresponds to (11.47). Comparing the two equations above, we see that T^* is covariant and simpler. When there are derivatives of field operators in the interaction term, the Hamiltonian density is not a scalar, but a tensor involving the time-like vector n. Although in this case both the Hamiltonian density and the contraction function depend on n, when we compute the S-matrix, the two types of n-dependence cancel out, so the final result does not depend on n. This is called *Matthew's theorem* [112].

11.4 Matthews' Theorem

If interactions do not include derivatives of the field operators, we have

$$\mathscr{H}_{\text{int}}(x) = -\mathscr{L}_{\text{int}}(x) . \tag{11.51}$$

Clearly, in this case,

$$S = T \exp\left[-\mathrm{i}\int \mathrm{d}^4x\, \mathscr{H}_{\text{int}}(x)\right] = T^* \exp\left[\mathrm{i}\int \mathrm{d}^4x\, \mathscr{L}_{\text{int}}(x)\right] . \tag{11.52}$$

This is because the difference between T and T^* does not appear anywhere here.

When derivatives are included in the interactions, the situation becomes more complicated. For example, when a charged scalar field and the electromagnetic field interact with each other,

$$\mathscr{L} = -\left[(\partial_\mu + \mathrm{i}eA_\mu)\varphi^\dagger \cdot (\partial_\mu - \mathrm{i}eA_\mu)\varphi + m^2\varphi^\dagger\varphi\right] + \mathscr{L}_{\text{em}} , \tag{11.53}$$

$$\mathscr{L}_{\text{int}} = -\mathrm{i}eA_\mu(\varphi^\dagger \cdot \partial_\mu\varphi - \partial_\mu\varphi^\dagger \cdot \varphi) - \mathrm{e}^2A_\mu^2\varphi^\dagger\varphi . \tag{11.54}$$

Thus, when derivatives of field operators are included, $\mathscr{H}_{\text{int}}$ differs from $-\mathscr{L}_{\text{int}}$.

First, we decompose the Lagrangian density:

$$\mathscr{L}(x) = \mathscr{L}_{\text{f}}(x) + \mathscr{L}_{\text{int}}(x) . \tag{11.55}$$

If φ_α is a real scalar field, the quantity canonically conjugate to φ_α in the free field case is

$$\pi_\alpha(x) = \frac{\partial\mathscr{L}_{\text{f}}(x)}{\partial\dot{\varphi}_\alpha(x)} = \dot{\varphi}_\alpha(x) . \tag{11.56}$$

If we now consider interactions, the canonically conjugate field is

$$\pi'_\alpha(x) = \frac{\partial\mathscr{L}(x)}{\partial\dot{\varphi}_\alpha(x)} = \dot{\varphi}_\alpha(x) + \frac{\partial\mathscr{L}_{\text{int}}(x)}{\partial\dot{\varphi}_\alpha(x)} . \tag{11.57}$$

By the Yang–Feldman equation (6.34), the relation between the Heisenberg picture and the interaction picture is

$$\varphi_\alpha(x) = U(x_0, -\infty)^{-1}\varphi_\alpha^{\text{in}}(x)U(x_0, -\infty) \,, \tag{11.58}$$

$$\pi'_\alpha(x) = U(x_0, -\infty)^{-1}\pi_\alpha^{\text{in}}(x)U(x_0, -\infty) \,. \tag{11.59}$$

The reason why π transforms into π' is that the transformation by U should not change the canonical commutation relation. In the following, we shall drop the superscript 'in' on operators in the interaction picture, and express Heisenberg operators in bold face. Therefore, the Hamiltonian density is

$$\boldsymbol{\mathscr{H}}(x) = \sum_\alpha \boldsymbol{\pi}'_\alpha(x)\dot{\boldsymbol{\phi}}_\alpha(x) - \mathscr{L}\big(\boldsymbol{\varphi}_\alpha(x), \dot{\boldsymbol{\varphi}}_\alpha(x)\big) \,, \tag{11.60}$$

where we have assumed that a spatial derivative is a linear combination of the $\boldsymbol{\varphi}_\alpha(x)$. Then,

$$\mathscr{H}(x) = U(x_0, -\infty)\boldsymbol{\mathscr{H}}(x)U(x_0, -\infty)^{-1} \,. \tag{11.61}$$

In this computation we use the inverse transformations of (11.58) and (11.59). Therefore,

$$\begin{aligned} U(x_0, -\infty)\dot{\boldsymbol{\varphi}}_\alpha(x)U(x_0, -\infty)^{-1} &= U(x_0, -\infty)\big[\boldsymbol{\pi}'_\alpha(x) - \boldsymbol{\sigma}_\alpha(x)\big]U(x_0, -\infty)^{-1} \\ &= \pi_\alpha(x) - \sigma_\alpha(x) \\ &= \dot{\varphi}_\alpha(x) - \sigma_\alpha(x) \,, \end{aligned} \tag{11.62}$$

where σ_α is defined by

$$\sigma_\alpha(x) = \frac{\partial \mathscr{L}_{\text{int}}(x)}{\partial \dot{\varphi}_\alpha(x)} = -n_\mu \frac{\partial \mathscr{L}_{\text{int}}(x)}{\partial \varphi_{\alpha,\mu}(x)} \,. \tag{11.63}$$

Thus,

$$\mathscr{H}(x) = \sum_\alpha \pi_\alpha(x)\big[\pi_\alpha(x) - \sigma_\alpha(x)\big] - \mathscr{L}\big(\varphi_\alpha(x), \dot{\varphi}_\alpha(x) - \sigma_\alpha(x)\big) \,. \tag{11.64}$$

We use a Taylor expansion for the second term. In this case, since $\mathscr{L}$ includes at most second order terms in $\dot{\varphi}_\alpha(x)$ and using

$$\frac{\partial^2}{\partial\dot{\varphi}_\alpha\partial\dot{\varphi}_\beta}\mathscr{L} = \delta_{\alpha\beta} \,, \quad \frac{\partial}{\partial\dot{\varphi}_\alpha}\mathscr{L} = \pi_\alpha + \sigma_\alpha \,, \tag{11.65}$$

we obtain the following power series expansion in σ :

$$\mathscr{L}(\varphi_\alpha, \dot{\varphi}_\alpha - \sigma_\alpha) = \mathscr{L}(\varphi_\alpha, \dot{\varphi}_\alpha) - \sum_\alpha \sigma_\alpha \pi_\alpha - \frac{1}{2}\sum_\alpha \sigma_\alpha^2(x) . \tag{11.66}$$

Inserting this into (11.64),

$$\mathscr{H}(x) = \sum_\alpha \pi_\alpha^2(x) - \mathscr{L}\big(\varphi_\alpha(x), \dot{\varphi}_\alpha(x)\big) + \frac{1}{2}\sum_\alpha \sigma_\alpha^2(x) . \tag{11.67}$$

Therefore,

$$\mathscr{H}_{\mathrm{int}}(x) = -\mathscr{L}_{\mathrm{int}}(x) + \frac{1}{2}\sum_\alpha n_\mu n_\nu \frac{\partial \mathscr{L}_{\mathrm{int}}(x)}{\partial \varphi_{\alpha,\mu}(x)} \frac{\partial \mathscr{L}_{\mathrm{int}}(x)}{\partial \varphi_{\alpha,\nu}(x)} . \tag{11.68}$$

It turns out that the second term on the right-hand side expresses a shift from (11.51). We now prove the equality (11.52) in this case, i.e., Matthews' theorem. To do so, we first specify the relation between the T-product and the normal product for the simple neutral scalar field theory.

We expand the T-product $T[AB\ldots Z]$ into normal products in the neutral scalar theory. In this case, we write the contraction function Δ_{F}, replaced by $\lambda\Delta_{\mathrm{F}}$, as $T_\lambda[AB\ldots Z]$. Since differentiating this with respect to λ is equivalent to contracting $\varphi(x)$ and $\varphi(y)$ and multiplying by Δ_{F}, we have

$$\frac{\partial}{\partial\lambda} T_\lambda[AB\ldots Z] = \frac{1}{2}\int \mathrm{d}^4x \int \mathrm{d}^4y \frac{\delta}{\delta\varphi(x)} \Delta_{\mathrm{F}}(x-y) \frac{\delta}{\delta\varphi(y)} T_\lambda[AB\ldots Z] , \tag{11.69}$$

where we have treated φ's in the normal product as c-numbers and differentiated with respect to them. Moreover, if we take $\lambda = 0$, then since this is exactly the same as that with no contraction, we have

$$T_0[AB\ldots Z] = :AB\ldots Z: . \tag{11.70}$$

Solving the differential equation (11.69) under the initial condition (11.70) and taking $\lambda = 1$, we find

$$T[AB\ldots Z] = \exp\left[\frac{1}{2}\int \mathrm{d}^4x \int \mathrm{d}^4y \frac{\delta}{\delta\varphi(x)} \Delta_{\mathrm{F}}(x-y) \frac{\delta}{\delta\varphi(y)}\right] :AB\ldots Z: . \tag{11.71}$$

Next we take the vacuum expectation value of this equation. Since the expectation value of the normal product vanishes, the right-hand side survives only if all field operators are contracted. Thus, setting to zero the operators φ which are not contracted on the right-hand side, this gives only those contributions which appear when all operators are contracted:

$$\langle 0|T[AB\ldots Z]|0\rangle = \exp\left[\frac{1}{2}\int \mathrm{d}^4x \int \mathrm{d}^4y \frac{\delta}{\delta\varphi(x)}\Delta_\mathrm{F}(x-y)\frac{\delta}{\delta\varphi(y)}\right] :AB\ldots Z:\Big|_{\varphi=0} \;. \tag{11.72}$$

In fact, (11.52) is only true if no derivative of φ is included in $AB\ldots Z$. This operator appearing in the argument of the exponential function is related to the field quantization. It tells us how to contract operators. Denoting this by D, the operation which converts the normal product including φ and its derivative to the T-product is

$$T[AB\ldots Z] = \mathrm{e}^D :AB\ldots Z: \;. \tag{11.73}$$

When the normal product includes φ, its derivative, and fermionic fields ψ and $\bar{\psi}$, we have

$$\begin{aligned} D = \frac{1}{2}\int \mathrm{d}^4x \int \mathrm{d}^4y \Bigg\{ & \frac{\delta}{\delta\varphi(x)}\Delta_\mathrm{F}(x-y)\frac{\delta}{\delta\varphi(y)} \\ & + 2\frac{\delta}{\delta\varphi_{,\mu}(x)}\left[\frac{\partial}{\partial x_\mu}\Delta_\mathrm{F}(x-y)\right]\frac{\delta}{\delta\varphi(y)} \\ & + \frac{\delta}{\delta\varphi_{,\mu}(x)}\left[\frac{\partial^2}{\partial x_\mu \partial y_\nu}\Delta_\mathrm{F}(x-y) - \mathrm{i}n_\mu n_\nu \delta^4(x-y)\right]\frac{\delta}{\delta\varphi_{,\nu}(y)}\Bigg\} \\ + \int \mathrm{d}^4x \int \mathrm{d}^4y & \frac{\delta}{\delta\bar{\psi}_\beta(y)} S_\mathrm{F}(x-y)_{\alpha\beta}\frac{\delta}{\delta\psi_\alpha(x)} \;, \end{aligned} \tag{11.74}$$

where we have used (11.49) as a contraction function. Regarding the functional derivatives with respect to the Dirac field, the reader is referred to the caution after (8.59). Likewise for the T*-product,

$$T^*[AB\ldots Z] = \mathrm{e}^{D^*} :AB\ldots Z: \,, \tag{11.75}$$

where D^* is defined so that (11.50) applies to contractions of the φ derivatives in (11.74), whence

$$D = D^* + \frac{1}{2}\int \mathrm{d}^4x \int \mathrm{d}^4y \frac{\delta}{\delta\varphi_{,\mu}(x)}\left[-\mathrm{i}n_\mu n_\nu \delta^4(x-y)\right]\frac{\delta}{\delta\varphi_{,\nu}(y)} \;. \tag{11.76}$$

In both cases, we have treated φ and its derivative as independent when taking functional derivatives.

We now choose the interaction Lagrangian density to have the form

$$\mathscr{L}_\mathrm{int}(x) = -j_\mu(x)\partial_\mu\varphi(x) + \mathscr{L}^{(0)}_\mathrm{int}(x) \;, \tag{11.77}$$

where $\mathscr{L}_{\text{int}}^{(0)}$ is a term which includes no derivatives. Therefore, from (11.68),

$$\mathscr{H}_{\text{int}}(x) = -\mathscr{L}_{\text{int}}(x) + \frac{1}{2}\left[n_\mu j_\mu(x)\right]^2 . \tag{11.78}$$

In order to compute the S-matrix, we consider

$$\begin{aligned} T\exp\left[-\mathrm{i}\int \mathrm{d}^4x\,\mathscr{H}_{\text{int}}(x)\right] &= \mathrm{e}^D :\exp\left[-\mathrm{i}\int \mathrm{d}^4x\,\mathscr{H}_{\text{int}}(x)\right]: \\ &= \mathrm{e}^{D^*}\mathrm{e}^{D-D^*} :\exp\left[-\mathrm{i}\int \mathrm{d}^4x\,\mathscr{H}_{\text{int}}(x)\right]: \end{aligned} \tag{11.79}$$

where $D - D^*$ is the second term on the right-hand side of (11.76). This is a functional derivative with respect to the derivative of φ. Hence, this acts only on the term $j_\mu \partial_\mu \varphi$ in $\mathscr{H}_{\text{int}}(x)$ which includes the derivative of φ. We must therefore compute

$$\mathrm{e}^{D-D^*} :\exp\left[-\mathrm{i}\int \mathrm{d}^4x\, j_\mu(x)\partial_\mu\varphi(x)\right]: . \tag{11.80}$$

To carry out this computation, we use

$$\exp\left(\lambda\frac{\mathrm{d}^2}{\mathrm{d}x^2}\right)\mathrm{e}^{ax} = \mathrm{e}^{ax+a^2\lambda} . \tag{11.81}$$

Generalizing this formula and using it to calculate (11.80), we find that (11.80) is equivalent to something of the form

$$:\exp\left[-\mathrm{i}\int \mathrm{d}^4x\left(j_\mu(x)\partial_\mu\varphi(x) - \frac{1}{2}\left[n_\mu j_\nu(x)\right]^2\right)\right]: . \tag{11.82}$$

Therefore, (11.79) can be written in the form

$$\begin{aligned} T\exp\left[-\mathrm{i}\int \mathrm{d}^4x\,\mathscr{H}_{\text{int}}(x)\right] &= \mathrm{e}^{D^*} :\exp\left[-\mathrm{i}\int \mathrm{d}^4x\left(\mathscr{H}_{\text{int}}(x) - \frac{1}{2}\left[n_\mu j_\nu(x)\right]^2\right)\right]: \\ &= \mathrm{e}^{D^*} :\exp\left[\mathrm{i}\int \mathrm{d}^4x\,\mathscr{L}_{\text{int}}(x)\right]: \\ &= T^*\exp\left[\mathrm{i}\int \mathrm{d}^4x\,\mathscr{L}_{\text{int}}(x)\right] . \end{aligned} \tag{11.83}$$

Dividing both sides of this by the vacuum expectation value yields the S-matrix:

$$S = \frac{T^* \exp\left[\mathrm{i} \int \mathrm{d}^4 x \mathscr{L}_{\mathrm{int}}(x)\right]}{\langle 0 | T^* \exp\left[\mathrm{i} \int \mathrm{d}^4 x \mathscr{L}_{\mathrm{int}}(x)\right] | 0 \rangle} . \tag{11.84}$$

This is known as *Matthews' theorem.* Note that this holds only if derivatives of the field operator are included linearly in (11.77).

11.5 Example of Matthews' Theorem with Modification

In the last section, we considered a situation where Matthews' theorem holds true and T and $\mathscr{H}_{\mathrm{int}}$ can be replaced by T^* and $-\mathscr{L}_{\mathrm{int}}$, respectively. In general, we have to replace $\mathscr{H}_{\mathrm{int}}$ by something slightly different from $-\mathscr{L}_{\mathrm{int}}$. We thus generalize Matthews' theorem to

$$T \exp\left[-\mathrm{i} \int \mathrm{d}^4 x \mathscr{H}_{\mathrm{int}}(x)\right] = T^* \exp\left[\mathrm{i} \int \mathrm{d}^4 x \mathscr{L}_{\mathrm{eff}}(x)\right] , \tag{11.85}$$

where $\mathscr{L}_{\mathrm{eff}}$ is an effective interaction Lagrangian density, which is equivalent to $\mathscr{L}_{\mathrm{int}}$ only when Matthews' theorem holds true. As an example of a situation where the theorem does not hold true, we consider the Lagrangian density

$$\mathscr{L} = -\frac{1}{2} D_{ab}(\varphi) \partial_\lambda \varphi_a \partial_\lambda \varphi_b - V(\varphi) , \tag{11.86}$$

where φ_a $(a = 1, 2, \ldots, N)$ are real scalar fields and D_{ab} is a real positive-definite matrix which is a function of φ. In addition, we assume that D and V do not include derivatives of φ. We shall now derive the effective interaction Lagrangian density for this theory.

When $x_0 = y_0$, the canonical commutation relations read

$$\left[\varphi_a(x), \varphi_b(y)\right] = 0 , \tag{11.87}$$

$$\left[\varphi_a(x), \dot{\varphi}_b(y)\right] = \mathrm{i} C_{ab}(\varphi) \delta^3(x - y) , \tag{11.88}$$

where C is the inverse matrix of D, i.e.,

$$\sum_b C_{ab} D_{bc} = \sum_b D_{ab} C_{bc} = \delta_{ac} . \tag{11.89}$$

With summing over repeated indices, the Euler–Lagrange equation gives

$$[\mathscr{L}]_{\varphi_a} = D_{ab}(\Box\varphi_b - j_b) = 0\,, \tag{11.90}$$

$$j_a = C_{ab}\left(\frac{1}{2}\frac{\partial}{\partial\varphi_b}D_{cd} - \frac{\partial}{\partial\varphi_d}D_{bc}\right)\partial_\lambda\varphi_c\partial_\lambda\varphi_d + C_{ab}\frac{\partial}{\partial\varphi_b}V\,. \tag{11.91}$$

We now introduce a generating functional for the Green's functions:

$$\mathscr{T}[J] = \left\langle\mathbf{0}\middle|T\exp\left[-\mathrm{i}\int\mathrm{d}^4x\,J_a(x)\varphi_a(x)\right]\middle|\mathbf{0}\right\rangle. \tag{11.92}$$

Therefore, combining (11.90) and the equal-time commutation relation (11.88),

$$\begin{aligned}\Box_x\langle\mathbf{0}|T\big[\varphi_a(x),\boldsymbol{A},\boldsymbol{B},\ldots\big]|\mathbf{0}\rangle &= \langle\mathbf{0}|T\big[j_a(x),\boldsymbol{A},\boldsymbol{B},\ldots\big]|\mathbf{0}\rangle \\ &\quad+\mathrm{i}\left\langle\mathbf{0}\middle|T\left[C_{ac}(\varphi(x))\frac{\delta}{\delta\varphi_b(x)},\boldsymbol{A},\boldsymbol{B},\ldots\right]\middle|\mathbf{0}\right\rangle+\cdots.\end{aligned} \tag{11.93}$$

Expressing this in terms of the generating functional,

$$\begin{aligned}\Box_x\mathrm{i}\frac{\delta}{\delta J_a(x)}\mathscr{T}[J] &= \Box_x\left\langle\mathbf{0}\middle|T\left[\varphi_a(x),\exp\left(-\mathrm{i}\int\mathrm{d}^4y\,J_c(y)\varphi_c(y)\right)\right]\middle|\mathbf{0}\right\rangle \\ &= \left\langle\mathbf{0}\middle|T\left[j_a(x),\exp\left(-\mathrm{i}\int\mathrm{d}^4y\,J_c(y)\varphi_c(y)\right)\right]\middle|\mathbf{0}\right\rangle \\ &\quad+J_b(x)\left\langle\mathbf{0}\middle|T\left[C_{ab}(\varphi(x)),\exp\left(-\mathrm{i}\int\mathrm{d}^4y\,J_c(y)\varphi_c(y)\right)\right]\middle|\mathbf{0}\right\rangle.\end{aligned} \tag{11.94}$$

Using $\mathscr{T}[J]$, the second term on the right-hand side can be expressed as

$$J_b(x)C_{ab}\left(\mathrm{i}\frac{\delta}{\delta J(x)}\right)\mathscr{T}[J]\,. \tag{11.95}$$

In the interaction picture, considering the T-product which includes only two derivatives of φ,

$$\begin{aligned}T\big[\partial_\mu\varphi(x),\partial_\nu\varphi(y),\varphi(z),\ldots\big] &= T^*\big[\partial_\mu\varphi(x),\partial_\nu\varphi(y),\varphi(z),\ldots\big] \\ &\quad-\mathrm{i}n_\mu n_\nu\delta^4(x-y)T\big[\varphi(z),\ldots\big]\,.\end{aligned} \tag{11.96}$$

Noting that, in this derivation, we have only used the equal-time commutation relations and we have not used the field equation, we can easily make the extension to the Heisenberg picture:

$$T\big[\partial_\mu \varphi_a(x), \partial_\nu \varphi_b(y), \varphi_c(z), \dots\big] = T^*\big[\partial_\mu \varphi_a(x), \partial_\nu \varphi_b(y), \varphi_c(z), \dots\big] \\ -\mathrm{i} n_\mu n_\nu \delta^4(x-y) T\big[C_{ab}(\varphi(x)), \varphi_c(z), \dots\big]\,. \tag{11.97}$$

In particular, if we take $y \to x$ and $\mu = \nu = \lambda$, then since $n_\lambda n_\lambda = -1$,

$$T\big[\partial_\lambda \varphi_a(x), \partial_\lambda \varphi_b(y), \varphi_c(z), \dots\big] = T^*\big[\partial_\lambda \varphi_a(x), \partial_\lambda \varphi_b(y), \varphi_c(z), \dots\big] \\ +\mathrm{i}\delta^4(0) T\big[C_{ab}(\varphi(x)), \varphi_c(z), \dots\big]\,. \tag{11.98}$$

Using this result, the first term on the right-hand side of (11.94) can be written

$$\left\langle \mathbf{0} \middle| T^*\left[j_a(x), \exp\left(-\mathrm{i}\int \mathrm{d}^4 y J_c(y)\varphi_c(y)\right)\right] \middle| \mathbf{0} \right\rangle \\ +\mathrm{i}\delta^4(0)\left\langle \mathbf{0} \middle| T\left[C_{ab}(x)\left(\frac{1}{2}\frac{\partial D_{cd}(x)}{\partial \varphi_b(x)} - \frac{\partial D_{bc}(x)}{\partial \varphi_d(x)}\right) C_{cd}(x) \right.\right. \\ \left.\left. \times \exp\left(-\mathrm{i}\int \mathrm{d}^4 y J_c(y)\varphi_c(y)\right)\right] \middle| \mathbf{0} \right\rangle , \tag{11.99}$$

where $C_{ab}(x)$ is an abbreviation for $C_{ab}(\varphi(x))$, and the same goes for $D_{cd}(x)$.

Thus, $\mathscr{T}[J]$ satisfies

$$\Box_x \mathrm{i}\frac{\delta}{\delta J_a(x)}\mathscr{T}[J] = \left[j_a\left(\mathrm{i}\frac{\delta}{\delta J(x)}\right) + \mathrm{i}\delta^4(0) F_a\left(\mathrm{i}\frac{\delta}{\delta J(x)}\right) + J_b(x) C_{ab}\left(\mathrm{i}\frac{\delta}{\delta J(x)}\right)\right]\mathscr{T}[J]\,, \tag{11.100}$$

where

$$F_a\big(\varphi(x)\big) = C_{ab}\left(\frac{1}{2}\frac{\partial D_{cd}(x)}{\partial \varphi_b(x)} - \frac{\partial D_{bc}(x)}{\partial \varphi_d(x)}\right) C_{cd}(x)\,. \tag{11.101}$$

Note that we can factorize the functional derivative j_a using the T*-product. Rewriting the third term on the right-hand side of (11.100), we have

$$J_b(x) C_{ab}\left(\mathrm{i}\frac{\delta}{\delta J(x)}\right) = C_{ab}\left(\mathrm{i}\frac{\delta}{\delta J(x)}\right) J_b(x) - \mathrm{i}\delta^4(0)\left.\frac{\partial C_{ab}(x)}{\partial \varphi_b(x)}\right|_{\varphi \to \mathrm{i}\delta/\delta J}\,. \tag{11.102}$$

Then the coefficients of $\mathrm{i}\delta^4(0)$ in (11.100) are

$$F_a - \frac{\partial C_{ab}}{\partial \varphi_b} = \frac{1}{2} C_{ab} \frac{\partial D_{cd}}{\partial \varphi_b} C_{cd} - C_{ab} \left(\frac{\partial D_{bc}}{\partial \varphi_d} C_{cd} + D_{bc} \frac{\partial C_{cd}}{\partial \varphi_d} \right)$$
$$= \frac{1}{2} C_{ab} \frac{\partial}{\partial \varphi_b} \ln(\det D) - C_{ab} \frac{\partial}{\partial \varphi_d} \delta_{bd}$$
$$= \frac{1}{2} C_{ab} \frac{\partial}{\partial \varphi_b} \ln(\det D)$$
$$= \frac{1}{2} G_a \,. \tag{11.103}$$

Consequently, (11.100) can be written as

$$\Box_x \mathrm{i} \frac{\delta}{\delta J_a(x)} \mathscr{T}[J] = \left[j_a \left(\mathrm{i} \frac{\delta}{\delta J(x)} \right) + \frac{1}{2} \mathrm{i}\delta^4(0) G_a \left(\mathrm{i} \frac{\delta}{\delta J(x)} \right) + C_{ab} \left(\mathrm{i} \frac{\delta}{\delta J(x)} \right) J_b(x) \right] \mathscr{T}[J] \,. \tag{11.104}$$

Multiplying this on the left by D, we obtain

$$D_{ab} \left(\mathrm{i} \frac{\delta}{\delta J(x)} \right) \Box_x \mathrm{i} \frac{\delta}{\delta J_b(x)} \mathscr{T}[J] \tag{11.105}$$
$$= \left[D_{ab} j_b \left(\mathrm{i} \frac{\delta}{\delta J(x)} \right) + J_a(x) + \frac{1}{2} \mathrm{i}\delta^4(0) \, \frac{\partial}{\partial \varphi_a(x)} \ln \big(\det D(x) \big) \bigg|_{\varphi \to \mathrm{i}\delta/\delta J} \right] \mathscr{T}[J] \,.$$

This is the equation satisfied by the generating functional $\mathscr{T}[J]$.

In order to derive a Feynman–Dyson-like formula in a theory like this, we must express the solution of the equation for $\mathscr{T}[J]$ in terms of φ in the interaction picture. We thus test the following quantity:

$$\text{“}\mathscr{T}[J]\text{”} = \Big\langle 0 \Big| T^* \exp \left[\mathrm{i} \int \mathrm{d}^4 x \big[\mathscr{L}_{\text{int}}(x) - J_a(x) \varphi_a(x) \big] \right] \Big| 0 \Big\rangle \,, \tag{11.106}$$

where

$$\mathscr{L}_{\text{int}} = \mathscr{L} + \frac{1}{2} (\partial_\lambda \varphi_a)^2 \,. \tag{11.107}$$

To obtain the functional equation satisfied by this “$\mathscr{T}[J]$”, we begin with

$$\text{“}\mathscr{T}[J]\text{”} = \exp \left[\mathrm{i} \int \mathrm{d}^4 x \mathscr{L}_{\text{int}} \left(\mathrm{i} \frac{\delta}{\delta J(x)} \right) \right] \Big\langle 0 \Big| \exp \left[-\mathrm{i} \int \mathrm{d}^4 x J_b(x) \varphi_b(x) \right] \Big| 0 \Big\rangle$$
$$= \exp \left[\mathrm{i} \int \mathrm{d}^4 x \mathscr{L}_{\text{int}} \left(\mathrm{i} \frac{\delta}{\delta J(x)} \right) \right] \exp \left[-\frac{1}{2} \int \mathrm{d}^4 x \int \mathrm{d}^4 y J_b(x) \Delta_{\mathrm{F}}(x-y) J_b(y) \right] , \tag{11.108}$$

where we replace φ by $\mathrm{i}\delta/\delta J$ in $\mathscr{L}_{\text{int}}$. For the transformation on the right-hand side, we have used (11.72). Hence,

$$\begin{aligned}
\Box_x \mathrm{i}\frac{\delta}{\delta J_a(x)}\text{“}\mathscr{T}[J]\text{”} &= \exp\left[\mathrm{i}\int \mathrm{d}^4x\,\mathscr{L}_{\text{int}}\left(\mathrm{i}\frac{\delta}{\delta J(x)}\right)\right] J_a(x) \\
&\quad \times \exp\left[-\frac{1}{2}\int \mathrm{d}^4x \int \mathrm{d}^4y\, J_b(x)\Delta_{\mathrm{F}}(x-y)J_b(y)\right] \\
&= J_a(x)\text{“}\mathscr{T}[J]\text{”} + \left[\mathrm{i}\int \mathrm{d}^4x'\,\mathscr{L}_{\text{int}}\left(\mathrm{i}\frac{\delta}{\delta J(x')}\right), J_a(x)\right]\text{“}\mathscr{T}[J]\text{”} \\
&= J_a(x)\text{“}\mathscr{T}[J]\text{”} - [\mathscr{L}_{\text{int}}]_{\varphi_a}\left(\mathrm{i}\frac{\delta}{\delta J(x)}\right)\text{“}\mathscr{T}[J]\text{”} \\
&= \left\{J_a(x) - D_{ab}\left(\mathrm{i}\frac{\delta}{\delta J(x)}\right)\left[\Box_x \mathrm{i}\frac{\delta}{\delta J_b(x)} - j_b\left(\mathrm{i}\frac{\delta}{\delta J(x)}\right)\right]\right. \\
&\quad \left. + \Box_x \mathrm{i}\frac{\delta}{\delta J_a(x)}\right\}\text{“}\mathscr{T}[J]\text{”}\,. \qquad (11.109)
\end{aligned}$$

Since the last term on the right-hand side is the same as the left-hand side, these terms cancel out:

$$D_{ab}\left(\mathrm{i}\frac{\delta}{\delta J(x)}\right)\Box_x \mathrm{i}\frac{\delta}{\delta J_b(x)}\text{“}\mathscr{T}[J]\text{”} = D_{ab}\, j_b\left(\mathrm{i}\frac{\delta}{\delta J(x)}\right)\text{“}\mathscr{T}[J]\text{”} + J_a(x)\text{“}\mathscr{T}[J]\text{”}\,. \qquad (11.110)$$

If we replace V in this equation by

$$V' = V + \frac{\mathrm{i}}{2}\delta^4(0)\ln(\det D)\,, \qquad (11.111)$$

it coincides with (11.105). Hence, $\mathscr{T}[J]$ can be obtained by replacing V by V' in "$\mathscr{T}[J]$" and normalizing in such a way that it is equal to unity when $J = 0$. Thus,

$$\mathscr{T}[J] = \frac{\langle 0|T^* \exp\left\{\mathrm{i}\int \mathrm{d}^4x\left[\mathscr{L}_{\text{eff}}(\varphi(x)) - J_a(x)\varphi_a(x)\right]\right\}|0\rangle}{\langle 0|T^* \exp\left[\mathrm{i}\int \mathrm{d}^4x\,\mathscr{L}_{\text{eff}}(\varphi(x))\right]|0\rangle}\,, \qquad (11.112)$$

where

$$\mathscr{L}_{\text{eff}} = \mathscr{L}_{\text{int}} - \frac{\mathrm{i}}{2}\delta^4(0)\ln(\det D)\,. \qquad (11.113)$$

It thus turns out that, in this example, Matthews' theorem has been modified.

The next question concerns properties of the additional term. First of all, this term is an imaginary number and includes the divergence $\delta^4(0)$. We have the following representation of the δ-function in the space-time coordinates:

$$\delta^4(x) = \frac{1}{(2\pi)^4}\int \mathrm{d}^4k\,\mathrm{e}^{\mathrm{i}k\cdot x} \rightarrow \delta^4(0) = \frac{1}{(2\pi)^4}\int \mathrm{d}^4k\,. \tag{11.114}$$

This therefore gives a fourth-order divergence in momentum space. The imaginary coefficient is connected with the fourth-order divergence. If we compute the closed loop without the additional term, a fourth-order divergence comes about in this theory. Introducing the cutoff Λ in momentum space, we obtain

$$\int \mathrm{d}^4k\frac{k^2}{k^2+m^2-\mathrm{i}\epsilon} \sim \frac{\pi^2}{2}\Lambda^4 - \mathrm{i}\pi^2 m^2\Lambda^2\,. \tag{11.115}$$

This implies that the divergences up to second order will give divergent contributions to the mass and coupling constants, as will be discussed later, but those are real numbers. On the other hand, the fourth-order divergence implied by the equation above, compared to the divergences up to second order, is an imaginary number. Such a contribution cannot be removed by renormalization and breaks the unitarity. Fortunately, the additional term mentioned above automatically cancels this fourth-order divergence, and in this sense a safety mechanism is automatically introduced into the theory.

Although the result mentioned above can also be derived by the path-integral method, this gives the additional term a different interpretation. The path-integral method produces the above result more easily than the method used here. This implies that, since the path-integral method is based on the Lagrangian, we can say that it is more suitable to derive the result including the Lagrangian and the T*-product.

11.6 Reduction Formula in the Interaction Picture

So far we have discussed the S-matrix computational method in the interaction picture. Combining the reduction formula given in Sect. 8.3 and the Gell-Mann–Low relation, we can also express the S-matrix elements in terms of the Green's function in the Heisenberg picture.

In Dyson's formula, the interaction Hamiltonian density or Lagrangian density appear when we express the S-matrix. We can ask ourselves whether it is possible to derive an equation which does not depend explicitly on the form of the interaction.

For simplicity, we consider the charged scalar field and analyze the S-matrix elements for the scattering process

$$a + b \rightarrow b + a\,. \tag{11.116}$$

To do so, we expand $U(\infty, -\infty)$ in normal products using Wick's theorem and determine the coefficient of the term

$$:\varphi_a^\dagger \varphi_b^\dagger \varphi_a \varphi_b: \, . \tag{11.117}$$

Since this requires us to read off the normal product from $U(\infty, -\infty)$ and contract the rest, we need to calculate

$$\left\langle 0 \right| \frac{\delta}{\delta\varphi_a^\dagger(x_1')} \frac{\delta}{\delta\varphi_b^\dagger(x_2')} \frac{\delta}{\delta\varphi_a(x_1)} \frac{\delta}{\delta\varphi_b(x_2)} U(\infty, -\infty) \left| 0 \right\rangle . \tag{11.118}$$

Writing the final state as $|a', b'\rangle$, the S-matrix element is given by

$$\begin{aligned}\langle a', b'|S - 1|a, b\rangle = & \int \mathrm{d}^4x_1'\mathrm{d}^4x_2'\mathrm{d}^4x_1\mathrm{d}^4x_2 \left\langle a' \right| \varphi_a^\dagger(x_1') \left| 0 \right\rangle \left\langle b' \right| \varphi_b^\dagger(x_2') \left| 0 \right\rangle \\ & \times \left\langle 0 \right| \frac{\delta}{\delta\varphi_a^\dagger(x_1')} \frac{\delta}{\delta\varphi_b^\dagger(x_2')} \frac{\delta}{\delta\varphi_a(x_1)} \frac{\delta}{\delta\varphi_b(x_2)} U(\infty, -\infty) \left| 0 \right\rangle \\ & \times \frac{\langle 0|\varphi_a(x_1)|a\rangle \langle 0|\varphi_b(x_2)|b\rangle}{\langle 0|U(\infty, -\infty)|0\rangle} \, . \end{aligned} \tag{11.119}$$

From the reduction formula, the functional derivative is given by

$$(\Box_x^2 - m_a^2) T[\varphi_a(x) \ldots] = \mathrm{i} \frac{\delta}{\delta\varphi_a^\dagger(x)} T[\ldots] \, . \tag{11.120}$$

Denoting the Klein–Gordon operator on the left-hand side by K_x^a and using (11.120), equation (11.119) becomes

$$\begin{aligned}\langle a', b'|S - 1|a, b\rangle = & \langle 0|U(\infty, -\infty)|0\rangle^{-1} \\ & \times \int \mathrm{d}^4x_1'\mathrm{d}^4x_2'\mathrm{d}^4x_1\mathrm{d}^4x_2 \left\langle a' \right| \varphi_a^\dagger(x_1') \left| 0 \right\rangle \left\langle b' \right| \varphi_b^\dagger(x_2') \left| 0 \right\rangle \\ & \times (-\mathrm{i})^4 K_{x_1'}^a K_{x_2'}^b K_{x_1}^a K_{x_2}^b \left\langle 0 \right| T\left[\varphi_a(x_1')\varphi_b(x_2')\varphi_a^\dagger(x_1)\varphi_b^\dagger(x_2) U(\infty, -\infty)\right] \left| 0 \right\rangle \\ & \times \langle 0|\varphi_a(x_1)|a\rangle \langle 0|\varphi_b(x_2)|b\rangle \, . \end{aligned} \tag{11.121}$$

Here $\langle 0|\varphi_a(x)|a\rangle$, which we should call a one-body wave function, has the same structure in both the interaction picture and the Heisenberg picture. The only difference would be a proportionality coefficient. Although it is not trivial to separate the whole Lagrangian density into the free part and the interaction part, the expressions are equal if we use the renormalized interaction picture discussed in the next chapter:

$$\langle 0|\varphi_a(x)|a\rangle = \langle \mathbf{0}|\boldsymbol{\varphi}_a(x)|\boldsymbol{a}\rangle = \frac{1}{\sqrt{2p_0 V}} e^{ip\cdot x} . \tag{11.122}$$

If we use the Gell-Mann–Low relation in this case, (11.121) can be expressed solely in terms of quantities in the Heisenberg picture:

$$\begin{aligned}\langle a', b'|S-1|a,b\rangle = &\int d^4x'_1 d^4x'_2 d^4x_1 d^4x_2 \langle \boldsymbol{a'}|\boldsymbol{\varphi}_a^\dagger(x'_1)|\mathbf{0}\rangle\langle \boldsymbol{b'}|\boldsymbol{\varphi}_b^\dagger(x'_2)|\mathbf{0}\rangle \\ &\times K^a_{x'_1} K^b_{x'_2} K^a_{x_1} K^b_{x_2} \langle \mathbf{0}|T\left[\boldsymbol{\varphi}_a(x'_1)\boldsymbol{\varphi}_b(x'_2)\boldsymbol{\varphi}_a^\dagger(x_1)\boldsymbol{\varphi}_b^\dagger(x_2)\right]|\mathbf{0}\rangle \\ &\times \langle \mathbf{0}|\varphi_a(x_1)|\boldsymbol{a}\rangle\langle \mathbf{0}|\varphi_b(x_2)|\boldsymbol{b}\rangle . \end{aligned} \tag{11.123}$$

Unlike Dyson's formula, in the above expression of the S-matrix element, the explicit form of the interaction does not appear. The problem in the interaction picture of separating the Lagrangian into the free part and the interaction part does not arise. However, in the process of deriving this formula, we have made the assumption (11.7), which is hard to justify. In fact, this result is justified only when we start with the renormalized interaction picture discussed above. Thus, we have to discuss the asymptotic conditions which lead to the above formula in the framework of the Heisenberg picture.

11.7 Asymptotic Conditions

The derivation of the S-matrix element in the Heisenberg picture in the previous section has been based on several assumptions. The question is whether or not we can derive the same result from clearer assumptions. In fact, this was done by Lehmann, Symanzik, and Zimmermann. The assumptions they made are called *asymptotic conditions* [113].

In Sect. 6.3, we introduced two kinds of asymptotic field in connection with the Yang–Feldman formalism. The asymptotic fields φ^{int} and φ^{out} for the real scalar field $\boldsymbol{\varphi}$ satisfy

$$(\Box - m^2)\varphi^{\text{in}}(x) = 0 = (\Box - m^2)\varphi^{\text{out}}(x) , \tag{11.124}$$

$$\left[\varphi^{\text{in}}(x), \varphi^{\text{in}}(y)\right] = i\Delta(x-y) = \left[\varphi^{\text{out}}(x), \varphi^{\text{out}}(y)\right] . \tag{11.125}$$

Since these two types of scalar field are engendered by the same scalar field $\boldsymbol{\varphi}$, we know that φ^{int} and φ^{out} are not independent of one another. This implies that they will not commute. Intuitively speaking, as in the case of the Yang–Feldman formalism,

$$\varphi(x) \longrightarrow \begin{cases} \varphi^{\mathrm{int}}(x)\,, & t \to -\infty\,, \\ \varphi^{\mathrm{out}}(x)\,, & t \to \infty\,. \end{cases} \tag{11.126}$$

As just described, the reason why the field asymptotes to the free field when $t \to \pm\infty$ is that since the particles are then far away from each other and there is no effect from other particles, so they behave like free particles. This fact is closely related to the issue of renormalization discussed in the next chapter, and to make this idea more rigorous we have to express the wave function of a particle, not by a plane wave, but by a wave packet.

We assume that some function $f(x)$ satisfies the conditions

$$(\Box - m^2) f(x) = 0\,, \tag{11.127}$$

$$-\mathrm{i}\int \mathrm{d}^3x \left(f\frac{\partial f^*}{\partial x_0} - f^*\frac{\partial f}{\partial x_0}\right) = 1\,. \tag{11.128}$$

Then corresponding to this f, we introduce the operators

$$\varphi_f(t) = -\mathrm{i}\int \mathrm{d}^3x \left[\varphi(x)\frac{\partial f^*(x)}{\partial x_0} - f^*(x)\frac{\partial \varphi(x)}{\partial x_0}\right]\,, \tag{11.129}$$

$$\varphi_f^{\dagger}(t) = \mathrm{i}\int \mathrm{d}^3x \left[\varphi(x)\frac{\partial f(x)}{\partial x_0} - f(x)\frac{\partial \varphi(x)}{\partial x_0}\right]\,, \tag{11.130}$$

where $t = x_0$. We then define the corresponding asymptotic fields φ_f^{in} and φ_f^{out} by

$$\lim_{\tau\to-\infty}\left(\Phi, \varphi_f(\tau)\Psi\right) = \left(\Phi, \varphi_f^{\mathrm{in}}\Psi\right)\,, \tag{11.131}$$

$$\lim_{\tau\to\infty}\left(\Phi, \varphi_f(\tau)\Psi\right) = \left(\Phi, \varphi_f^{\mathrm{out}}\Psi\right)\,. \tag{11.132}$$

For both states Φ and Ψ, we can define the same asymptotic field if we start with the normalized state vector $\varphi_f^{\dagger}(\tau)$. Note that the right-hand sides of (11.131) and (11.132) no longer depend on the time variable. Such a limit of an operator in the sense of the matrix element is called *weak convergence*, in contrast to strong convergence defined in the sense of the norm.

Next we consider an orthogonal system of wave functions. A wave function here is the matrix element with the vacuum of a field operator in a one-particle state. We consider the set of functions $\{f_\alpha(x)\}$ satisfying (11.127), (11.128), and the condition

$$-\mathrm{i}\int \mathrm{d}^3x \left(f_\alpha\frac{\partial f_\beta^*}{\partial x_0} - f_\beta^*\frac{\partial f_\alpha}{\partial x_0}\right) = \delta_{\alpha\beta}\,. \tag{11.133}$$

The completeness condition for this system of orthogonal functions is

$$\sum_{\alpha} f_\alpha(x) f_\alpha^*(y) = \mathrm{i}\Delta^{(+)}(x-y) . \tag{11.134}$$

We now introduce the complete system of state vectors $\{\Phi^{\text{in}}\}$. Assuming that Φ_0 is the vacuum,

$$\begin{aligned} &\Phi_0 , \\ &\Phi_\alpha^{\text{in}} = \varphi_\alpha^{\dagger\text{in}} \Phi_0 , \\ &\quad \vdots \\ \Phi_{\alpha_1 \ldots \alpha_k}^{\text{in}} &= (p_{\alpha_1 \ldots \alpha_k})^{-1/2} \varphi_{\alpha_1}^{\dagger\text{in}} \ldots \varphi_{\alpha_k}^{\dagger\text{in}} \Phi_0 , \end{aligned} \tag{11.135}$$

where $p_{\alpha_1 \ldots \alpha_k} = n_1! n_2! \ldots n_r!$ and n stands for the number of particles in the same one-particle state in $(\alpha_1, \ldots, \alpha_k)$. Replacing $\varphi^{\dagger\text{int}}$ by $\varphi^{\dagger\text{out}}$, we can also construct the complete system $\{\Phi^{\text{out}}\}$. The S-matrix can be defined as the unitary transformation between these two pairs of complete orthonormal systems. It will be shown later that this definition reproduces the S-matrix elements given in the previous section. In the next chapter, it will be shown that it also coincides with the definition of the S-matrix in the Lippmann–Schwinger theory, viz.,

$$S_{\beta\alpha} = (\Phi_\beta^{\text{out}}, \Phi_\alpha^{\text{in}}) . \tag{11.136}$$

An equivalent definition is

$$\Phi_\alpha^{\text{in}} = S \Phi_\alpha^{\text{out}} . \tag{11.137}$$

It is clear from the definition that, for the two asymptotic fields,

$$(\Phi_\beta^{\text{in}}, \varphi_f^{\text{in}} \Phi_\alpha^{\text{in}}) = (\Phi_\beta^{\text{out}}, \varphi_f^{\text{out}} \Phi_\alpha^{\text{out}}) . \tag{11.138}$$

Combining (11.137) and (11.138),

$$\varphi_f^{\text{out}} = S^{-1} \varphi_f^{\text{in}} S . \tag{11.139}$$

Similarly,

$$\varphi_f^{\dagger\text{out}} = S^{-1} \varphi_f^{\dagger\text{in}} S . \tag{11.140}$$

We introduce $\varphi^{\text{in}}(x)$ by

$$\varphi^{\text{in}}(x) = \sum_{\alpha} \left[f_\alpha^*(x) \varphi_\alpha^{\dagger\text{in}} + f_\alpha(x) \varphi_\alpha^{\text{in}} \right] , \tag{11.141}$$

and define φ^{out} by the same equation, viz.,

$$\varphi^{\text{out}}(x) = S^{-1}\varphi^{\text{in}}(x)S \,. \tag{11.142}$$

Since the latter coincides with (6.32), we see that this is the same as the S-matrix given previously. In relation to (11.122) in the previous section, we mentioned that we have not distinguished whether we take the in-state or the out-state for the one-particle state in the Heisenberg picture. This implies that, *for the stable one-particle state* α, we must have

$$\Phi_\alpha^{\text{in}} = \Phi_\alpha^{\text{out}} \,. \tag{11.143}$$

Combining this with (11.122) in the previous section,

$$(\Box - m^2)\big(\Phi_0, \varphi_0 \Phi_\alpha^{\text{in}}\big) = (\Box - m^2)\big(\Phi_0, \varphi_0 \Phi_\alpha^{\text{out}}\big) \,, \tag{11.144}$$

$$\big(\Phi_0, \varphi_0 \Phi_\alpha^{\text{in}}\big) = \big(\Phi_0, \varphi_0 \Phi_\alpha^{\text{out}}\big) = f_\alpha(x) \,. \tag{11.145}$$

We will discuss this requirement in the context of renormalization in the next chapter.

Equations (11.131) and (11.132), together with the assumption of the existence of the asymptotic field, are called *asymptotic conditions*. Starting from these conditions, we will derive the LSZ reduction formula in the Heisenberg picture, or as they called it, the magic formula (Zauberformel) [113]. We introduce a more concise notation:

$$T(x_1, \ldots, x_n) = T[\boldsymbol{\varphi}(x_1) \ldots \boldsymbol{\varphi}(x_n)] \,, \tag{11.146}$$

$$\tau(x_1, \ldots, x_n) = \big(\Phi_0, T[\boldsymbol{\varphi}(x_1) \ldots \boldsymbol{\varphi}(x_n)]\Phi_0\big) \,, \tag{11.147}$$

$$K_y = \Box_y - m^2 \,, \quad f \frac{\overleftrightarrow{\partial}}{\partial x} g = f \frac{\partial g}{\partial x} - \frac{\partial f}{\partial x} g \,. \tag{11.148}$$

To begin with, we prove the following equation:

$$\big(\Phi_0, T(x_1, \ldots, x_n)\Phi_\alpha^{\text{in}}\big) = -\mathrm{i} \int \mathrm{d}^4 y f_\alpha(y) K_y \tau(x_1, \ldots, x_n, y) \,. \tag{11.149}$$

The left-hand side is

$$\begin{aligned} \text{LHS} &= \lim_{y_0 \to -\infty} \mathrm{i} \int \mathrm{d}^3 y \big(\Phi_0, T(x_1, \ldots, x_n, y)\Phi_0\big) \frac{\overleftrightarrow{\partial}}{\partial y_0} f_\alpha(y) \\ &= \mathrm{i} \lim_{y_0 \to -\infty} \int \mathrm{d}^3 y \, \tau(x_1, \ldots, x_n, y) \frac{\overleftrightarrow{\partial}}{\partial y_0} f_\alpha(y) \,. \end{aligned} \tag{11.150}$$

Noting that $\varphi_\alpha^{\text{out}}$ is an annihilation operator, in the limit $y_0 \to \infty$, we have

$$\begin{aligned} \mathrm{i} \lim_{y_0 \to \infty} \int \mathrm{d}^3 y \, \tau(x_1, \ldots, x_n, y) \frac{\overleftrightarrow{\partial}}{\partial y_0} f_\alpha(y) &= \left(\Phi_0, \varphi_\alpha^{\dagger \text{out}} T(x_1, \ldots, x_n) \Phi_0 \right) \\ &= \left(\varphi_\alpha^{\text{out}} \Phi_0, T(x_1, \ldots, x_n) \Phi_0 \right) = 0 \,. \end{aligned} \tag{11.151}$$

Taking the difference between the two equations above,

$$\begin{aligned} &\left(\Phi_0, T(x_1, \ldots, x_n) \Phi_0^{\text{in}} \right) \\ &\quad = \mathrm{i} \left(\lim_{y_0 \to -\infty} - \lim_{y_0 \to \infty} \right) \int \mathrm{d}^3 y \, \tau(x_1, \ldots, x_n, y) \frac{\overleftrightarrow{\partial}}{\partial y_0} f_\alpha(y) \\ &\quad = -\mathrm{i} \int \mathrm{d}^4 y \frac{\partial}{\partial y_0} \left[\tau(x_1, \ldots, x_n, y) \frac{\overleftrightarrow{\partial}}{\partial y_0} f_\alpha(y) \right] \\ &\quad = -\mathrm{i} \int \mathrm{d}^4 y \left[\tau(x_1, \ldots, x_n, y) \frac{\partial^2 f_\alpha(y)}{\partial y_0^2} - \frac{\partial^2 \tau(x_1, \ldots, x_n, y)}{\partial y_0^2} f_\alpha(y) \right] . \end{aligned} \tag{11.152}$$

We combine this with Green's theorem:

$$\begin{aligned} &\int_V \mathrm{d}^3 y \left[\tau(x_1, \ldots, x_n, y) \cdot \Delta_y f_\alpha(y) - \Delta_y \tau(x_1, \ldots, x_n, y) \cdot f_\alpha(y) \right] \\ &\quad = \int_{S = \partial V} \mathrm{d}S \left[\tau(x_1, \ldots, x_n, y) \cdot \frac{\partial}{\partial y_n} f_\alpha(y) - \frac{\partial}{\partial y_n} \tau(x_1, \ldots, x_n, y) \cdot f_\alpha(y) \right] \\ &\quad \to 0 \,, \quad \partial V \to \infty \,. \end{aligned} \tag{11.153}$$

Here, we use the fact that, since $f_\alpha(y)$ is a wave packet and corresponds to a local wave, it vanishes at a long range. Therefore, combining (11.152) with (11.153),

$$\begin{aligned} &\left(\Phi_0, T(x_1, \ldots, x_n) \Phi_\alpha^{\text{in}} \right) \\ &\quad = \mathrm{i} \int \mathrm{d}^4 y \left[\tau(x_1, \ldots, x_n, y) \cdot \Box_y f_\alpha(y) - \Box_y \tau(x_1, \ldots, x_n, y) \cdot f_\alpha(y) \right] \\ &\quad = -\mathrm{i} \int \mathrm{d}^4 y \, f_\alpha(y) K_y \tau(x_1, \ldots, x_n, y) \,, \end{aligned} \tag{11.154}$$

where we have used $K_y f_\alpha(y) = 0$. The generalization of this equation is

$$\begin{aligned} \left(\Phi_0, T(x_1, \ldots, x_n) \Phi_{\alpha_1 \ldots \alpha_k}^{\text{in}} \right) = (-\mathrm{i})^k \int \mathrm{d}^4 y_1 \ldots \mathrm{d}^4 y_k f_{\alpha_1}(y_1) \ldots f_{\alpha_k}(y_k) K_{y_1} \ldots K_{y_k} \\ \times \tau(x_1, \ldots, x_n, y_1, \ldots, y_k) \end{aligned}$$

Generalizing further, we obtain

$$\begin{aligned}\left(\Phi_\alpha^{\text{out}}, T(x_1,\ldots,x_n)\Phi_\beta^{\text{in}}\right) &= -\mathrm{i}\int \mathrm{d}^4\eta f_{\beta_l}(\eta)K_\eta\left(\Phi_\alpha^{\text{out}}, T(x_1,\ldots,x_n,\eta)\Phi^{\text{in}}_{\beta_1\ldots\beta_{l-1}}\right)\\ &= -\mathrm{i}\int \mathrm{d}^4\zeta f^*_{\alpha_k}(\zeta)K_\zeta\left(\Phi^{\text{out}}_{\alpha_1\ldots\alpha_{k-1}}, T(x_1,\ldots,x_n,\zeta)\Phi_\beta^{\text{in}}\right),\end{aligned} \tag{11.155}$$

where $\alpha = \alpha_1\ldots\alpha_k$, $\beta = \beta_1\ldots\beta_l$. We have assumed that there is no common one-particle state between α and β. Under a similar assumption, the S-matrix element becomes

$$\begin{aligned}S_{\alpha\beta} &= (\Phi_\alpha^{\text{out}}, \Phi_\beta^{\text{in}})\\ &= (-1)^{k+l}\int \mathrm{d}^4\zeta_1\ldots\mathrm{d}^4\zeta_k\mathrm{d}^4\eta_1\ldots\mathrm{d}^4\eta_l f^*_{\alpha_1}(\zeta_1)\ldots f^*_{\alpha_k}(\zeta_k)\\ &\qquad\times f_{\beta_1}(\eta_1)\ldots f_{\beta_l}(\eta_l)K_{\zeta_1}\ldots K_{\zeta_k}K_{\eta_1}\ldots K_{\eta_l}\tau(\zeta_1\ldots\zeta_k\eta_1\ldots\eta_l).\end{aligned} \tag{11.156}$$

What we understand from this is that, when $k = l = 2$, the above expression is basically the same as (11.123).

Although we considered the matrix element in the above derivation, it also holds true for the operator, i.e.,

$$-\mathrm{i}\int \mathrm{d}^4y\, f_\alpha(y)K_yT(x_1,\ldots,x_n,y) = T(x_1,\ldots,x_n)\varphi_\alpha^{\dagger\text{in}} - \varphi_\alpha^{\dagger\text{out}}T(x_1,\ldots,x_n), \tag{11.157}$$

and

$$\mathrm{i}\int \mathrm{d}^4y\, f^*_\alpha(y)K_yT(x_1,\ldots,x_n,y) = T(x_1,\ldots,x_n)\varphi_\alpha^{\text{in}} - \varphi_\alpha^{\text{out}}T(x_1,\ldots,x_n). \tag{11.158}$$

Combining (11.134) and (4.18),

$$\int \mathrm{d}^4y\,\Delta(y-x)K_yT(x_1,\ldots,x_n,y) = T(x_1,\ldots,x_n)\varphi^{\text{in}}(x) - \varphi^{\text{out}}(x)T(x_1,\ldots,x_n). \tag{11.159}$$

This is the operator form of the LSZ reduction formula, which corresponds to (11.120). Putting together

$$S\varphi^{\text{out}}(x) = \varphi^{\text{in}}(x)S \tag{11.160}$$

and (11.159),

$$\int d^4y \Delta(y-x) K_y ST(x_1, \dots, x_n, y) = \left[ST(x_1, \dots, x_n), \varphi^{\text{in}}(x)\right], \tag{11.161}$$

$$\int d^4y \Delta(x-y) K_y ST(x_1, \dots, x_n, y) = \left[\varphi^{\text{in}}(x), ST(x_1, \dots, x_n)\right]. \tag{11.161'}$$

Using the above recursively,

$$\begin{aligned}\int d^4y_1 \dots d^4y_l \Delta(z_1 - y_1) \dots \Delta(z_l - y_l) K_{y_1} \dots K_{y_l} ST(x_1, \dots, x_n, y_1, \dots, y_l) \\ = [\varphi^{\text{in}}(z_1), [\varphi^{\text{in}}(z_2), [\dots [\varphi^{\text{in}}(z_l), ST(x_1, \dots, x_n)] \dots] .\end{aligned} \tag{11.162}$$

Taking the vacuum expectation value of this and using one of the renormalization conditions mentioned in the next chapter, viz.,

$$S\Phi_0 = \Phi_0 , \tag{11.163}$$

we obtain

$$\begin{aligned}\int d^4y_1 \dots d^4y_l \Delta(z_1 - y_1) \dots \Delta(z_l - y_l) K_{y_1} \dots K_{y_l} \tau(x_1, \dots, x_n, y_1, \dots, y_l) \\ = \left(\Phi_0, [\varphi^{\text{int}}(z_1), [\varphi^{\text{in}}(z_2), [\dots [\varphi^{\text{in}}(z_l), ST(x_1, \dots, x_n)] \dots]\Phi_0\right) .\end{aligned} \tag{11.164}$$

For $n = 0$,

$$\begin{aligned}\int d^4y_1 \dots d^4y_l \Delta(z_1 - y_1) \dots \Delta(z_l - y_l) K_{y_1} \dots K_{y_l} \tau(y_1, \dots, y_l) \\ = \left(\Phi_0, [\varphi^{\text{int}}(z_1), [\varphi^{\text{in}}(z_2), [\dots [\varphi^{\text{in}}(z_l), S] \dots]\Phi_0\right) .\end{aligned} \tag{11.165}$$

The operator form of the S-matrix is determined by (11.165). Expanding the S-matrix in the normal product form based on Wick's theorem, we have

$$S = \sum_{l=0}^{\infty} \frac{1}{l!} \int d^4y_1 \dots d^4y_l c(y_1, \dots, y_l) :\varphi^{\text{in}}(y_1) \dots \varphi^{\text{in}}(y_l): , \tag{11.166}$$

where we have assumed that c is symmetric with respect to $y_1, y_2, \dots, y_l$. When we insert (11.166) into (11.165), what is left on the right-hand side is only the term including the normal ordered product of l operators, whence

$$
\begin{aligned}
&(\Phi_0, [\varphi^{\text{in}}(z_1), [\dots[\varphi^{\text{in}}(z_l), S]\dots]\Phi_0) \\
&\qquad = \mathrm{i}^l \int \mathrm{d}^4 y_1 \dots \mathrm{d}^4 y_l \Delta(z_1 - y_1)\dots\Delta(z_l - y_l) c(y_1, \dots, y_l) \\
&\qquad = \int \mathrm{d}^4 y_1 \dots \mathrm{d}^4 y_l \Delta(z_1 - y_1)\dots\Delta(z_l - y_l) K_{y_1}\dots K_{y_l} \tau(y_1, \dots, y_l) .
\end{aligned}
\tag{11.167}
$$

Then c can be determined uniquely from (11.167), at least on the mass shell, i.e., for the Fourier components satisfying the Einstein energy–momentum dispersion relation. Moreover, since only the value of c on the mass shell contributes to (11.166),

$$
c(y_1, \dots, y_l) = (-\mathrm{i})^l K_{y_1}\dots K_{y_l} \tau(y_1, \dots, y_l) . \tag{11.168}
$$

Substituting this into (11.166),

$$
S = \sum_{l=0}^{\infty} \int \mathrm{d}^4 y_1 \dots \mathrm{d}^4 y_l K_{y_1}\dots K_{y_l} \tau(y_1, \dots, y_l) :\varphi^{\text{in}}(y_1)\dots\varphi^{\text{in}}(y_l): , \tag{11.169}
$$

where we have assumed that the term corresponding to $l = 0$ is equal to unity. This is the operator form of the S-matrix. In addition, going back to (11.164), we have

$$
\begin{aligned}
ST(x_1, \dots, x_n) = \sum_{l=0}^{\infty} \frac{(-\mathrm{i})^l}{l!} \int \mathrm{d}^4 y_1 \dots \mathrm{d}^4 y_l K_{y_1}\dots K_{y_l} \tau(x_1, \dots, x_n, y_1, \dots, y_l) \\
\times :\varphi^{\text{in}}(y_1)\dots\varphi^{\text{in}}(y_l): .
\end{aligned}
\tag{11.170}
$$

As just described, many reduction formulae can be obtained from the asymptotic conditions. Indeed, the last formula effectively defines the quantization method for fields.

11.8 Unitarity Condition on the Green's Function

The unitarity of the S-matrix is obvious as long as the S-matrix element is defined by (11.136) as a transition matrix between two complete orthonormal systems $\{\Phi^{\text{in}}\}$ and $\{\Phi^{\text{out}}\}$. When the asymptotic states form complete systems like this, we speak of *asymptotic completeness*. In this section, we extend the unitarity of the S-matrix from unitarity on the mass shell to unitarity off the mass shell. This can be expressed by the unitarity condition for the Green's functions.

To begin with, we consider the operator

$$T \exp\left[- \mathrm{i} \exp \int \mathrm{d}^4 x\, J(x) \varphi(x) \right] . \tag{11.171}$$

This operator is unitary, and denoting the operator for inverse time-ordering by $\tilde{T}$,

$$T \exp\left[- \mathrm{i} \int \mathrm{d}^4 x\; J(x) \varphi(x) \right] \tilde{T} \exp\left[\mathrm{i} \int \mathrm{d}^4 y\, J(y) \varphi(y) \right] = 1 . \tag{11.172}$$

Functionally differentiating this equation n times with respect to J and subsequently setting $J = 0$,

$$\sum_{\text{comb}} (-\mathrm{i})^k \mathrm{i}^{n-k} T(x'_1, \ldots, x'_k) \tilde{T}(x'_{k+1}, \ldots, x'_n) = 0 , \tag{11.173}$$

where $(x'_1, \ldots, x'_n)$ is a permutation of $(x_1, \ldots, x_n)$ and we sum over all ways of dividing a set of n variables into two complementary subsets. We take the vacuum expectation value of this equation. Inserting the complete system $\{\Phi^{\text{in}}\}$ between T and $\tilde{T}$, we use the equation

$$\begin{aligned} \left(\Phi_0, T(x_1, \ldots, x_k) \Phi^{\text{in}}_{\alpha_1 \ldots \alpha_l} \right) = (-\mathrm{i})^l \int \mathrm{d}^4 u_1 \ldots \mathrm{d}^4 u_l\, f_{\alpha_1}(u_1) \ldots f_{\alpha_l}(u_l) \\ \times K_{u_1} \ldots K_{u_l} \tau(x_1, \ldots, x_k, u_1, \ldots, u_l) \end{aligned} \tag{11.174}$$

and its complex conjugate

$$\begin{aligned} \left(\Phi^{\text{in}}_{\alpha_1 \ldots \alpha_l}, \tilde{T}(x_1, \ldots, x_k) \Phi_0 \right) = \mathrm{i}^l \int \mathrm{d}^4 v_1 \ldots \mathrm{d}^4 v_l\, f^*_{\alpha_1}(v_1) \ldots f^*_{\alpha_l}(v_l) \\ \times K_{v_1} \ldots K_{v_l} \tau(x_{k+1}, \ldots, x_n, v_1, \ldots, v_l). \end{aligned} \tag{11.175}$$

Then using (11.134), we sum over intermediate states. We use the notation

$$\bar{\tau}(x_1, \ldots, x_n) = (-\mathrm{i})^n K_{x_1} \ldots K_{x_n} \tau(x_1, \ldots, x_n) . \tag{11.176}$$

The Fourier transformation is the S-matrix element itself if all momenta are on the mass shell. Rewriting the expectation value of (11.173),

$$
\begin{aligned}
0 = \bar{\tau}(x_1, \ldots, x_n) + \bar{\tau}^*(x_1, \ldots, x_n) \qquad (11.177) \\
+ \sum_{\text{comb}}{}' \sum_{l=0}^{\infty} \frac{\mathrm{i}^l}{l!} \int \mathrm{d}^4 u_1 \ldots \mathrm{d}^4 u_l \mathrm{d}^4 v_1 \ldots \mathrm{d}^4 v_l \bar{\tau}(x_1', \ldots, x_k', u_1, \ldots, u_l) \\
\times \Delta^{(+)}(u_1 - v_1) \ldots \Delta^{(+)}(u_l - v_l) \bar{\tau}^*(x_{k+1}', \ldots, x_n', v_1, \ldots, v_l) \, ,
\end{aligned}
$$

where $l!$ in the denominator is a factor introduced to ensure that we do not count the same state more than once, and the prime on $\sum'$ indicates that we neglect $k = 0$ and $k = n$. Restricting all momenta to the mass shell in the Fourier transformation of this equation, it becomes the condition for unitarity. Hence, when the momenta lie outside the mass shell, the Fourier transformation can be taken as its generalization. We call (11.177) the *generalized unitarity condition*. In fact, it should be obvious from the following discussion that (11.177) yields the unitarity condition for the S-matrix on the mass shell. Using

$$
S = 1 + \sum_{l=1}^{\infty} \frac{1}{l!} \int \mathrm{d}^4 x_1 \ldots \mathrm{d}^4 x_l \bar{\tau}(x_1, \ldots, x_l) :\varphi^{\text{in}}(x_1) \ldots \varphi^{\text{in}}(x_l): \, , \qquad (11.178)
$$

$$
S^\dagger = 1 + \sum_{l=1}^{\infty} \frac{1}{l!} \int \mathrm{d}^4 x_1 \ldots \mathrm{d}^4 x_l \bar{\tau}^*(x_1, \ldots, x_l) :\varphi^{\text{in}}(x_1) \ldots \varphi^{\text{in}}(x_l): \, , \qquad (11.179)
$$

we expand $SS^\dagger$ as a sum of normal products:

$$
\begin{aligned}
&:\varphi^{\text{in}}(x_1) \ldots \varphi^{\text{in}}(x_l): :\varphi^{\text{in}}(y_1) \ldots \varphi^{\text{in}}(y_m): \\
&\quad = :\varphi^{\text{in}}(x_1) \ldots \varphi^{\text{in}}(x_l) \varphi^{\text{in}}(y_1) \ldots \varphi^{\text{in}}(y_m): \\
&\qquad + \sum_{\text{comb}} \mathrm{i} \Delta^{(+)}(x_1' - y_1') :\varphi^{\text{in}}(x_2') \ldots \varphi^{\text{in}}(x_l') \varphi^{\text{in}}(y_2') \ldots \varphi^{\text{in}}(y_m'): \\
&\qquad + \text{(terms with two or more contractions)}. \qquad (11.180)
\end{aligned}
$$

Therefore, looking at the coefficients of each of the normal products, we see that

$$
:\varphi^{\text{in}}(x_1) \ldots \varphi^{\text{in}}(x_n):
$$

is equal to the right-hand side of (11.177). Thus,

$$
SS^\dagger = 1 \, . \qquad (11.181)
$$

Similarly,

$$S^{\dagger} S = 1 \ , \tag{11.182}$$

using (11.177) with τ replaced by τ^*. As claimed, the unitarity for the S-matrix and for Green's function are consequences of asymptotic completeness.

In the above, we considered the T-product of the Heisenberg operator. Next, we introduce the Green's functions based on the retarded product, introduced in Sect. 6.2.

11.9 Retarded Green's Functions

If $A(x)$ is a local field, its retarded product is defined by

$$\begin{aligned} &R[\boldsymbol{A}(x) : \boldsymbol{\varphi}(x_1)\ldots\boldsymbol{\varphi}(x_n)] \\ &\quad = (-\mathrm{i})^n \sum_{p} \theta(x - x_1')\ldots\theta(x_{n-1}' - x_n')\big[\ldots[\boldsymbol{A}(x), \boldsymbol{\varphi}(x_1')]\ldots\boldsymbol{\varphi}(x_n')\big] \ , \end{aligned} \tag{11.183}$$

where $\theta(x)$ stands for $\theta(x_0)$, with $x_1', \ldots, x_n'$ a permutation of $x_1, \ldots, x_n$ and summation over all permutations. The only permutations to contribute are those satisfying $x_1' > x_2' > \ldots > x_n'$ for the time variables. We introduce the unitary operator (11.171), denoting it by U :

$$U = T \exp\left[-\mathrm{i} \int \mathrm{d}^4 x \, J(x)\boldsymbol{\varphi}(x)\right] . \tag{11.184}$$

Therefore,

$$U^{-1} = U^{\dagger} = \tilde{T} \exp\left[\mathrm{i} \int \mathrm{d}^4 x \, J(x)\boldsymbol{\varphi}(x)\right] . \tag{11.185}$$

We now introduce the generating functional

$$A_R[x, J] = U^{\dagger} T[U\boldsymbol{A}(x)] \ . \tag{11.186}$$

Therefore, it is easy to check that the R-product above can be expressed by

$$R[\boldsymbol{A}(x) : \boldsymbol{\varphi}(x_1)\ldots\boldsymbol{\varphi}(x_n)] = \left.\frac{\delta^n A_R[x, J]}{\delta J(x_1)\ldots\delta J(x_n)}\right|_{J=0} . \tag{11.187}$$

Directly from the definition,

$$\frac{\delta}{\delta J(y)} A_R[x, J] = -\mathrm{i}\theta(x-y)\big[A_R[x, J], \varphi_R[y, J]\big] . \tag{11.188}$$

In particular, taking $\boldsymbol{A} = \boldsymbol{\varphi}$, we have

$$\frac{\delta}{\delta J(y)} \varphi_R[x, J] - \frac{\delta}{\delta J(x)} \varphi_R[y, J] + \mathrm{i}\big[\varphi_R[x, J], \varphi_R[y, J]\big] = 0 . \tag{11.189}$$

This is called a *unitarity condition*. It corresponds to (11.173) in the case of the T-product. In addition, functionally differentiating A_R a total of n times with respect to J,

$$\begin{aligned} &R[\boldsymbol{A}(x) : \boldsymbol{\varphi}(x_1) \ldots \boldsymbol{\varphi}(x_n)] \\ &\quad = \sum_{\text{comb}} \mathrm{i}^k \tilde{T}\big[\boldsymbol{\varphi}(x_1') \ldots \boldsymbol{\varphi}(x_k')\big](-\mathrm{i})^{n-k} T\big[\boldsymbol{\varphi}(x_{k+1}') \ldots \boldsymbol{\varphi}(x_n')\boldsymbol{A}(x)\big] . \end{aligned} \tag{11.190}$$

Then taking the Hermitian conjugate of the reduction formula for the T-product, viz.,

$$\int \mathrm{d}^4 y \Delta(x-y) K_y S T(x_1, \ldots, x_n, y) = \big[\varphi^{\mathrm{int}}(x), S T(x_1, \ldots, x_n)\big] , \tag{11.191}$$

we obtain

$$\int \mathrm{d}^4 y \Delta(x-y) K_y \tilde{T}(x_1, \ldots, x_n, y) S^\dagger = -\big[\varphi^{\mathrm{int}}(x), \tilde{T}(x_1, \ldots, x_n) S^\dagger\big] . \tag{11.192}$$

Combining the three equations above,

$$\int \mathrm{d}^4 y \Delta(x-y) K_y R(w : x_1, \ldots, x_n, y) = -\mathrm{i}\big[\varphi^{\mathrm{int}}(x), R(w : x_1, \ldots, x_n)\big] , \tag{11.193}$$

where

$$R(w : x_1, \ldots, x_n, y) = R\big[\boldsymbol{A}(w) : \boldsymbol{\varphi}(x_1) \ldots \boldsymbol{\varphi}(x_n)\big] . \tag{11.194}$$

Then using (11.193) iteratively,

$$\int d^4y_1 \dots d^4y_l \Delta(z_1 - y_1) \dots \Delta(z_l - y_l) K_{y_1} \dots K_{y_l} R(w : x_1, \dots, x_n, y_1, \dots, y_l)$$
$$= (-i)^l [\varphi^{in}(z_1), \left[\dots [\varphi^{in}(z_l), R(w : x_1, \dots, x_n)] \dots \right] . \tag{11.195}$$

We now expand R as a sum of normal products:

$$R(w : x_1, \dots, x_n) = \sum_{l=0}^{\infty} \frac{1}{l!} \int d^4y_1 \dots d^4y_l f(w : x_1, \dots, x_n, y_1, \dots, y_l) :\varphi^{in}(y_1) \dots \varphi^{in}(y_l): , \tag{11.196}$$

where we have assumed that f is a symmetric function with respect to $y_1, \dots, y_l$. Inserting this into (11.195) and taking the vacuum expectation value,

$$(-i)^l \big(\Phi_0, [\varphi^{in}(z_1), [\dots [\varphi^{in}(z_l), R(w : x_1, \dots, x_n)] \dots] \Phi_0\big) \tag{11.197}$$
$$= \int d^4y_1 \dots d^4y_l \Delta(z_1 - y_1) \dots \Delta(z_l - y_l) f(w : x_1, \dots, x_n, y_1, \dots, y_l) .$$

However, from (11.195), the right-hand side is equivalent to

$$\int d^4y_1 \dots d^4y_l \Delta(z_1 - y_1) \dots \Delta(z_l - y_l) r(w : x_1, \dots, x_n, y_1, \dots, y_l) , \tag{11.198}$$

where

$$r(w : x_1, \dots, x_n, y_1, \dots, y_l) = K_{x_1} \dots K_{x_n} (\Phi_0, R(w : x_1, \dots, x_n) \Phi_0) . \tag{11.199}$$

Thus, if the momenta corresponding to $y_1, \dots, y_l$ are on the mass shell,

$$f(w : x_1, \dots, x_n, y_1, \dots, y_l) = r(w : x_1, \dots, x_n, y_1, \dots, y_l) . \tag{11.200}$$

Since only those on the mass shell exert any influence,

$$R(w : x_1, \dots, x_n) \tag{11.201}$$
$$= \sum_{l=0}^{\infty} \frac{1}{l!} \int d^4y_1 \dots d^4y_l r(w : x_1, \dots, x_n, y_1, \dots, y_l) :\varphi^{in}(y_1) \dots \varphi^{in}(y_l): .$$

In particular, for $n = 0$,

$$\boldsymbol{A}(w) = \sum_{l=0}^{\infty} \frac{1}{l!} \int \mathrm{d}^4 y_1 \ldots \mathrm{d}^4 y_l r(w : y_1, \ldots, y_l) :\varphi^{\mathrm{in}}(y_1) \ldots \varphi^{\mathrm{in}}(y_l): \ . \tag{11.202}$$

Moreover, if we take $\boldsymbol{A} = \boldsymbol{\varphi}$, then when $(\Phi_0, \boldsymbol{\varphi}(x)\Phi_0) = 0$, we have

$$\boldsymbol{\varphi}(x) = \varphi^{\mathrm{in}}(x) + \sum_{l=2}^{\infty} \frac{1}{l!} \int \mathrm{d}^4 y_1 \ldots \mathrm{d}^4 y_l r(x : y_1, \ldots, y_l) :\varphi^{\mathrm{in}}(y_1) \ldots \varphi(y_l): \ . \tag{11.203}$$

This gives the formal solution to the Yang–Feldman equation introduced in Sect. 6.3. In addition, from (11.189),

$$R(x : y, x_1, \ldots, x_n) - R(y : x, x_1, \ldots, x_n) + \mathrm{i} \sum_{\mathrm{comb}} \left[R(x : x'_1, \ldots, x'_k), R(y : x'_{k+1}, \ldots, x'_n) \right] = 0 \ . \tag{11.204}$$

Taking the vacuum expectation value of this equation and using the reduction formula obtained by inserting the complete system $\{\Phi^{\mathrm{in}}\}$, we obtain a non-linear equation for the system $(\Phi_0, R(x : x_1, \ldots, x_n)\Phi_0)$. This is also one of the generalized unitarity conditions.

Both the in- and the out-states appear in the reduction formula for the T-product, while only the in-states appear in the R-products.

Chapter 12
Renormalization Theory

So far we have shown only the lowest order calculations, but when computing higher order corrections, divergences must show up. A method for deducing finite consequences by a suitable interpretation is called a *renormalization theory*. As mentioned once or twice before, the basic idea of such a formalism is to specify a way of separating the Lagrangian density into the free part and the interaction part. This grouping is related to the definition of the interaction picture. We call the interaction picture defined by the correct grouping a *renormalized interaction picture*. Several properties of the Green's functions discussed in the previous chapter hold true in the renormalized interaction picture, while they may not hold true in other pictures.

In this chapter, in order to show that such a separation of the Lagrangian or the Hamiltonian is not necessarily trivial, we first review the scattering theory in non-relativistic quantum mechanics. The formal system discussed here has many similarities with the S-matrix theory based on the reduction formula given in the last chapter.

12.1 Lippmann–Schwinger Equation

In this section, we introduce the formal logic for the standard quantum mechanical system. This theory has a lot in common with the theory of Green's functions. What is important in the scattering problem is the way we formulate the boundary conditions. This issue is discussed in detail in my book, "Relativistic Quantum Mechanics" [78], but here we shall present it in a slightly different order.

In this section, we formulate the scattering problem using the notion of Stueckelberg causality:

> Assuming that the potential $V(t)$ is a function of time, if $V(t) = 0$ in the region $t < T$, then there exists no scattered wave for $t < T$.

K. Nishijima, *Quantum Field Theory*,
https://doi.org/10.1007/978-94-024-2190-3_12

The principle above has already been used for the quantization of free fields in Sect. 4.3 and for the derivation of the Yang–Feldman equation in Sect. 6.3. Here we apply it to the non-relativistic formulation. We consider the Schrödinger equation

$$\mathrm{i}\frac{\partial}{\partial t}\psi(t) = [H_0 + V(t)]\psi(t) , \tag{12.1}$$

where we have omitted the spatial coordinate for simplicity, and $V(t)$ is defined by

$$V(t) = \begin{cases} V , & t > T , \\ 0 , & t < T . \end{cases} \tag{12.2}$$

For $t < T$,

$$\mathrm{i}\frac{\partial}{\partial t}\psi(t) = H_0\psi(t) . \tag{12.3}$$

Since this is the equation for the incident wave, we write its solution as

$$\psi(t) = \psi_{\text{in}}(t) , \quad H_0\psi_{\text{in}}(t) = E\psi_{\text{in}}(t) . \tag{12.4}$$

Therefore, the scattered wave appears at a generic time t and its equation is

$$\psi(t) = \psi_{\text{in}}(t) + \psi_{\text{scatt}}(t) , \tag{12.5}$$

$$\left(\mathrm{i}\frac{\partial}{\partial t} - H_0\right)\psi_{\text{scatt}}(t) = V(t)\psi(t) . \tag{12.6}$$

To solve this equation, we introduce the Green's function, which is the solution of the equation

$$\left(\mathrm{i}\frac{\partial}{\partial t} - H_0\right)K_{\text{ret}}(t, \boldsymbol{x} : t', x') = \delta(t - t')\delta^3(\boldsymbol{x} - \boldsymbol{x}') , \tag{12.7}$$

$$K_{\text{ret}}(t, \boldsymbol{x} : t', \boldsymbol{x}') = 0 , \quad t < t' . \tag{12.8}$$

Therefore, the solution of (12.6) satisfying the causality condition is

$$\psi_{\text{scatt}}(t, \boldsymbol{x}) = \int \mathrm{d}t'\mathrm{d}^3x' K_{\text{ret}}(t, \boldsymbol{x} : t', \boldsymbol{x}')V(t', \boldsymbol{x}')\psi(t', x') . \tag{12.9}$$

Here, taking the limit $T \to -\infty$, the t-dependence of the potential disappears, whence

$$\psi_{\text{scatt}}(t, \boldsymbol{x}) = \int \mathrm{d}t'\mathrm{d}^3x' K_{\text{ret}}(t, \boldsymbol{x} : t', \boldsymbol{x}')V(\boldsymbol{x}')\psi(t', x') . \tag{12.10}$$

Then, if H_0 does not depend on the space-time coordinate,

$$K_{\text{ret}}(t, \boldsymbol{x} : t', \boldsymbol{x}') = K_{\text{ret}}(t - t', \boldsymbol{x} - \boldsymbol{x}') \ . \tag{12.11}$$

Thus,

$$\psi(t, \boldsymbol{x}) = \psi_{\text{in}}(t, \boldsymbol{x}) + \int \mathrm{d}t'\mathrm{d}^3x' K_{\text{ret}}(t - t', \boldsymbol{x} - \boldsymbol{x}')V(\boldsymbol{x}')\psi(t', \boldsymbol{x}') \ . \tag{12.12}$$

This is the integral equation governing the scattering, and corresponds to the Yang–Feldman equation. To remove the time variable, we take

$$\psi_{\text{in}}(t, \boldsymbol{x}) = \mathrm{e}^{-\mathrm{i}Et}\psi_{\text{in}}(\boldsymbol{x}) \ , \quad \psi(t, \boldsymbol{x}) = \mathrm{e}^{-\mathrm{i}Et}\psi(\boldsymbol{x}) \ , \tag{12.13}$$

leading to

$$\psi(\boldsymbol{x}) = \psi_{\text{in}}(\boldsymbol{x}) + \int \mathrm{d}^3x' G(\boldsymbol{x} - \boldsymbol{x}' : E)V(\boldsymbol{x}')\psi(\boldsymbol{x}') \ , \tag{12.14}$$

where

$$G(\boldsymbol{x} - \boldsymbol{x}' : E) = \int \mathrm{d}t' \mathrm{e}^{\mathrm{i}E(t-t')} K_{\text{ret}}(t - t', \boldsymbol{x} - \boldsymbol{x}') \ . \tag{12.15}$$

We now introduce the Fourier representation of the retarded Green's function:

$$K_{\text{ret}}(t, \boldsymbol{x}) = \frac{1}{(2\pi)^4} \int \mathrm{d}E\, \mathrm{d}^3p\, \mathrm{e}^{\mathrm{i}\boldsymbol{p}\boldsymbol{x} - \mathrm{i}Et} K(E, \boldsymbol{p}) \ . \tag{12.16}$$

Furthermore, we restrict H_0 to the form

$$H_0 = \frac{\boldsymbol{p}^2}{2m} \ . \tag{12.17}$$

Thus, the equation for $K(E, p)$ obtained from (12.7) can be written in the form

$$\left(E - \frac{\boldsymbol{p}^2}{2m}\right) K(E, \boldsymbol{p}) = 1 \ . \tag{12.18}$$

As will be shown later, the solution of this equation satisfying the boundary condition (12.8) is

$$K(E - \boldsymbol{p}) = \left(E - \frac{\boldsymbol{p}^2}{2m} + \mathrm{i}\epsilon\right)^{-1} \ , \tag{12.19}$$

Fig. 12.1 Path of E-integration in (12.22)

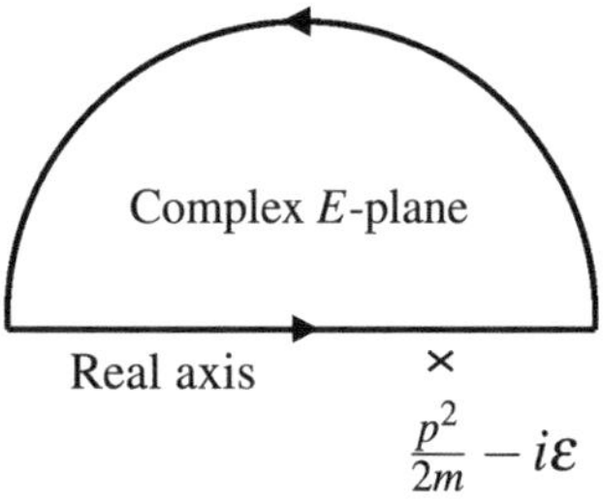

where ϵ is a positive infinitesimal. Taking $E = k^2/2m$, Eqs. (12.15) and (12.16) imply

$$G(\boldsymbol{x} - \boldsymbol{x}' : E) = -\frac{m}{2\pi} \frac{\mathrm{e}^{\mathrm{i}k|\boldsymbol{x}-\boldsymbol{x}'|}}{|\boldsymbol{x} - \boldsymbol{x}'|} . \tag{12.20}$$

Thus, (12.14) can be written in the form

$$\psi(\boldsymbol{x}) = \psi_{\mathrm{in}}(\boldsymbol{x}) - \frac{m}{2\pi} \int \mathrm{d}^3x' \frac{\mathrm{e}^{\mathrm{i}k|\boldsymbol{x}-\boldsymbol{x}'|}}{|\boldsymbol{x} - \boldsymbol{x}'|} V(\boldsymbol{x}')\psi(\boldsymbol{x}') . \tag{12.21}$$

This equation indicates that the scattered wave becomes an outward-directed spherical wave as a consequence of causality. When $t - t' < 0$,

$$K_{\mathrm{ret}}(t - t', \boldsymbol{x} - \boldsymbol{x}') = \frac{1}{(2\pi)^4} \int \mathrm{d}^3p\, \mathrm{e}^{\mathrm{i}\boldsymbol{p}(\boldsymbol{x}-\boldsymbol{x}')} \int \mathrm{d}E \frac{\mathrm{e}^{-\mathrm{i}E(t-t')}}{E - \frac{p^2}{2m} + \mathrm{i}\epsilon} . \tag{12.22}$$

Then, since $-\mathrm{i}E(t - t') = \mathrm{i}E|t - t'|$, if $\mathrm{Im}\, E > 0$, this exponential decreases exponentially with $\mathrm{Im}E$. Hence, selecting the integration path as in Fig. 12.1 and taking into account the fact that the pole is located outside the semicircle, it vanishes by Cauchy's theorem. Thus, if $t - t' < 0$, then $K_{\mathrm{ret}} = 0$ and we see that (12.8) is satisfied.

Clearly, using (12.22) and taking $\Psi = \psi$ and $\Phi = \psi_{\mathrm{in}}$, Eq. (12.21) can be written formally as

$$\Psi = \Phi + \frac{1}{E - H_0 + \mathrm{i}\epsilon} V\Psi \quad (E > 0) , \tag{12.23}$$

In fact, looking at the Fourier representation of the retarded Green's function (12.22), we see that this corresponds to the operator which gives the energy denominator in the equation above. From (12.23) and (12.3),

$$(E - H_0)\Psi = V\Psi , \quad (E - H_0)\Phi = 0 . \tag{12.24}$$

Moreover, for the bound state ($E < 0$), the inverse of $(E - H_0)$ is uniquely defined, so we obtain the homogenous equation

$$\Psi = \frac{1}{E - H_0} V \Phi \quad (E < 0) . \tag{12.25}$$

The pair of Eqs. (12.23) and (12.25) is known collectively as the *Lippmann–Schwinger equation* [114]. The solution of (12.23) consists of the incident wave and the outward-going spherical wave, and it can be written as

$$\Psi_a^{(+)} = \Phi_a + \frac{1}{E_a - H_0 + i\epsilon} V \Psi_a^{(+)} . \tag{12.26}$$

Mathematically, it is useful to consider the solution which consists of the incident wave and the inward-going spherical wave. The equation for such a state can be obtained by replacing $i\epsilon$ with $-i\epsilon$:

$$\Psi_a^{(-)} = \Phi_a + \frac{1}{E_a - H_0 - i\epsilon} V \Psi_a^{(-)} . \tag{12.27}$$

The two cases are solutions of the same Schrödinger equation under different boundary conditions:

$$(E_a - H_0)\Psi_a^{(\pm)} = V \Psi_a^{(\pm)} . \tag{12.28}$$

Thus, $\{\Psi_a^{(+)}\}$ and $\{\Psi_a^{(-)}\}$ form complete systems separately. Solving (12.26) sequentially,

$$\begin{aligned} \Psi_a^{(+)} &= \Phi_a + \frac{1}{E_a - H_0 + i\epsilon} \left(1 + V \frac{1}{E_a - H_0 + i\epsilon} + \dots \right) V \Phi_a \\ &= \Phi_a + \frac{1}{E_a - H_0 + i\epsilon} \left(1 - V \frac{1}{E_a - H_0 + i\epsilon}\right)^{-1} V \Phi_a . \end{aligned} \tag{12.29}$$

Then using $A^{-1}B^{-1} = (BA)^{-1}$,

$$\Psi_a^{(+)} = \Phi_a + \frac{1}{E_a - H + i\epsilon} V \Phi_a \quad (H = H_0 + V) . \tag{12.30}$$

We call this the *Chew–Goldberger formal solution* [115]. Another solution is

$$\Psi_a^{(-)} = \Phi_a + \frac{1}{E_a - H - i\epsilon} V \Phi_a . \tag{12.31}$$

Although this formal solution is in fact not useful for solving the problem, it is very useful for deriving the general properties of the S-matrix, and we will show this application below.

We assume that the state of the incident wave is suitably normalized, i.e.,

$$(\Phi_b, \Phi_a) = \delta_{ba} \ . \tag{12.32}$$

Therefore, combining the LS equation with the CG formal solution,

$$\begin{aligned}
\left(\Psi_b^{(+)}, \Psi_a^{(+)}\right) &= \left(\Phi_b + \frac{1}{E_b - H + \mathrm{i}\epsilon} V \Phi_b, \Psi_a^{(+)}\right) \\
&= \left(\Phi_b, \Psi_a^{(+)}\right) + \left(\Phi_b, V \frac{1}{E_b - H - \mathrm{i}\epsilon} \Psi_a^{(+)}\right) \\
&= \left(\Phi_b, \Phi_a + \frac{1}{E_a - H_0 + \mathrm{i}\epsilon} V \Psi_a^{(+)}\right) + \left(\Phi_b, V \frac{1}{E_b - H - \mathrm{i}\epsilon} \Psi_a^{(+)}\right) \\
&= (\Phi_b, \Phi_a) + \left(\frac{1}{E_a - E_b + \mathrm{i}\epsilon} + \frac{1}{E_b - E_a - \mathrm{i}\epsilon}\right) (\Phi_b, V \Psi_a^{(+)}) \\
&= (\Phi_b, \Phi_a) \ ,
\end{aligned} \tag{12.33}$$

so we see that $\{\Psi^{(+)}\}$ forms an orthonormal system just as $\{\Phi\}$ does. The same is true of $\{\Psi^{(-)}\}$. Thus, the transformation matrix between these two pairs of complete orthonormal systems, viz.,

$$S_{ab} = \left(\Psi_b^{(-)}, \Psi_a^{(+)}\right) \ , \tag{12.34}$$

is unitary. Comparing with (11.136), $\Psi^{(+)}$ and Ψ^{-} correspond to Φ^{in} and Φ^{out}, respectively. The unitarity condition is

$$S^\dagger S = S S^\dagger = 1 \ . \tag{12.35}$$

Starting with the definition and modifying it suitably, the S-matrix above can be written as

$$\begin{aligned}
S_{ba} &= \left(\Phi_b + \frac{1}{E_b - H - \mathrm{i}\epsilon} V \Phi_b, \Psi^{(+)}a\right) \\
&= \left(\Phi_b, \Psi_a^{(+)}\right) + \left(\Phi_b, V \frac{1}{E_b - H + \mathrm{i}\epsilon} \Psi_a^{(+)}\right) \\
&= (\Phi_b, \Phi_a) + \left(\Phi_b, \frac{1}{E_a - H_0 + \mathrm{i}\epsilon} V \Psi_a^{(+)}\right) + \left(\Phi_b, V \frac{1}{E_b - H + \mathrm{i}\epsilon} \Psi_a^{(+)}\right) \\
&= \delta_{ba} + \left(\frac{1}{E_a - E_b + \mathrm{i}\epsilon} + \frac{1}{E_b - E_a + \mathrm{i}\epsilon}\right) (\Phi_b, V \Psi_a^{(+)}) \\
&= \delta_{ba} - 2\pi \mathrm{i} \delta(E_b - E_a) \left(\Phi_b, V \Psi_a^{(+)}\right) \ .
\end{aligned} \tag{12.36}$$

Although in the transformation above we have expressed $\Psi^{(-)}$ in terms of Φ, expressing $\Psi^{(+)}$ in terms of Φ, we obtain

$$S_{ba} = \delta_{ba} - 2\pi \mathrm{i}\delta(E_b - E_a)\big(\Psi_b^{(-)}, V\Phi_a\big) . \tag{12.37}$$

Therefore, if $E_b = E_a$, the transition amplitude T_{ba} can be written in a symmetric form:

$$T_{ba} = \big(\Phi_b, V\Psi_a^{(+)}\big) = \big(\Psi_b^{(-)}, V\Phi_a\big) . \tag{12.38}$$

We now express the unitarity of the S-matrix in terms of T. Using T as above,

$$T^\dagger_{ba} = T^*_{ab} = \big(\Phi_a, V\Psi_b^{(+)}\big)^* = \big(\Psi_b^{(+)}, V\Phi_a\big) ,$$

$$\begin{aligned}
T^\dagger_{ba} - T_{ba} &= (\Psi_b^{(+)}, V\Phi_a) - (\Phi_b, V\Psi_a^{(+)}) \\
&= (\Phi_b, V\Phi_a) + \left(\frac{1}{E_b - H + \mathrm{i}\epsilon} V\Phi_b, V\Phi_a\right) \\
&\quad -(\Phi_b, V\Phi_a) - \left(\Phi_b, V\frac{1}{E_a - H + \mathrm{i}\epsilon} V\Phi_a\right) \\
&= \left(V\Phi_b, \left(\frac{1}{E_b - H - \mathrm{i}\epsilon} - \frac{1}{E_b - H + \mathrm{i}\epsilon}\right) V\Phi_a\right) \\
&= 2\pi \mathrm{i}(V\Phi_b, \delta(E_b - H)V\Phi_a) ,
\end{aligned} \tag{12.39}$$

then inserting the complete system $\{\Psi^{(-)}\}$, we obtain

$$\begin{aligned}
T^\dagger_{ba} - T_{ba} &= 2\pi \mathrm{i}\sum_n (V\Phi_b, \Psi_n^{(-)})\delta(E_b - E_n)\big(\Psi_n^{(-)}, V\Phi_a\big) \\
&= 2\pi \mathrm{i}\sum_n T^\dagger_{bn}\delta(E_b - E_n)T_{na} ,
\end{aligned} \tag{12.40}$$

where

$$T^\dagger_{bn} = T^*_{nb} = \big(\Psi_n^{(-)}, V\Phi_b\big)^* = \big(V\Phi_b, \Psi_n^{(-)}\big) .$$

If we inserted the other complete system $\{\Psi^{(+)}\}$ instead of $\{\Psi^{(-)}\}$, then instead of (12.40) we would get

$$2\pi \mathrm{i}\sum_n T_{bn}\delta(E_b - E_n)T^\dagger_{na} . \tag{12.41}$$

Thus, the unitarity condition can be written as

$$\begin{aligned} T_{ba}^{\dagger} - T_{ba} &= 2\pi \mathrm{i} \sum_n T_{bn}^{\dagger} \delta(E_b - E_n) T_{na} \\ &= 2\pi \mathrm{i} \sum_n T_{bn} \delta(E_b - E_n) T_{na}^{\dagger} \, . \end{aligned} \tag{12.42}$$

Using this unitarity condition, we can reproduce the optical theorem already discussed in Sect. 9.4. Although the content is exactly the same, let us express it in terms of the notation used in this section.

The probability per unit time for the transition $a \to b$ is

$$w_{ba} = 2\pi \delta(E_b - E_a) |T_{ba}|^2 \, . \tag{12.43}$$

If we take the sum over all probable final states, then from (12.42),

$$w_a = \sum_b w_{ba} = 2\pi \sum_b \delta(E_b - E_a) |T_{ba}|^2 = \frac{1}{\mathrm{i}} (T_{aa}^{\dagger} - T_{aa}) = -2 \mathrm{Im} T_{aa} \, . \tag{12.44}$$

Starting with the two-particle state a, the total cross-section can be obtained as

$$\sigma_a = \frac{\Omega}{v_{\mathrm{rel}}} w_a \, , \tag{12.45}$$

where Ω is the volume of quantization and v_{rel} is the relative speed of the two particles. Therefore, from (12.44),

$$\sigma_a = -\frac{2\Omega}{v_{\mathrm{rel}}} \mathrm{Im} T_{aa} \, . \tag{12.46}$$

Then, since the final result does not depend on Ω, we take $\Omega = 1$. The scattering amplitude $f(\theta)$ in the scattering potential is given by

$$\begin{aligned} f(\theta) &= -\frac{m}{2\pi} \int \mathrm{d}^3 x \, \mathrm{e}^{-\mathrm{i} \boldsymbol{k}_f \boldsymbol{x}} V(\boldsymbol{x}) \psi(\boldsymbol{x}) \\ &= -\frac{m}{2\pi} \left(\Phi_f, V \Psi_i^{(+)} \right) . \end{aligned} \tag{12.47}$$

Computing $w_{fi}/v_{\rm rel}$ for scattering into a constant solid angle, the cross-section becomes

$$
\begin{aligned}
\frac{d\sigma}{d\Omega} &= \frac{2\pi}{v}\frac{1}{(2\pi)^3}\int k_f^2 dk_f \delta\left(\frac{k_f^2}{2m}-\frac{k_i^2}{2m}\right)|T_{fi}|^2 \quad \left(v=\frac{k}{m}\right) \\
&= \left(\frac{m}{2\pi}\right)^2 |T_{fi}|^2 \\
&= |f(\theta)|^2 ,
\end{aligned}
\tag{12.48}
$$

which reproduces the well known result. In this regard, however, $T_{fi} = (\Phi_f, V\Psi_i^{(+)})$. In addition, from (12.46), the total cross-section becomes

$$
\sigma = -\frac{2}{v}\mathrm{Im}T_{aa} = -\frac{2}{v}\left(-\frac{2\pi}{m}\right)\mathrm{Im}f(0) = \frac{4\pi}{k}\mathrm{Im}f(0) ,
\tag{12.49}
$$

which also reproduces the optical theorem in its well-known form.

Although in the scattering potential the asymptotic form of the wave function has been obtained easily, it turns out that the asymptotic form of the abstract Lippmann–Schwinger state vector $\Psi^{(+)}$ is given by

$$
\Psi_a^{(+)} \sim S\Phi_a = \Phi_a - 2\pi i\delta(E_a - H_0)V\Psi_a^{(+)} .
\tag{12.50}
$$

Let us compare this asymptotic form with $\Psi_a^{(+)}$ itself:

$$
\Psi_a^{(+)} = \Phi_a + \frac{1}{E_a - H_0 + \mathrm{i}\epsilon}V\Psi_a^{(+)} .
$$

From this we understand that the asymptotic form can be derived if we make the following replacement for the scattered wave:

$$
\frac{1}{E_a - H_0 + \mathrm{i}\epsilon} \to -2\pi\mathrm{i}\delta(E_a - H_0) ,
\tag{12.51}
$$

or

$$
\Psi_a^{(+)} \sim \Phi_a - 2\pi\mathrm{i}\delta(E_a - H_0)\left[(E_a - H_0)\Psi_a^{(+)}\right] ,
\tag{12.52}
$$

where we multiply by $\delta(E_a - H_0)$ after multiplying by $(E_a - H_0)$. This operation reminds us of the Lehmann–Symanzik–Zimmermann (LSZ) asymptotic condition in Sect. 11.7. In (11.159), K_y corresponds to $(E_a - H_0)$, and $\Delta(y - x)$ corresponds to $\delta(E_a - H_0)$. Moreover, the asymptotic form satisfies the equation for the free particle, viz.,

$$(E_a - H_0)S\Phi_a = 0 \,. \tag{12.53}$$

This corresponds to (11.124).

Equation (12.52) is the reduction formula. It plays an important role in the discussion about recombination reactions. So far we have assumed that the separation of the Hamiltonian into the free part and the interaction part is unique, but in general, if a bound state appears, this uniqueness is lost. It is due to recombination reactions that this grouping in the initial state differs from that in the final state. For instance, consider a reaction such as

$$\mathrm{n} + \mathrm{d} \to \mathrm{n} + \mathrm{n}' + \mathrm{p} \,, \tag{12.54}$$

where d stands for the deuteron and n$'$ has been labeled by the prime in order to distinguish it from the other meson n. The total Hamiltonian is

$$H = T_\mathrm{p} + T_\mathrm{n} + T_{\mathrm{n}'} + V_\mathrm{np} + V_{\mathrm{n}'\mathrm{p}} + V_{\mathrm{nn}'} \,, \tag{12.55}$$

where T and V denote the kinetic energies and the potentials for the two-body forces, respectively. The decomposition of the Hamiltonian corresponding to the initial state is

$$H_0 = T_\mathrm{p} + T_\mathrm{n} + T_{\mathrm{n}'} + V_{\mathrm{n}'\mathrm{p}} \,, \quad V = V_\mathrm{np} + V_{\mathrm{nn}'} \,. \tag{12.56}$$

This is because, if we do not insert $V_{\mathrm{n}'\mathrm{p}}$ into H_0, there is no way to make d. On the other hand, the decomposition in the final state is

$$H_0' = T_\mathrm{p} + T_\mathrm{n} + T_{\mathrm{n}'} \,, \quad V' = V_\mathrm{np} + V_{\mathrm{n}'\mathrm{p}} + V_{\mathrm{nn}'} \,. \tag{12.57}$$

In general, we introduce two decompositions, one for the initial state and one for the final state:

$$H = H_a + V_a = H_b + V_b \,. \tag{12.58}$$

The free state vectors corresponding to each decomposition are Φ_a and Φ_b satisfying

$$(E_a - H_a)\Phi_a = 0 \,, \quad (E_b - H_b)\Phi_b = 0 \,. \tag{12.59}$$

Of course, in order for the transition $a \to b$ to occur, we must have $E_b = E_a$. Let us derive the transition amplitude T_{ba} in this case. Corresponding to the initial state, we have

$$\Psi_a^{(+)} = \Phi_a + \frac{1}{E_a - H + \mathrm{i}\epsilon} V_a \Phi_a \,. \tag{12.60}$$

Then, to construct the asymptotic form corresponding to the final state, we use the formula

$$\frac{1}{A} - \frac{1}{B} = \frac{1}{B}(B - A)\frac{1}{A} \,. \tag{12.61}$$

Therefore,

$$\frac{1}{E_a - H + \mathrm{i}\epsilon} - \frac{1}{E_a - H_b + \mathrm{i}\epsilon} = \frac{1}{E_a - H_b + \mathrm{i}\epsilon} V_b \frac{1}{E_a - H + \mathrm{i}\epsilon} \,, \tag{12.62}$$

which yields

$$\Psi_a^{(+)} = \Phi_a + \frac{1}{E_a - H_b + \mathrm{i}\epsilon}\left(1 + V_b \frac{1}{E_a - H + \mathrm{i}\epsilon}\right) V_a \Phi_a \,. \tag{12.63}$$

In the following, we write $E_a = E_b = E$ and make the replacement

$$\frac{1}{E - H_b + \mathrm{i}\epsilon} \longrightarrow -2\pi\mathrm{i}\delta(E - H_b) \,, \tag{12.64}$$

corresponding to (12.51). Then the asymptotic form corresponding to the final state is

$$\Psi_a^{(+)} \sim -2\pi\mathrm{i}\left(1 + V_b \frac{1}{E - H + \mathrm{i}\epsilon}\right) V_a \Phi_a \,. \tag{12.65}$$

Thus, T_{ba} can be given by the inner product of this asymptotic form and Φ_b :

$$\begin{aligned} T_{ba} &= \left(\Phi_b, \left(1 + V_b \frac{1}{E - H + \mathrm{i}\epsilon}\right) V_a \Phi_a\right) \\ &= \left(\left(1 + \frac{1}{E - H - \mathrm{i}\epsilon} V_b\right) \Phi_b, V_a \Phi_a\right) \\ &= \left(\Psi_b^{(-)}, V_a \Phi_a\right) \,. \end{aligned} \tag{12.66}$$

Note also that, corresponding to (12.38), there are also two ways to represent T_{ba}. Using $\Psi_a^{(+)}$ instead of $\Psi_b^{(-)}$,

$$T_{ba} = \left(\Psi_b^{(-)}, V_a \Phi_a\right) = \left(\Phi_b, V_b \Psi_a^{(+)}\right) \,. \tag{12.67}$$

This is because, subtracting one from the other,

$$
\begin{aligned}
\left(\Psi_b^{(-)}, V_a\Phi_a\right) - \left(\Phi_b, V_b\Psi_a^{(+)}\right) &= (\Phi_b, V_a\Phi_a) - (\Phi_b, V_b\Phi_a) \\
&\quad + (\Phi_b, V_a\Phi_a) - \left(\Phi_b, V_b\frac{1}{E - H + \mathrm{i}\epsilon}V_a\Phi_a\right) \\
&= (\Phi_b, (V_a - V_b)\Phi_a) \\
&= (\Phi_b, (H_b - H_a)\Phi_a) \\
&= (E_b - E_a)(\Phi_b, \Phi_a) = 0 \,.
\end{aligned}
$$

This shows that the two expressions in (12.67) are equal.

12.2 Renormalized Interaction Picture

In Chap. 8, we described the computational method for obtaining the S-matrix in the interaction picture based on a covariant perturbation theory. To define the interaction picture, we have to decompose the Hamiltonian or the Lagrangian into the free part and the interaction part. We took this decomposition to be trivial, but it is clear from the discussion about the recombination reaction in the previous section that this decomposition is not unique. Although at the lowest level of the perturbation it has not posed serious problems, it will turn out that this difference between decomposition methods has an important implication when computing higher order corrections.

We thus set down several conditions to determine the decomposition method. These conditions are called *renormalization conditions*. The interaction picture defined by the decomposition satisfying these conditions is called the *renormalized interaction picture*. In fact, it is in the renormalized interaction picture that the Gell-Mann–Low relation derived in Chap. 11, the related asymptotic conditions, and so on, all hold true, although we have not stated this clearly up to now. Another aspect of renormalization, and in general only this aspect is emphasized, is that it can remove the divergences appearing in higher order corrections. Indeed, it was through this that, in the period after World War II, there was a major development of QED. A theory in which divergences can be removed in this way is said to be *renormalizable*, and renormalizablity has been promoted as one of the guiding principles. It was also an important motivation for the more recent development of gauge theories.

Let us now go back to the problem of the decomposition of the Hamiltonian:

1. We have two complete orthonormal systems of eigenstates of the Hamiltonian:

$$
\left\{\Psi_a^{(+)}\right\} \quad \text{and} \quad \left\{\Psi_a^{(-)}\right\} . \tag{12.68}
$$

Furthermore, in relativistic quantum mechanics, these are eigenstates of the four-momentum:

$$P_\mu \Psi_a^{(\pm)} = (p_\mu)_a \Psi_a^{(\pm)} \,. \tag{12.69}$$

The S-matrix element is given by

$$S_{ab} = \left(\Psi_b^{(-)}, \Psi_a^{(+)}\right) . \tag{12.70}$$

It is clear from this expression for the S-matrix element that these two complete systems can be considered to be the same as

$$\{\Phi_a^{\text{in}}\} \quad \text{and} \quad \{\Phi_a^{\text{out}}\} \,. \tag{12.71}$$

2. In field theory, the vacuum state Φ_0 and the stable one-particle state Φ_α satisfy the conditions

$$\Phi_0^{\text{in}} = \Phi_0^{\text{out}} \,, \quad \Phi_\alpha^{\text{in}} = \Phi_\alpha^{\text{out}} \,. \tag{12.72}$$

If we express these conditions in terms of the S-matrix, then

$$S\Phi_0 = \Phi_0 \,, \quad S\Phi_\alpha = \Phi_\alpha \,, \tag{12.73}$$

where (12.72) implies that Φ_0 and Φ_α have been written without distinguishing between the in-state and the out-state. Equations (12.72) and (12.73) are called *renormalization conditions*.

Then, since the condition on the vacuum is satisfied by (8.70), this means that the S-matrix is defined by dropping all bubble diagrams. It turns out that we define the interaction picture in such a way as to satisfy these conditions. This requires as a consequence reintroducing several kinds of physical observable using the following two kinds of renormalization

- **Mass Renormalization.** In field theory, no elementary particles are in the bare state, because they have self-interactions. For example, even within the framework of classical theory, a charged particle carries the Coulomb field. Thus, considering an electron, the conditions (12.73) should hold true for the electron carrying its own field and not for the bare electron. Due to the existence of the self-interaction, the mass of an electron increases by what we call the *self-energy* δm. Thus, the observable mass of an electron is not just the originally given mass m, but changes to $m + \delta m$. We call m and $m + \delta m = m_{\text{obs}}$ the *bare mass* and the *observed mass*, respectively. We interpret this by saying that the one-electron state is the state with mass m_{obs}. We determine the self-energy δm from (12.73), i.e., we consider m_{obs} as the given mass rather than m. In the sense of absorbing δm into the mass, we call this reinterpretation a *renormalization of the mass*.

- **Charge Renormalization.** As discussed in Chap. 8, due to the phenomena of vacuum polarization, the vacuum behaves like a dielectric medium in field theory. Thus, taking the permittivity of the vacuum as ϵ, when the distance r between two charges e_1 and e_2 is large enough, the Coulomb potential between them is

$$V = \frac{e_1 e_2}{\epsilon} \frac{1}{4\pi r} . \tag{12.74}$$

However, since the permittivity of the vacuum is normalized to unity,

$$V = (e_1)_{\text{obs}} (e_2)_{\text{obs}} \frac{1}{4\pi r} . \tag{12.75}$$

Comparing (12.74) and (12.75),

$$(e_1)_{\text{obs}} = \frac{e_1}{\sqrt{\epsilon}} , \quad (e_2)_{\text{obs}} = \frac{e_2}{\sqrt{\epsilon}} , \tag{12.76}$$

where $(e)_{\text{obs}}$ is the experimental value. Thus, we must reinterpret $(e)_{\text{obs}}$ as the given charge, and not the bare charge e. Since the permittivity of the vacuum ϵ is absorbed into the definition of the electric charge, this is called a *renormalization of the electric charge*.

Renormalization can thus be additive, as for the mass, or multiplicative, as for the electric charge. Moreover, when we compute the self-energy δm or the permittivity ϵ using QED, we find that they diverge.

What we have discussed above is the renormalization of specific quantities due to the self-energy and the permittivity. There exists a procedure called *renormalization of the field operators*, which is a slightly abstract renormalization condition. This has already been required in Chap. 11 and expressed in (11.145). There exists a condition whereby the wave function of a given body is not varied by introducing interactions, i.e., the normalization is unchanged. What this condition means will be explained in detail later.

12.3 Mass Renormalization

We call a Feynman diagram which starts with and ends in a one-electron state a *self-energy diagram*. Any diagram which cannot be separated into two disconnected diagrams by cutting a single electron line is called an *irreducible self-energy diagram* (see Fig. 12.2 left). Others are said to be *reducible* (see Fig. 12.2 right).

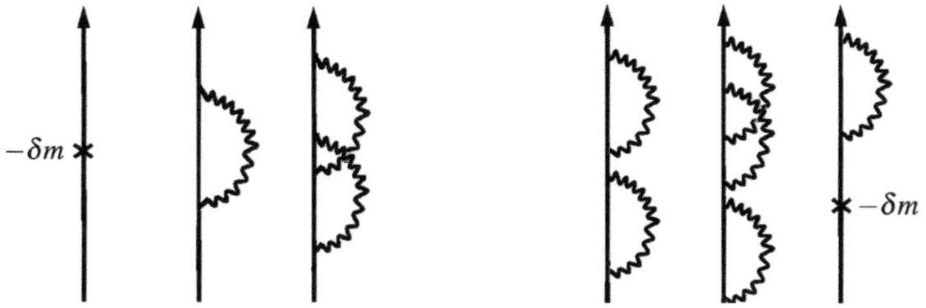

Fig. 12.2 *Left*: Irreducible self-energy diagram. *Right*: Reducible self-energy diagram

Considering the S-matrix element for the case where an electron with the four-momentum p enters and electron in the same state comes out, we have

$$\langle p|S|p\rangle = \langle p|p\rangle + \sum_{n=1}^{\infty} \frac{(-\mathrm{i})^n}{n!} \int \mathrm{d}^4x_1 \ldots \mathrm{d}^4x_n \langle p|T[\mathscr{H}_{\mathrm{int}}(x_1) \ldots \mathscr{H}_{\mathrm{int}}(x_n)]|p\rangle_{\mathrm{conn}}. \tag{12.77}$$

Here, in order to drop contributions from bubble diagrams, after the second term we keep only contributions from the connected parts in which bubble diagrams are omitted. The subscript 'conn' stands for dropping bubble diagrams which are not connected to any points in $x_1, \ldots, x_n$. The condition (12.73) requires the terms after the second term on the right-hand side of (12.77) to vanish. Taking m_{obs} as the mass in the free part, the difference from m, viz., δm, is included in the interaction part, so that in a first approximation where we consider only the renormalization of the mass, we can write

$$\mathscr{H}_{\mathrm{int}} = -\mathrm{i}e\bar{\psi}\gamma_\mu\psi A_\mu - \delta m\bar{\psi}\psi \,. \tag{12.78}$$

Computing the S-matrix element to order e^2, if we consider δm to be of order e^2, then from the Feynman–Dyson rule,

$$\langle p'|S^{(2)}|p\rangle = \int \mathrm{d}^4x \langle p'|\bar{\psi}(x)|0\rangle \left[\mathrm{i}\delta m^{(2)} + e^2 \int \mathrm{d}^4y\, \gamma_\mu S_{\mathrm{F}}(y)\gamma_\mu D_{\mathrm{F}}(y)\mathrm{e}^{-\mathrm{i}p\cdot y}\right] \langle 0|\psi(x)|p\rangle \,, \tag{12.79}$$

where we have taken into account only the first two irreducible self-energy diagrams shown above. This is because the third is of order e^4. We now introduce

$$\Sigma^{*(2)}(p) = \mathrm{i}e^2 \int \mathrm{d}^4y\, \gamma_\mu S_{\mathrm{F}}(y)\gamma_\mu D_{\mathrm{F}}(y)\mathrm{e}^{-\mathrm{i}p\cdot y} \,, \tag{12.80}$$

where the computation above is carried out in the Fermi–Feynman gauge corresponding to $\alpha = 1$. In general, the sum over contributions from all irreducible self-energy diagrams is called a *proper self-energy operator* or a *mass operator*.

Fig. 12.3 Self-energy diagram in momentum space

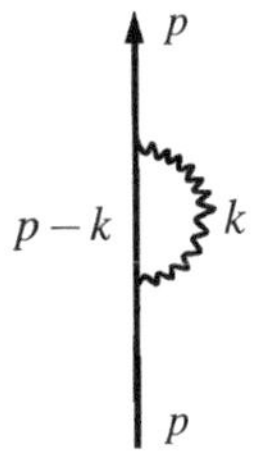

Hence,

$$\langle p'|S^{(2)}|p\rangle = \mathrm{i}\int \mathrm{d}^4x\langle p'|\bar{\psi}(x)|0\rangle\big[\delta m^{(2)} - \Sigma^{*(2)}(p)\big]\langle 0|\bar{\psi}(x)|p\rangle\ . \qquad (12.81)$$

The renormalization condition requires the expression above to vanish. Writing m_{obs} simply as m, we now have (Fig. 12.3)

$$\begin{aligned}\Sigma^{*(2)}(p) &= \frac{\mathrm{i}e^2}{(2\pi)^4}\int \mathrm{d}^4k\gamma_\mu\frac{\mathrm{i}(p-k)\cdot\gamma - m}{(p-k)^2+m^2-\mathrm{i}\epsilon}\gamma_\mu\frac{1}{k^2-\mathrm{i}\epsilon}\\ &= \frac{\mathrm{i}e^2}{(2\pi)^4}\int \mathrm{d}^4k\frac{-2\mathrm{i}(p-k)\cdot\gamma - 4m}{(p-k)^2+m^2-\mathrm{i}\epsilon}\frac{1}{k^2-\mathrm{i}\epsilon}\\ &= \frac{\mathrm{i}e^2}{(2\pi)^4}\int \mathrm{d}^4k\int \mathrm{d}x\frac{-2\mathrm{i}(p-k)\cdot\gamma - 4m}{(k^2+xp^2-2xp\cdot k+xm^2-\mathrm{i}\epsilon)^2}, \qquad (12.82)\end{aligned}$$

where we have used the fact that $\gamma_\mu\gamma_\mu = 4$ and $\gamma_\mu\gamma_\lambda\gamma_\mu = -2\gamma_\lambda$. Making the change of variable

$$k \ \rightarrow\ k' = k - xp\ , \qquad (12.83)$$

rewriting (12.82), and dropping odd-order terms in k', we obtain

$$\Sigma^{*(2)}(p) = \frac{-\mathrm{i}e^2}{(2\pi)^4}\int_0^1 \mathrm{d}x\int \mathrm{d}^4k'\frac{2\mathrm{i}(1-x)p\cdot\gamma + 4m}{\big[k'^2 + x(1-x)p^2 + xm^2 - \mathrm{i}\epsilon\big]^2}\ . \qquad (12.84)$$

Since $p^2 = -(\mathrm{i}p\cdot\gamma)^2$, $\Sigma^*(p)$ can be identified with a function of $\mathrm{i}p\cdot\gamma$, and expanding this as a power series in $(\mathrm{i}p\cdot\gamma + m)$, we have

$$\Sigma^{*(2)}(p) = A + B(\mathrm{i}p\cdot\gamma + m) + C(p)\ , \qquad (12.85)$$

where $C(p)$ is the sum over all terms higher than the second order in $(\mathrm{i}p\cdot\gamma + m)$. Therefore, A can be obtained if we set $\mathrm{i}p\cdot\gamma = -m$ in $\Sigma^{*(2)}(p)$. This yields

$$A = \frac{-\mathrm{i}e^2}{(2\pi)^4}\int_0^1 \mathrm{d}x\, 2m(1+x)\int \mathrm{d}^4k'\frac{1}{(k'^2 + x^2m^2 - \mathrm{i}\epsilon)^2}\ . \qquad (12.86)$$

To carry out this k'-integral, we use

$$\frac{1}{(k'^2 + x^2m^2 - \mathrm{i}\epsilon)^2} = 2 \lim_{\Lambda\to\infty} \int_{m^2}^{\Lambda^2} \frac{x^2 \mathrm{d}M^2}{(k'^2 + x^2M^2 - \mathrm{i}\epsilon)^3} \ . \tag{12.87}$$

Plugging this into (12.86) and carrying out the k'-integral, we obtain

$$\begin{aligned} A &= \frac{2\pi^2 e^2 m}{(2\pi)^4} \int_0^1 \mathrm{d}x(1+x) \int_{m^2}^{\Lambda^2} \frac{\mathrm{d}M^2}{M^2} \\ &= \frac{3}{2\pi}\alpha m \ln\frac{\Lambda}{m} \ , \quad \text{where } \alpha = \frac{e^2}{4\pi} \ . \end{aligned} \tag{12.88}$$

Going back to (12.81) and using the wave functions (3.191a), (3.191b), (3.192a) and (3.192b),

$$\langle p'|S^{(2)}|p\rangle = \frac{\mathrm{i}(2\pi)^4}{V}\delta^4(p'-p)\bar{u}(p')\big[\delta m^{(2)} - \Sigma^{*(2)}(p)\big]u(p) \ . \tag{12.89}$$

Using Dirac's equation for $u(p)$, we can replace $-p\cdot\gamma$ by $-m$ in the equation above. Thus, the condition that (12.89) should vanish can be written as

$$\delta m^{(2)} = A \ . \tag{12.90}$$

This stands for the self-energy of the electron up to order e^2. It is clear from (12.88) that this diverges logarithmically in the limit $\Lambda \to \infty$. However, what we should emphasize here is that the renormalization condition (12.73) requires a choice of interaction part of the form (12.78).

12.4 Renormalization of Field Operators

We have used (12.90) to understand the meaning of A in the expansion (12.85), but what about B? To answer this, we write down the condition for the normalization of the one-body wave function to remain unchanged when interactions are introduced:

$$\big\langle 0\big|T[\psi(x)U(\infty,-\infty)]_{\mathrm{conn}}\big|p\big\rangle = \langle 0|\psi(x)|p\rangle \ . \tag{12.91}$$

The Feynman diagram corresponding to the left-hand side is shown in Fig. 12.4.

Here, Σ^* is the contribution from all the irreducible self-energy diagrams except for δm. It should be clear from the diagram above that the contribution from the

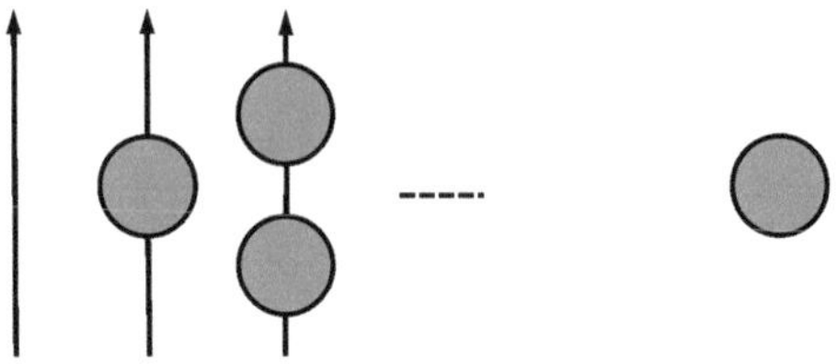

Fig. 12.4 Feynman diagram corresponding to the left-hand side of (12.91)

left-hand side of (12.91) is

$$\frac{\mathrm{i}p\cdot\gamma+m}{\mathrm{i}p\cdot\gamma+m+\Sigma^*(p)-\delta m}u(p)=\Bigg\{1-\big[\Sigma^*(p)-\delta m\big]\frac{1}{\mathrm{i}p\cdot\gamma+m} \tag{12.92}$$

$$+\big[\Sigma^*(p)-\delta m\big]\frac{1}{\mathrm{i}p\cdot\gamma+m}\big[\Sigma^*(p)-\delta m\big]\frac{1}{\mathrm{i}p\cdot\gamma+m}+\cdots\Bigg\}u(p)\,.$$

The discussion so far is based on the assumption that the interaction part is given by (12.78). Expanding Σ^* as suggested by (12.78), we can take $\mathrm{i}p\cdot\gamma=-m$ because $u(p)$ appears in (12.92). This yields

$$\lim_{\mathrm{i}p\cdot\gamma+m\to 0}\frac{\mathrm{i}p\cdot\gamma+m}{\mathrm{i}p\cdot\gamma+m+\Sigma^*(p)-\delta m}=\frac{1}{1+B}\,. \tag{12.93}$$

This is not equal to unity unless $B=0$, so (12.91) cannot be satisfied. Recall the origin of the condition that the normalization is unchanged. In fact, it originally arose from the asymptotic condition (11.126) for the operator in the Heisenberg picture. It was not assumed that the operator used in this case was the same as the original one. Hence, we may consider that the Heisenberg operator satisfying the asymptotic condition has a different normalization from the original Heisenberg operator appearing in the Lagrangian. We call operators satisfying the asymptotic condition *renormalized field operators*. They carry the subscript r and we assume the following multiplicative renormalization:

$$\boldsymbol{\psi}(x)=Z_2^{1/2}\boldsymbol{\psi}(x)_{\mathrm{r}}\,,\quad \bar{\boldsymbol{\psi}}(x)=Z_2^{1/2}\bar{\boldsymbol{\psi}}(x)_{\mathrm{r}}\,,\quad \mathbf{A}_\mu(x)=Z_3^{1/2}\mathbf{A}_\mu(x)_{\mathrm{r}}\,. \tag{12.94}$$

The multiplicative renormalization has already appeared in the renormalization of the electric charge, and we shall see that this is in fact closely related to the renormalization of operators mentioned above.

Going back to the general gauge, let us express the whole Lagrangian in terms of renormalized operators. Since in this case the operators are originally the Heisenberg operators, we should use bold face for them, but there should be no confusion if the usual type face is used:

$$\mathscr{L} = -Z_2\bar{\psi}_\mathrm{r}\Big[\gamma_\mu(\partial_\mu - \mathrm{i}eZ_3^{1/2}A_{\mu\mathrm{r}}) + m - \delta m\Big]\psi_\mathrm{r} - \frac{1}{4}Z_3F_{\mu\nu\mathrm{r}}F_{\mu\nu\mathrm{r}} - \frac{1}{2\alpha}Z_3(\partial_\mu A_{\mu\mathrm{r}})^2\,. \tag{12.95}$$

Although $m - \delta m$ should be written as $m_\mathrm{obs} - \delta m$, as already mentioned, for simplicity we have written m_obs as m. We decompose (12.95) into the free part and the interaction part as follows:

$$\mathscr{L}_\mathrm{f} = -\bar{\psi}_\mathrm{r}(\gamma_\mu\partial_\mu + m)\psi_\mathrm{r} - \frac{1}{4}F_{\mu\nu\mathrm{r}}F_{\mu\nu\mathrm{r}} - \frac{1}{2\alpha_\mathrm{r}}(\partial_\mu A_{\mu\mathrm{r}})^2\,, \tag{12.96}$$

$$\mathscr{L}_\mathrm{int} = (1 - Z_2)\bar{\psi}_\mathrm{r}(\gamma_\mu\partial_\mu + m)\psi_\mathrm{r} + (1 - Z_3)\frac{1}{4}F_{\mu\nu\mathrm{r}}F_{\mu\nu\mathrm{r}} + \mathrm{i}eZ_2Z_3^{1/2}A_{\mu\mathrm{r}}\bar{\psi}_\mathrm{r}\gamma_\mu\psi_\mathrm{r} + Z_2\delta m\bar{\psi}_\mathrm{r}\psi_\mathrm{r}\,, \tag{12.97}$$

where α_r is the renormalized gauge parameter. Hence, the gauge parameter changes under renormalization. The interaction picture corresponding to the partition above is called the *renormalized interaction picture*. The renormalized gauge parameter α_r is defined by

$$\alpha_\mathrm{r} = \alpha Z_3^{-1}\,. \tag{12.98}$$

Since we will only use the renormalized interaction picture in the following discussions, for simplicity we will drop the subscript r. In perturbation theory, several kinds of renormalization constant can be expanded as power series in e^2. We thus assume the expansions

$$\delta m = \delta m^{(2)} + \delta m^{(4)} + \cdots\,, \tag{12.99}$$

$$Z_2 = 1 + Z_2^{(2)} + \cdots\,, \tag{12.100}$$

$$Z_3 = 1 + Z_3^{(2)} + \cdots\,. \tag{12.101}$$

In the interaction picture, the interaction includes derivatives of field operators, so we use Matthew's theorem to express the S-matrix in the form

$$S = 1 + \sum_{n=1}^{\infty}\frac{\mathrm{i}^n}{n!}\int \mathrm{d}^4x_1\ldots\mathrm{d}^4x_nT^*\big[\mathscr{L}_\mathrm{int}(x_1)\ldots\mathscr{L}_\mathrm{int}(x_n)\big]_\mathrm{conn}\,. \tag{12.102}$$

We now repeat the discussion in the last section in this new interaction picture:

$$\langle p'|S^{(2)}|p\rangle = \int \mathrm{d}^4x\langle p'|\bar{\psi}(x)|0\rangle\mathscr{S}(p)\langle 0|\psi(x)|p\rangle\,. \tag{12.103}$$

The new quantity $\mathscr{S}(p)$ is given by the following equation, corresponding to $\mathrm{i}[\delta m^{(2)} - \Sigma^{*(2)}(p)]$ in (12.81):

$$\mathscr{S}(p) = \mathrm{i}Z_2\delta m + \mathrm{i}(1 - Z_2)(\mathrm{i}p \cdot \gamma + m) - \mathrm{i}\Sigma^*(p) . \tag{12.104}$$

In the lowest order approximation, Σ^* is given by

$$\Sigma^*(p) = \mathrm{i}e^2 Z_2^2 Z_3 \int \mathrm{d}^4 y\, \gamma_\mu S_\mathrm{F}(y)\gamma_\mu D_\mathrm{F}(y)\mathrm{e}^{-\mathrm{i}p\cdot y} . \tag{12.105}$$

Replacing Z_2 and Z_3 by 1, this coincides with the expression in the last section. Then in general, expanding as

$$\Sigma^*(p) = A + B(\mathrm{i}p \cdot \gamma + m) + C(p) , \tag{12.106}$$

the expression (12.103) vanishes. Using the fact that the normalization of the one-particle wave function is unchanged, we have

$$Z_2\delta m = A , \quad 1 - Z_2 = B . \tag{12.107}$$

It turns out that δm and Z_2 can be determined from this renormalization condition. Hence,

$$\mathscr{S}(p) = -\mathrm{i}C(p) \equiv -\mathrm{i}\Sigma^*_\mathrm{ren}(p) . \tag{12.108}$$

This equation defines the renormalized mass operator Σ^*_ren. Although $A^{(2)}$ and $B^{(2)}$ diverge logarithmically, $C^{(2)}(p)$ is finite.

12.5 Renormalized Propagators

In the renormalized interaction picture, the electron propagator is defined by

$$S_\mathrm{F}(x - y) = \langle 0|T[\psi(x)\bar{\psi}(y)]|0\rangle = \frac{-\mathrm{i}}{(2\pi)^4}\int \mathrm{d}^4 p\, \mathrm{e}^{\mathrm{i}p\cdot(x-y)} S_\mathrm{F}(p) ,$$

with

$$S_\mathrm{F}(p) = \frac{1}{\mathrm{i}p \cdot \gamma + m} . \tag{12.109}$$

A propagator including higher order corrections can be expressed in the Heisenberg picture by

$$S'_{\rm F} = \langle \mathbf{0} | T[\boldsymbol{\psi}(x)\bar{\boldsymbol{\psi}}(y)] | \mathbf{0} \rangle = \frac{\langle 0 | T^*[\psi(x), \bar{\psi}(y), U(\infty, -\infty)] | 0 \rangle}{\langle 0 | U(\infty, -\infty) | 0 \rangle} \,. \tag{12.110}$$

In in the renormalized interaction picture,

$$U(\infty, -\infty) = 1 + \sum_{n=1}^{\infty} \frac{\mathrm{i}^n}{n!} \int \mathrm{d}^4 x_1 \ldots \mathrm{d}^4 x_n T^* \big[\mathscr{L}_{\rm int}(x_1) \ldots \mathscr{L}_{\rm int}(x_n) \big] \,. \tag{12.111}$$

From the discussion about the Feynman diagram, the Fourier transform of $S'_{\rm F}$ is

$$\begin{aligned} S'_{\rm F}(p) &= S_{\rm F}(p) - S_{\rm F}(p) \Sigma^*_{\rm ren}(p) S_{\rm F}(p) + \cdots \\ &= S_{\rm F}(p) - S_{\rm F}(p) \Sigma^*_{\rm ren}(p) S'_{\rm F}(p) \,. \end{aligned} \tag{12.112}$$

This is called *Dyson's equation*. Its solution is

$$S'_{\rm F}(p) = S_{\rm F}(p) \big[1 + \Sigma^*_{\rm ren}(p) S_{\rm F}(p) \big]^{-1} = \big[\mathrm{i} p \cdot \gamma + m + \Sigma^*_{\rm ren}(p) \big]^{-1} \,. \tag{12.113}$$

Since $\Sigma^*_{\rm ren}(p)$ is a sum over terms higher than second order in $\mathrm{i} p \cdot \gamma + m$,

$$\lim_{\mathrm{i} p \cdot \gamma + m \to 0} (\mathrm{i} p \cdot \gamma + m) S'_{\rm F}(p) = 1 \,. \tag{12.114}$$

This is an important property of the renormalized propagator. Then in the computation in Sect. 12.3, and in particular in (12.84), we make the change of variables from the Feynman parameter x to M, where

$$m^2 = (1 - x) M^2 \,, \tag{12.115}$$

whence $S'_{\rm F}(p)$ can be written to order e^2 as

$$\begin{aligned} S'_{\rm F}(p) &= S_{\rm F}(p) - S_{\rm F}(p) \Sigma^{*(2)}_{\rm ren}(p) S_{\rm F}(p) \\ &= \frac{1}{\mathrm{i} p \cdot \gamma + m} + \frac{e^2}{16\pi^2} \int_m^{\infty} \frac{\mathrm{d}M}{M^3 (M^2 - m^2)} \left[\frac{(M + m)^2 (M^2 + m^2 - 4mM)}{\mathrm{i} p \cdot \gamma + M - \mathrm{i}\epsilon} \right. \\ &\qquad \left. + \frac{(M - m)^2 (M^2 + m^2 + 4mM)}{\mathrm{i} p \cdot \gamma - M + \mathrm{i}\epsilon} \right] . \end{aligned} \tag{12.116}$$

This integral diverges at $M = m$. In this case, we need to improve the approximation near the mass shell using some suitable method. We will return to this when discussing the renormalization group method in Chap. 20. However, what we call the ultraviolet divergence disappears completely.

Next, let us study the propagator of the electromagnetic field. Here we consider the renormalized Fermi–Feynman gauge $\alpha_\mathrm{r} = 1$. In the interaction picture,

$$\delta_{\mu\nu} D_\mathrm{F}(x-y) = \langle 0 | T[A_\mu(x) A_\nu(y)] | 0 \rangle = \frac{-\mathrm{i}}{(2\pi)^4} \delta_{\mu\nu} \int \mathrm{d}^4 k \mathrm{e}^{\mathrm{i}k\cdot(x-y)} D_\mathrm{F}(k) \,. \tag{12.117}$$

In the renormalized Heisenberg picture including higher order corrections, this is not proportional to $\delta_{\mu\nu}$:

$$D'_{F_{\mu\nu}}(x-y) = \langle \mathbf{0} | T[\boldsymbol{A}_\mu(x) \boldsymbol{A}_\nu(y)] | \mathbf{0} \rangle = \frac{\langle 0 | T^*[A_\mu(x) A_\nu(y) U(\infty, -\infty)] | 0 \rangle}{\langle 0 | U(\infty, -\infty) | 0 \rangle} \,. \tag{12.118}$$

Let us compute corrections up to order e^2. In this case, the diagrams which should be taken into account are shown in Fig. 12.5.

The terms in the interaction Lagrangian needed for this calculation are

$$\frac{1}{4}(1 - Z_3) F_{\mu\nu} F_{\mu\nu} + \mathrm{i} e Z_2 Z_3^{1/2} A_\mu \bar{\psi} \gamma_\mu \psi \,.$$

The first term and iterations of the second term correspond to the first diagram and the second diagram in Fig. 12.5, respectively. However, to order e^2, Z_2 and Z_3 in

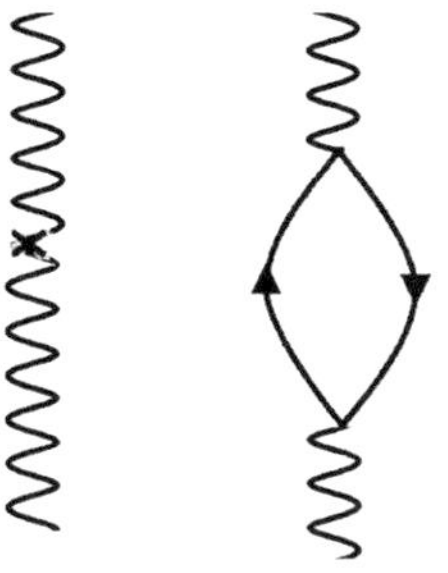

Fig. 12.5 Feynman diagrams for the electromagnetic field propagator up to order e^2

the second term can be set equal to 1:

$$
\begin{aligned}
D'_{F_{\mu\nu}}(x-y) &= \delta_{\mu\nu} D_{\mathrm{F}}(x-y) \\
&\quad -e^2 \int \mathrm{d}^4x'\mathrm{d}^4x'' D_{\mathrm{F}}(x-x') D_{\mathrm{F}}(x''-y) \mathrm{Tr}\big[\gamma_\mu S_{\mathrm{F}}(x'-x'')\gamma_\nu S_{\mathrm{F}}(x''-x')\big] \\
&\quad +\frac{\mathrm{i}}{2}(1-Z_3) \int \mathrm{d}^4x' \left(\delta_{\mu\nu}\frac{\partial}{\partial x'_\rho} - \delta_{\mu\rho}\frac{\partial}{\partial x'_\sigma}\right) D_{\mathrm{F}}(x-x') \,. \qquad (12.119)
\end{aligned}
$$

Taking the Fourier transform of this equation yields

$$
\begin{aligned}
D'_{F_{\mu\nu}}(k) &= \frac{\delta_{\mu\nu}}{k^2-\mathrm{i}\epsilon} \\
&\quad -\frac{\mathrm{i}e^2}{(2\pi)^4}\frac{1}{(k^2-\mathrm{i}\epsilon)^2}\int \mathrm{d}^4p \mathrm{Tr}\left[\gamma_\mu \frac{1}{\mathrm{i}p\cdot\gamma+m-\mathrm{i}\epsilon}\gamma_\nu \frac{1}{\mathrm{i}(p-k)\cdot\gamma+m-\mathrm{i}\epsilon}\right] \\
&\quad +(1-Z_3)\frac{1}{k^2-\mathrm{i}\epsilon}\left(\delta_{\mu\nu} - \frac{k_\mu k_\nu}{k^2-\mathrm{i}\epsilon}\right). \qquad (12.120)
\end{aligned}
$$

The second term is the Fourier transform of the expression

$$
\langle 0 | T^*[j_\mu(x), j_\nu(y)] | 0 \rangle \,.
$$

And formally, this satisfies the condition

$$
\frac{\partial}{\partial x_\mu}\langle 0 | T^*[j_\mu(x), j_\nu(y)] | 0 \rangle = 0 \,. \qquad (12.121)
$$

In momentum space, this condition becomes

$$
k_\mu \int \mathrm{d}^4 p \mathrm{Tr}[\ldots] = 0 \,. \qquad (12.122)
$$

The integral in the equation above, denoted by $f_{\mu\nu}(k)$, has the general form

$$
f_{\mu\nu}(k) = \delta_{\mu\nu} f(k^2) - k_\mu k_\nu g(k^2) \,. \qquad (12.123)
$$

From the condition (12.122),

$$
k_\mu f_{\mu\nu}(k) = k_\nu\big[f(k^2) - k^2 g(k^2)\big] = 0 \,. \qquad (12.124)
$$

Thus,

$$
f_{\mu\nu}(k) = (k^2\delta_{\mu\nu} - k_\mu k_\nu) g(k^2) \,. \qquad (12.125)
$$

Computing the trace in the integral in (12.122),

$$\mathrm{Tr} = \frac{4\big[\delta_{\mu\nu}(p^2 - p\cdot k + m^2) - 2p_\mu p_\nu + p_\mu k_\nu + p_\nu k_\mu\big]}{(p^2 + m^2 - \mathrm{i}\epsilon)\big[(p-k)^2 + m^2 - \mathrm{i}\epsilon\big]} \; .$$

Therefore,

$$\begin{aligned} f_{\mu\nu}(k) &= 4\int \mathrm{d}^4 p \frac{\delta_{\mu\nu}(p^2 - p\cdot k + m^2) - 2p_\mu p_\nu + p_\mu k_\nu + p_\nu k_\mu}{(p^2 + m^2 - \mathrm{i}\epsilon)\big[(p-k)^2 + m^2 - \mathrm{i}\epsilon\big]} \\ &= 4\int_0^1 \mathrm{d}x \int \mathrm{d}^4 p \frac{\delta_{\mu\nu}(p^2 - p\cdot k + m^2) - 2p_\mu p_\nu + p_\mu k_\nu + p_\nu k_\mu}{\big[p^2 + m^2 + x(k^2 - 2p\cdot k) - \mathrm{i}\epsilon\big]^2} \; . \end{aligned}$$

We make the change of variables

$$p \;\rightarrow\; p' = p - xk \; . \tag{12.126}$$

Thus,

$$f_{\mu\nu}(k) = 4\int_0^1 \mathrm{d}x \int \mathrm{d}^4 p' \frac{N}{\big[p'^2 + m^2 + x(1-x)k^2 - \mathrm{i}\epsilon\big]^2} \; , \tag{12.127}$$

where

$$N = \delta_{\mu\nu} p'^2 - 2p'_\mu p'_\nu + \delta_{\mu\nu}\big[m^2 + x(1-x)k^2\big] - 2x(1-x)(\delta_{\mu\nu}k^2 - k_\mu k_\nu) \; . \tag{12.128}$$

All terms except for the last produce terms proportional to $\delta_{\mu\nu}$. Hence, for this to coincide with (12.125), only the last term can survive. Thus,

$$f_{\mu\nu}(k) = -8(\delta_{\mu\nu}k^2 - k_\mu k_\nu)\int_0^1 \mathrm{d}x\, x(1-x)\int \frac{\mathrm{d}^4 p}{\big[p^2 + m^2 + x(1-x)k^2 - \mathrm{i}\epsilon\big]^2} \; , \tag{12.129}$$

and we can write

$$\begin{aligned} \frac{1}{\big[p^2 + m^2 + x(1-x)k^2 - \mathrm{i}\epsilon\big]^2} &= \frac{1}{(p^2 + m^2 - \mathrm{i}\epsilon)^2} \\ &\quad + \left\{ \frac{1}{\big[p^2 + m^2 + x(1-x)k^2 - \mathrm{i}\epsilon\big]^2} - \frac{1}{(p^2 + m^2 - \mathrm{i}\epsilon)^2} \right\} \; . \end{aligned}$$

Therefore, the expression for $D'_{F_{\mu\nu}}$ becomes

$$
\begin{aligned}
D'_{F_{\mu\nu}}(k) = {} & \frac{1}{k^2 - \mathrm{i}\epsilon} + \frac{8\mathrm{i}e^2}{(2\pi)^4}\frac{k^2\delta_{\mu\nu} - k_\mu k_\nu}{(k^2-\mathrm{i}\epsilon)^2}\int_0^1 x(1-x)\mathrm{d}x \\
& \times \int \mathrm{d}^4 p \left\{ \frac{1}{[p^2+m^2+x(1-x)k^2-\mathrm{i}\epsilon]^2} - \frac{1}{(p^2+m^2-\mathrm{i}\epsilon)^2} \right\} \\
& + \frac{k^2\delta_{\mu\nu} - k_\mu k_\nu}{(k^2-\mathrm{i}\epsilon)^2}\left[1 - Z_3 + \frac{8\mathrm{i}e^2}{(2\pi)^4}\int_0^1 x(1-x)\mathrm{d}x \int \frac{\mathrm{d}^4 p}{(p^2+m^2-\mathrm{i}\epsilon)^2}\right] .
\end{aligned}
\tag{12.130}
$$

Then to determine Z_3, corresponding to (12.114) in the case of the electron, or the equation

$$
S_{\mathrm{F}}(p)\Sigma^*_{\mathrm{ren}}(p)u(p) = 0 , \tag{12.131}
$$

we adopt the condition

$$
D_{\mathrm{F}}(k)\Pi^*_{\mathrm{ren}}(k)_{\mu\nu}e_\nu = 0 , \tag{12.132}
$$

where e_ν is the polarization vector of transverse photons and Π^* is called the *proper self-energy operator of the photon* or the *polarization operator*, which corresponds to Σ^* in the electron case. Thus, similarly to Dyson's equation (12.112) for Σ^* in the electron case, Π^* is defined by

$$
\begin{aligned}
D'_{F_{\mu\nu}} &= \delta_{\mu\nu}D_{\mathrm{F}}(k) - D_{\mathrm{F}}(k)\Pi^*_{\mathrm{ren}}(k)_{\mu\nu}D_{\mathrm{F}}(k) + \cdots \\
&= \delta_{\mu\nu}D_{\mathrm{F}}(k) - D_{\mathrm{F}}(k)\Pi^*_{\mathrm{ren}}(k)_{\mu\lambda}D'_{F_{\lambda\nu}}(k) .
\end{aligned}
\tag{12.133}
$$

To order e^2,

$$
\begin{aligned}
\Pi^*_{\mathrm{ren}}(k)_{\mu\nu} = {} & -\frac{8\mathrm{i}e^2}{(2\pi)^4}(k^2\delta_{\mu\nu} - k_\mu k_\nu)\int_0^1 x(1-x)\mathrm{d}x \\
& \times \int \mathrm{d}^4 p \left[\frac{1}{\left[p^2+m^2+x(1-x)k^2-\mathrm{i}\epsilon\right]^2} - \frac{1}{(p^2+m^2-\mathrm{i}\epsilon)^2} \right] \\
& + \frac{8\mathrm{i}e^2}{(2\pi)^4}(\delta_{\mu\nu}k^2 - k_\mu k_\nu)\left[1 - Z_3 + \frac{8\mathrm{i}e^2}{(2\pi)^4}\int_0^1 x(1-x)\mathrm{d}x \int \frac{\mathrm{d}^4 p}{(p^2+m^2-\mathrm{i}\epsilon)^2}\right] .
\end{aligned}
\tag{12.134}
$$

Then, in the one-photon state, taking into account $k^2 = 0$ and $k \cdot e = 0$, the condition (12.132) requires the coefficient of $\delta_{\mu\nu}k^2 - k_\mu k_\nu$ to vanish when $k^2 = 0$. Since the first term in (12.134) satisfies this condition, it implies that the second term vanishes, i.e.,

$$Z_3 = 1 + \frac{8\mathrm{i}e^2}{(2\pi)^4}\int_0^1 x(1-x)\mathrm{d}x \int \frac{\mathrm{d}^4 p}{(p^2+m^2-\mathrm{i}\epsilon)^2} \,. \tag{12.135}$$

Therefore, carrying out the Feynman integral,

$$\Pi^*_{\text{ren}}(k)_{\mu\nu} = \frac{e^2}{2\pi^2}(k^2\delta_{\mu\nu} - k_\mu k_\nu)\int_0^1 x(1-x)\ln\frac{m^2}{m^2+x(1-x)k^2-\mathrm{i}\epsilon}\mathrm{d}x \,. \tag{12.136}$$

In particular, when $|k^2| \ll m^2$,

$$\Pi^*_{\text{ren}}(k)_{\mu\nu} \approx \frac{e^2}{60\pi^2}(\delta_{\mu\nu}k^2 - k_\mu k_\nu)\left(-\frac{k^2}{m^2}\right) \,. \tag{12.137}$$

We see that, from the renormalization condition (12.132), Z_3 is uniquely determined by (12.135), and it turns out that D'_{F} is divergenceless.

12.6 Renormalization of Vertex Functions

We have seen that the propagator becomes finite in the renormalized interaction picture. However, there is one thing that does not become finite without multiplicative renormalization, namely the vertex function. We now consider its renormalization. Up to now, we have investigated the proper self-energy diagrams, but these are all related to two-point functions or propagators. We now consider the corrections to the vertex function γ_μ shown in Fig. 12.6.

These are diagrams in which the propagator, including the self-energy diagram, is removed from the Feynman diagrams of the three-point function. They give the correction to the vertex operator. With this correction, the vertex function γ_μ is replaced by the vertex function $\Gamma^{(0)}_\mu$. This is the same as replacing S_{F} and D_{F} by S'_{F} and $D'_{F_{\mu\nu}}$.

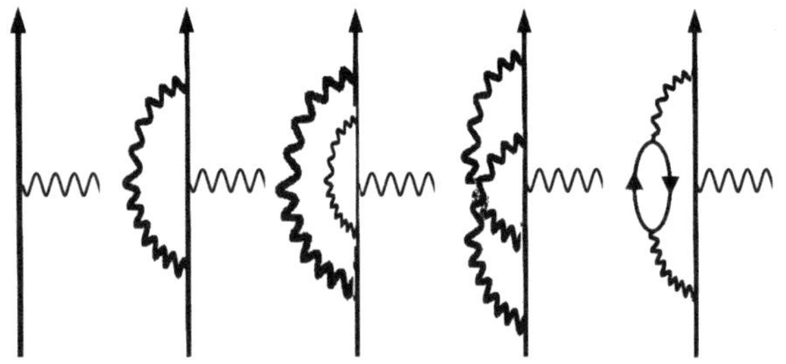

Fig. 12.6 Corrections to the vertex function

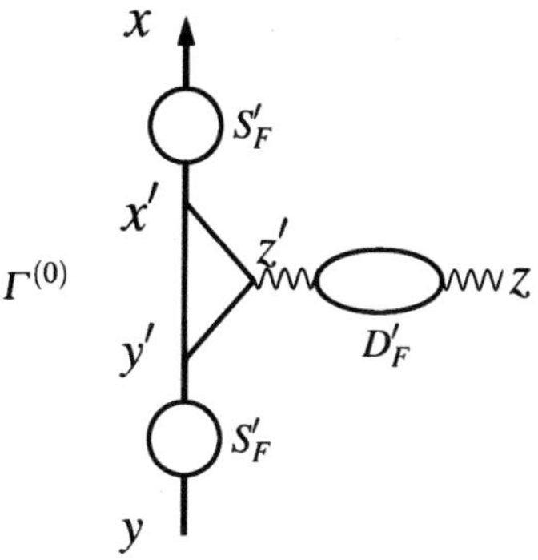

Fig. 12.7 Feynman diagram for the complete three-point function

Thus, the Feynman diagram corresponding to the complete three-point function (Fig. 12.7) can be written like the one above, i.e., the general three-point function can be expressed as a product of three propagators and one vertex function. Corresponding to this diagram, we have the expression

$$
\begin{aligned}
\langle \mathbf{0}|T^*[\boldsymbol{\psi}(x), \bar{\boldsymbol{\psi}}(y), \boldsymbol{A}_\nu(z)]|\mathbf{0}\rangle &= \frac{\langle 0|T^*[\psi(x), \bar{\psi}(y), A_\nu(z), U(\infty, -\infty)]|0\rangle}{\langle 0|U(\infty, -\infty)|0\rangle} \\
&= -eZ_2 Z_3^{1/2} \int \mathrm{d}^4x'\mathrm{d}^4y'\mathrm{d}^4z' S'_\mathrm{F}(x - x')\Gamma^{(0)}_\mu(x', y'; z') \\
&\quad \times S'_\mathrm{F}(y' - y) D'_{F_{\mu\nu}}(z' - z) \,. \qquad (12.138)
\end{aligned}
$$

Here, the Heisenberg operators and the interaction picture are renormalized. In the renormalized interaction picture, a propagator is automatically renormalized, but the vertex function is not. To the lowest order,

$$
\Gamma^{(0)}_\mu(x, y, z) = \gamma_\mu \delta^4(x - z)\delta^4(y - z) \,. \qquad (12.139)
$$

Since in general $\Gamma^{(0)}_\mu$ becomes a function of $(x - z)$ and $(y - z)$, we define its Fourier representation by

$$
\Gamma^{(0)}_\mu(x, y, z) = \frac{1}{(2\pi)^8} \int \mathrm{d}^4p\, \mathrm{d}^4q\, \mathrm{e}^{\mathrm{i}p\cdot(x-z)+\mathrm{i}q\cdot(z-y)} \Gamma^{(0)}_\mu(p, q) \,. \qquad (12.140)
$$

Therefore, according to (12.139), to lowest order,

$$
\Gamma^{(0)}_\mu(p, q) = \gamma_\mu \,. \qquad (12.141)
$$

We renormalize the vertex function in which higher-order corrections are included by the equations

$$\Gamma_\mu(p, q) = Z_1 \Gamma_\mu^{(0)}(p, q) \,, \tag{12.142}$$

$$\bar{u}(p)\Gamma_\mu(p, p)u(p) = \bar{u}(p)\gamma_\mu u(p) \,. \tag{12.143}$$

However, in (12.143), we have assumed that p is on the electron mass shell. Replacing $\Gamma_\mu^{(0)}$ on the right-hand side of (12.138) by Γ, it turns out that the coefficient in front of the integral on the right-hand side of that equation is given by

$$e_{\text{obs}} = eZ_1^{-1}Z_2Z_3^{1/2} \,. \tag{12.144}$$

In fact, this combination corresponds to the electric charge observed in experiments. This is manifested by several properties called *low-energy theorems*. Comparing (12.144) with (12.76), the permittivity of the vacuum is

$$\sqrt{\epsilon} = Z_1Z_2^{-1}Z_3^{-1/2} \,. \tag{12.145}$$

Although the definition of the renormalized interaction picture is unchanged, it is more useful to rewrite e as e_{obs}. Then we write

$$\begin{aligned} eZ_2Z_3^{1/2} &= eZ_1^{-1}Z_2Z_3^{1/2} - (1 - Z_1)eZ_1^{-1}Z_2Z_3^{1/2} \\ &= e_{\text{obs}} - (1 - Z_1)e_{\text{obs}} \\ &\equiv e_{\text{obs}} - \delta e \,. \end{aligned} \tag{12.146}$$

Thus, in the interaction part of the Lagrangian density, we rewrite as follows:

$$\mathrm{i}eZ_2Z_3^{1/2}\bar{\psi}\gamma_\mu\psi A_\mu = \mathrm{i}(e_{\text{obs}} - \delta e)\bar{\psi}\gamma_\mu\psi A_\mu \,. \tag{12.147}$$

So from now on, we use powers of e_{obs} rather than powers of e in the perturbation theory. We have

$$\delta e = (1 - Z_1)e_{\text{obs}} = \mathscr{O}(e_{\text{obs}}^3) \,. \tag{12.148}$$

The relation between Γ_μ and $\Gamma_\mu^{(0)}$ is

$$(e_{\text{obs}} - \delta e)\Gamma_\mu^{(0)}(x, y, z) = e_{\text{obs}}\Gamma_\mu(x, y, z) \,. \tag{12.149}$$

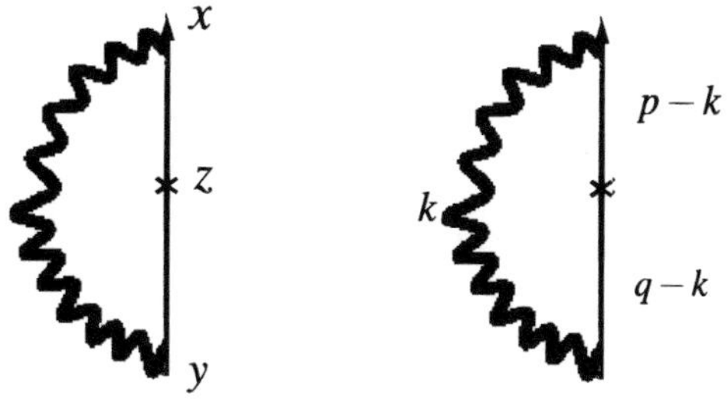

Fig. 12.8 Feynman diagram corresponding to the second term on the right-hand side of (12.150) in the x-space (*left*) and the p-space (*right*)

Computing the left-hand side up to order e_{obs}^3,

$$e_{\text{obs}}\Gamma_\mu(x, y, z) = (e_{\text{obs}} - \delta e)\gamma_\mu\delta^4(x - z)\delta^4(y - z) \qquad (12.150)$$
$$+e_{\text{obs}}^2\gamma_\lambda S_{\text{F}}(x - z)\gamma_\mu S_{\text{F}}(z - y)\gamma_\lambda D_{\text{F}}(x - y) \,.$$

The Feynman diagram corresponding to the second term on the right-hand side is shown in the x-space and the p-space in Fig. 12.8.

Taking the Fourier transform of (12.150), we obtain

$$e_{\text{obs}}\Gamma_\mu(p, q) = (e_{\text{obs}} - \delta e)\gamma_\mu \qquad (12.151)$$
$$+\frac{(-\text{i})^3}{(2\pi)^4}e_{\text{obs}}^3\int \text{d}^4k\gamma_\lambda\frac{1}{\text{i}(p - k)\cdot\gamma + m - \text{i}\epsilon}\gamma_\mu\frac{1}{\text{i}(q - k)\cdot\gamma + m - \text{i}\epsilon}\gamma_\lambda\frac{1}{k^2 - \text{i}\epsilon} \,.$$

To determine δe, we use (12.143). Rationalizing the denominator of (12.151), the numerator becomes

$$N = \gamma_\lambda\big[\text{i}(p - k)\cdot\gamma + m\big]\gamma_\mu\big[\text{i}(q - k)\cdot\gamma + m\big]\gamma_\lambda \,. \qquad (12.152)$$

Summing over λ, we can use the following formulas to calculate products of the γ-matrices:

$$(1) \quad \gamma_\lambda\gamma_a\gamma_\lambda = -2\gamma_a \,,$$
$$(2) \quad \gamma_\lambda\gamma_a\gamma_b\gamma_\lambda = 4\delta_{ab} \,,$$
$$(3) \quad \gamma_\lambda\gamma_a\gamma_b\gamma_c\gamma_\lambda = -2\gamma_c\gamma_b\gamma_a \,.$$

The numerator assumes the form

$$N = -2\text{i}(q-k)\cdot\gamma\gamma_\mu\text{i}(p-k)\cdot\gamma - 4\text{i}m\big[(p-k)_\mu + (q-k)_\mu\big] - 2m^2\gamma_\mu \,. \qquad (12.153)$$

Equation (12.151) then becomes

$$e_{\text{obs}}\Gamma_\mu(p,q) = (e_{\text{obs}} - \delta e)\gamma_\mu + \frac{(-\mathrm{i}e_{\text{obs}})^3}{(2\pi)^4}\int \mathrm{d}^4k \frac{N}{\left[(p-k)^2+m^2-\mathrm{i}\epsilon\right]\left[(p-k)^2+m^2-\mathrm{i}\epsilon\right](k^2-\mathrm{i}\epsilon)} . \tag{12.154}$$

In order to carry out the k-integral, we use the following formula and change of variables:

$$\frac{1}{abc} = 2\int_0^1 \mathrm{d}x_1 \int_0^1 \mathrm{d}x_2 \int_0^1 \mathrm{d}x_3 \delta(1-\Sigma x_i)\frac{1}{(x_1a+x_2b+x_3c)^3} , \tag{12.155}$$

$$P = \frac{1}{2}(p+q) , \quad \Delta = p-q . \tag{12.156}$$

Up to a factor of 2, the denominator of the integral in (12.154) can be written

$$\int_0^1 \mathrm{d}x_1 \int_0^1 \mathrm{d}x_2 \int_0^1 \mathrm{d}x_3 \delta\left(1-\sum x_i\right) f(x_1,x_2) = \int_D \mathrm{d}x_1 \mathrm{d}x_2 f(x_1,x_2) , \tag{12.157}$$

where

$$f(x_1,x_2) = \left[\left(P^2+\frac{\Delta^2}{4}+m^2\right)(x_1+x_2) + (P-k)\cdot\Delta(x_1-x_2) - 2k\cdot P(x_1+x_2) + k^2 - \mathrm{i}\epsilon\right]^{-3} . \tag{12.158}$$

The domain of integration D is shown in Fig. 12.9 (left). Changing the variables in (12.157) according to

$$x_1 + x_2 = u , \quad x_1 - x_2 = 2v , \tag{12.159}$$

the domain of integration for u and v is given by D', depicted in Fig. 12.9 (right). We now change the integration variable from k to k':

$$k' = k - uP - v\Delta . \tag{12.160}$$

The part corresponding to (12.155) becomes

$$2\int_{D'} \mathrm{d}u\,\mathrm{d}v\left[k'^2 + \left(P^2+\frac{\Delta^2}{4}+m^2\right)u + 2P\cdot\Delta v - (uP+v\Delta)^2 - \mathrm{i}\epsilon\right]^{-3} . \tag{12.161}$$

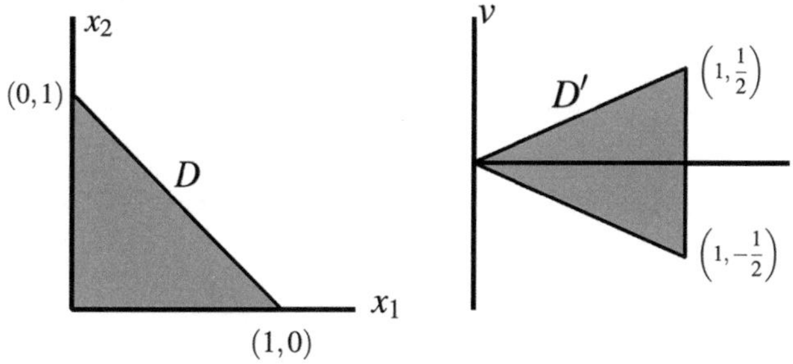

Fig. 12.9 Domains of integration for evaluating (12.154)

We now express the numerator in terms of k'. We can drop terms linear in k' because they give zero when we carry out the k'-integral. Since the integral is still complicated, we assume that p and q are on the mass shell, sandwich it between $\bar{u}(p)$ and $u(q)$, and use the relations

$$\bar{u}(p)(\mathrm{i}p \cdot \gamma + m) = \bar{u}(p)\left(\mathrm{i}P \cdot \gamma + \frac{1}{2}\mathrm{i}\Delta \cdot \gamma + m\right) = 0\,,$$

$$(\mathrm{i}q \cdot \gamma + m)u(p) = \left(\mathrm{i}P \cdot \gamma - \frac{1}{2}\mathrm{i}\Delta \cdot \gamma + m\right)u(q) = 0\,.$$

Moreover, when $\bar{u}(p)Au(q) = \bar{u}(p)Bu(q)$, we write $A \sim B$. From the mass-shell condition,

$$p^2 + m^2 = q^2 + m^2 = P^2 + \frac{\Delta^2}{4} + m^2 = 0\,, \quad P \cdot \Delta = 0\,. \tag{12.162}$$

Therefore, the term in square brackets in (12.161) simplifies as follows:

$$k'^2 + m^2u^2 + \Delta^2\left(\frac{u^2}{4} - v^2\right) - \mathrm{i}\epsilon\,. \tag{12.163}$$

Furthermore, assuming that Δ is small, we only keep terms linear in Δ in the denominator and the numerator, and drop those in Δ^2. Using the formula

$$2\mathrm{i}P_\mu \sim -\sigma_{\mu\nu}\Delta_\nu - 2m\gamma_\mu\,, \tag{12.164}$$

we obtain

$$e_{\mathrm{obs}}\Gamma_\mu(p,q) \sim (e_{\mathrm{obs}} - \delta e)\gamma_\mu - 4\frac{\mathrm{i}e_{\mathrm{obs}}^3}{(2\pi)^4}\int_{D'}\mathrm{d}u\,\mathrm{d}v\int \mathrm{d}^4k'\frac{N'}{(k'^2 + m^2u^2 - \mathrm{i}\epsilon)^3} + \mathscr{O}(\Delta^2)\,, \tag{12.165}$$

where

$$N' = -u(1-u)m\sigma_{\mu\nu}\Delta_\nu + m^2\gamma_\mu\big[(1-u)^2 - 4(1-u) + 1\big] + \frac{1}{2}\gamma_\mu k'^2 \,. \quad (12.166)$$

In the limit $\Delta \to 0$, the right-hand side should be equal to $e_{\text{obs}}\gamma_\mu$, so

$$\delta e = -\frac{4\mathrm{i}e_{\text{obs}}^3}{(2\pi)^4}\int_{D'} \mathrm{d}u\,\mathrm{d}v \int \mathrm{d}^4k' \frac{m^2(u^2+2u-2)+k'^2/2}{(k'^2+m^2u^2-\mathrm{i}\epsilon)^3} \,. \quad (12.167)$$

The right-hand side turns out to diverge logarithmically. Inserting (12.167) into (12.165),

$$\begin{aligned} e_{\text{obs}}\Gamma_\mu(p,q) &\sim e_{\text{obs}}\gamma_\mu + 4\frac{\mathrm{i}e_{\text{obs}}^3}{(2\pi)^4}\int_{D'} \mathrm{d}u\,\mathrm{d}v \int \mathrm{d}^4k' \frac{u(1-u)m\sigma_{\mu\nu}\Delta_\nu}{(k'^2+m^2u^2-\mathrm{i}\epsilon)^3} + \mathscr{O}(\Delta^2) \\ &= e_{\text{obs}}\gamma_\mu - \frac{\alpha}{2\pi}\frac{e_{\text{obs}}}{2m}\sigma_{\mu\nu}\Delta_\nu + \mathscr{O}(\Delta^2) \,. \end{aligned} \quad (12.168)$$

Here, α is, of course, equal to $e_{\text{obs}}^2/4\pi$. This equation is the one derived as the third-order perturbation. The effective Hamiltonian which yields the same result up to the first order in Δ is

$$\mathscr{H}_{\text{eff}} = -\frac{\alpha}{2\pi}\frac{e_{\text{obs}}}{4m}\bar{\psi}\sigma_{\mu\nu}\psi F_{\mu\nu} \,. \quad (12.169)$$

This term, which is gauge invariant, is called the *Pauli term*. When there is only a magnetic field, applying the non-relativistic approximation, we have

$$\mathscr{H}_{\text{eff}} = -\frac{\alpha}{2\pi}\frac{e_{\text{obs}}}{2m}\boldsymbol{\sigma}\cdot\boldsymbol{H} \,. \quad (12.170)$$

This tells us that the electron acquires a supplementary magnetic moment in addition to $e_{\text{obs}}/2m$ in the Dirac theory. This increase is called the *anomalous magnetic moment*. As a result, the magnetic moment of the electron up to this order is given by

$$\frac{e_{\text{obs}}}{2m}\left(1+\frac{\alpha}{2\pi}\right) \,. \quad (12.171)$$

This result was obtained by Schwinger and by Tomonaga et al. [116, 117], and it matches experimental values well. It is considered to be a great achievement of renormalization theory, which thus succeeded in explaining what we call the *Lamb shift* in the hydrogen atom, and has provided a foundation for the development of field theory.

So far, we have introduced Z_1, Z_2, and Z_3 as multiplicative renormalization constants. From (12.148) and (12.167), Z_1 is given by

$$Z_1 = 1 + \frac{4ie_{\text{obs}}^2}{(2\pi)^4} \int_{D'} du\, dv \int d^4k' \frac{m^2(u^2+2u-2)+k'^2/2}{(k'^2+m^2u^2-i\epsilon)^3} . \tag{12.172}$$

Although we have not provided an explicit computation of Z_2, it is given by (12.107), and computing to this order, it coincides with Z_1. We will discuss this equality in the next section.

12.7 Ward–Takahashi Identity

We now investigate in detail the relation (12.115) between the bare electric charge e and the observed electric charge e_{obs}. So far, we have only considered the electron. Let us consider the case where there is a wide variety of charged particles $a, b, \ldots$. Then,

$$(e_a)_{\text{obs}} = Z_{1a}^{-1} Z_{2a} Z_3^{1/2} e_a , \quad (e_b)_{\text{obs}} = Z_{1b}^{-1} Z_{2b} Z_3^{1/2} e_b , \ldots . \tag{12.173}$$

We consider the reaction

$$a + b \longrightarrow c + d . \tag{12.174}$$

In this case, it is the bare electric charge that is conserved by Noether's theorem:

$$e_a + e_b = e_c + e_d . \tag{12.175}$$

However, we know experimentally that charge conservation holds for the renormalized electric charges:

$$(e_a)_{\text{obs}} + (e_b)_{\text{obs}} = (e_c)_{\text{obs}} + (e_d)_{\text{obs}} . \tag{12.176}$$

In order for these two conservation laws to hold simultaneously, $Z_1^{-1}Z_2$ cannot depend on the type of charged particle, i.e.,

$$Z_{1a}^{-1} Z_{2a} = Z_{1b}^{-1} Z_{2b} = Z_{1c}^{-1} Z_{2c} = Z_{1d}^{-1} Z_{2d} . \tag{12.177}$$

Ward discovered that these equalities can be replaced by the following, which imply them [118]:

$$Z_{1a} = Z_{2a} , \ldots . \tag{12.178}$$

The equation $Z_1 = Z_2$ is referred to as the *Ward identity*. We shall now give its proof.

We take $\psi^{(0)}$ and $\bar{\psi}^{(0)}$ to be non-renormalized operators of the electric field. Setting

$$J_\mu = \mathrm{i}\bar{\psi}^{(0)}\gamma_\mu\psi^{(0)} = \mathrm{i}Z_2\bar{\psi}\gamma_\mu\psi \,, \tag{12.179}$$

and using the fact that non-renormalized Heisenberg operators satisfy the canonical commutation relations for $x_0 = y_0$, we have

$$[J_0(x), \psi^{(0)}(y)] = -\psi^{(0)}(y)\delta^3(x-y) \,, \quad [J_0(x), \bar{\psi}^{(0)}(y)] = \bar{\psi}^{(0)}(y)\delta^3(x-y) \,. \tag{12.180}$$

Thus, if J_μ is included in the T-product, then using $\partial_\mu J_\mu = 0$, we obtain

$$\begin{aligned}\partial_\mu T\big[J_\mu(z), \psi^{(0)}(x), \bar{\psi}^{(0)}(y)\big] &= T\big[[J_0(z), \psi^{(0)}(x)]\delta(z_0 - x_0), \bar{\psi}^{(0)}(y)\big] \\ &\quad + T\big[\psi^{(0)}(x), [J_0(z), \bar{\psi}^{(0)}(y)]\delta(z_0 - y_0)\big] \\ &= \big[\delta^4(z-y) - \delta^4(z-x)\big]T[\psi^{(0)}(x), \bar{\psi}^{(0)}(y)] \,.\end{aligned} \tag{12.181}$$

As we have seen before, at the point where the order of two time variables in the T-product are switched, a delta function in time shows up. Later, we will use a generalization of (12.181).

The renormalized interaction picture is defined by (12.96) and (12.97). On the other hand, expressing (12.181) in terms of renormalized operators,

$$\begin{aligned}\partial_\mu T\big[J_\mu(z), \psi(x), \bar{\psi}(y)\big] &= \big[\delta^4(z-y) - \delta^4(z-x)\big]T[\psi(x), \bar{\psi}(y)] \\ &\quad - \left[\psi(z)\frac{\delta}{\delta\psi(z)} - \bar{\psi}(z)\frac{\delta}{\delta\bar{\psi}(z)}\right]T[\psi(x), \bar{\psi}(y)] \,.\end{aligned} \tag{12.182}$$

We will use this formula later by generalizing it to a certain extent.

In order to apply the reduction formula for the electromagnetic field, we introduce the differential operator

$$D_{\mu\nu}(\partial) = \delta_{\mu\nu}\Box - \partial_\mu\partial_\nu + \frac{1}{\alpha}\partial_\mu\partial_\nu \,. \tag{12.183}$$

Applying the reduction formula to Green's function in the renormalized interaction picture, we compute the quantity

$$\begin{aligned}D_{\mu\nu}(\partial)T^*\big[A_\nu(x)\ldots, U(\infty,-\infty)\big] &= T^*\left[\mathrm{i}\frac{\delta}{\delta A_\mu(x)}\ldots, U(\infty,-\infty)\right] \\ &\quad - T^*\big[\ldots[\mathscr{L}_{\mathrm{int}}(x)]_{A_\mu}, U(\infty,-\infty)\big] \,,\end{aligned} \tag{12.184}$$

where an explicit expression for the Euler derivative is

$$[\mathscr{L}_{\rm int}(x)]_{A_\mu} = eZ_3^{1/2} J_\mu - (1 - Z_3)(\delta_{\mu\sigma}\Box - \partial_\mu \partial_\sigma) A_\sigma(x) \,. \tag{12.185}$$

Inserting this into (12.184) and taking the derivative in the second term of (12.185) outside T^*, we have

$$\begin{aligned} D_{\mu\nu}(\partial) T^*[A_\nu(x) \ldots, U(\infty, -\infty)] = T^*\left[\mathrm{i} \frac{\delta}{\delta A_\mu(x)} \ldots, U(\infty, -\infty) \right] \\ - eZ_3^{1/2} T^*\big[J_\mu(x), \ldots, U(\infty, -\infty)\big] \\ +(1 - Z_3)(\delta_{\mu\sigma} \Delta - \partial_\mu \partial_\sigma) T^*\big[A_\sigma(x), \ldots, U(\infty, -\infty)\big] \,. \end{aligned} \tag{12.186}$$

Here we use the relation

$$eZ_3^{1/2} = e(Z_1^{-1} Z_2 Z_3^{1/2}) Z_1 Z_2^{-1} = e_{\rm obs} Z_1 Z_2^{-1}. \tag{12.187}$$

Next, we differentiate (12.186) with respect to x_μ and use (12.182). Since the last term vanishes, we obtain

$$\begin{aligned} \partial_\mu D_{\mu\nu}(\partial) T^*\big[A_\nu(x), \ldots, U(\infty, -\infty)\big] = \partial_\mu T^*\left[\mathrm{i} \frac{\delta}{\delta A_\mu(x)} \ldots, U(\infty, -\infty) \right] \\ + e_{\rm obs} Z_1 Z_2^{-1} T^*\left[\left(\psi(x) \frac{\delta}{\delta \psi(x)} - \bar{\psi}(x) \frac{\delta}{\delta \bar{\psi}(x)} \right) \ldots, U(\infty, -\infty) \right] . \end{aligned} \tag{12.188}$$

Taking the vacuum expectation value of the above equation and using the Gell-Mann–Low formula, the relation between the Green's functions involving the renormalized Heisenberg operators is

$$\begin{aligned} \partial_\mu D_{\mu\nu}(\partial) \big\langle \mathbf{0} \big| T^*[\boldsymbol{A}_\nu(x) \ldots] \big| \mathbf{0} \big\rangle = \partial_\mu \left\langle \mathbf{0} \middle| \mathrm{i} \frac{\delta}{\delta \boldsymbol{A}_\mu(x)} T^*[\ldots] \middle| \mathbf{0} \right\rangle \\ + e_{\rm obs} Z_1 Z_2^{-1} \left\langle \mathbf{0} \middle| T^*\left[\left(\boldsymbol{\psi}(x) \frac{\delta}{\delta \boldsymbol{\psi}(x)} - \bar{\boldsymbol{\psi}}(x) \frac{\delta}{\delta \bar{\boldsymbol{\psi}}(x)} \right) \ldots \right] \middle| \mathbf{0} \right\rangle . \end{aligned} \tag{12.189}$$

In the equation above, dots stand for a suitable product of Heisenberg operators.

One feature of this equation is that only renormalized operators show up. Since $Z_1 Z_2^{-1} = 1$, as will be shown later, we can now obtain the *Ward–Takahashi identity* [119]:

$$\begin{aligned} \partial_\mu D_{\mu\nu}(\partial) \big\langle \mathbf{0} \big| T^*[\boldsymbol{A}_\nu(x) \ldots] \big| \mathbf{0} \big\rangle = \partial_\mu \left\langle \mathbf{0} \middle| \mathrm{i} \frac{\delta}{\delta \boldsymbol{A}_\mu(x)} T^*[\ldots] \middle| \mathbf{0} \right\rangle \\ + e_{\rm obs} \left\langle \mathbf{0} \middle| T^*\left[\left(\boldsymbol{\psi}(x) \frac{\delta}{\delta \boldsymbol{\psi}(x)} - \bar{\boldsymbol{\psi}}(x) \frac{\delta}{\delta \bar{\boldsymbol{\psi}}(x)} \right) \ldots \right] \middle| \mathbf{0} \right\rangle . \end{aligned} \tag{12.190}$$

To understand the relevance of (12.189), we first insert $A_\sigma(y)$ in place of the dots to obtain

$$\partial_\mu D_{\mu\nu}(\partial_x)\langle \mathbf{0}|T^*[\boldsymbol{A}_\nu(x), \boldsymbol{A}_\sigma(y)]|\mathbf{0}\rangle = \mathrm{i}\partial_\sigma \delta^4(x-y) . \tag{12.191}$$

Expressing this in momentum space and taking into account the equation

$$\partial_\mu D_{\mu\nu}(\partial) = \frac{1}{\alpha}\Box\partial_\nu , \tag{12.192}$$

we find

$$\frac{1}{\alpha}k^2 k_\nu D'_{F_{\nu\sigma}}(k) = k_\sigma . \tag{12.193}$$

We now write Dyson's equation for D'_{F} in an arbitrary gauge, viz.,

$$D'_{F_{\nu\sigma}}(k) = D_{F_{\nu\sigma}}(k) - D_{F_{\nu\lambda}}(k)\Pi^*_{\lambda\mu}(k)D'_{F_{\mu\sigma}}(k) , \tag{12.194}$$

where

$$D_{F_{\mu\sigma}}(k) = \left(\delta_{\nu\sigma} - \frac{k_\nu k_\sigma}{k^2-\mathrm{i}\epsilon}\right)\frac{1}{k^2-\mathrm{i}\epsilon} + \alpha\frac{k_{\nu\sigma}}{(k^2-\mathrm{i}\epsilon)^2} \tag{12.195}$$

and

$$\Pi^*_{\lambda\mu}(k) = \Pi^*_{\mathrm{ren}}(k)_{\lambda\mu} = (k_\lambda k_\mu - \delta_{\lambda\mu}k^2)\Pi^*(-k^2) . \tag{12.196}$$

From this, we obtain the solution of (12.194) in the form

$$D'_{F_{\nu\sigma}}(k) = \left(\delta_{\nu\sigma} - \frac{k_{\nu\sigma}}{k^2-\mathrm{i}\epsilon}\right)\frac{1}{k^2-\mathrm{i}\epsilon}\frac{1}{1-\Pi^*(-k^2)} + \alpha\frac{k_\nu k_\sigma}{(k^2-\mathrm{i}\epsilon)^2} . \tag{12.197}$$

This certainly satisfies (12.193).

Another example is

$$\begin{aligned}\partial_\mu D_{\mu\nu}(\partial_z)\langle 0|T^*[\boldsymbol{\psi}(x), \bar{\boldsymbol{\psi}}(y), \boldsymbol{A}_\nu(z)]|\mathbf{0}\rangle\\ = -e_{\mathrm{obs}}Z_1Z_2^{-1}\big[\delta^4(y-z) - \delta^4(x-z)\big]\langle \mathbf{0}|T^*[\boldsymbol{\psi}(x), \bar{\boldsymbol{\psi}}(y)]|\mathbf{0}\rangle.\end{aligned} \tag{12.198}$$

We combine (12.191) and this equation with the following equation:

$$\begin{aligned}\langle \mathbf{0}|T^*[\boldsymbol{\psi}(x), \bar{\boldsymbol{\psi}}(y), \boldsymbol{A}_\nu(z)]|\mathbf{0}\rangle\\ = -e_{\mathrm{obs}}\int \mathrm{d}^4x'\mathrm{d}^4y'\mathrm{d}^4z' S'_{\mathrm{F}}(x-x')\Gamma_\mu(x',y';z')S'_{\mathrm{F}}(y'-y)D'_{F_{\mu\nu}}(z'-z).\end{aligned} \tag{12.199}$$

Inserting the right-hand side of (12.199) into the left-hand side of (12.198) and using (12.191) to cancel D'_{F}, we obtain the following equation in momentum space:

$$-\,\mathrm{i}(p-q)_\nu S'_{\mathrm{F}}(p)\Gamma_\nu(p,q)S'_{\mathrm{F}}(q) = Z_1 Z_2^{-1}\left[S'_{\mathrm{F}}(p) - S'_{\mathrm{F}}(q)\right] . \qquad (12.200)$$

Alternatively, differentiating both sides with respect to p_μ and setting $q = p$,

$$\Gamma_\mu(p,p) = -\mathrm{i}Z_1 Z_2^{-1}\frac{\partial}{\partial p_\mu}S'_{\mathrm{F}}(p)^{-1} . \qquad (12.201)$$

We sandwiching this between $\bar{u}(p)$ and $u(p)$ and use the relations

$$S'_{\mathrm{F}}(p)^{-1} = \mathrm{i}p\cdot\gamma + m + \mathscr{O}\big((\mathrm{i}p\cdot\gamma + m)^2\big) ,$$

$$\bar{u}(p)(\mathrm{i}p\cdot\gamma + m) = (\mathrm{i}p\cdot\gamma + m)u(p) = 0 .$$

This leads finally to

$$\bar{u}(p)\gamma_\mu u(p) = Z_1 Z_2^{-1}\bar{u}(p)\gamma_\mu u(p) , \qquad (12.202)$$

which yields Ward's identity

$$Z_1 = Z_2 . \qquad (12.203)$$

Thus, in the Ward–Takahashi identity (12.189), we can replace $Z_1 Z_2^{-1}$ by 1 to obtain the Ward–Takahashi identity given by (12.190). In general, an equation obtained by subtracting a divergence of the Green's function is called a Ward–Takahashi identity.

Equations (12.201) and (12.203) were found by Ward [118], while (12.200) was derived by Takahashi [119]. Equation (12.190) for the Green's function was proved by the author in the Fermi gauge [120], but it was eventually shown that it can be extended to an arbitrary gauge.

12.8 Integral Representation of the Propagator

The propagator with higher-order corrections can be expressed by a certain kind of integral representation as the superposition of free-field propagators with different masses. This was established by Umezawa et al. in the early 1950s [121–123].

12.8.1 Integral Representation

We consider a neutral scalar field and make the following general assumptions:

1. There exists a four-vector operator P_μ, which stands for the energy–momentum, satisfying

$$[P_\mu, P_\nu] = 0 \,, \tag{12.204}$$

$$[\varphi(x), P_\mu] = \frac{1}{\mathrm{i}}\partial_\mu \varphi(x) \,. \tag{12.205}$$

2. There exists a set $\{\Phi_{k,\alpha}\}$ of eigenstates of P_μ which form a complete system. Here, k_μ is an eigenvalue of P_μ and α is another quantity specifying the states:

$$P_\mu \Phi_{k,\alpha} = k_\mu \Phi_{k,\alpha} \,. \tag{12.206}$$

Of course, as $\{\Phi\}$, we can choose either $\{\Phi^{\mathrm{in}}\}$ or $\{\Phi^{\mathrm{out}}\}$. In what follows, we assume the existence of the vacuum and write it as $|\mathbf{0}\rangle$. Affixing a prime to the invariant function in the presence of interactions,

$$\begin{aligned} \langle \mathbf{0}|\varphi(x)\varphi(y)|\mathbf{0}\rangle &= \mathrm{i}\Delta^{(+)\prime}(x-y) \,, \\ \langle \mathbf{0}|[\varphi(x), \varphi(y)]|\mathbf{0}\rangle &= \mathrm{i}\Delta'(x-y) \,, \\ \langle \mathbf{0}|T[\varphi(x)\varphi(y)]|\mathbf{0}\rangle &= \Delta'_{\mathrm{F}}(x-y) \,. \end{aligned} \tag{12.207}$$

Then, to carry the Fourier expansion, we introduce the matrix element

$$\langle \mathbf{0}|\varphi(x)|\boldsymbol{k}, \boldsymbol{\alpha}\rangle = a_{k,\alpha} \mathrm{e}^{\mathrm{i}k\cdot x} \,. \tag{12.208}$$

Hence,

$$\begin{aligned} \langle \mathbf{0}|\varphi(x)\varphi(y)|\mathbf{0}\rangle &= \sum_{k,\alpha} \langle \mathbf{0}|\varphi(x)|\boldsymbol{k}, \boldsymbol{\alpha}\rangle\langle \boldsymbol{k}, \boldsymbol{\alpha}|\varphi(y)|\mathbf{0}\rangle \\ &= \sum_{k,\alpha} |a_{k,\alpha}|^2 \mathrm{e}^{\mathrm{i}k\cdot(x-y)} \,. \end{aligned} \tag{12.209}$$

We now introduce the Lorentz invariant function

$$\rho(-p^2) = (2\pi)^3 \sum_{k,\alpha} |a_{k,\alpha}|^2 \delta^4(p-k) \geq 0 \,. \tag{12.210}$$

Inserting this into (12.209), for the time-like four-momentum p, we have

$$\begin{aligned} \mathrm{i}\Delta^{(+)\prime}(x-y) &= \frac{1}{(2\pi)^3}\int \mathrm{d}^4 p\,\theta(p_0)\rho(-p^2)\mathrm{e}^{\mathrm{i}p\cdot(x-y)} \qquad (-p^2 \geq 0) \\ &= \int_0^\infty \mathrm{d}\kappa^2 \rho(\kappa^2)\frac{1}{(2\pi)^3}\int \mathrm{d}^4 p\,\theta(p_0)\delta(p^2+\kappa^2)\mathrm{e}^{\mathrm{i}p\cdot(x-y)} \\ &= \int_0^\infty \mathrm{d}\kappa^2 \rho(\kappa^2)\mathrm{i}\Delta^{(+)}(x-y;\kappa^2)\,. \end{aligned} \tag{12.211}$$

Generalizing this to an arbitrary function, we obtain

$$\Delta^{(+)\prime}(x) = \int_0^\infty \mathrm{d}\kappa^2 \rho(\kappa^2)\Delta^{(+)}(x;\kappa^2)\,, \tag{12.212}$$

so in the presence of interactions, the invariant two-point function can be expressed by the integral representation as a superposition of invariant two-point functions for free fields with different masses.

The spectral function $\rho(\kappa^2)$ for a free field with the mass m is

$$\rho(\kappa^2) = \delta(\kappa^2 - m^2)\,. \tag{12.213}$$

If interactions are introduced, we can decompose into contributions from the one-particle intermediate states and the multi-particle intermediate states:

$$\rho(\kappa^2) = c\delta(\kappa^2 - m^2) + \sigma(\kappa^2)\,. \tag{12.214}$$

We note, however, that σ vanishes below the lowest invariant mass $2m$ in the two-particle system, i.e.,

$$\sigma(\kappa^2) = \theta(\kappa^2 - 4m^2)\sigma(\kappa^2)\,. \tag{12.215}$$

If we apply the renormalization condition, namely that the normalization of the wave function becomes the same as that of the free field in the one-particle state, then c in (12.214) becomes unity, i.e.,

$$\rho(\kappa^2) = \delta(\kappa^2 - m^2) + \sigma(\kappa^2)\,. \tag{12.216}$$

We now introduce the Fourier representation of the propagator:

$$\Delta_{\mathrm{F}}'(x) = \frac{-\mathrm{i}}{(2\pi)^4}\int \mathrm{d}^4 k\,\mathrm{e}^{\mathrm{i}k\cdot x}\,\Delta_{\mathrm{F}}'(-k^2)\,. \tag{12.217}$$

Hence,

$$\Delta_F'(-k^2) = \frac{1}{k^2 + m^2 - i\epsilon} + \int_{4m^2}^{\infty} d\kappa^2 \frac{\sigma(\kappa^2)}{k^2 + \kappa^2 - i\epsilon} . \quad (12.218)$$

We introduce the renormalization constant Z_φ, noting that, for the electron field $Z_\psi = Z_2$ and for the electromagnetic field $Z_A = Z_3$:

$$\boldsymbol{\varphi}^{(0)}(x) = Z_\varphi^{1/2} \boldsymbol{\varphi}(x) , \quad (12.219)$$

where the superscript (0) denotes unrenormalized quantities. Therefore,

$$\langle \mathbf{0} | [\boldsymbol{\varphi}^{(0)}(x), \boldsymbol{\varphi}^{(0)}(y)] | \mathbf{0} \rangle = Z_\varphi \int d\kappa^2 \rho(\kappa^2) i\Delta(x - y; \kappa^2) . \quad (12.220)$$

Differentiating both sides of this equation with respect to x_0 and then setting $x_0 = y_0$, the left-hand side becomes $-i\delta^3(x-y)$ from the canonical commutation relation. In addition, on the right-hand side, using (4.13),

$$\left. \frac{\partial}{\partial x_0} \Delta(x - y; \kappa^2) \right|_{x_0 = y_0} = -\delta^3(x - y) , \quad (12.221)$$

whence

$$- i\delta^3(x - y) = -i\delta^3(x - y) Z_\varphi \int d\kappa^2 \rho(\kappa^2) . \quad (12.222)$$

The integral representation of the renormalization constant can now be obtained immediately as

$$Z_\varphi^{-1} = \int d\kappa^2 \rho(\kappa^2) = 1 + \int d\kappa^2 \sigma(\kappa^2) \geq 1 . \quad (12.223)$$

This inequality originates from the positive-definite metric assumed for the scalar field. Thus,

$$1 \geq Z_\varphi \geq 0 . \quad (12.224)$$

12.8.2 Self-Energy

As shown above, the renormalization constant is written in terms of the integral representation via the spectral function. But what about the self-energy? To discuss this problem, we consider the following simple model:

$$\mathscr{L} = -\left(\partial_\lambda \Phi^\dagger \partial_\lambda \Phi + M_0^2 \Phi^\dagger \Phi\right) - \frac{1}{2}\left[(\partial_\lambda \varphi)^2 + m_0^2 \varphi^2\right] - g_0 \Phi^\dagger \Phi \varphi \, . \qquad (12.225)$$

Here φ stands for a neutral scalar field, Φ and $\Phi^\dagger$ for charged scalars, and m_0 and M_0 for the bare masses associated with each field, which are related to the observed masses m and M via

$$M^2 = M_0^2 + \delta M^2 \, , \quad m^2 = m_0^2 + \delta m^2 \, . \qquad (12.226)$$

We introduce multiplicative renormalizations via the relations

$$\boldsymbol{\Phi}^{(0)} = Z_2^{1/2} \boldsymbol{\Phi} \, , \quad \boldsymbol{\varphi}^{(0)} = Z_3^{1/2} \boldsymbol{\varphi} \, , \quad g_0 = Z_1 Z_2^{-1} Z_3^{-1/2} g \, . \qquad (12.227)$$

The renormalized field equations are

$$(\Box - m^2)\boldsymbol{\varphi} = Z_1 Z_3^{-1} g \boldsymbol{\Phi}^\dagger \boldsymbol{\Phi} - \delta m^2 \boldsymbol{\varphi} \, , \qquad (12.228)$$

$$(\Box - M^2)\boldsymbol{\Phi} = Z_1 Z_2^{-1} g \boldsymbol{\Phi} \boldsymbol{\varphi} - \delta M^2 \boldsymbol{\Phi} \, . \qquad (12.229)$$

Using (12.228), we can derive an integral representation of δm^2. Likewise, we can use (12.229) to derive an integral representation of δM^2. Expressing (12.228) in terms of m_0,

$$(\Box_x - m_0^2)\langle \mathbf{0} | [\boldsymbol{\varphi}(x), \boldsymbol{\varphi}(y)] | \mathbf{0} \rangle = Z_1 Z_3^{-1} g \langle \mathbf{0} | [\boldsymbol{\Phi}^\dagger(x) \boldsymbol{\Phi}(x), \boldsymbol{\varphi}(y)] | \mathbf{0} \rangle \, . \qquad (12.230)$$

Differentiating this equation with respect to y_0 and setting $y_0 = x_0$, the right-hand side becomes

$$\langle \mathbf{0} | [\boldsymbol{\Phi}^\dagger(x) \boldsymbol{\Phi}(x), \dot{\boldsymbol{\varphi}}(y)] | \mathbf{0} \rangle = 0 \qquad (x_0 = y_0) \, . \qquad (12.231)$$

It thus turns out that the left-hand side also vanishes. The left-hand side should still vanish if we carry out the operation discussed above in the integral representation:

$$(\Box^2 - m_0^2)\langle \mathbf{0} | [\boldsymbol{\varphi}(x), \boldsymbol{\varphi}(y)] | \mathbf{0} \rangle = \mathrm{i} \int \mathrm{d}\kappa^2 \rho_\varphi(\kappa^2)(\kappa^2 - m_0^2) \Delta(x - y; \kappa^2) \, .$$

Differentiating this with respect to y_0 and setting $y_0 = x_0 = 0$, Eq. (12.221) implies

$$\mathrm{i}\delta^3(x - y) \int \mathrm{d}\kappa^2 (\kappa^2 - m_0^2) \rho_\varphi(\kappa^2) = 0 \, ,$$

and hence,

$$\int \mathrm{d}\kappa^2(\kappa^2 - m^2)\rho_\varphi(\kappa^2) = -\delta m^2 \int \mathrm{d}\kappa^2 \rho_\varphi(\kappa^2) = -Z_3^{-1}\delta m^2 \,. \tag{12.232}$$

Thus,

$$\delta m^2 = -Z_3 \int \mathrm{d}\kappa^2(\kappa^2 - m^2)\rho_\varphi(\kappa^2) = -Z_3 \int \mathrm{d}\kappa^2(\kappa^2 - m^2)\sigma(\kappa^2) < 0 \,. \tag{12.233}$$

This equation implies that δm^2 is always negative. This is, of course, a predictable result. If we assume that, in (12.226), m^2 is positive and finite and m_0^2 is positive and infinite, then δm^2 can only be negative and infinite.

12.8.3 Integral Representation of the Electromagnetic Field Propagator

We now apply the previous discussion for the scalar field to the electromagnetic field. We first introduce a propagator for the electromagnetic field:

$$\langle \mathbf{0}|T^*[\boldsymbol{A}_\mu(x)\boldsymbol{A}_\nu(y)]|\mathbf{0}\rangle = \frac{-\mathrm{i}}{(2\pi)^4}\int \mathrm{d}^4k\, \mathrm{e}^{\mathrm{i}k\cdot(x-y)} D'_{F_{\mu\nu}}(k) \,. \tag{12.234}$$

For the free electromagnetic field, $D_{F_{\mu\nu}}$ is given by (12.195). Furthermore, it is clear from (12.197) that the effects of the interaction appear only in the transverse part of the wave. Recalling that

$$D_{\mu\nu}(\partial)\boldsymbol{A}_\nu = -\boldsymbol{j}_\mu \,, \tag{12.235}$$

$$\boldsymbol{j}_\mu = -\mathrm{i}e_{\mathrm{obs}}Z_2\bar{\boldsymbol{\psi}}\gamma_\mu\boldsymbol{\psi} - (1 - Z_3)(\delta_{\mu\sigma}\Box - \partial_\mu\partial_\sigma)\boldsymbol{A}_\sigma \,, \tag{12.236}$$

we introduce the current $\boldsymbol{j}_\mu$ and define the function

$$\langle \mathbf{0}|T^*[\boldsymbol{j}_\mu(x)\boldsymbol{j}_\nu(y)]|\mathbf{0}\rangle = \frac{-\mathrm{i}}{(2\pi)^4}\int \mathrm{d}^4k\, \mathrm{e}^{\mathrm{i}k\cdot(x-y)} \Pi_{\mu\nu}(k) \,. \tag{12.237}$$

Therefore,

$$k^4 \mathrm{Im} D'_{F_{\mu\nu}}(k) = \mathrm{Im}\Pi_{\mu\nu}(k) \,. \tag{12.238}$$

We now seek an integral representation of $\Pi_{\mu\nu}$. From the Lorentz covariance,

$$\langle \mathbf{0}|\boldsymbol{j}_\mu(x)\boldsymbol{j}_\nu(y)|\mathbf{0}\rangle = \mathrm{i}\delta_{\mu\nu}\int \mathrm{d}\kappa^2\sigma_1(\kappa^2)\Delta^{(+)}(x-y;\kappa^2) \qquad (12.239)$$

$$+\mathrm{i}\partial_\mu\partial_\nu\int \mathrm{d}\kappa^2\sigma_2(\kappa^2)\Delta^{(+)}(x-y;\kappa^2)\,.$$

Note also that, since $\partial_\mu \boldsymbol{j}_\mu = 0$,

$$\sigma_1(\kappa^2)+\kappa^2\sigma_2(\kappa^2)=0\,. \qquad (12.240)$$

We thus have

$$\sigma_1(\kappa^2)=\kappa^2\sigma(\kappa^2)\,,\quad \sigma_2(\kappa^2)=-\sigma(\kappa^2)\,, \qquad (12.241)$$

$$\langle\mathbf{0}|\boldsymbol{j}_\mu(x)\boldsymbol{j}_\nu(y)|\mathbf{0}\rangle=\mathrm{i}\int \mathrm{d}\kappa^2\sigma(\kappa^2)(\delta_{\mu\nu}\Box-\partial_\mu\partial_\nu)\Delta^{(+)}(x-y;\kappa^2)\,. \qquad (12.242)$$

Taking $\mu=\nu=4$,

$$-\langle\mathbf{0}|\boldsymbol{j}_0(x)\boldsymbol{j}_0(y)|\mathbf{0}\rangle=\mathrm{i}\int \mathrm{d}\kappa^2\sigma(\kappa^2)\nabla^2\Delta^{(+)}(x-y;\kappa^2)\,. \qquad (12.243)$$

Because $\boldsymbol{j}_\mu$ is gauge invariant, the indefinite metric cannot appear on the left-hand side. Hence,

$$(2\pi)^3\sum_n\left|\langle \boldsymbol{n}|\boldsymbol{j}_0(0)|\mathbf{0}\rangle\right|^2\delta^4(p_n-k)=\boldsymbol{k}^2\sigma(-k^2)\geq 0\,, \qquad (12.244)$$

i.e., $\sigma(\kappa^2)$ is positive-definite as in the case of the scalar field:

$$\sigma(\kappa^2)\geq 0\,. \qquad (12.245)$$

Using this result, for the integral representation of $\Pi_{\mu\nu}$, we have

$$\Pi_{\mu\nu}(k)=(k_\mu k_\nu-\delta_{\mu\nu}k^2)\Pi(-k^2)\,, \qquad (12.246)$$

$$\Pi(-k^2)=\int \mathrm{d}\kappa^2\frac{\sigma(\kappa^2)}{k^2+\kappa^2-\mathrm{i}\epsilon}+\text{const.} \qquad (12.247)$$

Since the renormalization condition requires $\Pi(0)$ to vanish,

$$\Pi(-k^2)=\Pi(-k^2)-\Pi(0)=-k^2\int\frac{\mathrm{d}\kappa^2}{\kappa^2}\frac{\sigma(\kappa^2)}{k^2+\kappa^2-\mathrm{i}\epsilon}\,. \qquad (12.248)$$

At the lowest order in the perturbation theory,

$$\sigma(\kappa^2) = \frac{e_{\mathrm{obs}}^2}{12\pi^2}\left(1 + \frac{2m^2}{\kappa^2}\right)\sqrt{1 - \frac{4m^2}{\kappa^2}}\theta(\kappa^2 - 4m^2) . \tag{12.249}$$

The integral representation of the electromagnetic field propagator is

$$D'_{F_{\mu\nu}}(k) = \left(\delta_{\mu\nu} - \frac{k_\mu k_\nu}{k^2 - \mathrm{i}\epsilon}\right)\left[\frac{1}{k^2 - \mathrm{i}\epsilon} + \int \frac{\mathrm{d}\kappa^2}{\kappa^2}\frac{\sigma(\kappa^2)}{k^2 + \kappa^2 - \mathrm{i}\epsilon}\right] + \alpha\frac{k_\mu k_\nu}{(k^2 - \mathrm{i}\epsilon)^2} . \tag{12.250}$$

12.8.4 Goto–Imamura–Schwinger Term

The integral representation of the vacuum expectation value of the commutator of $\boldsymbol{j}_\mu$ and $\boldsymbol{j}_\nu$ is

$$\langle \mathbf{0}|[\boldsymbol{j}_\mu(x), \boldsymbol{j}_\nu(y)]|\mathbf{0}\rangle = \mathrm{i}\int \mathrm{d}\kappa^2\sigma(\kappa^2)(\delta_{\mu\nu}\Box - \partial_\mu\partial_\nu)\Delta(x - y;\kappa^2) . \tag{12.251}$$

To obtain this result, we have only used the gauge invariance of $\partial_\mu \boldsymbol{j}_\mu = 0$ and $\boldsymbol{j}_0$.
Assuming that $\mu = 0$ and $\nu = k$ (1, 2, 3) and setting $x_0 = y_0$,

$$\begin{aligned} \mathrm{i}\langle \mathbf{0}|[j_0(x), j_k(y)]|\mathbf{0}\rangle &= -\int \mathrm{d}\kappa^2\sigma(\kappa^2)\big[\partial_0\partial_k\Delta(x - y;\kappa^2)\big]_{y_0 = x_0} \\ &= \frac{\partial}{\partial x_k}\delta^3(x - y)\int \mathrm{d}\kappa^2\sigma(\kappa^2) . \end{aligned} \tag{12.252}$$

Since $\sigma(\kappa^2) \geq 0$, the right-hand side does not vanish. Since $j_\mu = -\mathrm{i}e\bar{\psi}\gamma_\mu\psi$ at the lowest order, this contradicts the result obtained from the canonical commutation relation, i.e., using the commutation relation,

$$[j_0(x), j_k(y)] = 0 \quad (x_0 = y_0) . \tag{12.253}$$

The term which survives on the right-hand side of (12.252) is called the *Goto–Imamura–Schwinger term* [124, 125]. We carry out the following computation in the interaction picture using $\partial_\mu j_\mu = 0$:

$$\partial_\mu T[j_\mu(x), j_\nu(y)] = \delta(x_0 - y_0)[j_0(x), j_\nu(y)] \neq 0 . \tag{12.254}$$

The reason why this vanished in (12.121) is that we used the T*-product instead of the T-product, i.e.,

$$\langle \mathbf{0}|T^*[j_\mu(x), j_\nu(y)]|\mathbf{0}\rangle = \int \mathrm{d}\kappa^2 \sigma(\kappa^2)(\delta_{\mu\nu}\Box - \partial_\mu\partial_\nu)\Delta_{\mathrm{F}}(x-y;\kappa^2). \quad (12.255)$$

In this case, it turns out that (12.121) obviously holds. As shown above, the representation which includes a product of field operators at the same point often has a singularity. A different result is often obtained from the one derived by a simple computation.

As a similar example, if we consider a neutral vector field and assume that the mass is zero, then since $\partial_\mu \varphi_\mu = 0$,

$$\begin{aligned}[\varphi_\mu(x), \varphi_\nu(y)] &= \mathrm{i}\left(\delta_{\mu\nu} - \frac{1}{m^2}\frac{\partial^2}{\partial x_\mu \partial x_\nu}\right)\Delta(x-y;m^2) \\ &= \mathrm{i}\int \mathrm{d}\kappa^2 \frac{\delta(\kappa^2 - m^2)}{m^2}(\delta_{\mu\nu}\Box - \partial_\mu\partial_\nu)\Delta(x-y;\kappa^2)\,. \quad (12.256)\end{aligned}$$

Therefore,

$$\sigma(\kappa^2) = \frac{1}{m^2}\delta(\kappa^2 - m^2)\,, \quad (12.257)$$

$$\mathrm{i}\langle 0|[\varphi_0(x), \varphi_k(y)]|0\rangle = \frac{\partial}{\partial x_k}\delta^3(x-y)\frac{1}{m^2}\,. \quad (12.258)$$

The derivation here is exactly the same as for the Goto–Imamura–Schwinger term.

In this chapter, we have discussed the concept of renormalization by asking how we can separate the Lagrangian into the free part and the interaction part. We have shown that, at the next to lowest order, the divergence disappears, or rather that the renormalization constant can be chosen so as to make the divergence vanish. In the renormalization theory, the following stance is taken: the observed parameters can be replaced by finite experimental values, while the bare parameters which are not observed directly are allowed to be divergent. But if we now compute higher order corrections, can we be sure there will still be no divergence? Although this issue has been discussed by many people, we shall not pursue it here. We will discuss this aspect from a slightly different point of view in the last chapter.

Chapter 13
Classification of Hadrons and Models

Up to and including Chap. 12, we have described the fundamental framework of quantum field theory, focusing mainly on issues related to QED. However, as mentioned in Chap. 1, an important part of elementary particle theory concerns the strong interactions. Hence, we need to consider the classification, structure, and modelling of hadrons, and in this chapter, we shall turn to a more phenomenological discussion about such issues, including the group-theoretical classification of hadrons, the quark model, the parton model, and so on.

13.1 Unitary Groups

We start with a simple review of group theory. In a set G consisting of elements $a, b, \ldots$, we define a product denoted ab: if a and b are elements of G, then the product ab is also an element of G. If this product satisfies the following four axioms, then we call G a *group*:

1. If $a, b \in G$, then their product ab is uniquely defined and $ab \in G$.
2. There is a unique identity element e in G, such that, for an arbitrary element a in G,

$$ae = ea = a\ .$$

3. For an arbitrary element a in G, there is a unique inverse element, denoted a^{-1}, such that

$$aa^{-1} = a^{-1}a = e\ .$$

K. Nishijima, *Quantum Field Theory*,
https://doi.org/10.1007/978-94-024-2190-3_13

4. The product is associative, i.e., for arbitrary elements a, b, c,

$$a(bc) = (ab)c\ .$$

13.1.1 Representations of a Group

If there exists a matrix $D(a)$ for each element a in G such that

$$D(a)D(b) = D(ab)\ , \tag{13.1}$$

$$D(a^{-1}) = D(a)^{-1}\ , \tag{13.2}$$

$$D(e) = E \quad \text{(unit matrix)}\ , \tag{13.3}$$

then we call this set of matrices a *representation* of the group G. The vector space on which the matrix operators act is called a *representation space* and the dimension of this vector space is called the *dimension of the representation*.

The transformation of a vector in the n-dimensional representation is described by

$$\xi_a \;\rightarrow\; \xi'_a = D_a{}^b \xi_b\ , \quad a, b = 1, 2, \ldots, n\ . \tag{13.4}$$

In addition, the contravariant representation $\bar{\xi}^a$ associated with ξ_a is specified by the transformation

$$\bar{\xi}^a \;\rightarrow\; \bar{\xi}^{a\prime} = \bar{\xi}^b (D^{-1})_b{}^a\ . \tag{13.5}$$

The quantity

$$\bar{\xi}^a \xi_a \quad \text{(summing over } a\text{)} \tag{13.6}$$

is then an invariant.

Two representations D_1 and D_2 are said to be equivalent if there exists a matrix U such that, for every element a in G,

$$D_2(a) = U D_1(a) U^{-1}\ . \tag{13.7}$$

If, for an arbitrary element a, the matrix $D(a)$ can be diagonalized to the form

$$D(a) = \begin{pmatrix} D_1(a) & 0 & 0 \\ 0 & D_2(a) & 0 \\ 0 & 0 & \ddots \end{pmatrix}, \tag{13.8}$$

the representation is said to be *reducible* and we write

$$D = D_1 \oplus D_2 \oplus \cdots \oplus D_k \ . \tag{13.9}$$

A representation which does not permit such a decomposition is said to be *irreducible*.

13.1.2 Direct Product Representation

Suppose there are two representations D^α and D^β, with dimensions α and β, respectively. If we now use the bases $\xi^\alpha(i)$ and $\xi^\beta(j)$ of the two representation spaces to define

$$\xi^{\alpha\beta}(ij) = \xi^\alpha(i)\xi^\beta(j) \ , \tag{13.10}$$

we can construct an $\alpha\beta$-dimensional representation. This is called a *direct product representation*, and we write it as

$$D^{\alpha\beta} = D^\alpha \otimes D^\beta \ . \tag{13.11}$$

In general, the direct product representation is reducible.

13.1.3 Lie Groups

We call a transformation group parameterized by n continuous variables a *Lie group*, and n is called the *order* of the group.

13.1.4 Orthogonal Group $O(n)$

For n real variables $x_1, \ldots, x_n$, the set of linear transformations $O(n)$ of those variables which leave the quadratic form

$$\sum_{i=1}^{n} x_i^2 = x_1^2 + \cdots + x_n^2 \tag{13.12}$$

invariant is called the *orthogonal group*.

13.1.5 Unitary Group U(n)

Similarly, the set of linear transformations $U(n)$ of the complex variables $\xi_1, \ldots, \xi_n$ which leave the quantity

$$\sum_{i=1}^{n} |\xi_i|^2 = |\xi_1|^2 + \cdots + |\xi_n|^2 \tag{13.13}$$

invariant is called the *unitary group*. This is the group of n-dimensional unitary matrices. Assuming that U is one of these matrices,

$$U^\dagger U = U U^\dagger = E \,, \quad U^{-1} = U^\dagger \,. \tag{13.14}$$

The order N of this group is n^2. Considering the transformation generated by a composition of infinitesimal transformations in this group, as we did for the proper Lorentz transformation, a general element of $U(n)$ can be written as

$$U = \exp(\mathrm{i}\epsilon^a F_a) \,, \quad a = 1, 2, \ldots, N \,, \tag{13.15}$$

where we have summed over a, and F_a is a Hermitian matrix, i.e.,

$$F_a^\dagger = F_a \,. \tag{13.16}$$

Moreover, an infinitesimal transformation has the form

$$U = 1 + \mathrm{i}\epsilon^a F_a \,, \tag{13.17}$$

where ϵ^a is an infinitesimal parameter. The group $SU(n)$ comprises all unitary matrices with unit determinant, i.e., matrices U such that

$$\det U = 1 \,. \tag{13.18}$$

This is called the *special unitary group*. For the matrix written in the form (13.15), the condition (13.18) can be expressed by

$$\mathrm{Tr}\, F_a = 0 \,. \tag{13.19}$$

There are in fact $N = n^2 - 1$ independent matrices F_a. The F_a are called *generators*. The set F of these generators satisfies the commutation relations

$$[F_i, F_j] = \mathrm{i} f_{ijk} F_k \,. \tag{13.20}$$

This makes it a Lie algebra. The f_{ijk} are called *structure constants*. The commutators in (13.20) satisfy the following two conditions:

$$\text{Antisymmetry:} \qquad [F_i, F_j] = -[F_j, F_i] \,, \tag{13.21}$$

$$\text{Jacobi identity:} \qquad \big[F_i, [F_j, F_k]\big] + \big[F_j, [F_k, F_i]\big] + \big[F_k, [F_i, F_j]\big] = 0 \,. \tag{13.22}$$

In addition, the maximum number of the $F_1, F_2, \ldots, F_N$ which can be simultaneously diagonalized, in the quantum mechanical sense, is called the *rank* of the algebra. For instance, in $SU(2)$, choosing the three Pauli matrices $\sigma_i/2$ ($i = 1, 2, 3$) to be the F's, we have

$$[J_i, J_j] = \mathrm{i}\epsilon_{ijk} J_k \,, \tag{13.23}$$

where $J_i = \sigma_i/2$. Since only J_3 is in general diagonal, the rank of the algebra is unity.

We consider a representation of $U(n)$ as in (13.4),

$$\xi'_a = U_a{}^b \xi_b \,. \tag{13.24}$$

Taking the complex conjugate, since $U(n)$ is unitary,

$$\xi^{*\prime}_a = \xi^*_b (U_a{}^b)^* = \xi^*_b (U^{-1})_b{}^a \,.$$

Thus, comparing with (13.5), we can choose the contravariant vector to be

$$\xi^*_a = \bar{\xi}^a \,. \tag{13.25}$$

13.1.6 Special Unitary Group $SU(2)$

A two-dimensional representation of $U(2)$ is given by

$$U = \mathrm{e}^{\mathrm{i}\epsilon_0} \mathrm{e}^{\mathrm{i}\hat{\epsilon}\hat{\sigma}/2} \,. \tag{13.26}$$

The first factor is a phase and the second factor rotates in the isospin space. Moving from $U(2)$ to $SU(2)$, the general form is

$$U = \exp(\mathrm{i}\hat{\epsilon}\hat{\sigma}/2) \,. \tag{13.27}$$

This is a representation of the isospin group for the nucleon. In this case, if we take the three components of the quantized operators which express the isospin to be I_1,

I_2, and I_3, then they satisfy

$$[I_i, I_j] = \mathrm{i}\epsilon_{ijk} I_k \ . \tag{13.28}$$

The two-dimensional representation of I_i is $\sigma_i/2$. Since this group is isomorphic to the rotation group in three-dimensional space, the direct product representation has the decomposition

$$D_I \otimes D_{I'} = D_{I+I'} \oplus D_{I+I'-1} \oplus \cdots \oplus D_{|I-I'|} \ . \tag{13.29}$$

Although $SU(2)$ shows up in many topics in physics and is of fundamental importance, we will now discuss $SU(3)$, which is an extension of $SU(2)$.

13.2 The Group *SU(3)*

The group $SU(2)$ was introduced in relation to the isospin, and all its representations were obtained from the fundamental representation using p and n (the proton and neutron, respectively) as its basis. If we consider only I_3 as a hadron quantum number, the hadrons can be classified using $SU(2)$ with rank 1, but if we include strangeness, as discussed in Chap. 1, a group of rank 2 is required. And hence, $SU(3)$ comes on the scene.

The group $SU(3)$ has $3^2 - 1 = 8$ generators, and we write them as $F_1, F_2, \ldots, F_8$. Since a traceless three-dimensional diagonal matrix has only two independent components, the group has rank two. And so, as in the case of p and n, we consider a three-component basis:

$$\text{Sakata's model} \qquad \mathrm{p}\,,\ \mathrm{n}\,,\ \Lambda\,, \tag{13.30}$$

$$\text{Gell-Mann's quark model} \qquad \mathrm{u}\,,\ \mathrm{d}\,,\ \mathrm{s}\,. \tag{13.31}$$

Although we discussed both Sakata's model [64] and the quark model [74, 75] in Chap. 1, in the following we will only describe the quark model. Just like p and n, u and d correspond to the up-state and the down-state of the isospin, respectively, while s corresponds to the quark with strangeness. The electric charge Q of the hadron is connected with the third component of its isospin by the relations

$$Q = e\left(I_3 + \frac{Y}{2}\right) , \tag{13.32}$$

$$Y = B + S \ . \tag{13.33}$$

Here Y is the *hypercharge*, an integer specific to each isospin multiplet. B and S are the *baryon number* and the *strangeness*, respectively. The baryon number is $+1$ for

Table 13.1 Quantum numbers of the three different quarks

	B	I_3	S	Q/e
u	1/3	1/2	0	2/3
d	1/3	−1/2	0	−1/3
s	1/3	0	−1	−1/3

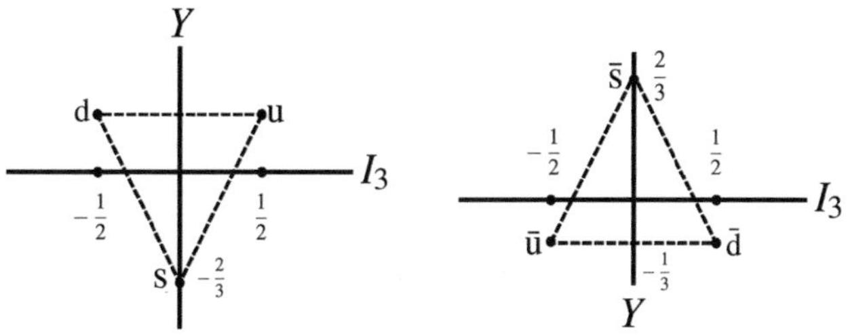

Fig. 13.1 Weight diagram for quarks (*left*) and anti-quarks (*right*)

the nucleon and -1 for the anti-nucleon. It is also $+1$ for Λ^0, Σ^+, Σ^0, Σ^-, Ξ^0, Ξ^-, and so on, which decay into the nucleon by emitting a π-meson or a photon.

Although I_3 and S are conserved for the strong interaction and the electromagnetic interaction, they are not conserved for the weak interaction and can vary by 1. If we denote quarks such as u, d, and s generically by q, we can consider that, in the quark model, mesons correspond to bound states of the form $\mathrm{q\bar{q}}$, and baryons to bound states of the form qqq. In this model, the electric charges and the baryon numbers of the quarks are not integers but fractions, being integer multiples of 1/3. Their values are shown in Table 13.1.

Now the quark, anti-quark, and their bound states form multiplets. Each member of a multiplet is distinguished by the values of I_3 and Y. In order to make the structure of the multiplets clear, we define a so-called *weight diagram* in the I_3–Y plane (see Fig. 13.1). The isospin distinguishes members in the horizontal direction, such as u and d. We simply call this I-spin. On the other hand, we sometimes introduce the U-spin distinguishing d and s and the V-spin distinguishing s and u. We will return to this when discussing the mass formula.

The weight diagram of the representation created by the direct product of two representations can be made by superimposing the center of one weight diagram on each particle point in the other weight diagram. For instance, from the direct product of q and $\bar{\mathrm{q}}$, we obtain the weight diagram shown in Fig. 13.2. The center of this diagram is occupied by three particles.

The different states $q_i\bar{q}^j$ can be expressed in a matrix form:

$$D = (q_i\bar{q}^j) = \begin{pmatrix} \mathrm{u\bar{u}} & \mathrm{u\bar{d}} & \mathrm{u\bar{s}} \\ \mathrm{d\bar{u}} & \mathrm{d\bar{d}} & \mathrm{d\bar{s}} \\ \mathrm{s\bar{u}} & \mathrm{s\bar{d}} & \mathrm{s\bar{s}} \end{pmatrix} . \tag{13.34}$$

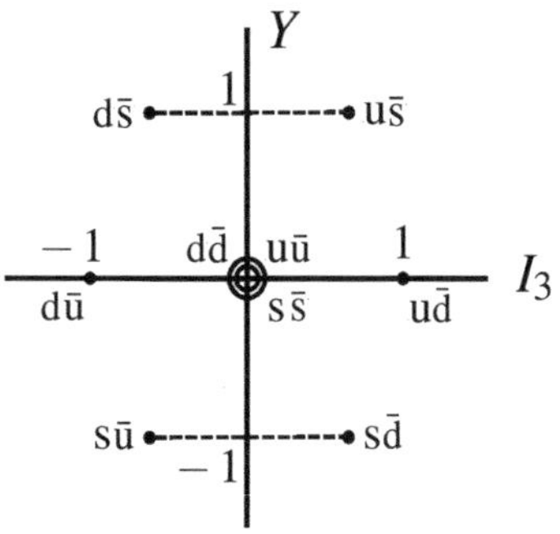

Fig. 13.2 Weight diagram for the quark–anti-quark system

We decompose this direct product as follows:

$$D_i{}^j = \frac{1}{3}\delta_i{}^j D_k{}^k + \left(D_i{}^j - \frac{1}{3}\delta_i{}^j D_k{}^k \right) . \tag{13.35}$$

The first term is a scalar singlet and the second term is a traceless tensor octet. The coefficient of the first term is

$$\frac{1}{3}(\mathrm{u\bar{u}} + \mathrm{d\bar{d}} + \mathrm{s\bar{s}}) = \frac{1}{\sqrt{3}} \left(\frac{\mathrm{u\bar{u}} + \mathrm{d\bar{d}} + \mathrm{s\bar{s}}}{\sqrt{3}} \right) . \tag{13.36}$$

This belongs to the one-dimensional representation of $SU(3)$. Since this meson is 0^-, it corresponds to the particle called $\eta'(958)$, where 958 indicates the mass in units of MeV. We thus replace the matrix elements of D by the 0^- mesons. For example,

$$(13.36) \longrightarrow \frac{1}{\sqrt{3}}\eta' .$$

We now consider the octet, in which the off-diagonal terms are replaced directly by suitable mesons without the need for any adjustments. On the other hand, the diagonal terms are first decomposed in terms of the isospin. For instance, the second term in (13.35), corresponding to $i = j = 1$, is

$$\begin{aligned} \frac{1}{3}(2\mathrm{u\bar{u}} - \mathrm{d\bar{d}} - \mathrm{s\bar{s}}) &= \frac{1}{6}(\mathrm{u\bar{u}} + \mathrm{d\bar{d}} - 2\mathrm{s\bar{s}}) + \frac{1}{2}(\mathrm{u\bar{u}} - \mathrm{d\bar{d}}) \\ &\to \frac{1}{\sqrt{6}}\eta^0 + \frac{1}{\sqrt{2}}\pi^0 . \end{aligned} \tag{13.37}$$

The off-diagonal terms are replaced as follows:

$$\mathrm{u\bar{s}} \to \mathrm{K}^+ , \quad \mathrm{d\bar{s}} \to \mathrm{K}^0 , \quad \mathrm{s\bar{u}} \to \mathrm{K}^- , \quad \mathrm{s\bar{d}} \to \bar{\mathrm{K}}^0 . \tag{13.38}$$

These replacements are useful when discussing only group-theoretic properties, without any relation to the details of the interactions. Thus, the part of D corresponding to the 0^- mesons is

$$D(0^-) = \frac{1}{\sqrt{3}}\eta' + \begin{pmatrix} \frac{\eta^0}{\sqrt{6}} + \frac{\pi^0}{\sqrt{2}} & \pi^+ & K^+ \\ \pi^- & \frac{\eta^0}{\sqrt{6}} - \frac{\pi^0}{\sqrt{2}} & K^0 \\ K^- & \bar{K}^0 & -\frac{2}{\sqrt{6}}\eta^0 \end{pmatrix} . \tag{13.39}$$

The decomposition of the direct product into irreducible representations is

$$\mathbf{3} \otimes \bar{\mathbf{3}} = \mathbf{1} \oplus \mathbf{8} , \tag{13.40}$$

where $\mathbf{3}$ and $\bar{\mathbf{3}}$ are the three-dimensional representations to which the quark and the anti-quark belong, respectively. A similar decomposition can be carried out for the 1^- mesons. Thus, the part of D corresponding to the 1^- mesons is

$$D(1^-) = \frac{1}{\sqrt{3}}\omega + \begin{pmatrix} \frac{\phi^0}{\sqrt{6}} + \frac{\rho^0}{\sqrt{2}} & \rho^+ & K^{*+} \\ \rho^- & \frac{\phi^0}{\sqrt{6}} - \frac{\rho^0}{\sqrt{2}} & K^{*0} \\ K^{*-} & \bar{K}^{*0} & -\frac{2}{\sqrt{6}}\phi^0 \end{pmatrix} . \tag{13.41}$$

13.2.1 Generators of *SU*(3)

The generators of $SU(2)$ in the two-dimensional spinor representation are

$$I_i \rightarrow \frac{1}{2}\sigma_i . \tag{13.42}$$

What about $SU(3)$? Considering the 2×2 part of $D(0^-)$, the isospin-1 part can be written as

$$\begin{pmatrix} \frac{\pi^0}{\sqrt{2}} & \pi^+ \\ \pi^- & -\frac{\pi^0}{\sqrt{2}} \end{pmatrix} = \frac{1}{\sqrt{2}} \begin{pmatrix} \pi^3 & \pi^1 - i\pi^2 \\ \pi^1 + i\pi^2 & -\pi^3 \end{pmatrix} = \frac{1}{\sqrt{2}}\hat{\tau}\hat{\pi} . \tag{13.43}$$

For a Hermitian field, we write the octet part in the form

$$\frac{1}{\sqrt{2}}(\lambda_i)_a{}^b M^i = \frac{1}{\sqrt{2}}\hat{\lambda}\hat{M}\ , \tag{13.44}$$

where M stands for the eight-component Hermitian meson field.

We use the following numbering for, say, 0^- :

$$\begin{pmatrix} \frac{\pi^0}{\sqrt{2}}+\frac{\eta^0}{\sqrt{6}} & \pi^+ & \mathrm{K}^+ \\ \pi^- & -\frac{\pi^0}{\sqrt{2}}+\frac{\eta^0}{\sqrt{6}} & \mathrm{K}^0 \\ \mathrm{K}^- & \bar{\mathrm{K}}^0 & -\frac{2\eta^0}{\sqrt{6}} \end{pmatrix} \equiv \frac{1}{\sqrt{2}} \begin{pmatrix} M^3+\frac{M^8}{\sqrt{3}} & M^1-\mathrm{i}M^2 & M^4-\mathrm{i}M^5 \\ M^1+\mathrm{i}M^2 & -M^3+\frac{M^8}{\sqrt{3}} & M^6-\mathrm{i}M^7 \\ M^4+\mathrm{i}M^5 & M^6+\mathrm{i}M^7 & -\frac{2}{\sqrt{3}}M^8 \end{pmatrix} . \tag{13.45}$$

Thus, corresponding to (13.42) in the three-dimensional representation, the generators λ_i generalize σ_i and replace the generators F_i for $SU(3)$:

$$F_i \longrightarrow \frac{\lambda_i}{2}\ . \tag{13.46}$$

Following Gell-Mann, these eight matrices λ_i can be chosen as follows:

$$\lambda_1 = \begin{pmatrix} 0&1&0\\1&0&0\\0&0&0 \end{pmatrix},\quad \lambda_2 = \begin{pmatrix} 0&-\mathrm{i}&0\\ \mathrm{i}&0&0\\0&0&0 \end{pmatrix},\quad \lambda_3 = \begin{pmatrix} 1&0&0\\0&-1&0\\0&0&0 \end{pmatrix},$$
$$\lambda_4 = \begin{pmatrix} 0&0&1\\0&0&0\\1&0&0 \end{pmatrix},\quad \lambda_5 = \begin{pmatrix} 0&0&-\mathrm{i}\\0&0&0\\ \mathrm{i}&0&0 \end{pmatrix},\quad \lambda_6 = \begin{pmatrix} 0&0&0\\0&0&1\\0&1&0 \end{pmatrix}, \tag{13.47}$$
$$\lambda_7 = \begin{pmatrix} 0&0&0\\0&0&-\mathrm{i}\\0&\mathrm{i}&0 \end{pmatrix},\quad \lambda_8 = \frac{1}{\sqrt{3}}\begin{pmatrix} 1&0&0\\0&1&0\\0&0&-2 \end{pmatrix}.$$

Using this choice for the λ_i, the structure constants can be determined from

$$\left[\frac{\lambda_i}{2}, \frac{\lambda_j}{2}\right] = \mathrm{i} f_{ijk}\frac{\lambda_k}{2}\ . \tag{13.48}$$

The λ_i are all Hermitian matrices and normalized so that

$$\mathrm{Tr}(\lambda_i\lambda_j) = 2\delta_{ij}\ . \tag{13.49}$$

This choice is important. The symmetries of the structure constants can be derived from this equation. Furthermore, using (13.49), the anti-commutation relations can be written in the form

$$\left\{\frac{\lambda_i}{2}, \frac{\lambda_j}{2}\right\} = \frac{1}{3}\delta_{ij} + d_{ijk}\frac{\lambda_k}{2} \,. \tag{13.50}$$

This introduces another set of structure constants.

The two sets of structure constants f_{ijk} and d_{ijk} above are completely anti-symmetric and completely symmetric in the indices i, j, and k, respectively. To see this, we use (13.49):

$$4\mathrm{i} f_{ijk} = \mathrm{Tr}([\lambda_i, \lambda_j]\lambda_k) = \mathrm{Tr}(\lambda_j[\lambda_k, \lambda_i]) = \mathrm{Tr}(\lambda_i[\lambda_j, \lambda_k]) \,. \tag{13.51}$$

This shows that f_{ijk} is completely anti-symmetric. Similarly, the symmetry of d_{ijk} is obvious from the equation

$$4d_{ijk} = \mathrm{Tr}(\{\lambda_i, \lambda_j\}\lambda_k) = \mathrm{Tr}(\lambda_j\{\lambda_k, \lambda_i\}) = \mathrm{Tr}(\lambda_i\{\lambda_j, \lambda_k\}) \,. \tag{13.52}$$

13.2.2 *I*-, *U*-, and *V*-Spin

The $SU(3)$ algebra has several kinds of $SU(2)$ algebras as subalgebras, and many useful consequences can be derived from this. Firstly, examining the content of (13.32), we obtain

$$F_3 = I_3 \,, \quad F_8 = \frac{\sqrt{3}}{2}Y \,. \tag{13.53}$$

Thus, for the quarks,

$$Q = e\left(F_3 + \frac{F_8}{\sqrt{3}}\right) \,. \tag{13.54}$$

This relation holds for all hadrons composed of quarks and anti-quarks. We define the I-, U-, and V-spin as follows:

$$\begin{aligned} &I_\pm = F_1 \pm \mathrm{i}F_2 \,, \quad U_\pm = F_6 \pm \mathrm{i}F_7 \,, \quad V_\pm = F_4 + \mathrm{i}F_5 \,, \\ &I_3 = F_3 \,, \quad U_3 = -\frac{1}{2}F_3 + \frac{\sqrt{3}}{2}F_8 \,, \quad V_3 = -\frac{1}{2}F_3 - \frac{\sqrt{3}}{2}F_8 \,. \end{aligned} \tag{13.55}$$

Hence, I, U, and V generate three $SU(2)$ subalgebras. Moreover, among these three algebras, the following condition is satisfied:

$$I_3 + U_3 + V_3 = 0 . \tag{13.56}$$

It is straightforward to show that the commutation relations among these are

$$\begin{aligned}
&[I_3, I_\pm] = \pm I_\pm , \quad [I_3, U_\pm] = \mp\frac{1}{2}U_\pm , \quad [I_3, V_\pm] = \mp\frac{1}{2}V_\pm , \\
&[U_3, I_\pm] = \mp\frac{1}{2}I_\pm , \quad [U_3, U_\pm] = \pm U_\pm , \quad [U_3, V_\pm] = \mp\tfrac{1}{2}V_\pm , \\
&[[V_3, I_\pm] = \mp\frac{1}{2}I_\pm , \quad [V_3, U_\pm] = \mp\frac{1}{2}U_\pm , \quad [V_3, V_\pm] = \pm V_\pm , \\
&[Y, I_\pm] = 0 , \quad [Y, U_\pm] = \pm U_\pm , \quad [Y, V_\pm] = \pm V_\pm , \\
&[I_+, I_-] = 2I_3 , \quad [U_+, U_-] = 2U_3 , \quad [V_+, V_-] = 2V_3 , \\
&[I_\pm, U_\pm] = \pm V_\pm , \quad [U_\pm, V_\pm] = \pm I_\mp , \quad [V_\pm, I_\pm] = \pm U_\mp .
\end{aligned} \tag{13.57}$$

13.2.3 Three-Body Quark Systems

As we have seen before, a meson is a quark–anti-quark bound state $q\bar{q}$, whereas a baryon (or heavy particle) is a three-quark bound state qqq. We thus begin with the weight diagram for a two-quark state qq.

We can carry out the following direct-product decomposition corresponding to (13.40):

$$\mathbf{3} \otimes \mathbf{3} = \bar{\mathbf{3}} \oplus \mathbf{6} . \tag{13.58}$$

Since three of the six points appearing in Fig. 13.3 are double points, we have to make a suitable linear combination of the relevant states. To do this, we have to

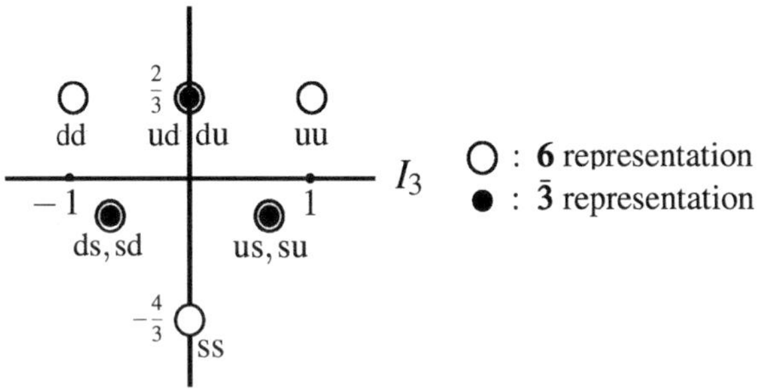

Fig. 13.3 Weight diagram for the two-quark system

make symmetric and anti-symmetric combinations:

$$q_i q_j = T_{ij} = \frac{1}{2}(T_{ij} - T_{ji}) + \frac{1}{2}(T_{ij} + T_{ji}) = \frac{1}{2}T_{[ij]} + \frac{1}{2}T_{(ij)} \ . \tag{13.59}$$

The anti-symmetric one is equal to the contravariant vector

$$\epsilon^{ijk} T_{[ij]} = T^k \ . \tag{13.60}$$

This is a three-dimensional representation and transforms like $\bar{q}^k$, so it is the representation $\bar{\mathbf{3}}$. The symmetric part, on the other hand, has six components and corresponds to the representation **6**. This confirms (13.58).

We now look at the transformation of ϵ under $SU(3)$:

$$\epsilon'_{ijk} = U_i{}^l U_j{}^m U_k{}^n \epsilon_{lmn} = \epsilon_{ijk} \det U \ .$$

Thus, since $\det U = 1$,

$$\epsilon'_{ijk} = \epsilon_{ijk} \ . \tag{13.61}$$

Similarly,

$$\epsilon'^{ijk} = \epsilon^{ijk} \ . \tag{13.62}$$

We can now list the states belonging to the $\bar{\mathbf{3}}$ representation and the **6** representation:

$$\{\bar{\mathbf{3}}\} \qquad \begin{cases} I = 0 & \frac{1}{\sqrt{2}}(\mathrm{ud} - \mathrm{du}) \ , \\ I = \frac{1}{2} & \frac{1}{\sqrt{2}}(\mathrm{us} - \mathrm{su}) \ , \quad \frac{1}{\sqrt{2}}(\mathrm{ds} - \mathrm{sd}) \ , \end{cases} \tag{13.63}$$

$$\{\mathbf{6}\} \qquad \begin{cases} I = 0 & \mathrm{ss} \\ I = \frac{1}{2} & \frac{1}{\sqrt{2}}(\mathrm{us} + \mathrm{su}) \ , \quad \frac{1}{\sqrt{2}}(\mathrm{ds} + \mathrm{sd}) \ , \\ I = 1 & \mathrm{uu} \ , \quad \frac{1}{\sqrt{2}}(\mathrm{ud} + \mathrm{du}) \ , \quad \mathrm{dd} \ . \end{cases} \tag{13.64}$$

We now consider a general tensor:

$$T^{(p)}_{(q)} = T^{\alpha\beta\ldots\gamma}_{ij\ldots k} \ , \quad \text{such as} \ \ \bar{q}^\alpha \bar{q}^\beta \ldots \bar{q}^\gamma q_i q_j \ldots q_k \ , \tag{13.65}$$

and an irreducible representation, i.e., one where the tensor cannot be reduced to a lower-dimensional tensor by contractions with the tensors

$$\delta_i{}^j\ , \quad \epsilon^{ijk}\ , \quad \epsilon_{ijk}\ . \tag{13.66}$$

Reducible tensors can be reduced to lower-dimensional tensors by contractions, as in the following examples:

$$\delta_\alpha{}^i T^{\alpha\beta\ldots\delta}_{ij\ldots l} = B^{\beta\ldots\delta}_{j\ldots l}\ , \quad \epsilon_{\mu\alpha\beta} T^{\alpha\beta\ldots\delta}_{ij\ldots l} = C^{\gamma\ldots\delta}_{\mu ij\ldots l}\ , \quad \epsilon^{mij} T^{\alpha\beta\ldots\delta}_{ij\ldots l} = D^{m\alpha\beta\ldots\delta}_{k\ldots l}\ .$$

If T is irreducible, it turns out that the right-hand sides of the equations above vanish. This is the condition that T be symmetric in $\alpha, \beta, \ldots$ and $i, j, \ldots$, and the trace vanishes, i.e.,

$$T^{i\beta\ldots\delta}_{ij\ldots l} = 0\ . \tag{13.67}$$

If $T^{(p)}_{(q)}$ satisfies the condition above, there is one irreducible representation $D(p,q)$. Here, p and q are the numbers of superscripts and subscripts, respectively. In this case, the dimension N of $D(p,q)$ is given by

$$N = (1+p)(1+q)\left[1 + \frac{1}{2}(p+q)\right]\ . \tag{13.68}$$

Proof First we ask how many tensors there are which are symmetric in p indices. We consider as indices the case where the numbers 1, 2, and 3 are x, y, and z. We must therefore count how many choices of x, y, and z there are. This is equivalent to counting the number of solutions to $x+y+z=p$. x can take any of the values $x = 0, 1, \ldots, p$. Once x is fixed, we need to find the number of solutions to

$$y + z = p - x\ . \tag{13.69}$$

This yields $p - x + 1$. The total will thus be

$$\sum_{x=0}^{p}(p-x+1) = \frac{1}{2}(p+1)(p+2)\ . \tag{13.70}$$

Now, for the p superscripts and the q subscripts, the number of symmetric tensors is

$$\frac{1}{4}(p+1)(p+2)(q+1)(q+2)\ . \tag{13.71}$$

We now take into account the vanishing trace condition. To do this, we reduce the number of symmetric tensors by changing p into $p-1$ and q into $q-1$ and subtracting the result:

$$N = \frac{1}{4}(p+1)(p+2)(q+1)(q+2) - \frac{1}{4}p(p+1)q(q+1)$$
$$= (1+p)(1+q)\left[1+\frac{1}{2}(p+q)\right] ,$$

which proves (13.68).

We can now specify the following examples of $D(p, q)$:

$$\begin{aligned} &D(0,0) = \{\mathbf{1}\} , \quad D(1,1) = \{\mathbf{8}\} , \quad D(0,3) = \{\mathbf{10}\} , \\ &D(0,1) = \{\mathbf{3}\} , \quad D(0,2) = \{\mathbf{6}\} , \quad D(3,0) = \{\bar{\mathbf{10}}\} , \\ &D(1,0) = \{\bar{\mathbf{3}}\} , \quad D(2,0) = \{\bar{\mathbf{6}}\} , \quad \ldots . \end{aligned} \tag{13.72}$$

And so we come back to the first issue, i.e., investigating the representation of the three-quark system:

$$\mathbf{3} \otimes \mathbf{3} \otimes \mathbf{3} = (\mathbf{3} \oplus \mathbf{6}) \otimes \mathbf{3} = (\mathbf{1} \oplus \mathbf{8}) \oplus (\mathbf{8} \oplus \mathbf{10}) . \tag{13.73}$$

Nucleons such as p and n belong to the eight-dimensional representation here. An excited state of the nucleon known as Δ, which arises in π-meson–nucleon scattering, belongs to the ten-dimensional representation.

In the same vein as (13.39) and (13.41), we now express the eight-dimensional representation which includes the nucleon in the form of a matrix:

$$\mathscr{B} = \begin{pmatrix} \dfrac{\Sigma^0}{\sqrt{2}} + \dfrac{\Lambda^0}{\sqrt{6}} & \Sigma^+ & \mathrm{p} \\ \Sigma^- & -\dfrac{\Sigma^0}{\sqrt{2}} + \dfrac{\Lambda^0}{\sqrt{6}} & \mathrm{n} \\ \Xi^- & \Xi^0 & -\dfrac{2}{\sqrt{6}}\Lambda^0 \end{pmatrix} . \tag{13.74}$$

We can now derive the anti-baryon octet from the baryon octet using the charge conjugation operator $\mathscr{C}$, whence

$$\mathscr{C}^{-1}\mathscr{B}^{\mathrm{T}}\mathscr{C} = C\bar{\mathscr{B}} , \tag{13.75}$$

where T stands for the transposition of the 3×3 matrix. This is a generalization of (3.198). Finally, imitating (13.74), we denote the matrix representations of $D(0^-)$ and $D(1^-)$ by $\mathscr{P}$ and $\mathscr{V}$, respectively.

In the treatment above, the idea of using the quark as the fundamental representation is a mathematical tool for deriving other representations. Physically, even if there are no quarks, we can derive all observable representations by considering eight-dimensional baryon representations and eight-dimensional meson representations. We call this the *eightfold way* [52, 67, 68], in contrast with Sakata's model, which starts with the three-dimensional representation of the baryon,

13.2.4 Mass Formulas

$SU(3)$ is not a symmetry group in the strict sense. In the real world, this symmetry is broken. However, by treating this symmetry-breaking perturbatively, meaningful consequences can be obtained, both qualitatively and semi-quantitatively. In the following, we investigate the mass difference between hadrons and also the magnetic moment, using a perturbative computation in the context of the quark model. Later in this chapter, we will show the corresponding computation based on the full quark model.

The most important assumption introduced here is that, when we split the Hamiltonian into the invariant part under $SU(3)$ and the other part, viz.,

$$H = H_{\text{inv}} + H_{\text{SB}} , \tag{13.76}$$

the symmetry-breaking part H_{SB} has the same transformation property as F_8 and we can treat this term as a perturbation. Now,

$$\frac{\sqrt{3}}{2} F_8 = U_3 + \frac{1}{2} I_3 . \tag{13.77}$$

We introduce the following U-spin triplet:

$$\mathrm{n} , \quad -\frac{1}{2}\Sigma^0 + \frac{\sqrt{3}}{2}\Lambda^0 , \quad \Xi^0 . \tag{13.78}$$

The matrix elements of U_3 and I_3 in the eight-dimensional representation are

$$\begin{aligned}
&\langle \mathrm{n}|U_3|\mathrm{n}\rangle = 1 , \\
&\left\langle -\frac{1}{2}\Sigma^0 + \frac{\sqrt{3}}{2}\Lambda^0 \middle| U_3 \middle| -\frac{1}{2}\Sigma^0 + \frac{\sqrt{3}}{2}\Lambda^0 \right\rangle = 0 , \\
&\left\langle \Xi^0 \middle| U_3 \middle| \Xi^0 \right\rangle = -1 ,
\end{aligned} \tag{13.79}$$

$$\begin{aligned}
&\langle \mathrm{n}|I_3|\mathrm{n}\rangle = -\frac{1}{2} , \\
&\left\langle -\frac{1}{2}\Sigma^0 + \frac{\sqrt{3}}{2}\Lambda^0 \middle| I_3 \middle| -\frac{1}{2}\Sigma^0 + \frac{\sqrt{3}}{2}\Lambda^0 \right\rangle = 0 , \\
&\left\langle \Xi^0 \middle| I_3 \middle| \Xi^0 \right\rangle = \frac{1}{2} .
\end{aligned} \tag{13.80}$$

If H_{SB} transforms in the same way as (13.77),

$$\langle \mathrm{n}|H_{\mathrm{SB}}|\mathrm{n}\rangle + \langle \Xi^0|H_{\mathrm{SB}}|\Xi^0\rangle = 2\Big\langle -\frac{1}{2}\Sigma^0 + \frac{\sqrt{3}}{2}\Lambda^0\Big|H_{\mathrm{SB}}\Big| -\frac{1}{2}\Sigma^0 + \frac{\sqrt{3}}{2}\Lambda^0\Big\rangle$$
$$= \frac{1}{2}\langle \Sigma^0|H_{\mathrm{SB}}|\Sigma^0\rangle + \frac{3}{2}\langle \Lambda^0|H_{\mathrm{SB}}|\Lambda^0\rangle\,. \qquad (13.81)$$

Thus, adding the common mass of the invariant part to each diagonal element, which is taken as a shift in the mass, we deduce the relation

$$m(\mathrm{n}) + m(\Xi) = \frac{1}{2}m(\Sigma) + \frac{3}{2}m(\Lambda)\,. \qquad (13.82)$$

Additionally, for the 10-dimensional representation, considering the spin quartet,

$$\Delta^-\,,\quad \Sigma^{*-}\,,\quad \Xi^{*-}\,,\quad \Omega^-\,, \qquad (13.83)$$

and carrying out a similar computation,

$$m(\Delta) - m(\Sigma^*) = m(\Sigma^*) - m(\Xi^*) = m(\Sigma^*) - m(\Omega)\,. \qquad (13.84)$$

These are called the *Gell-Mann mass formulas* [67]. The weight diagram of the ten-dimensional representation is shown in Fig. 13.4. It was using these formulas that Ω^- was found in [71].

Okubo also showed that, using the isospin I and hypercharge Y, the mass of a particle in one of the $SU(3)$ multiplets can be expressed in the form [70]

$$m = a + bY + c\left[I(I+1) - \frac{Y^2}{4}\right]\,. \qquad (13.85)$$

Fig. 13.4 Weight diagram for the decuplet

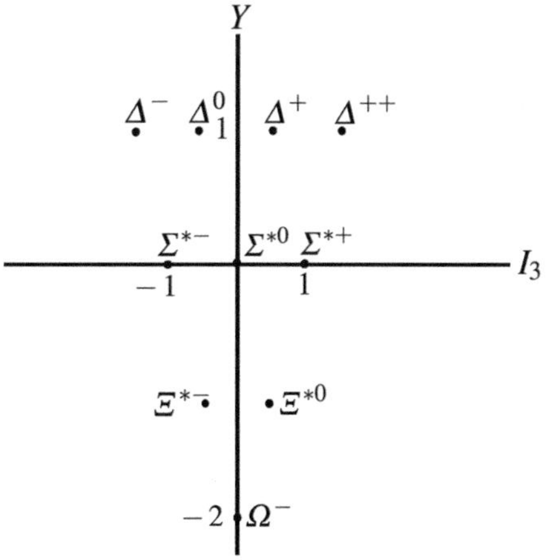

Although this equation holds for both bosons and fermions, a better agreement with experiment is obtained for bosons by writing the equation for m^2.

13.2.5 Baryon Magnetic Moments

In the example above, we have assumed that the term H_{SB} which breaks the symmetry transforms in the same way as Y (which is proportional to F_8). It may be assumed that the electromagnetic interaction, which was not considered there, has the same transformation property as Q. Consequently, the magnetic moment can be considered to have the same transformation property. Note that Q commutes with the U-spin:

$$[U_{\pm}, Q] = 0\ , \quad [U_3, Q] = 0\ . \tag{13.86}$$

Moreover, since the electromagnetic interaction is a sum of the isospin singlet and triplet, we have

$$\mu\left(\Sigma^{+}\right) + \mu\left(\Sigma^{-}\right) = 2\mu\left(\Sigma^{0}\right)\ , \tag{13.87}$$

where $\mu\ (\Sigma)$ stands for the magnetic moment of Σ, and Σ^+, Σ^0, and Σ^- belong to the isospin triplet.

Since the magnetic moments of particles belonging to the same multiplet of the U-spin are all equal, we can classify the baryon octet by the U-spin:

$$\begin{array}{ll} U - \text{spin singlet} & U_0^0 = \frac{1}{2}(\sqrt{3}\Sigma^0 + \Lambda^0)\ , \\ U - \text{spin doublet} & \Sigma^+\ , \quad \mathrm{p}\ , \quad \text{and} \quad \Xi^-\ , \quad \Sigma\ , \\ U - \text{spin triplet} & \Xi^0\ , \quad U_1^0 = \frac{1}{2}(\Sigma^0 - \sqrt{3}\Lambda^0)\ , \quad \mathrm{n}\ . \end{array}$$

From these, we immediately obtain the following equalities:

$$\mu(\Sigma^+) = \mu(\mathrm{p})\ , \quad \mu(\Xi^-) = \mu(\Sigma^-)\ , \quad \mu(\Xi^0) = \mu(U_1^0) = \mu(\mathrm{n})\ . \tag{13.88}$$

Furthermore, the sum of the eigenvalues of Q over all members of the octet vanishes:

$$\sum_{\{\mathbf{8}\}} \mu = 0\ . \tag{13.89}$$

Since (13.87), (13.88), and (13.89) amount to six conditions, we can express the eight magnetic moments in terms of just two magnetic moments, say $\mu(\mathrm{p})$ and $\mu(\mathrm{n})$. It is more useful to use Σ^0 and Λ^0 as particles, rather than U_0^0 and U_1^0, so we consider the latter as eigenstates of μ and use the condition

$$\langle U_0^0 | \mu | U_1^0 \rangle = 0 \,. \tag{13.90}$$

This yields the relations

$$2\mu(\Sigma^0\Lambda^0) \equiv 2\langle \Sigma^0 | \mu | \Lambda^0 \rangle = \sqrt{3}\big[\mu(\Sigma^0) - \mu(\Lambda^0)\big] \,, \tag{13.91}$$

$$\mu(U_1^0) = \frac{3}{2}\mu(\Lambda^0) - \frac{1}{2}\mu(\Sigma^0) \,. \tag{13.92}$$

Here, $\mu(\Sigma^0\Lambda^0)$ is called a *transition magnetic moment*. It contributes to the decay $\Sigma^0 \to \Lambda^0 + \gamma$. We can now obtain the magnetic moments of all members of the octet:

$$\begin{aligned} &\mu(\Sigma^+) = \mu(\mathrm{p}) \,, \quad \mu(\Sigma^0) = -\frac{1}{2}\mu(\mathrm{n}) \,, \quad \mu(\Sigma^-) = -\mu(\mathrm{p}) - \mu(\mathrm{n}) \,, \\ &\mu(\Lambda^0) = \frac{1}{2}\mu(\mathrm{n}) \,, \quad \mu(\Xi^0) = \mu(\mathrm{n}) \,, \quad \mu(\Xi^-) = -\mu(\mathrm{p}) - \mu(\mathrm{n}) \,. \end{aligned} \tag{13.93}$$

Taking $e/2m_\mathrm{p}$ as the basic unit, known as the *nuclear magneton*, and using

$$\mu(\mathrm{p}) = 2.79 \,, \quad \mu(\mathrm{n}) = -1.91 \,, \tag{13.94}$$

we can compute the magnetic moments of all members of the baryon octet and compare them with experimental data. The results are displayed in Table 13.2.

Although the agreement of theoretical values and experimental values is qualitatively good, it is not so good quantitatively. Later, we will return to a more quantitative discussion using the quark model. In any case, the discussions so far are group theoretical, and not based directly on the quark model.

Table 13.2 Choosing C_j and O_j

	$\mu(\Lambda^0)$	$\mu(\Sigma^+)$	$\mu(\Sigma^-)$	$\mu(\Xi^-)$	$\mu(\Xi^0)$
Theoretical value	-0.96	2.79	-0.88	-0.88	-1.91
Experimental value	-0.613 ± 0.004	2.38 ± 0.02	-1.10 ± 0.05	-1.85 ± 0.75	-1.250 ± 0.014

13.2.6 $SU(3)$-Invariant Interactions

So far, we have treated the octets of baryons, pseudo-scalar mesons, and vector mesons. These are expressed by 3×3 matrices like $\mathscr{B}$, $\mathscr{P}$, and $\mathscr{V}$. We now consider the transformation under $SU(3)$ (13.24). For this transformation, the matrix D given by (13.34) transforms by

$$D \to U D U^{-1} . \tag{13.95}$$

Thus, $\mathscr{B}$, $\mathscr{P}$, and $\mathscr{V}$, and so on, all transform in the same way as (13.95). This means that any trace of the product of these matrices will be invariant under the $SU(3)$ transformations. Moreover, under charge conjugation,

$$\mathscr{C}^{-1} \mathscr{P}^T \mathscr{C} = \mathscr{P} , \quad \mathscr{C}^{-1} \mathscr{V}^T \mathscr{C} = -\mathscr{V} . \tag{13.96}$$

These correspond to (13.75) in the case of baryons.

An $SU(3)$-invariant Yukawa-type interaction is

$$\mathscr{H}_{\text{int}} = \sqrt{2} D \operatorname{Tr}\big(\bar{\mathscr{B}}\{\mathscr{M}, \mathscr{B}\}\big) + \sqrt{2} F \operatorname{Tr}\big(\bar{\mathscr{B}}[\mathscr{M}, \mathscr{B}]\big) , \tag{13.97}$$

for $\mathscr{M} = \mathscr{P}$ or $\mathscr{V}$. The reason why we use the notation D and F is that d and f appear as the Clebsch–Gordan coefficients connecting different kinds of members when we compute the trace. In the case of the pseudo-scalar meson, the Yukawa-type coupling constant G appearing in (6.96) is given by

$$G = D + F . \tag{13.98}$$

Note also that, if there are interactions corresponding to

$$\mathscr{V} \to \mathscr{P} + \mathscr{P} , \quad \mathscr{V} \to \mathscr{V} + \mathscr{P} , \tag{13.99}$$

it is easy to see that only type F or type D can occur, respectively.

13.2.7 Casimir Operator

Up to now, we have introduced matrices corresponding to the octets $\mathscr{B}$, $\mathscr{P}$, and $\mathscr{V}$. Considering a member corresponding to their matrix elements, say the one such that M in (13.45) is replaced by F, we write it as

$$\mathscr{F} = \sum_{i=1}^{8} \frac{\lambda_i}{\sqrt{2}} F_i . \tag{13.100}$$

Although the F_i transform under $SU(3)$, their transformation rules are once again given by (13.95). Hence,

$$\mathrm{Tr}\,\mathscr{F}^2\,, \quad \mathrm{Tr}\,\mathscr{F}^3\,, \tag{13.101}$$

are invariant under the $SU(3)$ transformations, and therefore,

$$[F_i, \mathrm{Tr}\,\mathscr{F}^n] = 0\,. \tag{13.102}$$

An operator which is a polynomial in the generators F_i and commutes with all the F_i as just described is called a *Casimir operator*. It is clear from the definition that the Casimir operator has the same eigenvalue for all members of the same irreducible representation. The operator

$$\boldsymbol{J}^2 = J_x^2 + J_y^2 + J_z^2\,, \tag{13.103}$$

for the rotation group in three-dimensional space is an example of a Casimir operator, and its eigenvalue $J(J+1)$ specifies the irreducible representation.

In the case of $SU(3)$, there are only two independent quantities in $\mathrm{Tr}\,\mathscr{F}^n$. If we take

$$C_n = \mathrm{Tr}\,\mathscr{F}^n\,, \tag{13.104}$$

then it is straightforward to show that the following recursion equation holds:

$$C_{n+3} = \frac{1}{2}C_2C_{n+1} + \frac{1}{3}C_3C_n\,. \tag{13.105}$$

Using $C_1 = 0$, all the C_n can be expressed in terms of C_2 and C_3. Thus, we can choose C_2 and C_3 as independent Casimir operators. For instance, C_4 is expressed in terms of C_2 by

$$\mathrm{Tr}\,\mathscr{F}^4 = \frac{1}{2}\left(\mathrm{Tr}\,\mathscr{F}^2\right)^2\,, \quad C_4 = \frac{1}{2}C_2^2\,. \tag{13.106}$$

We can express C_2 and C_3 in terms of the F_i :

$$C_2 = \mathrm{Tr}\,\mathscr{F}^2 = \sum_i F_i^2\,, \quad C_3 = \mathrm{Tr}\,\mathscr{F}^3 = \frac{1}{\sqrt{2}}\sum_{i,j,k} d_{ijk}F_iF_jF_k\,. \tag{13.107}$$

We write the eigenvalue of C_2 for a given irreducible representation R as $C_2(R)$. In particular, we denote the eigenvalue for the adjoint representation by $C_2(G)$. Now we know that, for the $SU(3)$ group, the irreducible representations can be expressed as $D(p,q)$, and it can be shown that, for $R = D(p,q)$,

$$C_2(R) = \frac{1}{3}(p^2 + pq + q^2) + p + q \,. \tag{13.108}$$

This result will be used when discussing the gauge theory later on.

In this section we have used quarks as a mathematical tool and discussed how to apply the $SU(3)$ group to classify hadrons. Later, we will explain why quarks should be considered as real physical entities.

13.3 Universality of π-Meson Decay Interactions

So far, we have discussed the classification of hadrons. We shall now consider leptons. As mentioned in Chap. 1, leptons are fermions which are not affected by the strong interaction. These include those carrying electric charges, such as the electron and the muon, as well as the neutral neutrino. The latter has several forms, e.g., the electron neutrino ν_e, which forms a doublet with the electron, and the muon neutrino ν_μ, which forms a doublet with the muon. We call the pairs (e, ν_e) and (μ, ν_μ) the *first generation* and the *second generation*, respectively. It seems that the lepton number in each generation is conserved. M.L. Perl found a third charged lepton in [126]. This is called the *tauon* or *tau-lepton* and denoted by τ. Of course, there is also a tau neutrino which is paired with the tauon. We thus know of three generations of leptons:

$$(e, \nu_e) \,, \quad (\mu, \nu_\mu) \,, \quad (\tau, \nu_\tau) \,. \tag{13.109}$$

The masses of the charged leptons are

$$m_e = 0.5110034 \pm 0.0000014\,\text{MeV} \,, \quad m_\mu = 105.65943 \pm 0.00018\,\text{MeV} \,,$$
$$m_\tau = 1776.86 \pm 0.12\,\text{MeV} \,. \tag{13.110}$$

Since the mass of the π-meson or pion is

$$m_\pi = 139.5637 \pm 0.0007\,\text{MeV} \,, \tag{13.111}$$

the pion can decay into lepton pairs of the first and second generations:

$$\pi^+ \to \mu^+ + \nu_\mu \,, \qquad \pi^+ \to e^+ + \nu_e \,. \tag{13.112}$$

However, it is the tau-lepton that can decay into the pion:

$$\tau^+ \to \pi^+ + \bar{\nu}_\tau \,. \tag{13.113}$$

We assume that the interaction between the lepton l (l = e, μ, τ), the l-neutrino ν_l, and π has the form

$$\mathscr{H}_{\text{decay}} = \frac{\mathrm{i}g}{m_\pi}\bar{\psi}_{\nu_l}\gamma_\lambda(1+\gamma_5)\psi_l\partial_\lambda\varphi + \text{h.c.} \tag{13.114}$$

In Sect. 9.2, assuming that the coupling constant g has the same value for the first and second generations, and computing the branching ratio of (13.112), we obtained (9.61), i.e.,

$$\frac{\Gamma(\pi^+ \to \mathrm{e}^+ + \nu_\mathrm{e})}{\Gamma(\pi^+ \to \mu^+ + \nu_\mu)} = 1.29 \times 10^{-4}\,. \tag{13.115}$$

This agrees well with the experimental value of $(1.266 \pm 0.023) \times 10^{-4}$. Let us assume that it is the same for the third generation. The decay width for the reaction (13.113) is

$$\Gamma(\tau^+ \to \pi^+ + \bar{\nu}_\tau) = \frac{g^2}{8\pi}m_\tau\left(\frac{m_\tau}{m_\pi}\right)^2\left[1-\left(\frac{m_\pi}{m_\tau}\right)^2\right]. \tag{13.116}$$

In this case, since we have not summed over the spins for τ^+, but averaged over them because the reaction differs from (13.112) (in the sense that the charged lepton τ is in the initial, and not the final state), $g^2/4\pi$ has been altered to $g^2/8\pi$. Then, taking the ratio with (9.59), g^2 drops out. If we compute the ratio, the value expected theoretically is

$$\frac{\Gamma(\tau^+ \to \pi^+ + \bar{\nu}_\tau)}{\Gamma(\pi^+ \to \mu^+ + \nu_\mu)} \approx 1.0 \times 10^4\,. \tag{13.117}$$

In order to investigate this ratio experimentally, we write

$$\begin{aligned}\frac{\Gamma(\tau^+ \to \pi^+ + \bar{\nu}_\tau)}{\Gamma(\pi^+ \to \mu^+ + \nu_\mu)} &= \frac{\Gamma(\tau^+ \to \text{all})}{\Gamma(\pi^+ \to \text{all})}r(\tau^+ \to \pi^+ + \nu_\tau)\\ &= \frac{\tau(\pi^+)}{\tau(\tau^+)}r(\tau^+ \to \pi^+ + \nu_\tau)\,,\end{aligned} \tag{13.118}$$

where $r(\tau^+ \to \pi^+ + \nu_\tau)$ stands for the branching ratio for τ^+ to decay as in (13.113), and $\tau(\pi^+)$ and $\tau(\tau^+)$ are the mean lifetimes of π^+ and τ^+, respectively:

$$\begin{aligned}\tau(\pi^+) &= (2.6030 \pm 0.0023) \times 10^{-8}\,\mathrm{s}\,,\\ \tau(\mu^+) &= (2.19714 \pm 0.00007) \times 10^{-6}\,\mathrm{s}\,,\\ \tau(\tau^+) &= (3.4 \pm 0.5) \times 10^{-13}\,\mathrm{s}\,.\end{aligned} \tag{13.119}$$

Inserting these values into (13.118) and equating it with (13.117), we obtain

$$r(\tau^+ \to \pi^+ + \nu_\tau) = (13 \pm 2)\% . \qquad (13.120)$$

In the case of π^+, we have assumed that they almost all decay into $\mu^+ + \nu_\mu$. Equation (13.120) does not contradict the experimental value $(10.7 \pm 1.6)\%$. This confirms that the coupling constant g does not depend on the generation in the lepton-decay interactions of π.

13.4 Beta-Decay

The interaction term (13.114) given in the previous section is in fact phenomenological. We may consider that this interaction derives from a more fundamental one. Before the birth of gauge theory, the fundamental weak interaction theory was Fermi's, and the oldest and most typical example of such an interaction was the beta-decay of an atomic nucleus.

A theory of beta-decay, viz.,

$$\mathrm{n} \to \mathrm{p} + \mathrm{e}^- + \bar{\nu}_\mathrm{e} , \qquad (13.121)$$

was proposed by Fermi in 1934 [13]. The interaction term was considered to be

$$\mathscr{H}_\beta = \sum_j C_j \bar{\psi}_\mathrm{p} O_j \psi_\mathrm{n} \cdot \bar{\psi}_\mathrm{e} O_j \psi_\nu + \sum_j C'_j \bar{\psi}_\mathrm{p} O_j \psi_\mathrm{n} \cdot \bar{\psi}_\mathrm{e} O_j \gamma_5 \psi_\nu + \mathrm{h.c.} , \qquad (13.122)$$

where the O_j are the Dirac matrices corresponding to the scalar, the vector, the tensor, the axial vector, and the pseudo-scalar (SVTAP), introduced in Sect. 3.9. From the point of view of the Pauli–Gürsey group introduced in Sect. 7.7, this is not general enough. A term with ψ_ν replaced by $C\bar{\psi}_\nu$ should be added to (13.122). However, we consider here only the interaction with the restricted form above. Therefore, we are actually considering only a subgroup of the Pauli–Gürsey group:

$$\mathscr{G}_1 \times \mathscr{G}_2 \in \mathscr{G}_\mathrm{F} . \qquad (13.123)$$

Phase transformations under $\mathscr{G}_\mathrm{F}$ are

$$\psi \to \mathrm{e}^{\mathrm{i}\alpha + \mathrm{i}\beta\gamma_5} \psi , \quad \bar{\psi} \to \bar{\psi} \mathrm{e}^{-\mathrm{i}\alpha + \mathrm{i}\beta\gamma_5} . \qquad (13.124)$$

These transformations lead to the following transformations of the coupling constants:

$$C_j \to \mathrm{e}^{\mathrm{i}\alpha} (C_j \cos\beta + \mathrm{i}C'_j \sin\beta) , \quad C'_j \to \mathrm{e}^{\mathrm{i}\alpha} (C'_j \cos\beta + \mathrm{i}C_j \sin\beta) . \qquad (13.125)$$

An invariant combination of the coupling constants under these transformations is

$$C_i^* C_j + C_i'^* C_j' \,. \tag{13.126}$$

Experimentally, only this combination is measured. It was Lee and Yang who introduced a term including C_j' into (13.122) and considered the possibility of non-conservation of parity [45]. That was in 1956. They pointed out that it was only by making the replacement

$$C_i^* C_j \;\to\; C_i^* C_j + C_i'^* C_j' \tag{13.127}$$

in several kinds of calculation in the previous theory without C_j', that agreement could be obtained with previously obtained experimental data for beta-decay.

Furthermore, restricting to the transformation (13.124), fixed points exist for C_i and C_j' under the transformation (13.125):

$$\text{when } \alpha = -\beta\,, \quad C_j = C_j'\,, \text{ and when } \alpha = \beta\,, \quad C_j = -C_j'\,. \tag{13.128}$$

Fixed points imply that the interaction (13.122) is invariant under this restricted transformation, and as discussed in Sect. 7.7, this leads to invariance under the restricted transformation (7.146). As shown in Sect. 3.7, experiment implies that

$$C_j = C_j'\,. \tag{13.129}$$

Thus, (13.122) can be written in the form

$$\mathscr{H}_\beta = \sum_j C_j \bar{\psi}_p O_j \psi_n \cdot \bar{\psi}_e O_j (1 + \gamma_5) \psi_\nu + \text{h.c.} \tag{13.130}$$

As shown above, the neutrino is left-handed. Lee and Yang chose C_j and O_j as in Table 13.3.

We shall now determine what kinds of coupling constants exist. In beta-decay, there is one for the emission of e^- and one for the emission of e^+. We can then obtain the following equation for the dependence of the helicity $\boldsymbol{\sigma} \cdot \boldsymbol{p}/p$ of the emitted $e^\mp$ on the coupling constants:

$$h(e^\pm) = \mp \frac{|\langle 1 \rangle|^2 (|C_S|^2 - |C_V|^2) + |\langle \boldsymbol{\sigma} \rangle|^2 (|C_T|^2 - |C_A|^2)}{|\langle 1 \rangle|^2 (|C_S|^2 + |C_V|^2) + |\langle \boldsymbol{\sigma} \rangle|^2 (|C_T|^2 + |C_A|^2)} \frac{v}{c}\,, \tag{13.131}$$

Table 13.3 Choice of C_j and O_j made by Lee and Yang

Type	Scalar	Vector	Tensor	Axial vector	Pseudo-scalar
C_j	C_S	C_V	C_T	C_A	C_P
O_j	1	γ_μ	$\frac{1}{\sqrt{2}}\sigma_{\mu\nu}$	$-i\gamma_\mu\gamma_5$	γ_5

where $\langle \mathbf{1} \rangle$ and $\langle \boldsymbol{\sigma} \rangle$ are the nuclear matrix elements of Fermi type and Gamow–Teller type. That is, assuming that $\boldsymbol{\psi}$ is a non-relativistic Heisenberg operator,

$$\langle \mathbf{1} \rangle = \langle \boldsymbol{B} | \boldsymbol{\psi}_{\mathrm{p}}^{\dagger}(0) \boldsymbol{\psi}_{\mathrm{n}}(0) | \boldsymbol{A} \rangle \ , \quad \langle \boldsymbol{\sigma} \rangle = \langle \boldsymbol{B} | \boldsymbol{\psi}_{\mathrm{p}}^{\dagger}(0) \boldsymbol{\sigma} \boldsymbol{\psi}_{\mathrm{n}}(0) | \boldsymbol{A} \rangle \ . \tag{13.132}$$

In deriving (13.131), we have omitted the so-called *Fierz interference term* between the VA-type and the ST-type, and we have also omitted the Coulomb correction. We now prove this formula, assuming for simplicity that there is only a scalar-type interaction.

We start with the state $|i\rangle$ and end up with the state $S|i\rangle$. Therefore,

$$S|i\rangle = \sum_{n} |n\rangle S_{ni} \ . \tag{13.133}$$

Depending on experimental conditions, we may or may not have to take a sum over all states $|n\rangle$. We write observable states as (n). The expectation value of the physical quantity A in the final state is given by

$$\langle A \rangle = \frac{\sum_{(m)} \sum_{(n)} \langle i | T^{\dagger} | m \rangle \langle m | A | n \rangle \langle n | T | i \rangle}{\sum_{n} \langle i | T^{\dagger} | n \rangle \langle n | T | i \rangle} \ , \tag{13.134}$$

where S is replaced by T and common factors in the numerator and denominator have been eliminated. We now choose the eigenstates of A as the partially complete system (n):

$$A|n\rangle = a_n |n\rangle \ . \tag{13.135}$$

If we choose $A = \boldsymbol{p} \cdot \boldsymbol{\sigma}$ for the electron, then $|n\rangle$ can be an eigenstate of the Dirac Hamiltonian $H(p)$. This is because

$$[\boldsymbol{\sigma} \cdot \boldsymbol{p}, H(p)] = 0 \ . \tag{13.136}$$

In order to compute $\langle A \rangle$, we only have to make the following replacement for the Casimir operator of the electron in a molecule:

$$-\mathrm{i}p \cdot \gamma + m \rightarrow (-\mathrm{i}p \cdot \gamma + m)(\boldsymbol{\sigma} \cdot \boldsymbol{p}) = (\boldsymbol{\sigma} \cdot \boldsymbol{p})(-\mathrm{i}p \cdot \gamma + m) \ . \tag{13.137}$$

We thus start with

$$\mathscr{T}_{fi} \propto \bar{u}_{\mathrm{e}}(p)(1 + \gamma_5) v_{\nu}(q) \langle \mathbf{1} \rangle \ . \tag{13.138}$$

Therefore,

$$\langle \boldsymbol{\sigma} \cdot \boldsymbol{p} \rangle = \frac{\sum \mathrm{Tr}\big[(1 + \gamma_5)(-\mathrm{i}q \cdot \gamma)(-\mathrm{i}p \cdot \gamma + m)(\boldsymbol{\sigma} \cdot \boldsymbol{p})\big] \cdot |\langle 1 \rangle|^2}{\sum \mathrm{Tr}\big[(1 + \gamma_5)(-\mathrm{i}q \cdot \gamma)(-\mathrm{i}p \cdot \gamma + m)\big] \cdot |\langle 1 \rangle|^2} \ , \tag{13.139}$$

where the four-momentum p of the electron has been fixed. The sum over the final states can for all practical purposes be replaced by the following phase space integral:

$$\sum \longrightarrow \int \frac{d^3 q}{2q_0} \int \frac{d^3 P_B}{2P_{B0}} \delta^4(P_f - P_i) \dots . \tag{13.140}$$

Since an atomic nucleus B is heavy, P_{B0} can be treated as constant. Then, $d^3 P_B$ cancels $\delta^3(P_f - P_i)$ for the momentum. Regarding $\boldsymbol{q}$, its direction becomes free only if energy is conserved, as in the case of scattering in a potential:

$$\sum \longrightarrow \int \frac{d^3 q}{2q_0} \delta(E - p_0 - q_0) \dots . \tag{13.141}$$

The energy can be approximated by the mass difference with the atomic nucleus, i.e., $E = P_{A0} - P_{B0} \approx M_A - M_B$.

Regarding the traces,

$$\mathrm{Tr}\,(\text{numerator}) = 4(q_0 \boldsymbol{p}^2 - p_0 \boldsymbol{p} \cdot \boldsymbol{q}) , \quad \mathrm{Tr}\,(\text{denominator}) = 4(q_0 p_0 - \boldsymbol{p} \cdot \boldsymbol{q}) . \tag{13.142}$$

Hence, carrying out the q-integral in polar coordinates, we have

$$\int \frac{d^3 q}{2q_0} \delta(E - p_0 - q_0)(q_0 \boldsymbol{p}^2 - p_0 \boldsymbol{p} \cdot \boldsymbol{q}) = 2\pi \int q dq \delta(E - p_0 - q_0) q_0 \boldsymbol{p}^2 ,$$

$$\int \frac{d^3 q}{2q_0} \delta(E - p_0 - q_0)(q_0 p_0 - \boldsymbol{p} \cdot \boldsymbol{q}) = 2\pi \int q dq \delta(E - p_0 - q_0) p_0 q_0 .$$

Taking into account the fact that $q_0 = q$ and carrying out the q-integral, it turns out that $q = q_0$ is replaced by $E - p_0$. Therefore, the ratio of the two terms yields

$$\langle \boldsymbol{\sigma} \cdot \boldsymbol{p} \rangle = \frac{q_0 \boldsymbol{p}^2}{p_0 q_0} = p \frac{p}{p_0} = p \frac{v}{c} . \tag{13.143}$$

Thus, the helicity is

$$\langle h \rangle = \langle \boldsymbol{\sigma} \cdot \boldsymbol{p}/p \rangle = \frac{v}{c} . \tag{13.144}$$

A generalization of this derivation yields (13.131). The helicity of e^- was measured for

$$^{60}\mathrm{Co} \rightarrow {}^{60}\mathrm{Ni} + e^- + \bar{\nu}_e , \tag{13.145}$$

by Frauenfelder in [127, 128]. The result was

$$h(e^-) = -\left(\frac{v}{c}\right) \pm 10\% \,. \tag{13.146}$$

Comparing this with (13.131), we see that beta-decay is a VA combination. Thus, the interaction term is

$$\mathscr{H}_\beta = (C_V \bar{\psi}_p \gamma_\lambda \psi_n - C_A \bar{\psi}_p \gamma_\lambda \gamma_5 \psi_n) \bar{\psi}_e \gamma_\lambda (1 + \gamma_5) \psi_\nu + \text{h.c.} \tag{13.147}$$

Taking

$$C_V = \frac{G}{\sqrt{2}} \,, \quad C_A = \frac{G}{\sqrt{2}} x \,, \tag{13.148}$$

G itself is determined from

$$^{14}\text{O} \to {}^{14}\text{N}^* + e^+ + \nu_e \,, \tag{13.149}$$

corresponding to the transition $0^+ \to 0^+$, which is not affected by C_A. This is because $\langle \mathbf{1} \rangle$ is obtained group-theoretically in this case, using the fact that ^{14}C, ^{14}N*, and ^{14}O form an isospin triplet. We will mention an issue regarding this intermediate process in the next section.

In fact, x can be determined from the beta-decay of a polarized meson, for example:

$$G = \frac{1.01}{m_p^2} \times 10^{-5} \,, \quad x = -1.254 \pm 0.006 \,. \tag{13.150}$$

Since x is negative, we call this the V − A theory. Explicitly, the interaction term is

$$\mathscr{H}_\beta = \frac{G}{\sqrt{2}} \bar{\psi}_p \gamma_\lambda (\mathbf{1} + 1.25\gamma_5) \psi_n \cdot \bar{\psi}_e \gamma_\lambda (\mathbf{1} + \gamma_5) \psi_\nu + \text{h.c.} \tag{13.151}$$

The factor of $1/\sqrt{2}$ originates from the replacement in (13.127). For this reason, when using the old coupling constant G, we need to write it in this form.

13.5 Universality of the Fermi Interaction

We have already mentioned the universality of the lepton decay of the π-meson with interaction (13.114), since it does not depend on the lepton generation. The interaction term specified in (13.114) is phenomenological. The Fermi interaction discussed above can be considered to cover more phenomena and is therefore more fundamental. We will discuss this idea step by step. So we begin this section by interpreting the universality of (13.114) from the point of view of the universality of the Fermi interaction. Equation (13.151) specifies the interaction between nucleons and leptons in the first generation, and we assume that the same form holds universally for leptons in all generations:

$$\mathscr{H}_{\mathrm{F}} = \frac{G}{\sqrt{2}}\bar{\psi}_{\mathrm{p}}\gamma_{\lambda}(\mathbf{1}+1.25\gamma_5)\psi_{\mathrm{n}} \cdot \bar{\psi}_{l}\gamma_{\lambda}(\mathbf{1}+\gamma_5)\psi_{\nu_l} + \mathrm{h.c.} \tag{13.152}$$

If we assume that $\pi^+ \rightarrow \mu^+ + \nu_\mu$ occurs through this interaction, then

$$\begin{aligned}&\langle \mu^+, \nu_\mu | \mathscr{H}_{\mathrm{F}}(x) | \boldsymbol{\pi}^+ \rangle \\ &\quad = 1.25\frac{G}{\sqrt{2}}\langle 0 | \bar{\boldsymbol{\psi}}_{\mathrm{n}}(x)\gamma_\lambda\gamma_5\boldsymbol{\psi}_{\mathrm{p}}(x) | \boldsymbol{\pi}^+ \rangle\langle \mu^+, \nu_\mu | \bar{\psi}_{\nu_\mu}(x)\gamma_\lambda(1+\gamma_5)\psi_\mu(x) | 0 \rangle.\end{aligned} \tag{13.153}$$

From its transformation property, we may write the first factor as

$$1.25\frac{G}{\sqrt{2}}\langle \mathbf{0} | \bar{\boldsymbol{\psi}}_{\mathrm{n}}(x)\gamma_\lambda\gamma_5\boldsymbol{\psi}_{\mathrm{p}}(x) | \boldsymbol{\pi}^+ \rangle = -\frac{g}{m_\pi}p_\lambda\frac{\mathrm{e}^{\mathrm{i}p\cdot x}}{\sqrt{2p_0 V}}, \tag{13.154}$$

where p is the four-momentum of π^+. Therefore, phenomenologically, as far as this process is concerned, we can make the following replacement:

$$1.25\frac{G}{\sqrt{2}}\bar{\psi}_{\mathrm{n}}\gamma_\lambda\gamma_5\psi_{\mathrm{p}} \;\rightarrow\; \frac{\mathrm{i}g}{m_\pi}\partial_\lambda\varphi\,. \tag{13.155}$$

The fact that g does not depend on the generation in (13.114) can be interpreted as a consequence of the universality wherein G does not depend on the generation in (13.152). We can thus consider (13.114) as a phenomenological expression of (13.152) as far as this process is concerned. Let us go a little further with this universality.

As far as the lepton part goes, the universal form

$$\bar{\psi}_{l}\gamma_{\lambda}(1+\gamma_5)\psi_{\nu_l} \tag{13.156}$$

appears in the Fermi interaction, but the nucleon part still has a complicated form. It may thus be better to consider the nucleon part as still being a phenomenological expression. Let us treat this part slightly more abstractly. From (7.30), we write down the isospin current density for the nucleon–meson system as

$$\hat{j}_\lambda = \mathrm{i}\bar{\psi}\gamma_\lambda\frac{\hat{\tau}}{2}\psi - \hat{\varphi}\times\partial_\lambda\hat{\varphi}\,, \tag{13.157}$$

where we have considered nucleons and π-mesons as elementary particles here. As long as the theory preserves charge independence, the isospin current density is always expressed in terms of the fundamental field operators as the Noether current density. However, the main properties of isospin do not depend on what field we choose as the fundamental field, being determined by group-theoretical considerations. This is similar in the scattering theory: the scattering amplitude for stable particles obeys the scattering formula regardless of whether the particles are elementary or composite.

The vector part of the nucleon term in the Fermi interaction is

$$\mathrm{i}\bar{\psi}_{\mathrm{p}}\gamma_\lambda\psi_{\mathrm{n}} = \mathrm{i}\left(\bar{\psi}\gamma_\lambda\frac{\tau}{2}\psi + \mathrm{i}\bar{\psi}\gamma_\lambda\frac{\tau_2}{2}\psi\right) . \tag{13.158}$$

Note that this is the nucleon part of $j_{1\lambda}+\mathrm{i}j_{2\lambda}$ in the isospin current density (13.157). To give the phenomenological vector part of the nucleon greater generality, we replace it with the isospin current density, which can be written without specifying the fundamental fields. Hence, we rewrite the vector part in (13.152) as

$$\mathscr{H}_{\mathrm{V}} = -\mathrm{i}\frac{G}{\sqrt{2}}(j_{1\lambda}+\mathrm{i}j_{2\lambda})\bar{\psi}_l\gamma_\lambda(\mathbf{1}+\gamma_5)\psi_{\nu_l} + \text{h.c.} \tag{13.159}$$

The isospin current density satisfies the conservation law, so Gell-Mann called this replacement the CVC hypothesis [49], where CVC is an abbreviation for the conserved vector current. Since this vector current can be used for all hadrons with fixed isospins, it becomes a truly universal expression. For example, consider the beta-decay of π^+:

$$\pi^+ \to \pi^0 + \mathrm{e}^+ + \nu_{\mathrm{e}} . \tag{13.160}$$

If the momentum is conserved, the matrix elements of the isospin are

$$\langle\mathbf{p}|\boldsymbol{I}_1+\mathrm{i}\boldsymbol{I}_2|\mathbf{n}\rangle = 1 , \qquad \langle\boldsymbol{\pi}^+|\boldsymbol{I}_1+\mathrm{i}\boldsymbol{I}_2|\boldsymbol{\pi}^0\rangle = -\sqrt{2} . \tag{13.161}$$

Regarding their densities when the momentum transfer is almost zero,

$$\langle\mathbf{p}|\boldsymbol{j}_{10}+\mathrm{i}\boldsymbol{j}_{20}|\mathbf{n}\rangle = \frac{1}{V}\langle\mathbf{p}|\boldsymbol{I}_1+\mathrm{i}\boldsymbol{I}_2|\mathbf{n}\rangle = \frac{1}{V} , \tag{13.162}$$

$$\langle\boldsymbol{\pi}^+|\boldsymbol{j}_{10}+\mathrm{i}\boldsymbol{j}_{20}|\boldsymbol{\pi}^0\rangle = \frac{1}{V}\langle\boldsymbol{\pi}^+|\boldsymbol{I}_1+\mathrm{i}\boldsymbol{I}_2|\boldsymbol{\pi}^0\rangle = -\frac{\sqrt{2}}{V} . \tag{13.163}$$

This result does not depend on whether nucleons and π-mesons are elementary or composite particles. Generalizing the above equation,

$$\langle\boldsymbol{\pi}^+|\boldsymbol{j}_{1\lambda}+\mathrm{i}\boldsymbol{j}_{2\lambda}|\boldsymbol{\pi}^0\rangle = -\frac{\sqrt{2}}{V}\frac{q_\lambda}{q_0} , \tag{13.164}$$

where we assume that π^+ and π^0 have almost the same momentum q_λ, and neglect the difference in their momenta. From this, we can immediately compute the decay width of the beta-decay of π^+:

$$\frac{\Gamma(\pi^+ \to \pi^0 + e^+ + \nu_e)}{\Gamma(\pi^+ \to \mu^+ + \nu_\mu)} = 1.0 \times 10^{-8} . \tag{13.165}$$

This agrees extremely well with the experimental value of $(1.033 \pm 0.034) \times 10^{-8}$. Assuming that the CVC hypothesis is correct, it thus turns out that the interaction (13.159) is extremely general and can be used for all hadrons.

So far in the discussion, we have only appealed to the properties of the isospin, and in particular the commutation relation

$$[I_i, I_j] = i\epsilon_{ijk} I_k . \tag{13.166}$$

This commutation relation therefore characterizes the universality. It is thus tempting to consider the same thing for the axial-vector part of the interaction. We thus make the substitution

$$1.25\frac{G}{\sqrt{2}} i\bar{\psi}_p \gamma_\lambda \gamma_5 \psi_n \rightarrow \frac{G}{\sqrt{2}}\left(j^5_{1\lambda} + i j^5_{2\lambda}\right) . \tag{13.167}$$

The factor of 1.25 can be considered as phenomenological, i.e., it is given by

$$\langle \mathbf{p} | j^5_{1\lambda} + i j^5_{2\lambda} | \mathbf{n} \rangle = 1.25\frac{i}{V}\bar{u}(p)\gamma_\lambda\gamma_5 u(n) . \tag{13.168}$$

As will soon be shown, this is a calculable factor. We can now define the pseudoscalar isospin by

$$\hat{I}^5 = \int d^3x\, \hat{j}^5_0 . \tag{13.169}$$

Since $\hat{I}^5$ is itself an isospin vector, the commutation relation with $\hat{I}$ is given by

$$[I_i, I^5_j] = i\epsilon_{ijk} I^5_k . \tag{13.170}$$

The remaining problems here are the normalization of $\hat{I}^5$ and, in association with that, the closure of the commutation relations between $\hat{I}$ and $\hat{I}^5$. Gell-Mann assumed the following equation:

$$[I^5_i, I^5_j] = i\epsilon_{ijk} I_k . \tag{13.171}$$

The algebra closed in the way just described is called a *current algebra* [52, 53].

To test this algebra, starting with (13.171), we have to see whether we can prove that $-C_A/C_V = 1.25$. To do this, we use the partially conserved axial-vector current (PCAC) hypothesis [129], according to which $\partial_\lambda \hat{j}^5_\lambda$ can express the π-meson field and the dispersion formula. The result is the Adler–Weisberger formula [54–56]:

$$\left(\frac{C_V}{C_A}\right)^2 = 1 + \frac{2}{\pi}\left(\frac{m_p}{G}\right)^2 \int_{m_\pi}^{\infty} \frac{k\,d\omega}{\omega^2}\big[\sigma_-(\omega) - \sigma_+(\omega)\big] , \tag{13.172}$$

where ω and k are the incident energy and momentum of π in the laboratory frame, and $\sigma_-(\omega)$ and $\sigma_+(\omega)$ are the total cross-sections for the following processes:

$$\sigma_-(\omega) :\ \pi^- + p \ \rightarrow\ \text{all} , \quad \sigma_+(\omega) :\ \pi^+ + p \ \rightarrow\ \text{all} , \tag{13.173}$$

G is the coupling constant appearing in the phenomenological interaction between nucleons and π-mesons given in (6.96). Its approximate value is $G^2/4\pi \approx 15$. Inserting an experimental value into the right-hand side of (13.172), the result almost coincides with $C_A/C_V = -1.25$, and this therefore supports the current algebra hypothesis. For a proof of (13.172), the reader is referred to my book *Fields and Particles* [2].

So far, regarding the Fermi interaction, the following universality has been established. First, for the lepton part, the interactions are common to all generations. Next, for the hadron part, it can be expressed by the current density satisfying the current algebra hypothesis. Moreover, it is easy to check that exactly the same commutation relations hold for the lepton part as for the hadron part, i.e., taking

$$i\bar{\psi}_l \gamma_\lambda (1+\gamma_5)\psi_{\nu_l} \ \rightarrow\ j'_{1\lambda} - i j'_{2\lambda} , \tag{13.174}$$

and writing spatial integrals of j'_{10} and j'_{20} as I'_1 and I'_2, respectively, we find that I'_1 and I'_2 satisfy the same commutation relations as $I_1 + I_1^5$ and $I_2 + I_2^5$. Thus, algebraically, the lepton part and the hadron part have the universal properties.

In order to investigate the Fermi interaction among leptons, such as in the reaction

$$\mu^+ \rightarrow\ e^+ + \nu_e + \bar{\nu}_\mu , \tag{13.175}$$

we assume an interaction with the same form:

$$\mathscr{H}_{\mu e} = \frac{G_F}{\sqrt{2}}\big[\bar{\psi}_{\nu_e}\gamma_\lambda(1+\gamma_5)\psi_e\big]\big[\bar{\psi}_\mu\gamma_\lambda(1+\gamma_5)\psi_{\nu_\mu}\big] + \text{h.c.} \tag{13.176}$$

Since it was explained in detail how to compute this decay width in Sect. 9.5, we only show the result:

$$\Gamma(\mu^+ \rightarrow\ e^+ + \nu_e + \bar{\nu}_\mu) = \frac{1}{192\pi^2}G_F^2 m_\mu^5 . \tag{13.177}$$

This coupling constant does not have a correction from the strong interaction, so it can be determined from the lifetime and the mass of the muon:

$$G_{\mathrm{F}} = \frac{1.03}{m_{\mathrm{p}}^2} \times 10^{-5} \,. \tag{13.178}$$

Similarly, in the beta-decay of the neutron,

$$\Gamma(\mathrm{n} \rightarrow \mathrm{p} + \mathrm{e}^- + \bar{\nu}_{\mathrm{e}}) = \frac{0.94}{60\pi^3}\left(|C_{\mathrm{V}}|^2 + 3|C_{\mathrm{A}}|^2\right)\Delta^5 \,, \tag{13.179}$$

where $\Delta = m_{\mu} - m_{\mathrm{p}}$. Comparing (13.178) with (13.150), there is a slight difference between G and G_{F}. Cabibbo provided an interpretation of this difference. The current algebra is based on the isospin group $SU(2)$, and the weak interaction wherein a hadron decays into a lepton pair has a small component which changes the strangeness. This extends the current algebra to $SU(3)$. Cabibbo's interpretation was that the hadron part of the Fermi interaction is the isospin density but slightly altered by an $SU(3)$ rotation [57]. The angle of rotation is known as the *Cabibbo angle*. Writing it as θ_{c},

$$G = G_{\mathrm{F}} \cos\theta_{\mathrm{c}} \,. \tag{13.180}$$

Cabibbo's theory will be discussed in detail in the next section. Here we discuss the quark model. Expressing the commutation relation in terms of field operators at equal times $x_0 = y_0$, we can use the density to write

$$\left[j^5_{i0}(x), j^5_{j0}(y)\right] = \mathrm{i}\epsilon_{ijk} j_{k0}(x)\delta^3(x - y) + \text{GIS term} \,, \tag{13.181}$$

where the GIS term is the term which vanishes, like the Goto–Imamura–Schwinger term, when we carry out the spatial integral. Assuming that the fields of the π-meson and the nucleon are fundamental fields, the first term on the right-hand side, $j_{k\lambda}(x)$, or $\hat{j}_{\lambda}(x)$, is given by (13.157). It remains to express $\hat{j}^5_{\lambda}$. We shall not give an explicit expression here, as $\hat{j}^5_{\lambda}$ satisfying (13.181) is highly non-linear and hence complicated. The dynamics in a system with the symmetry corresponding to such a non-linear current expression is called *chiral dynamics*. This gives a phenomenological representation of the current algebra, but it would be too complicated to consider the fundamental theory. It would be utterly impossible to implement the renormalization. We thus turn to the expression in terms of quarks.

13.6 Quark Model in Weak Interactions

In Sect. 13.2, we introduced three types of quarks, u, d, and s. These quarks were invented as a mathematical tool and as carriers of the fundamental representation, rather than as real particles. However, once we recognize that the current which satisfies the current algebra exists in the weak interaction as a physical entity, the way we express the current density in terms of field operators becomes a real physical problem, i.e., a problem of what the fundamental fields are in real life.

Denoting the quark field with three components u, d, and s as q, and using the Gell-Mann matrices λ_i which mix these three components, we introduce the current density

$$\mathscr{F}_{j\mu} = \mathrm{i}\bar{q}\gamma_\mu \frac{\lambda_j}{2} q \,, \quad \mathscr{F}^5_{j\mu} = \mathrm{i}\bar{q}\gamma_\mu\gamma_5 \frac{\lambda_j}{2} q \,. \tag{13.182}$$

We also define the quantities

$$F_j = \int \mathrm{d}^3x \mathscr{F}_{j0} \,, \quad F^5_j = \int \mathrm{d}^3x \mathscr{F}^5_{j0} \,. \tag{13.183}$$

These are not necessarily conserved. Their equal-time commutation relations are

$$[F_i, F_j] = \mathrm{i} f_{ijk} F_k \,, \quad [F_i, F^5_j] = \mathrm{i} f_{ijk} F^5_k \,, \quad [F^5_i, F^5_j] = \mathrm{i} f_{ijk} F_k \,. \tag{13.184}$$

Thus, it turns out that, using quarks, it is a straightforward matter to express the current algebra.

Here we have constructed the $SU(3)$ current algebra, which has $SU(2)$ as a subalgebra. To understand the algebraic structure of (13.184) in detail, we build the combination

$$F^{(\pm)}_i = \frac{1}{2}\left(F_i \pm F^5_i\right) , \tag{13.185}$$

which yields the commutation relations

$$\left[F^{(+)}_i, F^{(-)}_j\right] = 0 \,, \quad \left[F^{(+)}_i, F^{(+)}_j\right] = \mathrm{i} f_{ijk} F^{(+)}_k \,, \quad \left[F^{(-)}_i, F^{(-)}_j\right] = \mathrm{i} f_{ijk} F^{(-)}_k \,. \tag{13.186}$$

This algebraic structure is thus a direct product of the algebras $SU(3)_\mathrm{L}$ and $SU(3)_\mathrm{R}$, generated by $F^{(+)}$ corresponding to the left-handed current and $F^{(-)}$ corresponding to the right-handed current, respectively:

$$SU(3)_\mathrm{L} \times SU(3)_\mathrm{R} \,. \tag{13.187}$$

As described above, the quark model is the simplest way to satisfy the current algebra and the best expression from the point of view of high-energy behaviour and renormalizability. Here, we no longer consider the quark model as a simple mathematical tool, but take the quarks to be real physical entities. Cabibbo defined the current carrying electric charge which appeared in the Fermi interaction in [57] by

$$J_\mu = \left[\mathscr{F}^{(\mu)}_{1\mu} - \mathrm{i}\mathscr{F}^{(+)}_{2\mu}\right]\cos\theta_{\mathrm{C}} + \left[\mathscr{F}^{(+)}_{4\mu} - \mathrm{i}\mathscr{F}^{(+)}_{5\mu}\right]\sin\theta_{\mathrm{C}} + \sum_l \mathrm{i}\bar{\psi}_l\gamma_\mu(1+\gamma_5)\psi_{\nu_l}\ , \tag{13.188}$$

and wrote the universal Fermi interaction as

$$\mathscr{H}_{\mathrm{F}} = \frac{G_{\mathrm{F}}}{\sqrt{2}} J^+_\mu J_\mu\ . \tag{13.189}$$

From these, for instance, the beta-decay interaction, which does not change the strangeness, is

$$\mathscr{H}_\beta = \frac{G_{\mathrm{F}}}{\sqrt{2}}\cos\theta_{\mathrm{C}}\left[\mathscr{F}^{(+)}_{1\mu} + \mathrm{i}\mathscr{F}^{(+)}_{2\mu}\right]\mathrm{i}\bar{\psi}_{\mathrm{e}}\gamma_\mu(\mathbf{1}+\gamma_5)\psi_{\nu_{\mathrm{e}}} + \text{h.c.} \tag{13.190}$$

Taking into account (13.180), this coincides with (13.152). In addition, it turns out that the 0-components of J_μ defined by (13.188) and those of its Hermitian conjugate $J^\dagger_\mu$ satisfy the algebra $SU(2)_{\mathrm{L}}$.

The interaction (13.189) consists of the charged current, and the neutral current is missing. However, this will be added later, in the gauge theory. The quantum number discriminating between the three types of quarks u, d, and s is called *flavour*. Isospin and strangeness are examples of flavours. In fact, there is further discrimination between quarks, known as *colour*. This will be discussed in the next section.

To explain contributions changing the strangeness in the weak interaction, Cabibbo suggested that d′ and s′ might differ from d and s as mass eigenstates, rewriting the hadron part of the Fermi interaction as

$$\bar{\psi}_{\mathrm{p}}\gamma_\lambda(\mathbf{1}+1.25\gamma_5)\psi_{\mathrm{n}} \to \bar{\mathrm{u}}\gamma_\lambda(\mathbf{1}+\gamma_5)d'\ . \tag{13.191}$$

Since both d and s have electric charge $-e/3$, he proposed the transformation

$$\begin{pmatrix}\mathrm{d}'\\ \mathrm{s}'\end{pmatrix} = \begin{pmatrix}\cos\theta_{\mathrm{C}} & \sin\theta_{\mathrm{C}}\\ -\sin\theta_{\mathrm{C}} & \cos\theta_{\mathrm{C}}\end{pmatrix}\begin{pmatrix}\mathrm{d}\\ \mathrm{s}\end{pmatrix}\ . \tag{13.192}$$

This is the origin of the form (13.188). In 1970, Glashow et al. introduced a fourth quark called charm, denoted by c. Assuming that it carried an electric charge of $+2e/3$, this completed a second generation of quarks [130]:

$$\begin{pmatrix}\mathrm{u}\\ \mathrm{d}\end{pmatrix}\ ,\quad \begin{pmatrix}\mathrm{c}\\ \mathrm{s}\end{pmatrix}\ . \tag{13.193}$$

Then, in order to explain the violation of CP invariance, Kobayashi and Maskawa predicted a third generation of quarks, replacing (13.193) by [131]

$$\begin{pmatrix} \mathrm{u} \\ \mathrm{d} \end{pmatrix} , \quad \begin{pmatrix} \mathrm{c} \\ \mathrm{s} \end{pmatrix} , \quad \begin{pmatrix} \mathrm{t} \\ \mathrm{b} \end{pmatrix} . \tag{13.194}$$

The quarks t and b are called the *top quark* and the *bottom quark*, respectively. Extending (13.192), they introduced the mixing matrix

$$\begin{pmatrix} \mathrm{d}' \\ \mathrm{s}' \\ \mathrm{b}' \end{pmatrix} = \begin{pmatrix} c_1 & s_1 c_2 & s_1 s_2 \\ -s_1 c_3 & c_1 c_2 c_3 - s_2 s_3 \mathrm{e}^{-\mathrm{i}\delta} & c_1 s_2 c_3 + c_2 s_3 \mathrm{e}^{-\mathrm{i}\delta} \\ -s_1 s_3 & c_1 c_2 s_3 + s_2 c_3 \mathrm{e}^{-\mathrm{i}\delta} & c_1 s_2 s_3 - c_2 c_3 \mathrm{e}^{-\mathrm{i}\delta} \end{pmatrix} \begin{pmatrix} \mathrm{d} \\ \mathrm{s} \\ \mathrm{b} \end{pmatrix} . \tag{13.195}$$

This is called the *Kobayashi–Maskawa matrix*. The factor of $\mathrm{e}^{-\mathrm{i}\delta}$ leads to time reversal or CP violation, and accordingly, the third-generation quarks are needed here. Moreover,

$$s_i = \sin\theta_i , \quad c_i = \cos\theta_i , \tag{13.196}$$

where θ_i $(i = 1, 2, 3)$ are three angles, extending the Cabibbo angle.

We have introduced the quark model by considering the weak interactions, but what about the strong interactions?

13.7 Quark Model in Strong Interactions

In Sect. 13.2, we discussed the classification of hadrons, the mass formula, the magnetic moment, and so on, based on the $SU(3)$ algebra. In this section, we use the quark model to show that these results can be further improved.

The most important thing here is that quarks have, besides flavour, another degree of freedom called *colour*. Let us see why a new degree of freedom is needed.

The quark model is supported by the following group theoretical considerations. Since each meson is a bound state $\mathrm{q\bar{q}}$, for flavour $SU(3)$,

$$\mathbf{3} \otimes \bar{\mathbf{3}} = \mathbf{1} \oplus \mathbf{8} , \tag{13.197}$$

which means that a singlet and an octet are possible. There is experimental support for this, namely, both the singlet and the octet exist, and there is nothing else. Each baryon has the form qqq and we have

$$\mathbf{3} \otimes \mathbf{3} \otimes \mathbf{3} = \mathbf{1} \oplus \mathbf{8} \oplus \mathbf{10} . \tag{13.198}$$

Experimentally, it is precisely these multiplets that have been detected. The quarks forming the basic triplet have not been detected. The only thing that has been detected experimentally are states belonging to **8**, $\mathbf{8} \otimes \mathbf{8}$, and so on. This is what the eightfold way means.

Now let us consider a member Δ^{++} of the decuplet. This particle is a bound state of three u quarks, i.e., uuu, and its spin is $3/2$. Assuming that the orbital angular momenta are each in the S-state, it can be considered that the three $1/2$-spins are parallel and the composite spin is $3/2$. If so, both the spatial wave function and the spin wave function are symmetric, which leads to a contradiction, since the wave function represents three identical fermions. Thus, there must be another, unknown degree of freedom. The wave function here should be completely antisymmetric with respect to this degree of freedom, which we shall call colour for the time being. We assume the existence of three kinds of colour states 1, 2, and 3. We note, however, that this colour degree of freedom has never been observed at the level of hadrons. Assuming that hadrons do have an extra degree of freedom, this yields several contradictions, unless we consider that hadrons are the singlets for this new degree of freedom.

In this case, apart from the spatial part, the wave function of Δ^{++} may be written as

$$\epsilon_{\alpha\beta\gamma} u_{\uparrow}^{\alpha} u_{\uparrow}^{\beta} u_{\uparrow}^{\gamma} , \tag{13.199}$$

where u stands for the flavour, $\uparrow$ indicates spin up, and α, β, and γ indicate the colour degree of freedom. When we carry out an $SU(3)$ transformation on the three kinds of colour state, (13.199) is invariant, so this is a colour singlet. All hadrons will thus be invariant under this colour $SU(3)$ transformation. This means that the quark model must be a theory which is invariant under the $SU(3)$ group.

13.7.1 Mass Formula

We will show that we can improve the mass formula in Sect. 13.2 by using the quark model. To do so, we first express the hadrons as quark bound states:

$$\begin{aligned} &\pi^+ \sim \mathrm{u\bar{d}} , \quad \pi^- \sim \mathrm{d\bar{u}} , \quad \mathrm{K}^+ \sim \mathrm{u\bar{s}} , \quad \mathrm{K}^0 \sim \mathrm{d\bar{s}} , \\ &\mathrm{p} \sim \mathrm{uud} , \quad \mathrm{n} \sim \mathrm{udd} , \quad \Lambda \sim \mathrm{uds} , \quad \Sigma^+ \sim \mathrm{uus} , \\ &\Delta^{++} \sim \mathrm{uuu} , \quad \Omega^- \sim \mathrm{sss} , \quad \text{etc.} \end{aligned}$$

The quarks have the following charges:

$$Q_{\mathrm{u}} = \frac{2}{3}e , \quad Q_{\mathrm{d}} = Q_{\mathrm{s}} = -\frac{1}{3}e . \tag{13.200}$$

In some cases, we have to consider mixed states in order to distinguish the flavour singlet from the flavour octet:

$$\eta_1 \sim \frac{1}{\sqrt{3}}(u\bar{u} + d\bar{d} + s\bar{s}) , \quad \eta_8 \sim \frac{1}{\sqrt{6}}(u\bar{u} + d\bar{d} - 2s\bar{s}) . \tag{13.201}$$

A rough formula for the mass can be obtained by taking a sum over the quark masses. The following formulas can be derived from both the quark model and the previous $SU(3)$ method. Particle names stand for masses here:

$$\underset{496\,\text{MeV}}{\text{K}} = \underset{446\,\text{MeV}}{(\pi + 3\eta)/4} , \quad \underset{776\,\text{MeV}}{\rho} = \underset{783\,\text{MeV}}{\omega} , \quad \underset{127\,\text{MeV}}{\phi - \text{K}^*} = \underset{116\,\text{MeV}}{\text{K}^* - \rho} ,$$

$$\underset{152\,\text{MeV}}{\Sigma^* - \Delta} = \underset{150\,\text{MeV}}{\Xi^* - \Sigma^*} = \underset{139\,\text{MeV}}{\Omega^- - \Xi^*} , \quad \underset{1129\,\text{MeV}}{(\text{N} + \Xi)/2} = \underset{1135\,\text{MeV}}{(\Sigma + 3\Lambda)/4} . \tag{13.202}$$

In the quark model, we can include further corrections from the interaction. However, to do this, we will have to borrow a result from the gauge theory discussed later. The short-range potential among $q\bar{q}$ or qq is given as follows, assuming that the composite system belongs to the **1**-colour representation or the $\bar{\mathbf{3}}$-colour representation:

$$V_{q\bar{q}}(\mathbf{1}) = -\frac{4}{3}\alpha_s/r , \quad V_{qq}(\bar{\mathbf{3}}) = -\frac{2}{3}\alpha_s/r , \tag{13.203}$$

where α_s is a coupling constant characterizing the interaction among particles carrying colour. Its expectation value ΔE is given up to a multiplicative constant by

$$\Delta E \propto \frac{\boldsymbol{\sigma}_1 \cdot \boldsymbol{\sigma}_2}{m_1 m_2} . \tag{13.204}$$

We can thus add a correction term to the equation wherein the mass M of the composite system is considered to be a sum of masses m_i of the components:

$$M(q\bar{q}) = m_1 + m_2 + a(\boldsymbol{\sigma}_1 \cdot \boldsymbol{\sigma}_2)/m_1 m_2 , \tag{13.205}$$

$$M(qqq) = \sum_{i=1}^{3} + a' \sum_{i<j} (\boldsymbol{\sigma}_1 \cdot \boldsymbol{\sigma}_2)/m_i m_j . \tag{13.206}$$

These can be computed if the spin wave functions of the hadrons are known. The results are shown in Tables 13.4 and 13.5. Here, both the meson case and the baryon case involve three parameters, but given the simplicity of the formula, it cannot be helped that the quarks have to be attributed slightly different masses to fit the mesons and to fit the baryons. Even so, the qualitative agreement is good.

Table 13.4 Meson masses in MeV. Choice of parameters: $m_u = m_d = 310$ MeV, $m_s = 483$ MeV, $a/m_u^2 = 160$ MeV

Meson	π(138)	K(496)	η(549)	ρ(776)	ω(783)	K*(892)	φ(1020)
Mass	140	485	559	780	780	896	1032

Table 13.5 Baryon masses in MeV. Choice of parameters: $m_u = m_d = 363$ MeV, $m_s = 538$ MeV, $a'/m_u^2 = 50$ MeV

Baryon	N(939)	Λ(1116)	Σ(1193)	Ξ(1318)	Δ(1232)	Σ*(1384)	Ξ*(1533)	Ω(1672)
Mass	939	1114	1179	1327	1239	1381	1529	1682

13.7.2 Magnetic Moments

Just as we have done for mass, we can use the spin wave function to express the magnetic moments of baryons by

$$\mu_i = \frac{Q_i}{2m_i}, \quad i = \mathrm{u, d, s}\,. \tag{13.207}$$

For instance, the wave function of the proton is

$$|\mathrm{p}\uparrow\rangle = \sqrt{\frac{2}{3}}\,\big|\mathrm{uu}, S_z = 1\big\rangle \times \big|\mathrm{d}\downarrow\big\rangle - \sqrt{\frac{1}{3}}\,\big|\mathrm{uu}, S_z = 0\big\rangle \times \big|\mathrm{d}\uparrow\big\rangle, \tag{13.208}$$

where the symbols have the obvious meaning. The magnetic moment can then be computed from

$$\mu(\mathrm{p}) = \frac{2}{3}(2\mu_\mathrm{u}) + \left(-\frac{2}{3} + \frac{1}{3}\right)\mu_\mathrm{d} = \frac{4}{3}\mu_\mathrm{u} - \frac{1}{3}\mu_\mathrm{d}\,. \tag{13.209}$$

We can express the magnetic moments in units of the nuclear magneton, as in Sect. 13.2. In baryons, the values of the μ_i in (13.207) are

$$\mu_\mathrm{u} = 1.863\ \text{n.m.}\,, \quad \mu_\mathrm{d} = -0.931\ \text{n.m.}\,, \quad \mu_\mathrm{s} = -0.583\ \text{n.m.} \tag{13.210}$$

The magnetic moments of the baryons are shown in Table 13.6. In the table, the magnetic moments of the quarks in (13.210) are expressed directly by the quark symbols u, d, and s. Apart from Ξ^-, where there is a large error, the agreement with experiment is much improved over Table 13.2.

So far we have discussed three types of quark. It took much longer for other kinds of quark to be "detected". This is because c and b are much more massive than u, d, and s. As will be discussed later, an isolated quark cannot be observed, so what is observed is always a hadron. In 1974, Ting and Richter independently found a vector particle with a mass of about 3100 MeV, which they called J and

Table 13.6 Baryon magnetic moments

Baryon	Expression for magnetic moment	Theoretical value		Experimental value
p	$\frac{4}{3}\mathrm{u} - \frac{1}{3}\mathrm{d}$	2	0.73	2.793
n	$\frac{4}{3}\mathrm{d} - \frac{1}{3}\mathrm{u}$	−1	0.86	−1.913
Λ^0	s	−0	0.58	-0.613 ± 0.004
$\Sigma^0 \to \Lambda^0$	$\frac{1}{\sqrt{3}}(\mathrm{d} - \mathrm{u})$	−1	0.61	$-\left(1.82^{+0.25}_{-0.18}\right)$
Σ^+	$\frac{4}{3}\mathrm{u} - \frac{1}{3}\mathrm{s}$	2	0.68	2.379 ± 0.020
Σ^0	$\frac{4}{3}(\mathrm{u} + \mathrm{d}) - \frac{1}{3}\mathrm{s}$	0	0.82	
Σ^-	$\frac{4}{3}\mathrm{d} - \frac{1}{3}\mathrm{s}$	−1	0.05	-1.10 ± 0.50
Ξ^0	$\frac{4}{3}\mathrm{s} - \frac{1}{3}\mathrm{u}$	−1	0.4	-1.250 ± 0.014
Ξ^-	$\frac{4}{3}\mathrm{s} - \frac{1}{3}\mathrm{d}$	−0	0.47	-1.85 ± 0.75

ψ, respectively [132, 133]. This was interpreted as the bound state $c\bar{c}$, where c was identified with the c quark introduced by Glashow et al. in 1970. Regarding the fourth quark, its existence was considered by Maki and Hara from a different point of view in about 1963 [134, 135]. Then a particle called Υ with a mass of about 9460 MeV was detected by Lederman and Yamanouchi [136]. It was interpreted as the bound state $b\bar{b}$.

Hence, it gradually became clear that there are also three generations of quarks. The next problem was therefore to find the t quark. The detection of this particle is (now we should use 'was' instead of 'is') one of the aims of the accelerator TRISTAN, constructed in Tsukuba. Moreover, bound states of these heavy quarks have many excited states, and they have been explained by assuming the potential

$$V(r) = -\frac{4}{3}\frac{\alpha_s}{r} + kr \ , \tag{13.211}$$

which has proved extremely successful. It is assumed that, the more massive the quark, the better the approximation provided by such a non-relativistic computation.

13.8 Parton Model

We have been discussing the quark model, according to which each hadron is a bound state of quarks. A key issue is the way the strong interaction glues quarks together. The *parton model* characterizes this property of the strong interaction. If we translate this literally, it should be called the *partial-particle model*. In 1969, Feynman assumed that "the nucleon consists of almost free point particles, called partons." He dubbed this the parton model [137]. Thus, for instance, if the four-momentum of a proton is P and the four-momenta of the partons making it up are p_i, then we require

$$p_i \approx x_i P \ , \quad 1 \geq x_i \geq 0 \ , \tag{13.212}$$

because without such a proportionality relation, almost free partons will fly away on their own and the proton will fall to pieces.

In order to make this idea more explicit, assuming that the parton is in fact a quark, the proton has the composition uud. If the probabilities for the u quark and d quark to exist in the region $(x, x + \Delta x)$ are $u(x)\Delta x$ and $d(x)\Delta x$, respectively, then we must have

$$\int_0^1 \mathrm{d}x \ u(x) = 2 \ , \quad \int_0^1 \mathrm{d}x \ d(x) = 1 \ . \tag{13.213}$$

We therefore compute the structure function for the inelastic scattering of electrons discussed in Sect. 9.3 using this model. To ensure consistency with (13.212), if the proton mass is denoted by M, we take the parton mass to be

$$m_i = x_i M \ . \tag{13.214}$$

We must then compute contributions corresponding to the structure function for each parton. If we write a contribution from one parton as $w_{\mu\nu}$, we have

$$2\pi e^2 w_{\mu\nu} = \int \mathrm{d}^4 x \frac{p_0}{m} \langle p | j_\mu(x) j_\nu(0) | p \rangle \mathrm{e}^{-\mathrm{i}Q\cdot x} \ , \tag{13.215}$$

where we have dropped the subscript i. Actually, we have to use a matrix element in the Heisenberg picture, but since we use free quarks here, we compute in the interaction picture. Averaging over possible spins of the quark, the equation above becomes

$$\begin{aligned} 2\pi e^2 w_{\mu\nu} &= -e_q^2 \int \frac{\mathrm{d}^3 p'}{(2\pi)^3} \frac{p_0}{m} \frac{1}{2} \mathrm{Tr}\left(\gamma_\mu \frac{-\mathrm{i}p' \cdot \gamma + m}{2p'_0} \gamma_\nu \frac{-\mathrm{i} \cdot \gamma + m}{2p_0} \right) (2\pi)^4 \delta^4(p' - p - Q) \\ &= -2\pi e_q^2 \delta(p'^2 + m^2) \frac{1}{4m} \mathrm{Tr}\Big[\gamma_\mu \big[-\mathrm{i}(P + Q) \cdot \gamma + m \big] \gamma_\nu (-\mathrm{i}p \cdot \gamma + m) \Big] \ , \end{aligned}$$

where we have assumed that the parton absorbs the momentum Q, and as a result it has the momentum $p' = p + Q$. Therefore,

$$w_{\mu\nu} = -\left(\frac{e_q}{e} \right)^2 \delta\big[(p + Q)^2 + m^2 \big] \frac{1}{m} (2p_\mu p_\nu + p_\mu Q_\nu + p_\nu Q_\mu - \delta_{\mu\nu} p \cdot Q) \ . \tag{13.216}$$

So, using $p^2 + m^2 = 0$,

$$w_{\mu\nu} = \left(\frac{e_q}{e}\right)^2 \delta(2p \cdot Q + Q^2)\left[\left(\delta_{\mu\nu} - \frac{Q_\mu Q_\nu}{Q^2}\right)\nu \right. \\ \left. + \left(P_\mu - \frac{P \cdot Q}{Q^2}Q_\mu\right)\left(P_\nu - \frac{P \cdot Q}{Q^2}Q_\nu\right)\frac{2xM}{M^2}\right], \tag{13.217}$$

where, as in Chap. 9, ν has been defined by

$$p \cdot Q = xP \cdot Q = -xM\nu . \tag{13.218}$$

Using (9.82), we can now obtain the structure functions of the parton:

$$w_1 = \left(\frac{e_q}{e}\right)^2 \delta(2p \cdot Q + Q^2)\nu = \nu\left(\frac{e_q}{e}\right)^2 \delta(Q^2 - 2M\nu x) , \tag{13.219}$$

$$w_2 = \left(\frac{e_q}{e}\right)^2 2xM\delta(2p \cdot Q + Q^2) = 2xM\left(\frac{e_q}{e}\right)^2 \delta(Q^2 - 2M\nu x) . \tag{13.220}$$

To derive the structure functions of the proton from these, we have to multiply by $u(x)$ and $d(x)$ for $q = \mathrm{u}$ and $q = \mathrm{d}$, respectively, and integrate with respect to x. If we take

$$P(x) = \left(\frac{e_\mathrm{u}}{e}\right)^2 u(x) + \left(\frac{e_\mathrm{d}}{e}\right)^2 d(x) = \frac{4}{9}u(x) + \frac{1}{9}d(x) , \tag{13.221}$$

then

$$F_1 = MW_1 = \int \mathrm{d}x\ P(x)\delta\left(\frac{Q^2}{M\nu} - 2x\right) = \frac{1}{2}P\left(\frac{Q^2}{2M\nu}\right) , \tag{13.222}$$

$$F_2 = \nu W_2 = \int \mathrm{d}x\ P(x)x\delta\left(\frac{Q^2}{2M\nu} - x\right) = \frac{Q^2}{2M\nu}P\left(\frac{Q^2}{2M\nu}\right) . \tag{13.223}$$

This establishes the Bjorken scaling law [105], according to which, in the limits $Q^2 \to \infty$ and $M\nu \to \infty$, MW_1 and νW_2 become functions of $Q^2/2M\nu = x$ alone. Although in the computation above we have not taken these limits explicitly, in the free parton picture we are assuming these limits implicitly because we neglect the binding energy. In this limit, inelastic scattering is called *deep-inelastic scattering*. We thus conclude that the parton model satisfies Bjorken's scaling law for deep inelastic scattering. Moreover, since we have assumed that the spin of the parton is $1/2$ in the computation above, it follows that

$$F_2(x) = 2xF_1 . \tag{13.224}$$

This relation, which agrees with experiment, is known as the Callan–Gross relation [138].

If the spin of the parton is zero, we may ask how the relation above changes. So, assuming that the parton is a scalar, let us compute $w_{\mu\nu}$:

$$2\pi e^2 w_{\mu\nu} = 2\pi e_{\mathrm{p}}^2 \int \frac{\mathrm{d}^3 p'}{2p'_0} \frac{1}{2m} (p + p')_\mu (p + p')_\nu \delta^4(p' - p - Q)$$
$$= 2\pi e_{\mathrm{p}}^2 \delta(2p \cdot Q + Q^2)(2p + Q)_\mu (2p + Q)_\nu / m \,, \tag{13.225}$$

where e_{p} is the electric charge of the parton. In this case,

$$F_1(x) = 0 \,, \quad F_2(x) = x P(x) \,. \tag{13.226}$$

In other words, F_1 vanishes. This contradicts the experiment. Hence, it is impossible to construct the proton using only scalar particles, and the parton must have spin $1/2$. Note that, in the calculation above, the interference term has been dropped, but this may be justified to some extent because of the large momentum transfer.

To conclude this section, we can assert that "quarks in the hadron are almost free." In other words, the interaction among quarks becomes weak at short range. If this holds for the strong interaction in general, this is a remarkable feature of the strong interaction which clearly restricts its structure.

If we consider the same thing in momentum space, it turns out that, in the strong interaction, the probability of a process involving a large momentum transfer is very small. As will be discussed later, only a non-Abelian gauge interaction can satisfy this condition.

Chapter 14
What Is Gauge Theory?

So far we have introduced a variety of interactions among elementary particles. If we arrange them in order of decreasing strength, we find the strong interaction, the electromagnetic interaction, the weak interaction, and finally the gravitational interaction. Typical in each category are Yukawa's nucleon–meson interaction, the interaction between the electron and the electromagnetic field, the Fermi interaction, and so on. For elementary particles with their very small masses, gravity is not generally considered in elementary particle theory, because it is many orders of magnitude weaker than the others.

These interactions often have various kinds of symmetry. For instance, the nucleon–meson interaction is invariant under rotations in isospin space, and as a consequence, there is a conserved quantity, namely isospin. It is thus natural to ask whether there is a general principle for determining the forms of different interactions and whether there exists a relationship between the form of an interaction and the symmetries of the theory.

The Coulomb potential and the Newtonian potential, already known to classical physics, are both inversely proportional to the distance. Their strengths are proportional to the product of the particles' electric charges and the product of the bodies' masses, respectively. Both the electric charge and the mass are conserved quantities in non-relativistic theories. If we consider that such potentials emerge from fields, then we can say that the origins of these fields are the current densities corresponding to conserved quantities. We call these fields and interactions *gauge fields* and *gauge interactions*. It turns out that in this sort of interaction the gauge field and the conserved quantity, or the symmetry inducing it, are in a one-to-one correspondence, which gives one solution to the question mentioned above.

The best-known example of a gauge field is the electromagnetic field. The origin of the electromagnetic field is a four-dimensional electric current. Its representation can be determined immediately by designating what particles are fundamental. So far we have written the Lagrangian by treating hadrons and leptons as fundamental particles, but it is also easy to write down the Lagrangian by considering quarks

K. Nishijima, *Quantum Field Theory*,
https://doi.org/10.1007/978-94-024-2190-3_14

and leptons as fundamental. This involves the same procedure as when we take the electric current in terms of hadrons and rewrite it in terms of quarks. Moreover, when two hadrons are far enough apart, the Coulomb potential depends only on the electric charges of the hadrons and not on whether the hadrons are fundamental particles or composite particles. This is because the electric charge is an additive quantum number.

Yang and Mills proposed in 1954 the existence of a gauge field corresponding to the conserved isospin [139]. Since the isospin symmetry is described by a non-Abelian algebra, this is called a *non-Abelian gauge field*. In 1956, Utiyama introduced the non-Abelian gauge field corresponding to a more general symmetry. He also described the gravitational field as the gauge field corresponding to the conservation of energy–momentum [140]. The assumption that all interactions are of the gauge type, i.e., that there is a one-to-one correspondence between any conserved quantity and a gauge field, is called the *gauge principle*. Essentially, this was proposed by Utiyama. Let us give an example of the kind of result that can be obtained from this principle. Consider the hypothesis of baryon number conservation. If this is correct, there should be a gauge field that corresponds to it. This will be a vector field, and its interaction strength should be proportional to the baryon number. Since the mass is approximately proportional to the baryon number in macroscopic matter, this force looks very much like gravity. However, its effect should have been observed as a deviation from Newton's law of gravitation. And since the law of gravitation holds to very good accuracy, it is hard to imagine the existence of such a force. Therefore, the basis for assuming the conservation of baryon number is flimsy. This turns out to be connected with the possibility of proton decay.

In the following, we shall present the formalism needed to treat gauge fields.

14.1 Gauge Transformations of the Electromagnetic Field

As mentioned above, the electromagnetic field was the first known gauge field. In the mathematical formulation of a gauge field, the gauge transformations have a fundamental importance, not yet discussed above. We thus begin by studying the gauge transformations of the electromagnetic gauge field, and extend the concept to other, non-Abelian gauge fields.

We express a quantum of electric charge of a field ψ_a as $e_a = eq_a$, where e has units of electric charge and q_a is a dimensionless number. If eQ is the electric charge operator, then

$$[\psi_a(x), Q] = q_a \psi_a(x) , \tag{14.1}$$

so q_a is one of the eigenvalues of Q. We consider the unitary operator

$$\mathscr{R}(\lambda) = \exp(\mathrm{i}\lambda Q) , \tag{14.2}$$

where λ is a real parameter, and we introduce a phase transformation for $\psi_a(x)$:

$$\psi_a(x) \longrightarrow \psi'_a(x) = \mathscr{R}^{-1}(\lambda)\psi_a(x)\mathscr{R}(\lambda) = \mathrm{e}^{\mathrm{i}\lambda q_a}\psi_a(x) . \tag{14.3}$$

In a system where the electric charge is conserved, the Lagrangian density is invariant under this transformation. This is called a *global symmetry*. The transformation (14.3) is an example of (14.2), and we write it as

$$U(\lambda) = \mathrm{e}^{\mathrm{i}\lambda q_a} . \tag{14.4}$$

We now replace the parameter λ by a function of the coordinate, viz.,

$$\lambda \to \lambda(x) . \tag{14.5}$$

When we insert this $\lambda(x)$, we call (14.2), (14.3), and (14.4) a *local gauge transformation*, in contrast to the one with constant λ, which is called a *global gauge transformation*. Hence, (14.3) becomes

$$\psi_a(x) \to U\big(\lambda(x)\big)\psi_a(x) . \tag{14.6}$$

Introducing the gauge field A_μ, we define the covariant derivative D_μ by

$$D_\mu = \partial_\mu - \mathrm{i}e_a A_\mu(x) , \tag{14.7}$$

which is intended to act on $\psi_a(x)$. To determine how A_μ transforms under the gauge transformation, we require $D_\mu\psi_a$ to transform in the same way as (14.6):

$$D'_\mu\psi'_a(x) = U\big(\lambda(x)\big)\big(D_\mu\psi_a(x)\big) . \tag{14.8}$$

Combining (14.6) with (14.8),

$$D'_\mu = U\big(\lambda(x)\big)D_\mu U\big(\lambda(x)\big)^{-1} . \tag{14.9}$$

Inserting (14.7) into this equation, the transformation property of A_μ is determined to be

$$A_\mu(x) \to A_\mu(x) + \frac{1}{e}\partial_\mu\lambda(x) . \tag{14.10}$$

Note that the transformation above no longer depends on how we choose a. Moreover, from (14.9),

$$\big[D'_\mu, D'_\nu\big] = U\big(\lambda(x)\big)[D_\mu, D_\nu]U\big(\lambda(x)\big)^{-1} . \tag{14.11}$$

Since this commutator does not include derivatives, it commutes with U. It therefore becomes an invariant quantity under the gauge transformation:

$$[D_\mu, D_\nu] = -\mathrm{i}e_a(\partial_\mu A_\nu - \partial_\nu A_\mu) = -\mathrm{i}e_a F_{\mu\nu} \ . \tag{14.12}$$

We can also write down the Jacobi identity for D:

$$\big[D_\mu, [D_\nu, D_\sigma]\big] + \big[D_\nu, [D_\sigma, D_\mu]\big] + \big[D_\sigma, [D_\mu, D_\nu]\big] = 0 \ . \tag{14.13}$$

Inserting (14.12) into this equation, we find

$$\partial_\mu F_{\nu\sigma} + \partial_\nu F_{\sigma\mu} + \partial_\sigma F_{\mu\nu} = 0 \ . \tag{14.14}$$

This is called the *Bianchi identity*.

14.2 Non-Abelian Gauge Fields

In the previous section, we only considered one conserved quantity, namely Q. Let us now consider the case where there are N conserved charges $Q_1, Q_2, \dots, Q_N$, which form a Lie algebra, i.e., the commutation relations among them take the form

$$[Q_a, Q_b] = \mathrm{i} f_{abc} Q_c \ . \tag{14.15}$$

We consider a field $\psi(x)$ that carries an n-dimensional irreducible representation of this algebra, i.e.,

$$[\psi_r(x), Q_a] = (t_a)_{rs}\psi_s(x) \ , \quad r, s = 1, 2, \dots, n \ . \tag{14.16}$$

The matrices t_a form an n-dimensional representation of the charges Q_a. Just as representations of Q for different fields were given by different values of q in the case of the electromagnetic field, so the t_a representing the Q_a will also be different for different fields. By analogy with the last section, we define

$$\mathscr{R}(\lambda) = \exp\left(\mathrm{i}\sum_{i=1}^{N} \lambda^a Q_a\right) \ , \tag{14.17}$$

and specify a phase transformation for $\psi(x)$ by

$$\psi(x) \ \to \ \psi'(x) = \mathscr{R}^{-1}(\lambda)\psi(x)\mathscr{R}(\lambda) = U(\lambda)\psi(x) \ , \tag{14.18}$$

where

$$U(\lambda) = \exp\left(\mathrm{i}\sum_{a=1}^{N} \lambda^a t_a\right) . \tag{14.19}$$

We now make this transformation local, viz.,

$$\lambda^a \to \lambda^a(x) ,$$

and define the covariant derivative D_μ of $\psi(x)$ by

$$D_\mu = \partial_\mu - \mathrm{i}g \sum_{a=1}^{N} t_a A^a_\mu(x) \equiv \partial_\mu - \mathrm{i}g\mathscr{A}_\mu(x) . \tag{14.20}$$

Of course, $\mathscr{A}_\mu$ depends explicitly on the representation t. As in the last section, the gauge transformation of D_μ is

$$D'_\mu = U\big(\lambda(x)\big) D_\mu U\big(\lambda(x)\big)^{-1} . \tag{14.21}$$

This determines the form of the gauge transformation of $\mathscr{A}_\mu$:

$$\mathscr{A}'_\mu(x) = U\big(\lambda(x)\big)\mathscr{A}_\mu(x)U\big(\lambda(x)\big)^{-1} - \frac{1}{\mathrm{i}g}U\big(\lambda(x)\big)\partial_\mu U\big(\lambda(x)\big)^{-1} . \tag{14.22}$$

If we write down this transformation for the component A^a_μ, we see that it does not in fact depend on how we choose the representation t. For instance, considering an infinitesimal $\lambda(x)$,

$$A^{a\prime}_\mu(x) = A^a_\mu(x) + f_{abc}A^b_\mu(x)\lambda^c(x) + \frac{1}{g}\partial_\mu\lambda^a(x) , \tag{14.23}$$

and the representation t drops out completely. Additionally, the commutators of the components of the covariant derivative are

$$[D_\mu, D_\nu] = -\mathrm{i}g\big(\partial_\mu\mathscr{A}_\mu - \partial_\mu\mathscr{A}_\nu - \mathrm{i}g[\mathscr{A}_\mu, \mathscr{A}_\nu]\big) \equiv -\mathrm{i}g\mathscr{F}_{\mu\nu} , \tag{14.24}$$

$$\mathscr{F}_{\mu\nu} = \sum_{a=1}^{N} t_a \left(\partial_\mu A^a_\nu - \partial_\nu A^a_\mu + g f_{abc} A^b_\mu A^c_\nu\right) . \tag{14.25}$$

Once again, we see that the components of $\mathscr{F}_{\mu\nu}$ do not depend on how we choose t. The gauge transformation of $\mathscr{F}_{\mu\nu}$ is given by

$$\mathscr{F}'_{\mu\nu}(x) = U\big(\lambda(x)\big)\mathscr{F}_{\mu\nu}(x)U\big(\lambda(x)\big)^{-1} . \tag{14.26}$$

Since the t_a do not commute in this case, $\mathscr{F}_{\mu\nu}$ is not gauge invariant. However,

$$\mathrm{Tr}\left(\mathscr{F}_{\mu\nu}(x)\mathscr{F}_{\mu\nu}(x)\right) \tag{14.27}$$

is clearly gauge invariant, and can be used as the Lagrangian density of the gauge field. $A^a_\mu(x)$ is called a *non-Abelian gauge field*. Once again, we can derive the Bianchi identity from (14.13).

14.3 Gravitational Field as a Gauge Field

Utiyama was the first to note that the gravitational field is a sort of gauge field [140]. Since the symmetry of the gravitational field under a space-time transformation is different from an internal symmetry that is independent of space-time, many complications arise. So far we have introduced the gauge field from the condition that a given field and its covariant derivative should transform in the same way. However, for a space-time coordinate transformation, it is the rank of the tensor that determines the irreducible representation, and the covariant derivative alters this rank, so the given field and its covariant derivative clearly transform in different ways. For example, for a scalar field ϕ, if we consider $D_\mu D_\nu \phi$, the first covariant derivative D_ν is the covariant derivative of the scalar field, which is just ∂_ν, while the second covariant derivative D_μ is the covariant derivative of the vector field $\phi_\nu \equiv \partial_\nu \phi$, i.e.,

$$D_\mu \phi_\nu = \partial_\mu \phi_\nu - \Gamma^\lambda_{\mu\nu}\phi_\lambda . \tag{14.28}$$

If we consider $\Gamma^\lambda_{\mu\nu}$ as a matrix element of the matrix Γ_μ , viz.,

$$\left(\Gamma_\mu\right)_\nu{}^\lambda = \Gamma^\lambda_{\mu\nu} , \tag{14.29}$$

it turns out that Γ_μ corresponds to $\mathrm{i}g\mathscr{A}_\mu$ in the previous section. $\Gamma^\lambda_{\mu\nu}$ is the Christoffel symbol of the second kind and is defined by

$$\Gamma^\lambda_{\mu\nu} = \frac{1}{2} g^{\lambda\sigma}\left(\partial_\mu g_{\sigma\nu} + \partial_\nu g_{\sigma\mu} - \partial_\sigma g_{\mu\nu}\right) . \tag{14.30}$$

For a general coordinate transformation, Γ does not transform as a tensor. An additional term arises, characteristic of the transformation of a gauge field, involving a derivative of the transformation matrix.

We now construct a quantity corresponding to $\mathscr{F}_{\mu\nu}$. As mentioned before, the representation of D depends on what D acts on. For instance, if we consider a scalar field ϕ, then

$$[D_\mu, D_\nu]\phi = \phi_{;\nu;\mu} - \phi_{;\mu;\nu} , \tag{14.31}$$

where ; μ indicates the covariant derivative. However, for a vector field A_ν,

$$[D_\sigma, D_\rho]A_\nu = A_{\nu;\rho;\sigma} - A_{\nu;\sigma;\rho} = R^\lambda_{\nu\rho\sigma} A_\lambda . \tag{14.32}$$

Hence, using the matrix Γ, what corresponds to $\mathscr{F}_{\mu\nu}$ in the previous section is

$$\begin{aligned} R^\lambda_{\nu\mu\sigma} &= \left(\mathscr{F}_{\rho\sigma}\right)_\nu{}^\lambda = \left(\partial_\rho \Gamma_\sigma - \partial_\sigma \Gamma_\rho - [\Gamma_\rho, \Gamma_\sigma]\right)_\nu{}^\lambda \\ &= \partial_\rho \Gamma^\lambda_{\sigma\nu} - \partial_\sigma \Gamma^\lambda_{\rho\nu} - \Gamma^\alpha_{\rho\nu}\Gamma^\lambda_{\sigma\alpha} + \Gamma^\alpha_{\sigma\nu}\Gamma^\lambda_{\rho\alpha} . \end{aligned} \tag{14.33}$$

This has essentially the same form as $\mathscr{F}_{\mu\nu}$ in the previous section. Moreover, the Jacobi identity for the scalar field ϕ, viz.,

$$\left(\left[D_\nu, [D_\rho, D_\sigma]\right] + \left[D_\rho, [D_\sigma, D_\nu]\right] + \left[D_\sigma , [D_\nu, D_\rho]\right]\right)\phi = 0 , \tag{14.34}$$

is equivalent to the following equation (multiplied by $\partial_\lambda \phi$):

$$R^\lambda_{\nu\rho\sigma} + R^\lambda_{\rho\sigma\nu} + R^\lambda_{\sigma\nu\rho} = 0 . \tag{14.35}$$

Similarly, for the vector field A_μ, the equation

$$\left(\left[D_\nu, [D_\rho, D_\sigma]\right] + \left[D_\rho, [D_\sigma, D_\nu]\right] + \left[D_\sigma, [D_\nu, D_\rho]\right]\right) A_\mu = 0 \tag{14.36}$$

implies

$$R^\lambda_{\mu\rho\sigma;\nu} + R^\lambda_{\mu\sigma\nu;\rho} + R^\lambda_{\mu\nu\rho;\sigma} = 0 . \tag{14.37}$$

Equations (14.35) and (14.37) are the Bianchi identities in this case.

Chapter 15
Spontaneous Symmetry Breaking

In the last chapter we discussed gauge theory. As far as it goes, we remain unable to eliminate other interactions and replace them by gauge interactions. It may happen that the symmetry in the Lagrangian is broken for some reason and is not apparent. This is called a *broken or hidden symmetry*. In this case, a quantum of the gauge field corresponding to the broken or hidden symmetry acquires mass, in contrast to the photon. In this chapter, we will discuss a mechanism whereby the symmetry in the Lagrangian is broken.

15.1 Nambu–Goldstone Particles

We assume that a given Lagrangian is invariant under the Lie group G. We call the ground state of this system the *vacuum* and denote it by $|\mathbf{0}\rangle$. In this case, there are two possibilities:

1. $|\mathbf{0}\rangle$ is invariant under G, whence it belongs to a one-dimensional representation of G;
2. $|\mathbf{0}\rangle$ is not invariant under G, and it belongs to a multi-dimensional representation of G.

The former case is a manifest symmetry and the latter is a hidden symmetry. In the latter case, we say that *the symmetry is spontaneously broken*, and $|\mathbf{0}\rangle$ is clearly degenerate. There exist many examples like this outside field theory. A well known example is the ferromagnet. In the ground state, the electron spins align in one direction. The direction in which the spins align depends on how the ground state has been prepared, but originally the system is rotationally symmetric, so this ground state is degenerate with respect to spin directions.

K. Nishijima, *Quantum Field Theory*,
https://doi.org/10.1007/978-94-024-2190-3_15

Theorem 15.1 *Nambu–Goldstone theorem. When the Lagrangian is invariant under the Lie group G and its invariance is spontaneously broken, there exist particles whose masses and spins are both zero.*

This was first discovered by Nambu and Jona-Lasinio [141, 142], and then a simpler example was given by Goldstone [143]. The particles in question are called *Nambu–Goldstone particles*.

If $|\mathbf{0}\rangle$ does not belong to the one-dimensional representation, there exist local scalar operators satisfying

$$\langle \mathbf{0}|\rho_a(x)|\mathbf{0}\rangle \neq 0 \,. \tag{15.1}$$

Although the ρ_a themselves belong to the multi-dimensional representation, how they are represented depends on the representation of $|\mathbf{0}\rangle$. The expectation value of operators with non-zero spins always disappears because of Lorentz invariance. In the following, we consider the case where we can choose the scalar field operators themselves as the ρ_a. The original Nambu–Jona-Lasinio model chose a bilinear form of the Fermi field as ρ. This is because no scalar field was included in their model.

We express an infinitesimal transformation of the group G in the form

$$\delta\varphi_a = \epsilon_\alpha T^\alpha_{ab}\varphi_b \,. \tag{15.2}$$

If we rewrite this with $\epsilon_\alpha \rightarrow \lambda^a$, $T^\alpha \rightarrow it_a$, it coincides with (14.18). The corresponding Noether current is

$$J^\alpha_\mu = \frac{\partial \mathscr{L}}{\partial \varphi_{a,\mu}} T^\alpha_{ab}\varphi_b \,. \tag{15.3}$$

It satisfies the local conservation law

$$\partial_\mu J^\alpha_\mu = 0 \,. \tag{15.4}$$

We thus analyze the global conservation law in the two cases mentioned at the beginning of this section:

1. When a symmetry is manifest, there are global conserved quantities

$$Q^\alpha = \int \mathrm{d}^3x \, J^\alpha_0(x) \,, \tag{15.5}$$

and the following global conservation law holds:

$$\frac{\mathrm{d}Q^\alpha}{\mathrm{d}t} = 0 \,. \tag{15.6}$$

Moreover, since $|\mathbf{0}\rangle$ belongs to the one-dimensional representation of G, the generators Q^α of G annihilate the vacuum:

$$Q^\alpha |\mathbf{0}\rangle = 0 . \tag{15.7}$$

It is easily shown from the definition of (15.3) that, because the Q^α are generators of the group G, we must have

$$\mathrm{i}[Q^\alpha, \boldsymbol{\varphi}_a] = T^\alpha_{ab}\boldsymbol{\varphi}_b . \tag{15.8}$$

In this case, it is clear from (15.7) that

$$\mathrm{i}\big\langle \mathbf{0}\big|[Q^\alpha, \boldsymbol{\varphi}_a]\big|\mathbf{0}\big\rangle = T^\alpha_{ab}\langle \mathbf{0}|\boldsymbol{\varphi}_b|\mathbf{0}\rangle = 0 . \tag{15.9}$$

In addition, the generators Q^α are the same both with and without renormalization, i.e., (15.8) and (15.9) hold for the field operators both with and without renormalization, and they have the same form. However, for the discussion below, it will be more useful to consider the renormalized case.

2. For the broken symmetry, there is a component $\boldsymbol{\varphi}_b$ satisfying

$$T^\alpha_{ab}\langle \boldsymbol{\varphi}_b\rangle \neq 0 , \quad \langle \boldsymbol{\varphi}_b\rangle = \langle \mathbf{0}|\boldsymbol{\varphi}_b|\mathbf{0}\rangle . \tag{15.10}$$

In this case, we will prove the Nambu–Goldstone theorem. We first study the structure of the two-point function

$$\mathrm{i}\big\langle \mathbf{0}\big|[\boldsymbol{J}^\alpha_\mu(x), \boldsymbol{\varphi}_\alpha(y)]\big|\mathbf{0}\big\rangle \tag{15.11}$$

under the *explicit Lorentz covariance*. First, from the discussion about the integral representation in Sect. 12.8, we can write down the following integral representation as the most general form of (15.11):

$$\mathrm{i}\big\langle \mathbf{0}\big|[\boldsymbol{J}^\alpha_\mu(x), \boldsymbol{\varphi}_a(y)]\big|\mathbf{0}\big\rangle = \partial_\mu \int \mathrm{d}m^2 \rho^\alpha_a(m^2)\Delta(x-y; m^2) . \tag{15.12}$$

In addition, the spectral function $\rho^\alpha_a(m^2)$ is given by

$$p_\mu\theta(p_0)\rho^\alpha_a(-p^2) = (2\pi)^3\mathrm{i}\sum_n \delta^4(p-p_n)\big\langle \mathbf{0}\big|\boldsymbol{J}^\alpha_\mu(0)\big|\boldsymbol{n}\big\rangle\big\langle \boldsymbol{n}\big|\boldsymbol{\varphi}_a(0)\big|\mathbf{0}\big\rangle . \tag{15.13}$$

Extracting a divergence in (15.12) and using the local conservation law (15.4),

$$\int \mathrm{d}m^2 \rho^\alpha_a(m^2) m^2 \Delta(x-y; m^2) = 0 , \tag{15.14}$$

$$m^2 \rho^\alpha_a(m^2) = 0 . \tag{15.15}$$

A general from of ρ satisfying these equations is

$$\rho_a^\alpha(m^2) = N_a^\alpha \delta(m^2) . \tag{15.16}$$

Thus, denoting the Δ-function for $m^2 = 0$ by D,

$$\mathrm{i}\langle \mathbf{0} | [\boldsymbol{J}_a^\alpha(x), \boldsymbol{\varphi}_a(y)] | \mathbf{0} \rangle = N_a^\alpha \partial_\mu D(x-y) . \tag{15.17}$$

Now, taking $\mu = 4$ and using (15.8),

$$N_a^\alpha = T_{ab}^\alpha \langle \boldsymbol{\varphi}_b \rangle \neq 0 . \tag{15.18}$$

Hence,

$$\mathrm{i}\langle \mathbf{0} | [\boldsymbol{J}_\mu^\alpha(x), \boldsymbol{\varphi}_a(y)] | \mathbf{0} \rangle = T_{ab}^\alpha \langle \boldsymbol{\varphi}_b \rangle \partial_\mu D(x-y) . \tag{15.19}$$

This equation implies that the scalar field $\boldsymbol{\varphi}_a$ yields a massless one-particle state. This particle is the Nambu–Goldstone particle. Moreover, in this case the term proportional to $\partial_\mu \boldsymbol{\varphi}_a$ is included in $\boldsymbol{J}_\mu^\alpha$:

$$\int \mathrm{d}^3 x \, \partial_\mu \boldsymbol{\varphi}_a \sim V \frac{1}{\sqrt{V}} = \sqrt{V} \to \infty , \quad V \to \infty . \tag{15.20}$$

Hence, the integral (15.18) defining Q^α diverges, so (15.8) can be rewritten as

$$\mathrm{i} \int \mathrm{d}^3 x \big[\boldsymbol{J}_0^\alpha(x), \boldsymbol{\varphi}_a(y) \big] = T_{ab}^\alpha \boldsymbol{\varphi}_b(y) , \quad x_0 = y_0 . \tag{15.21}$$

Changing the order of taking the commutator and doing the integration makes no difference. In this case, the global conserved quantities Q^α are zero. As a consequence, this broken symmetry becomes a hidden symmetry.

No particle whose mass and spin are both zero has yet been observed experimentally. Does this mean that such a broken symmetry cannot be realized in the real world? Or does the Nambu–Goldstone particle disappear as a result of some other mechanism? Solving this issue will be the aim of the present chapter.

15.2 Sigma Model

In the last section, we investigated the relationship between the broken symmetry and the Nambu–Goldstone particle. In this section, we will discuss a simple model which realizes this. We consider the system consisting of a pseudo-scalar triplet π

and a scalar singlet σ:

$$\pi_1\,,\quad \pi_2\,,\quad \pi_3\,,\quad \sigma\,. \tag{15.22}$$

To distinguish between the scalar and the pseudo-scalar, we have to introduce fermions and interactions such as

$$\mathrm{i}\bar{\psi}\gamma_5\frac{\hat{\tau}}{2}\psi\hat{\varphi}\,,\quad \bar{\psi}\psi\sigma\,. \tag{15.23}$$

For simplicity, we will only consider bosons. We introduce the notation

$$\varphi_a^2=\varphi_1^2+\varphi_2^2+\varphi_3^2+\sigma^2=\hat{\varphi}^2+\sigma^2\,, \tag{15.24}$$

where the φ_i are the field operators for the π_i and σ is the field operator for σ.

For the Lagrangian density, we take

$$\mathscr{L}=-\frac{1}{2}(\partial_\mu\varphi_a)^2+\frac{1}{4}m^2\varphi_a^2-\frac{1}{8}f^2(\varphi_a^2)^2\,, \tag{15.25}$$

which has $O(4)$ symmetry. Note that the mass term has the opposite sign to the usual one. Minus the sum of the second and third terms is the potential, viz.,

$$U(\varphi)=-\frac{1}{4}m^2\varphi_a^2+\frac{1}{8}f^2(\varphi_a^2)^2\,. \tag{15.26}$$

Then $\varphi_a=0$ gives an extremum of $U(\varphi)$, but it does not give the minimum. Put another way, the ordinary symmetric vacuum is not stable. Before introducing non-symmetric vacua, let us investigate the algebraic structure of this model. In this system, there are six local conservation laws. We thus introduce the following vector and axial-vector currents:

$$\hat{\boldsymbol{V}}_\mu=-\hat{\boldsymbol{\varphi}}\times\partial_\mu\hat{\boldsymbol{\varphi}}\,,\quad \hat{\boldsymbol{A}}_\mu=\hat{\boldsymbol{\varphi}}\partial_\mu\boldsymbol{\sigma}-\boldsymbol{\sigma}\partial_\mu\hat{\boldsymbol{\varphi}}\,. \tag{15.27}$$

Both these current densities satisfy the local conservation laws

$$\partial_\mu\hat{\boldsymbol{V}}_\mu=0\,,\quad \partial_\mu\hat{\boldsymbol{A}}_\mu=0\,. \tag{15.28}$$

If the global conservation laws hold, then taking $\hat{\boldsymbol{\pi}}=\dot{\hat{\boldsymbol{\varphi}}}$ and $\boldsymbol{\pi}_0=\dot{\boldsymbol{\sigma}}$, the conserved quantities are

$$\hat{I}=\int\mathrm{d}^3x\,\left(\hat{\boldsymbol{\varphi}}\times\hat{\boldsymbol{\pi}}\right)\,,\quad \hat{I}^5=\int\mathrm{d}^3x\,\left(\boldsymbol{\sigma}\hat{\boldsymbol{\pi}}-\hat{\boldsymbol{\varphi}}\boldsymbol{\pi}_0\right). \tag{15.29}$$

Comparing with (13.166), (13.170), and (13.171), the commutation relations among them are

$$[I_i, I_j] = \mathrm{i}\epsilon_{ijk} I_k\ , \quad \left[I_i, I_j^5\right] = \mathrm{i}\epsilon_{ijk} I_k^5\ , \quad \left[I_i^5, I_j^5\right] = \mathrm{i}\epsilon_{ijk} I_k\ . \tag{15.30}$$

Even if the global conservation laws do not hold, the commutation relations among the densities still do. Neglecting the Goto–Imamura–Schwinger term discussed in Sect. 12.8 and setting $x_0 = y_0$,

$$\left[V_{i0}(x), V_{j0}(y)\right] = \mathrm{i}\epsilon_{ijk}\delta^3(x-y)V_{k0}(x)\ . \tag{15.31}$$

Taking

$$\mathscr{F}_{i\mu}^{(\pm)} = \frac{1}{2}\left(V_{i\mu} \pm A_{i\mu}\right) \tag{15.32}$$

for the density, as discussed in Sect. 13.6, we have the current algebra

$$SU(2)_{\mathrm{L}} \times SU(2)_{\mathrm{R}}\ . \tag{15.33}$$

Adding further fermions to this model, a representation of the current algebra can be made. Indeed, it leads to chiral dynamics and we can construct a simple representation for the current algebra. However, in the real world there is no particle corresponding to σ, so representations without σ have been proposed. Unfortunately, since the representation without σ is a rather non-linear theory, work eventually shifted to the quark model. In the following, we will consider the model in which σ survives. Coming back to the topic of broken symmetry, we write the field equation in the form

$$\Box\varphi_a + \frac{1}{2}\left(m^2 - f^2\varphi_b^2\right)\varphi_a = 0\ . \tag{15.34}$$

Considering this as a classical equation, the requirement of causality cannot be satisfied unless

$$\frac{1}{2}\left(m^2 - f^2\varphi_b^2\right) \leq 0\ , \tag{15.35}$$

whence $\langle\varphi_b\rangle^2 \neq 0$. Corresponding to the minimum of $U(\langle\varphi\rangle)$, we take

$$\langle\hat{\varphi}\rangle = 0\ , \quad \langle\sigma\rangle = \eta \neq 0\ , \tag{15.36}$$

where the parity and the isospin are conserved. Hence, if we take

$$\sigma = \sigma_0 + \eta\ , \tag{15.37}$$

then η is determined by the condition minimizing the potential, viz.,

$$m^2 = f^2\eta^2 \,. \tag{15.38}$$

Since we have introduced symmetry breaking, if we investigate what kinds of equations $\hat{\varphi}$ and σ_0 satisfy, we find

$$\Box\hat{\boldsymbol{\varphi}} = \text{non-linear terms}\,, \quad (\Box - m^2)\boldsymbol{\sigma}_0 = \text{non-linear terms}\,. \tag{15.39}$$

What we have understood from these is that the mass of π vanishes, σ has a finite mass, and π is the Nambu–Goldstone particle. Although the discussion about symmetry breaking has been classical in this section, we will give a more detailed quantum mechanical discussion in the next.

Interpreting π as the pion, since it actually has a small mass, we add a small term to introduce a slight symmetry breaking artificially. This term can be considered to determine the direction of symmetry breaking. If a is a small quantity,

$$\mathscr{L} \to \mathscr{L}' = \mathscr{L} + a\sigma \,. \tag{15.40}$$

In this case, η is shifted by order a and becomes η':

$$\sigma = \sigma_0 + \eta' \,. \tag{15.41}$$

Minimizing the potential classically in this case,

$$m^2 = f^2\eta'^2 - \frac{2a}{\eta'} \,. \tag{15.42}$$

Therefore, if a is small, η' differs from η only by a term of order a. Although the mass of σ is given once again by m, the mass of π becomes finite, and if we write it as μ, we have

$$\mu^2 = \frac{1}{2}\left(f^2\eta'^2 - m^2\right) = \frac{a}{\eta'} \,. \tag{15.43}$$

Moreover, in this case, the local conservation law of the axial vector current is no longer observed. If we write η' as f_π, then

$$\partial_\mu \hat{\boldsymbol{A}}_\mu = -\mu^2 f_\pi \hat{\boldsymbol{\varphi}} \,. \tag{15.44}$$

This is called the partially conserved axial-vector current (PCAC) hypothesis [129]. Considering π as a composite particle, we define its phenomenological field operator by this equation.

Since the above symmetry breaking argument has been classical and phenomenological, we now formulate this using Green's functions in order to give a rigorous quantum mechanical discussion.

15.3 The Mechanism of Spontaneous Symmetry Breaking

As an example of spontaneous symmetry breaking, we previously mentioned the ferromagnet. When there is no magnetic field, the directions of the electron spins are random at high temperature because of their thermal motions, whence there is rotational symmetry. If we turn on a magnetic field, the ferromagnet is magnetized in the direction of the magnetic field. If we now gradually decrease the temperature a phase transition occurs. It turns out that even after removing the magnetic field, it remains magnetized. In order to translate this phenomenon into the case for spontaneous symmetry breaking in field theory, we use a Green's function. The necessary formalism has been given in Sect. 11.2.

The generating function $\mathscr{R}[J]$ for connected Green's functions is a functional of the external field J, where J corresponds to the magnetic field imposed from the outside in the ferromagnet case. The vacuum expectation value of the field $\varphi(x)$ produced as a consequence corresponds to the magnetization at low temperature. We wish to investigate whether magnetization occurs even in the case where there is no external field. Choosing the magnetic field as an independent variable, we ask whether the magnetization can vanish when the magnetic field takes a suitable value. This suitable value corresponds to the magnetization of the permanent magnet.

Although in Sect. 11.2, we introduced the functional $\mathscr{R}[J]$ of $J(x)$ and the functional $\mathscr{F}[\langle\varphi(x)\rangle]$ of $\langle\varphi(x)\rangle$, since it is useful to slightly change the phases, we write

$$\mathscr{R} = -\mathrm{i}W\,, \quad \mathscr{F} = -\mathrm{i}\Gamma\,, \tag{15.45}$$

whence (11.36) becomes

$$\Gamma = W - \int \mathrm{d}^4x\, \langle\varphi(x)\rangle J(x)\,. \tag{15.46}$$

The role of the new variables as generating functionals is expressed by

$$\frac{\delta W}{\delta J(x)} = \langle\varphi(x)\rangle\,, \quad \frac{\delta\Gamma}{\delta\langle\varphi(x)\rangle} = -J(x)\,. \tag{15.47}$$

W and Γ are generating functionals of the connected Green's function and of the one-particle irreducible Green's function, respectively. The former is the Green's function corresponding to connected Feynman diagrams and the latter is the Green's function without a pole corresponding to the one-particle state. This is because in the latter the one-particle propagator is absorbed into $\langle \boldsymbol{\varphi}(x) \rangle$.

The connected and one-particle irreducible Green's functions in the momentum space are defined by

$$\left.\frac{\delta^n W[J]}{\delta J(x_1)\ldots\delta J(x_n)}\right|_{J=0} = \frac{1}{(2\pi)^{4(n-1)}}\int (\mathrm{d}p)\delta(p_1+\cdots+p_n)\mathrm{e}^{\mathrm{i}p_1\cdot x_1+\cdots+\mathrm{i}p_n\cdot x_n}G^{(n)}(p_1,\ldots,p_n) \tag{15.48}$$

and

$$\left.\frac{\delta^n \Gamma[\langle \boldsymbol{\varphi}\rangle]}{\delta\langle \boldsymbol{\varphi}(x_1)\rangle\ldots\delta\langle \boldsymbol{\varphi}(x_n)\rangle}\right|_{\langle \boldsymbol{\varphi}(x)\rangle=\varphi} = \frac{1}{(2\pi)^{4(n-1)}}\int (\mathrm{d}p)\delta(p_1+\cdots+p_n)\mathrm{e}^{\mathrm{i}p_1\cdot x_1+\cdots+\mathrm{i}p_n\cdot x_n}\Gamma^{(n)}(p_1,\ldots,p_n)\,, \tag{15.49}$$

respectively, where $(\mathrm{d}p) = \mathrm{d}^4 p_1 \ldots \mathrm{d}^4 p_n$ and φ is a constant. In the theory in which the symmetry is spontaneously broken, it turns out that φ takes a value different from zero. We now define the effective potential $V(\varphi)$ by

$$V(\varphi) = -\Gamma[\varphi]\,. \tag{15.50}$$

At the lowest order, $\Gamma[\varphi]$ coincides with the value when the field operator is replaced by the constant φ in the Lagrangian density. Therefore, it coincides with $-U(\varphi)$ in the previous section, i.e.,

$$V^{(0)}(\varphi) = -\mathscr{L}(\varphi) = U(\varphi)\,. \tag{15.51}$$

However, since there are differences when we include higher-order quantum corrections, this is called an *effective potential*. In the following, we consider the case where φ is multi-component. We assume that the theory is invariant under a Lie group G, as considered in Sect. 15.1. Thus, if we have invariance under the transformation (15.2),

$$V(\varphi_a + \varepsilon_\alpha T^\alpha_{ab}\varphi_b) = V(\varphi_a)\,, \quad \text{or} \quad \frac{\partial V}{\partial \varphi_a}T^\alpha_{ab}\varphi_b = 0\,. \tag{15.52}$$

Corresponding to (11.38) and (11.40), we have

$$\frac{\partial V}{\partial \varphi_a} = J_a\,, \qquad \frac{\partial^2 V}{\partial \varphi_a \partial \varphi_b} = -\Delta_{\mathrm{F}}^{-1}(0)_{ab}\,, \tag{15.53}$$

where J_a is the external field corresponding to φ_a, while $\Delta_F^{-1}(0)$ corresponds to $p^2 = 0$ and expresses the inverse matrix of the multi-component propagator when the expectation value φ does not necessarily vanish. Differentiating (15.52),

$$\frac{\partial^2 V}{\partial\varphi_a\partial\varphi_c} T^\alpha_{ab}\varphi_b + \frac{\partial V}{\partial\varphi_a} T^\alpha_{ac} = 0 . \tag{15.54}$$

Combining this with (15.53),

$$\Delta_F^{-1}(0)_{ca} T^\alpha_{ab}\varphi_b = J_a T^\alpha_{ac} . \tag{15.55}$$

The conditions for φ to have finite components even if the external field J vanishes, i.e., the conditions for the symmetry to be spontaneously broken, are

$$J_a = \frac{\partial V}{\partial\varphi_a} = 0 , \quad \Delta_F^{-1}(0)_{ca} T^\alpha_{ab}\varphi_b = 0 . \tag{15.56}$$

The second equation means that there will be massless Nambu–Goldstone particles. It is a necessary condition for this equation to have solutions except for $\varphi = 0$. A sufficient condition is that $V(\varphi)$ should have a minimum for the value of φ.

We consider the $SO(n)$ symmetry as an example, and take

$$V(\varphi) = F(\varphi_a^2) , \tag{15.57}$$

where $\varphi_a^2 = \varphi_1^2 + \cdots + \varphi_n^2$. Assuming that $F(x)$ has a minimum at $x = x_0$,

$$F'(x_0) = 0 , \quad F''(x_0) > 0 . \tag{15.58}$$

In this case, J_a vanishes for $\varphi_a^2 = x_0$:

$$J_a = 2\varphi_a F'(x_0) = 0 . \tag{15.59}$$

It turns out that the second equation in (15.56) then also holds. The solutions for $\varphi_a^2 = x_0$ are now degenerate. Thus, the ground states of this system turn out to be degenerate. In this case,

$$\Delta_F^{-1}(0)_{ab} = -\frac{\partial^2 V}{\partial\varphi_a\partial\varphi_b} = -2\delta_{ab}F'(\varphi_a^2) - 4\varphi_a\varphi_b F''(\varphi_a^2) . \tag{15.60}$$

Taking $\varphi_a^2 = x_0$,

$$\Delta_F^{-1}(0)_{ab} = -4\varphi_a\varphi_b F''(x_0) . \tag{15.61}$$

The eigenvalues of this matrix are $-4x_0 F''(x_0)$ and 0 for all others. The eigenvector for the first of these eigenvalues is parallel to the n-dimensional vector $(\varphi_1, \ldots, \varphi_n)$, while the eigenvector for the latter eigenvalue is a vector normal to it. Hence, by a proper rotation in $SO(n)$, we can write $\Delta_F^{-1}(0)$ in diagonal form:

$$\Delta_F^{-1}(0) = \begin{pmatrix} D & & & \\ & 0 & & \\ & & \ddots & \\ & & & 0 \end{pmatrix}. \tag{15.62}$$

It turns out that, in this new coordinate system, the components of φ can be written as $(\sqrt{x_0}, 0, \ldots, 0)$. In this case, if we take $c = 1$ in (15.56), then from the antisymmetry of T, we obtain $T^{\alpha}_{ab}\varphi_b = 0$. If we take $c \neq 1$, then we obtain $\Delta_F^{-1}(0)_{ca} = 0$, so (15.56) holds in every case. Since the component corresponding to $c = 1$ has a finite mass and components corresponding to $c \neq 1$ give massless poles for $\Delta_F(p^2)$, Nambu–Goldstone particles turn out to be included. Additionally, since for $c \neq 1$, $T^{\alpha}_{cb}\varphi_b \neq 0$ for a proper α, the discussion in Sect. 15.1 also leads to the conclusion that Nambu–Goldstone particles exist.

The discussion about the sigma model in the previous section shows that the computation above has been carried out explicitly based on the approximation (15.51). The components for $c = 1$ and for $c \neq 1$ above correspond to σ and π, respectively.

15.4 Higgs Mechanism

So far we have considered the gauge principle. If the symmetry is not broken, there must be quanta of massless gauge fields for the number of conserved quantities. On the other hand, if the symmetry is spontaneously broken, then there must be massless scalar particles. However, in reality, no massless boson has yet been observed, except for the photon. So is it possible for massless bosons to be eliminated by an as-yet-unknown mechanism? It was Englert, Brout, and Higgs in 1964 who gave the first interesting discussion of this issue [144–146]. In the following, we will present their approach.

We consider the simplest symmetric group, $U(1)$ or $SO(2)$. We assume that a given system includes two real scalar fields φ_1 and φ_2, and is invariant under the transformation

$$\begin{pmatrix} \varphi_1 \\ \varphi_2 \end{pmatrix} \to \begin{pmatrix} \cos\alpha & \sin\alpha \\ -\sin\alpha & \cos\alpha \end{pmatrix} \begin{pmatrix} \varphi_1 \\ \varphi_2 \end{pmatrix}. \tag{15.63}$$

If the Lagrangian density of this system is invariant under this transformation, then we can introduce the corresponding Noether current $j_\mu(x)$. We obtain the commutation relations

$$i\int d^3x\big[\boldsymbol{j}_0(x), \boldsymbol{\varphi}_1(y)\big] = \boldsymbol{\varphi}_2(y)\ , \tag{15.64}$$

$$i\int d^3x\big[\boldsymbol{j}_0(x), \boldsymbol{\varphi}_2(y)\big] = -\boldsymbol{\varphi}_1(y)\ , \tag{15.65}$$

where $x_0 = y_0$. We assume the following spontaneous symmetry breaking:

$$\langle \boldsymbol{\varphi}_2(x)\rangle \neq 0\ . \tag{15.66}$$

In the proof of the Nambu–Goldstone theorem,

$$\text{FT } i\big\langle \mathbf{0}\big|[\boldsymbol{j}_\mu(x), \boldsymbol{\varphi}_1(y)]\big|\mathbf{0}\big\rangle \propto \langle \boldsymbol{\varphi}_2\rangle \epsilon(k_0) k_\mu \delta(k^2)\ , \tag{15.67}$$

where FT indicates the Fourier transform, so we could conclude that there are massless scalar particles. However, in this proof we assumed *explicit Lorentz covariance*. But what happens if this assumption does not hold true, as for example in non-relativistic theories, where this theorem does not necessarily apply. If we do not have explicit covariance and a particular unit vector along the time axis appears in a theory, then the general form of the Fourier transform in (15.67) above becomes

$$\text{FT} \sim k_\mu \rho_1(k^2, n\cdot k) + n_\mu \rho_2(k^2, n\cdot k) + C n_\mu \delta^4(k)\ . \tag{15.68}$$

The general case within this form which satisfies the conservation law, i.e., the one whose internal product with k_μ vanishes, becomes

$$\text{FT} \sim k_\mu \delta(k^2)\rho_4(n\cdot k) + \big[k^2 n_\mu - k_\mu(n\cdot k)\big]\rho_5(k^2, n\cdot k) + C n_\mu \delta^4(k)\ . \tag{15.69}$$

Thus, a counterexample for the Nambu–Goldstone theorem occurs when

$$\rho_4 = 0\ . \tag{15.70}$$

But even though there exists no explicit covariance, could the theory be Lorentz covariant? The example given by Higgs is QED in a non-covariant gauge, i.e., with the gauge condition in the form

$$n_\mu A_\mu = 0\ . \tag{15.71}$$

Maxwell's equations are then

$$\partial_\mu F_{\mu\nu} = -j_\nu\ , \quad F_{\mu\nu} = \partial_\mu A_\nu - \partial_\nu A_\mu\ . \tag{15.72}$$

We consider the following general form of the Fourier transform:

$$\text{FT}\ \mathrm{i}\langle\mathbf{0}|[\boldsymbol{A}_\mu(x),\boldsymbol{\varphi}_1(y)]|\mathbf{0}\rangle = k_\mu\sigma_1(k^2,n\cdot k)+n_\mu\sigma_2(k^2,n\cdot k)+Dn_\mu\delta^4(k)\ . \tag{15.73}$$

Therefore, from the gauge condition (15.71), taking $n^2=-1$, we obtain

$$\sigma_2=(n\cdot k)\sigma_1\ ,\quad D=0\ . \tag{15.74}$$

Thus, combining this result with (15.72),

$$\text{FT};\ \mathrm{i}\langle\mathbf{0}|[\boldsymbol{j}_\mu(x),\boldsymbol{\varphi}_1(y)]|\mathbf{0}\rangle = \big[k^2n_\mu-k_\mu(n\cdot k)\big]\sigma_2(k^2,n\cdot k)\ . \tag{15.75}$$

Comparing this with (15.69), it turns out that (15.70) is satisfied. Choosing a non-covariant gauge like this, the massless Nambu–Goldstone particle does not appear. The reason why we could do this is that there is a gauge field. Needless to say, the transformations (15.64) and (15.65) appearing at the beginning of this section correspond to global gauge transformations.

Bearing this in mind, we introduce the gauge field corresponding to the $U(1)$ symmetry and consider what happens when the symmetry is spontaneously broken. The following discussion was given by Higgs in [147]. Since this gauge field can be considered to be the same as the electromagnetic field, we introduce the Lagrangian density

$$\mathscr{L}=-\frac{1}{4}F_{\mu\nu}F_{\mu\nu}-\frac{1}{2}D_\mu\varphi_aD_\mu\varphi_a+\frac{1}{4}m_0^2\varphi_a\varphi_a\ . \tag{15.76}$$

The scalar part is the same as that of the sigma model, but φ_a has two components labelled by $a=1,2$. The covariant derivative D_μ is defined by

$$D_\mu\varphi_a=\partial_\mu\varphi_1-eA_\mu\varphi_2\ ,\quad D_\mu\varphi_2=\partial_\mu\varphi_2+eA_\mu\varphi_1\ . \tag{15.77}$$

In QED, we use complex fields, φ and $\varphi^\dagger$, and their covariant derivatives, i.e.,

$$\varphi=\frac{1}{\sqrt{2}}(\varphi_1-\mathrm{i}\varphi_2)\ ,\quad \varphi^\dagger=\frac{1}{\sqrt{2}}(\varphi_1-\mathrm{i}\varphi_2)\ , \tag{15.78}$$

$$D_\mu\varphi=(\partial_\mu-\mathrm{i}eA_\mu)\varphi\ ,\quad D_\mu\varphi^\dagger=(\partial_\mu+\mathrm{i}eA_\mu)\varphi^\dagger\ . \tag{15.79}$$

However, to introduce spontaneous symmetry breaking, it is more useful to use a representation with real scalar fields. The field equations are

$$\partial_\nu F_{\mu\nu}=j_\mu=e(\varphi_2D_\mu\varphi_1-\varphi_1D_\mu\varphi_2)\ , \tag{15.80}$$

$$D_\mu^2\varphi_a+\frac{1}{2}(m_0^2-f^2\varphi_b\varphi_b)\varphi_a=0\ . \tag{15.81}$$

In this example, as in the case of the sigma model, the Lagrangian density has been chosen so that the symmetry is spontaneously broken. Regarding solutions which give an extremum of the effective potential in the lowest order approximation, in addition to $\langle \boldsymbol{A}_\mu \rangle = 0$ and $\langle \boldsymbol{\varphi}_b \rangle = 0$, there is also

$$\langle \boldsymbol{A}_\mu \rangle = 0 \, , \quad \langle \boldsymbol{\varphi}_b \rangle^2 = \frac{m_0^2}{f^2} \equiv \eta^2 \, . \tag{15.82}$$

This corresponds to (15.38).

Note that the Lagrangian density, and therefore the field equations, are invariant under the gauge transformation

$$A_\mu(x) \;\rightarrow\; A_\mu(x) + \frac{1}{e}\partial_\mu \lambda(x) \, , \tag{15.83}$$

$$\begin{pmatrix} \varphi_1(x) \\ \varphi_2(x) \end{pmatrix} \rightarrow \begin{pmatrix} \cos\lambda(x) & \sin\lambda(x) \\ -\sin\lambda(x) & \cos\lambda(x) \end{pmatrix} \begin{pmatrix} \varphi_1(x) \\ \varphi_2(x) \end{pmatrix} . \tag{15.84}$$

If the symmetry is spontaneously broken as in (15.82), this solution is degenerate and its degeneracy is generated by the transformation (15.84), for constant λ, i.e., the solution to (15.82) has the following form which involves an arbitrary parameter α :

$$\langle \boldsymbol{\varphi}_1 \rangle = \eta \cos\alpha \, , \quad \langle \boldsymbol{\varphi}_2 \rangle = \eta \sin\alpha \, . \tag{15.85}$$

In the following discussion, we fix $\alpha = \pi/2$. Therefore, in the chosen ground state,

$$\langle \boldsymbol{A}_\mu \rangle = 0 \, , \quad \langle \boldsymbol{\varphi}_1 \rangle = 0 \, , \quad \langle \boldsymbol{\varphi}_2 \rangle = \eta \, . \tag{15.86}$$

We then expand the field operators around this solution, viz.,

$$\varphi_1 = g \, , \quad \varphi_2 = \eta + h \, , \tag{15.87}$$

and introduce a vector field B_μ with field strength $F_{\mu\nu}$:

$$B_\mu = A_\mu - (e\eta)^{-1}\partial_\mu g \, , \tag{15.88}$$

$$F_{\mu\nu} = \partial_\mu B_\nu - \partial_\nu B_\mu = \partial_\mu A_\nu - \partial_\nu A_\mu \, . \tag{15.89}$$

Neglecting interactions, we write down the field equations in a linear approximation with respect to the differences with the classical solution:

$$(\Box - m_0^2)h = 0 \, , \tag{15.90}$$

$$\partial_\nu F_{\mu\nu} = -m_1^2 B_\mu \, , \quad \partial_\mu B_\mu = 0 \, , \tag{15.91}$$

where m_0 and m_1 can be expressed using η as

$$m_0 = f\eta\,, \quad m_1 = e\eta\,. \tag{15.92}$$

Hence, the scalar field has mass m_0, the vector field B_μ has mass m_1, and the massless particle disappears. Although the original vector field only has transverse wave components and its mass is 0, B_μ absorbs the Nambu–Goldstone field g as a longitudinal wave and it becomes a vector field with finite mass. This is called the *Higgs mechanism*. Because of this, the massless Nambu–Goldstone particle disappears and the gauge quantum obtains mass.

In order to analyze the interactions, we write down the Lagrangian density. The free part, corresponding to (15.90) and (15.91), is

$$\mathscr{L}_f = -\frac{1}{4}F_{\mu\nu}F_{\mu\nu} - \frac{1}{2}m_1^2 B_\mu B_\mu - \frac{1}{2}\left(\partial_\mu h \partial_\mu h + m_0^2 h^2\right)\,. \tag{15.93}$$

The interaction part is

$$\begin{aligned}\mathscr{L}_{\text{int}} = eA_\mu(h\partial_\mu g - g\partial_\mu h) - em_1 h A_\mu A_\mu - \frac{1}{2}f m_0 h(g^2 + h^2) \\ -\frac{1}{2}e^2 A_\mu A_\mu(g^2 + h^2) - \frac{f^2}{8}(g^2 + h^2)^2\,.\end{aligned} \tag{15.94}$$

The commutation relations in the interaction picture can be derived from the free part as

$$\left[B_\mu(x), B_\nu(y)\right] = \mathrm{i}\left(\delta_{\mu\nu} - \frac{\partial_\mu\partial_\nu}{m_1^2}\right)\Delta(x - y; m_1^2)\,, \tag{15.95}$$

$$\left[h(x), h(y)\right] = \mathrm{i}\Delta(x - y; m_0^2)\,. \tag{15.96}$$

This finite-mass scalar field h is called the *Higgs field* and its quantum is called the *Higgs particle*. To quantize the gauge field, we need gauge conditions. We thus need to impose constraints on the gauge field classically or add additional terms in the Lagrangian to determine the gauge. In this section, we adopt the former approach and impose the Coulomb gauge condition

$$\partial \cdot A + (n \cdot \partial)(n \cdot A) = 0\,. \tag{15.97}$$

In this gauge, we can obtain the field g from (15.88):

$$g = -m_1\left[\Box + (n \cdot \partial)^2\right]^{-1}\left[\partial \cdot B + (n \cdot \partial)(n \cdot B)\right]\,. \tag{15.98}$$

Thus, we can also express A_μ in terms of B_μ. Using this,

$$\begin{aligned}
\big[A_\mu(x), A_\nu(y)\big] &= \mathrm{i}\Big\{\delta_{\mu\nu} - \big[(n_\mu\partial_\nu + n_\nu\partial_\mu)(n\cdot\partial) + \partial_\mu\partial_\nu\big]\big[\Box + (n\cdot\partial)^2\big]^{-1}\Big\} \\
&\qquad \times \Delta(x-y; m_1^2)\,, \\
\big[A_\mu(x), g(y)\big] &= \mathrm{i}m_1 n_\mu(n\cdot\partial)\big[\Box + (n\cdot\partial)^2\big]^{-1}\Delta(x-y; m_1^2)\,, \\
\big[g(x), g(y)\big] &= \mathrm{i}(n\cdot\partial)^2\big[\Box + (n\cdot\partial)^2\big]^{-1}\Delta(x-y; m_1^2)\,, \\
\big[B_\mu(x), g(y)\big] &= \mathrm{i}m_1^{-1}\big[n_\mu\Box - (n\cdot\partial)\partial_\mu\big](n\cdot\partial)\big[\Box + (n\cdot\partial)^2\big]^{-1}\Delta(x-y; m_1^2)\,.
\end{aligned} \tag{15.99}$$

The local charge conservation law holds true:

$$\partial_\mu \boldsymbol{j}_\mu = 0\,. \tag{15.100}$$

However, when the $U(1)$ symmetry is broken, it is clear from (15.80) and (15.91) that

$$\boldsymbol{j}_\mu(x) = -m_1^2 \boldsymbol{B}_\mu(x) + \text{non-linear term}\,. \tag{15.101}$$

In this case, as discussed in Sect. 15.1, since a term proportional to $1/\sqrt{V}$ is included in $\boldsymbol{j}_0(x)$, its spatial integral Q diverges when $V \to \infty$:

$$Q = \int \mathrm{d}^3x\, \boldsymbol{j}_0(x) \sim \sqrt{V} \;\to\; \infty\,. \tag{15.102}$$

Moreover, the integral itself is the Fourier transform of $B_0(x)$ and corresponds to the component of $p = 0$, so it has the time-dependence $\exp(\pm \mathrm{i}m_1 t)$. It cannot therefore be a global conserved quantity. This means that the global conservation law is lost.

We shall now check whether the discussion at the beginning of this section is valid, i.e., whether the massless boson disappears under the covariant gauge condition. We compute the following commutator to lowest order:

$$\begin{aligned}
\mathrm{i}\langle \boldsymbol{0}|[\boldsymbol{j}_\mu(x), \boldsymbol{\varphi}_1(y)]|\boldsymbol{0}\rangle &\approx -\mathrm{i}m_1^2\langle 0|[B_\mu(x), g(y)]|0\rangle \\
&= m_1\big[n_\mu\Box - (n\cdot\partial)\partial_\mu\big](n\cdot\partial)\big[\Box + (n\cdot\partial)^2\big]^{-1}\Delta(x-y; m_1^2)\,.
\end{aligned}$$

Its Fourier transform is thus proportional to the quantity

$$\big[(n\cdot k)k_\mu - k^2 n_\mu\big](n\cdot k)\big[k^2 + (n\cdot k)^2\big]^{-1}\epsilon(k_0)\delta(k^2 + m_1^2)\,. \tag{15.103}$$

Comparing this with the general form (15.69) given at the beginning of this section, this term corresponds to the case where we only have ρ_5, while ρ_4 disappears. This is therefore a counterexample to the Nambu–Goldstone theorem (15.70). If the symmetry is not broken spontaneously, $\eta \to 0$ and $m_1 = e\eta \to 0$. In this case, (15.103) once more has the form

$$k_\mu \epsilon\big(k_0 \delta(k^2)\big) , \tag{15.104}$$

whence the Nambu–Goldstone theorem is valid again, and g becomes a massless field.

In the treatment above, the explicit Lorentz covariance of the theory has been lost. Additionally, an extra field g has come in. We shall now present a formalism which conserves the Lorentz covariance and does not include any extra field. Higgs applied a transformation from the Cartesian coordinate system to the polar coordinate system to the field operators:

$$\varphi_1 = R\cos\Theta , \quad \varphi_2 = R\sin\Theta , \quad A_\mu = B_\mu - \frac{1}{e}\partial_\mu \Theta . \tag{15.105}$$

The gauge transformation is

$$\Theta(x) \to \Theta(x) - \lambda(x) . \tag{15.106}$$

With this transformation, the Lagrangian can be written in the form

$$\mathscr{L} = -\frac{1}{4}F_{\mu\nu}F_{\mu\nu} - \frac{1}{2}\mathrm{e}^2 B_\mu^2 R^2 - \frac{1}{2}\partial_\mu R \partial_\mu R + \frac{1}{4}m_0^2 R^2 - \frac{1}{8}f^2 R^4 . \tag{15.107}$$

The extra degree of freedom has clearly been eliminated. R is the Higgs field, and the solution which spontaneously breaks the symmetry is

$$\langle \boldsymbol{B}_\mu \rangle = 0 , \quad \langle \boldsymbol{R} \rangle = \eta . \tag{15.108}$$

These show that $\boldsymbol{B}_\mu$ once again becomes a vector field with mass m_1, while $\boldsymbol{R}$ becomes a scalar field with mass m_0. Obviously, the massless Nambu–Goldstone particle disappears completely. Although this proof seems to be very clear, it is quite classical. Quantum-mechanically, the transformation (15.105) is quite non-linear and has a strong specificity. Additionally, in the form (15.107), we cannot apply the renormalization theory, and computations of higher-order corrections become impossible. Thus, although we have lots of gauge choices, every case has both advantages and disadvantages. Since an indefinite metric does not appear in the gauge mentioned here, this is called the *unitarity gauge*.

15.5 Higgs Mechanism with Covariant Gauge Condition

In the last section, we saw that the Nambu–Goldstone particle does not appear under a non-covariant gauge condition. If the physical contents of the gauge theory do not depend on how we choose the gauge, then under the covariant gauge condition, it turns out that the appearance of this massless particle cannot be observed, even though it is inevitable, as proven in Sect. 15.1. So, by what kind of mechanism is this particle concealed? Finding a solution to this problem will be the main theme of this section.

In order to discuss this issue, we begin with a formal discussion. It sometimes happens that the problem is simplified in the gauge theory by choosing the Landau gauge ($\alpha = 0$). However, the Lagrangian density (5.121) diverges if we take $\alpha = 0$. Hence, following Utiyama and Nakanishi, we introduce an auxiliary field B, and rewrite (5.121) in the form

$$\mathscr{L}_{\text{em}} = -\frac{1}{4}F_{\mu\nu}F_{\mu\nu} - \partial_\mu B A_\mu + \frac{\alpha}{2}B^2 . \tag{15.109}$$

Varying B, we obtain

$$\partial_\mu A_\mu = \alpha B , \tag{15.110}$$

and inserting this into (15.109), we see that it is equivalent to the original (5.121). Taking this form, there is no confusion if we choose the Landau gauge and set $\alpha = 0$.

Although in the last section, in order to quantize the vector field, we eliminated the extra components by imposing the Coulomb gauge condition (15.97) in the classical theory, in the present section, we shall add a further term called a *gauge-fixing term* to (15.109) and quantize in a covariant way. Therefore, the free part of the Lagrangian density (15.93) is

$$\mathscr{L}_{\text{f}} = \mathscr{L}_1 + \mathscr{L}_2 , \tag{15.111}$$

$$\mathscr{L}_1 = -\frac{1}{4}F_{\mu\nu}F_{\mu\nu} - \frac{1}{2}(m_1 A_\mu - \partial_\mu g)(m_1 A_\mu - \partial_\mu g) + B\partial_\mu A_\mu , \tag{15.112}$$

$$\mathscr{L}_2 = -\frac{1}{2}(\partial_\mu h \partial_\mu h + m_0^2 h^2) , \tag{15.113}$$

and the interaction part is again given by (15.94). Hence, quantizing the independent fields A_μ, B, g, and h appearing in the free part, in the interaction picture, and expressing each of them in terms of the irreducible fields, we have from (15.112),

$$A_\mu^{\text{in}} = U_\mu^{\text{in}} - \frac{1}{m_1}\partial_\mu g_{(-)}^{\text{in}} , \quad B^{\text{in}} = m_1\big[g_{(+)}^{\text{in}} + g_{(-)}^{\text{in}}\big] , \quad g^{\text{in}} = g_{(+)}^{\text{in}} , \tag{15.114}$$

where h is already irreducible. These asymptotic fields have the properties

$$(\Box - m_1^2)H_\mu^{\text{in}} = 0\,, \quad \partial_\mu U_\mu^{\text{in}} = 0\,,$$

$$\langle 0|T^*[U_\mu^{\text{in}}(x), U_\nu^{\text{in}}(y)]|0\rangle = \left(\delta_{\mu\nu} - \frac{\partial_\mu\partial_\nu}{m_1^2}\right)\Delta_{\text{F}}(x-y;m_1^2)\,. \tag{15.115}$$

This means that the asymptotic field U_μ^{in} has mass m_1 and is a spin-1 field. Moreover, it is clear from

$$\Box g_{(+)}^{\text{in}} = \Box g_{(-)}^{\text{in}} = 0\,,$$

$$\langle 0|T[g_{(\pm)}^{\text{in}}(x), g_{(\pm)}^{\text{in}}(y)]|0\rangle = \pm D_{\text{F}}(x-y)\,, \tag{15.116}$$

that the asymptotic fields $g_{(\pm)}^{\text{in}}$ are the fields of the Nambu–Goldstone particles, and are a mixture of the one with the positive-definite metric and the one with the indefinite metric. Finally, the asymptotic field h^{in} satisfies

$$(\Box - m_0^2)h^{\text{in}} = 0\,, \quad \langle 0|T[h^{\text{in}}(x), h^{\text{in}}(y)]|0\rangle = \Delta_{\text{F}}(x-y;m_0^2)\,. \tag{15.117}$$

This corresponds to the Higgs particle.

What is important here is that

$$\Box \boldsymbol{B} = 0\,, \tag{15.118}$$

as a consequence of the equations of motion, as in the case of QED. Therefore, we can make the identification

$$\boldsymbol{B}(x) = B^{\text{in}}(x) = B^{\text{out}}(x)\,. \tag{15.119}$$

In this case, as in the case of QED, $\boldsymbol{B}$ commutes with the S-matrix:

$$[S, \boldsymbol{B}(x)] = 0\,. \tag{15.120}$$

Since this theory has an indefinite metric, following the argument in Chap. 5, we introduce the physical state by the subsidiary condition

$$B^{(+)}(x)|\text{phys}\rangle = 0\,. \tag{15.121}$$

At different space-time points, there is no contradiction here because

$$[\boldsymbol{B}(x), \boldsymbol{B}(y)] = 0\,. \tag{15.122}$$

To make a physical state from a different physical state including the vacuum, we have to act with operators that commute with $\boldsymbol{B}$, which leaves only

$$U_\mu^{\text{in}}, \quad h^{\text{in}}, \quad \boldsymbol{B}. \tag{15.123}$$

If we act with $\boldsymbol{B}$, a zero norm state is created. After all, considering the first two operators in (15.123), only the positive-definite and finite-mass particles survive. This is the Higgs mechanism which conceals the Nambu–Goldstone particle.

Although we have discussed the non-renormalized interaction picture here, the situation is essentially the same as in the renormalized interaction picture. Such an argument has been given by Nakanishi [148], but it has been slightly modified by the author here.

So far in this chapter, we have considered only field operators whose vacuum expectation values do not disappear, such as (15.1). Basically, if (15.10) holds, the Nambu–Goldstone theorem holds for $\boldsymbol{\varphi}$, whether it be the field operator itself or one of its local polynomials. If (15.1) does not hold for the field operator itself, but holds for a local polynomial, we call this *dynamical symmetry breaking*. In this case, there are some examples where the Higgs mechanism works. It should be noted that, in 1963, Anderson made the same argument as Higgs for plasma oscillations [149].

15.6 Kibble's Theorem

In the last section, we discussed the gauge theory corresponding to the Abelian group $U(1)$. We will now present Kibble's theorem, which is an extension of the Higgs theorem for the Abelian group to the case of a non-Abelian group.

To investigate the Higgs mechanism, we first considered ways to set the gauge condition within the classical theory and to apply the unitarity gauge. This is more like a classical theoretical method. In contrast, in the last section, we quantized the field in the covariant gauge, and introducing an additional condition, we eliminated the indefinite metric. From the point of view of quantum field theory, the latter method is preferable, but it is then very difficult to quantize the non-Abelian gauge theory in the covariant gauge, something we shall discuss in a later chapter. Therefore, in this section we apply the former method and accept the limitations of a quasi-classical treatment. Here we present the demonstration due to Kibble [150], in which the unitarity gauge is introduced for the non-Abelian gauge field.

15.6.1 Adjoint Representation

An infinitesimal transformation of field operators for a given group G is given by (15.2). In the following, distinguishing upper indices from lower indices, we shall write

$$\delta\varphi^A = \epsilon^a T^A_{aB}\varphi^B \ . \tag{15.124}$$

We also define the structure constants f^c_{ab} by

$$[T_a, T_b]^A{}_B = f^c_{ab} T^A_{cB} \ . \tag{15.125}$$

Thus, from the anti-commutativity of commutators and the Jacobi identity, the structure constants should satisfy

$$f^c_{ab} = -f^c_{ba} \ , \tag{15.126}$$

$$f^m_{ab} f^l_{mc} + f^m_{bc} f^l_{ma} + f^m_{ca} f^l_{mb} = 0 \ . \tag{15.127}$$

We introduce the matrix M by

$$(M_c)^a{}_b = f^a_{cb} \ . \tag{15.128}$$

Therefore, from (15.127)

$$[M_a, M_b]^l{}_c = f^m_{ab} (M_m)^l{}_c \ . \tag{15.129}$$

Hence, M is a representation of the algebra corresponding to the group G, called the *adjoint representation*. To prove Kibble's theorem, we need to know certain things about this representation, so we list some relevant formulas:

1. Taking λ as infinitesimal and writing $\lambda^a T_a$ simply as λT,

$$(1 + \lambda T) T_d (1 - \lambda T) = T_b (1 + \lambda M)^b{}_d \ . \tag{15.130}$$

   This is because the coefficient of λ^c on the left-hand side is

$$[T_c, T_d] = f^b_{cd} T_b = T_b (M_c)^b{}_d \ ,$$

   which is equal to the coefficient of λ^c on the right-hand side. Thus, if λ is finite,

$$\mathrm{e}^{\lambda T} T_d \mathrm{e}^{-\lambda T} = T_b (\mathrm{e}^{\lambda M})^b{}_d \ . \tag{15.131}$$

2. Next, using Feynman's formula (without proof),

$$\begin{aligned}\partial_\mu \mathrm{e}^{\theta T} &= \int_0^1 \mathrm{d}\alpha\, \mathrm{e}^{(1-\alpha)\theta T} T_d \mathrm{e}^{\alpha\theta T} \partial_\mu \theta^d \\ &= \mathrm{e}^{\theta T} \int_0^1 \mathrm{d}\alpha\, \mathrm{e}^{-\alpha\theta T} T_d \mathrm{e}^{\alpha\theta T} \partial_\mu \theta^d \\ &= \mathrm{e}^{\theta T} T_d \left(\int_0^1 \mathrm{d}\alpha\, \mathrm{e}^{-\alpha\theta M} \right)^b{}_d \partial_\mu \theta^d \\ &= \mathrm{e}^{\theta T} T_d \Lambda^b{}_d \partial_\mu \theta^d \,, \end{aligned} \tag{15.132}$$

where

$$\Lambda = \frac{1 - \mathrm{e}^{-\theta M}}{\theta M} \,. \tag{15.133}$$

We now define the covariant derivative by

$$D_\mu = \partial_\mu - \mathscr{A}_\mu = \partial_\mu - T A_\mu = \partial_\mu - T_a A^a_\mu \,. \tag{15.134}$$

Note that the coupling constant has been included in T, and therefore in the structure constants and also M, and $\mathscr{A}$ is equal to what we wrote as $\mathrm{i}g\mathscr{A}_\mu$ in our previous notation. We can now implement the gauge transformation:

$$\mathscr{A}_\mu(x) \;\rightarrow\; U\big(\lambda(x)\big)\mathscr{A}_\mu(x)U^{-1}\big(\lambda(x)\big) - U\big(\lambda(x)\big)\partial_\mu U^{-1}\big(\lambda(x)\big) \,. \tag{15.135}$$

In the following, replacing $\exp(\theta T)$ by $U\big(\lambda(x)\big)$, the *inverse* of the transformation above can be written as

$$\begin{aligned}\mathscr{A}_\mu &= \mathrm{e}^{\theta T} T_a A'^a_\mu \mathrm{e}^{-\theta T} - \mathrm{e}^{\theta T} \partial_\mu \mathrm{e}^{-\theta T} \\ &= T_b (\mathrm{e}^{\theta M} A'_\mu)^b - T_b \left(\frac{1 - \mathrm{e}^{\theta M}}{\theta M} \partial_\mu \theta \right)^b \\ &= T_b \big(\mathrm{e}^{\theta M} A'_\mu + \mathrm{e}^{\theta M} \Lambda \partial_\mu \theta \big)^b \,. \end{aligned} \tag{15.136}$$

Thus,

$$A^A_\mu = (\mathrm{e}^{\theta M})^A{}_B A'^B_\mu + (\mathrm{e}^{\theta M} \Lambda)^A{}_b \partial_\mu \theta^b \,. \tag{15.137}$$

In the equation above, we have distinguished capital letters and small letters, and we will discuss this difference later.

15.6.2 Kibble's Theorem

The Lagrangian of the scalar field which is invariant under a group G is written in the form

$$\begin{aligned}\mathscr{L} &= \frac{1}{2}(\varphi_\mu - \partial_\mu\varphi)^2 - \frac{1}{2}(\partial_\mu\varphi)^2 - U(\varphi) \\ &= -\varphi_\mu\partial_\mu\varphi + \frac{1}{2}\varphi_\mu\varphi_\mu - U(\varphi)\ , \end{aligned} \tag{15.138}$$

without specifying the components explicitly.[1] Writing it in this form, the Lagrangian is linear in the derivatives, and the quantity canonically conjugate to φ is $-\varphi_0$. Using the effective potential $V(\varphi)$, the formula which determines the spontaneous symmetry breaking is

$$\frac{\partial V(\varphi)}{\partial\varphi} = 0\ . \tag{15.139}$$

To lowest order, $V(\varphi)$ can be replaced by $U(\varphi)$. We assume that one of the solutions is

$$\varphi = \eta\ . \tag{15.140}$$

Therefore, in general, $e^{\lambda T}\eta$ is also a solution and the solutions are degenerate. However, λ is a constant. If we write the subgroup of G which keeps η invariant as G_η, i.e.,

$$e^{\lambda T}\eta = \eta\ , \tag{15.141}$$

then it turns out that the symmetry corresponding to G_η is not broken for this solution. The number of invariants which are algebraically independent and constructed by η is denoted ν, which we call the *canonical number*. For instance, for $O(3)$, the invariant

$$\boldsymbol{\eta}^2 = \eta_1^2 + \eta_2^2 + \eta_3^2$$

is the unique invariant for the three-dimensional η and in this case $\nu = 1$. We now assume that T is an n-dimensional representation and therefore that both φ and η are n-dimensional vectors. In this case, we choose the canonical form such that only ν components of η are non-zero. That is, we eliminate the other components by a transformation in G. For instance, in the Higgs model, $G = U(1)$, $n = 2$, and

[1] In this section, the notation ϕ_μ means $\partial_\mu\phi$, following the original article by T. Kibble [150].

$\nu = 1$. We consider the manifold constructed by the set of solutions, viz.,

$$\left\{ e^{\lambda T} \eta | e^{\lambda T} \in G \right\} . \tag{15.142}$$

We shall now determine its dimension.

Although η is n-dimensional, since the values of the ν invariants made from η cannot be changed, the number of parameters r which can be chosen freely is given by $r = n - \nu$. Therefore, the dimension of the above set is r. Note also that the number of components of λ is equal to the number of the generators T, so that it becomes the order of g. Since r parameters among them play the role of designating η, it turns out that $g - r$ do not change η. Thus, this is also the dimension of G_η.

In the following, fixing the form of η, the general form of the vacuum expectation value that solves (15.139) is

$$\varphi = \underset{(r)}{e^{\mu T}} \underset{(g-r)}{e^{\nu T}} = \eta = e^{\mu T} , \qquad e^{\nu T} \in G_\eta , \tag{15.143}$$

where the dimensions are indicated under the two factors. Introducing the angular variable θ and the radial variable ρ, as a generalization of the unitarity gauge in the Higgs model, we take

$$\varphi = e^{\theta T} \rho = e^{\theta T} (\eta + \rho') , \tag{15.144}$$

where θ is r-dimensional. For the canonical form, we have

$$\theta = \begin{pmatrix} \theta_1 \\ \vdots \\ \theta_r \\ 0 \\ \vdots \\ 0 \end{pmatrix} , \qquad \rho = \begin{pmatrix} 0 \\ \vdots \\ 0 \\ \rho_1 \\ \vdots \\ \rho_\nu \end{pmatrix} , \tag{15.145}$$

where the total number of components of θ is g and $r = n - \nu$, while ρ has altogether n components, of which the first r are equal to zero.

For the generators T_A , $A = 1, 2, \dots, g - r$, of G_η, we assume that ρ satisfies

$$T_A \rho = 0 , \tag{15.146}$$

while for the other T_a , $a = 1, 2, \dots, r$,

$$T_a \rho \neq 0 . \tag{15.147}$$

In the following, we express the r dimensions of θ by small Roman letters and ν dimensions of ρ by Greek letters. We define the r-dimensional matrix X by

$$X_a{}^b = (T_a\rho)^b = T^b_{a\alpha}\rho^\alpha \,. \tag{15.148}$$

This matrix is invertible, because if $\det X = 0$, then there exist c^a satisfying

$$c^a T_a \rho = 0 \,,$$

whence $c^a T_a$ becomes a generator of G_η, which contradicts our assumptions.

We now rewrite φ in the Lagrangian density in terms of θ and ρ using (15.144). With (15.132), we obtain

$$\partial_\mu \varphi^A = (\mathrm{e}^{\theta T} T_b \rho)^A \Lambda^b{}_d \partial_\mu \theta^d + (\mathrm{e}^{\theta T})^A{}_\beta \partial_\mu \rho^\beta \,, \tag{15.149}$$

$$\varphi_\mu \partial_\mu \varphi = (\varphi_\mu \mathrm{e}^{\theta T} T_b \rho) \Lambda^b{}_d \partial_\mu \theta^d + \varphi_\mu \mathrm{e}^{\theta T} \partial_\mu \rho \,. \tag{15.150}$$

We denote the coefficients of the derivatives on the right-hand side of (15.150) as follows:

$$\theta_{\mu d} \equiv \left(\varphi_\mu \mathrm{e}^{\theta T} T_b \rho\right) \Lambda^b{}_d \,, \quad \rho_{\mu\alpha} \equiv \varphi_{\mu A} \left(\mathrm{e}^{\theta T}\right)^A{}_\alpha \,. \tag{15.151}$$

Now φ has n components, but ρ has only ν components. Thus,

$$\rho_{\mu\alpha} \rho_\mu{}^\alpha = \varphi_{\mu A} \varphi_\mu{}^A - \varphi_{\mu A} \left(\mathrm{e}^{\theta T}\right)^A{}_a \left(\mathrm{e}^{-\theta T}\right)^A{}_a \left(\mathrm{e}^{-\theta T}\right)^a{}_B \varphi_\mu{}^B \,. \tag{15.152}$$

However,

$$\begin{aligned} \theta_{\mu d} &= \varphi_{\mu A} \left(\mathrm{e}^{\theta T}\right)^A{}_\alpha T^\alpha_{b\beta} \rho^\beta \Lambda^b{}_d + \varphi_{\mu A} \left(\mathrm{e}^{\theta T}\right)^A{}_a T^a_{b\beta} \rho^\beta \Lambda^b{}_d \\ &= \rho_{\mu\alpha} T^\alpha_{b\beta} \rho^\beta \Lambda^b{}_d + \varphi_{\mu A} \left(\mathrm{e}^{\theta T}\right)^A{}_a X_b{}^a \Lambda^b{}_d \,. \end{aligned} \tag{15.153}$$

Therefore,

$$\theta_{\mu d} \left(\Lambda^{-1} X^{-1}\right)^d{}_a = \left(\rho_\mu T_b \rho\right) \left(X^{-1}\right)^b{}_a + \varphi_{\mu A} \left(\mathrm{e}^{\theta T}\right)^A{}_a \,. \tag{15.154}$$

If we express the difference between $\rho\rho$ and $\varphi\varphi$ in (15.152) using (15.154),

$$\begin{aligned} \mathscr{L} = -\rho_\mu \partial_\mu \rho + \frac{1}{2} \rho_\mu \rho_\mu - U(\rho) - \theta_\mu \partial_\mu \theta \\ + \frac{1}{2} \left\{ \left[\theta_{\mu a} (\Lambda^{-1})^a{}_b - \rho^\mu T_b \rho \right] (X^{-1})^b{}_c \right\}^2 \,. \end{aligned} \tag{15.155}$$

In the Lagrangian density above, θ always appears together with θ_μ and never alone, so we see that θ becomes a massless Nambu–Goldstone particle. Now, ρ and ρ_μ are gauge invariant, and the gauge transformations of θ and θ_μ will be derived below.

From (15.144), φ and $\mathrm{e}^{\theta T}$ transform in the same way under the gauge transformation, so considering δ as the gauge transformation corresponding to the infinitesimal gauge function $\delta\lambda^A$, we have

$$\delta\left(\mathrm{e}^{\theta T}\right) = T_A \delta\lambda^A \mathrm{e}^{\theta T} \ . \tag{15.156}$$

Therefore, from (15.132),

$$\delta\left(\mathrm{e}^{\theta T}\right) = \mathrm{e}^{\theta T} T_b {\Lambda^b}_d \delta\theta^d \ . \tag{15.157}$$

Combining the two equations above,

$$T_b {\Lambda^b}_d \delta\theta^d = \mathrm{e}^{-\theta T} T_A \mathrm{e}^{\theta T} \delta\lambda^A = T_b \left(\mathrm{e}^{-\theta M}\right)^b{}_A \delta\lambda^A \ .$$

Therefore,

$$\delta\theta^a = \left(\Lambda^{-1}\right)^a{}_b \left(\mathrm{e}^{-\theta M}\right)^b{}_A \delta\lambda^A \ . \tag{15.158}$$

Although the gauge transformation for $\theta_{\mu a}$ is not necessary for the following discussion, we shall give it for the record. Because θ^a and $-\theta_{0a}$ are canonical conjugates,

$$\delta\left[\theta_{0a}, \theta^b\right] = \left[\delta\theta_{0a}, \theta^b\right] + \left[\theta_{0a}, \delta\theta^b\right] .$$

Alternatively,

$$\mathrm{i}\frac{\partial}{\partial\theta_{0b}}\delta\theta_{0a} + \mathrm{i}\frac{\partial}{\partial\theta^a}\delta\theta^b = 0 \ , \qquad \delta\theta_{\mu a} = -\theta_{\mu b}\frac{\partial}{\partial\theta^a}\delta\theta^b \ . \tag{15.159}$$

We now introduce a gauge field into the Lagrangian density in polar form. In this case, the covariant derivative is obtained by the replacement

$$D_\mu\varphi = \partial_\mu\varphi - \delta\varphi(\lambda \to A_\mu) \ . \tag{15.160}$$

In other words, we replace the infinitesimal gauge function λ by A_μ to obtain the second term on the right-hand side. As already noted, the coupling constant is included in the structure constant. So writing

$$f^A_{BC} = -g f_{ABC} \ , \tag{15.161}$$

the Lagrangian density including the gauge field is

$$\begin{aligned}\mathscr{L} = &-\frac{1}{2}F_{\mu\nu,A}\big(\partial_\mu A_\nu^A - \partial_\nu A_\mu^A + g f_{ABC} A_\mu^B A_\nu^C\big) + \frac{1}{4}F_{\mu\nu}F_{\mu\nu} \\ &-\rho_\mu \partial_\mu \rho + \frac{1}{2}\rho_\mu \rho_\mu - U(\rho) \\ &+\frac{1}{2}\Big\{\big[\theta_{\mu a}(\Lambda^{-1})^a{}_b - \rho_\mu T_b \rho\big](X^{-1})^b{}_c\Big\}^2 \\ &-\theta_\mu\big(\partial_\mu \theta - \Lambda^{-1}\mathrm{e}^{-\theta M} A_\mu\big)\,, \end{aligned} \tag{15.162}$$

where the last term comes from introducing the covariant derivative for θ by the replacement (15.160).

Now, from the gauge transformation (15.137), we express the gauge field A_μ in terms of a new gauge field A'_μ:

$$A_\mu^A = \big(\mathrm{e}^{\theta M}\big)^A{}_B A'^B_\mu + \big(\mathrm{e}^{\theta M}\Lambda\big)^A{}_b \partial_\mu \theta^b\,, \quad F = \mathrm{e}^{\theta M} F'\,. \tag{15.163}$$

Making this replacement, the last term in (15.162) becomes

$$\partial_\mu \theta - \Lambda^{-1}\mathrm{e}^{-\theta M} A_\mu = \partial_\mu \theta - \Lambda^{-1}\mathrm{e}^{-\theta M}\mathrm{e}^{\theta M}(A'_\mu + \Lambda \partial_\mu \theta) = -\Lambda^{-1} A'_\mu\,, \tag{15.164}$$

whence the term in $\partial_\mu \theta$ vanishes! This is exactly the same as in the Higgs mechanism. As a result, (15.162) becomes

$$\begin{aligned}\mathscr{L} = &-\frac{1}{2}F'_{\mu\nu,A}\big(\partial_\mu A'^A_\nu - \partial_\nu A'^A_\mu + g f_{ABC} A'^B_\mu A'^C_\nu\big) + \frac{1}{4}F'_{\mu\nu}F'_{\mu\nu} \\ &-\rho_\mu \partial_\mu \rho + \frac{1}{2}\rho_\mu \rho_\mu - U(\rho) \\ &+\frac{1}{2}\Big\{\big[\theta_{\mu a}(\Lambda^{-1})^a{}_b - \rho_\mu T_b \rho\big](X^{-1})^b{}_c\Big\}^2 + \theta_\mu \Lambda^{-1} A'_\mu\,. \end{aligned} \tag{15.165}$$

Since $\partial_\mu \theta$ does not appear here, θ_μ is expressed algebraically in terms of the other quantities. If we now set

$$\theta_{\mu a}\big(\Lambda^{-1}X^{-1}\big)^a{}_b \equiv h_{\mu b}\,, \quad \rho_\mu T_a \rho \big(X^{-1}\big)^a{}_b \equiv k_{\mu b}\,, \tag{15.166}$$

then only $\theta_{\mu a}$ involves $h_{\mu a}$, and the sum of the last two terms in (15.165) becomes

$$\frac{1}{2}\big(h_\mu - k_\mu\big)^2 + h_\mu X A'_\mu = \frac{1}{2}\big(h_\mu - k_\mu + X A'_\mu\big)^2 + k_\mu X A'_\mu - \frac{1}{2}\big(X A'_\mu\big)^2\,.$$

Hence, considering h_μ as an independent variable and using the variation principle, the first term on the right-hand side in the above equation can be dropped. Thus, the final Lagrangian density takes the form

$$\mathscr{L} = -\frac{1}{2}F'_{\mu\nu,A}\left(\partial_\mu A'^A_\nu - \partial_\nu A'^A_\mu + g f_{ABC} A'^B_\mu A'^C_\nu\right) + \frac{1}{4}F'_{\mu\nu}F'_{\mu\nu}$$
$$-\rho_\mu \partial_\mu \rho + \frac{1}{2}\rho_\mu \rho_\mu - U(\rho) + (\rho_\mu T_a \rho) A'^a_\mu - \frac{1}{2}\left(X^a{}_b A'^b_\mu\right)^2 . \quad (15.167)$$

This means that θ has been completely eliminated and the scalar field is described solely by ρ. As in the case of the Abelian Higgs model, it turns out that the unitarity gauge has been realized.

Although the number of θ components corresponding to Nambu–Goldstone particles was r, they were absorbed into the massless gauge particles as longitudinal modes and gave finite masses to the gauge particles. Moreover, ρ became ν scalar particles with finite masses, the so-called *Higgs particles*. It turns out that massless gauge particles are left, corresponding to $g - r$ generators for G_η. We summarize this as follows:

ρ (ν components)	Higgs particles with finite masses
θ (r components)	gauge particles with finite masses
A (r components)	gauge particles with finite masses
A ($g - r$ components)	massless gauge particles

The above argument is the proof of Kibble's theorem, which is an extension of Higgs' theorem to non-Abelian gauge fields. As an example, Kibble discussed the model in which three-component complex scalar fields are introduced for an $SU(2) \times U(1)$ gauge group. In the Weinberg–Salam model, the three-component fields are replaced by two-component complex scalar fields. This is the subject of the next chapter.

Chapter 16
Weinberg–Salam Model

As discussed in Chap. 13, the weak interaction is described phenomenologically by Fermi's universal four-fermion theory. One property of this interaction is that it can be expressed as a product of the current densities which form the Lie algebra. However, since this theory is not renormalizable, the cross-section increases without limit at high energies and also in higher order corrections, i.e., the higher the order, the greater the degree of divergence. Thus, we cannot avoid interpreting Fermi's theory as purely phenomenological, expressing the S-matrix element only for low energies and only at the lowest order. Many have tried to interpret Fermi's interaction, by introducing intermediate vector fields, as an iteration of Yukawa interactions. The vector field is necessary in order to reproduce a (V − A)-type interaction, but the theory is still not renormalizable. This led to the model by Weinberg and Salam, combining gauge fields with the Higgs–Kibble mechanism and describing the weak interaction and the electromagnetic interaction in a single theory. These authors proposed the model in [151] and in [152], just after the publication of Kibble's work. In fact, as explained at the end of the last section, Kibble himself considered almost the same model as this.

16.1 Weinberg–Salam Model

What kind of Lie algebra is needed to describe the weak interaction according to a gauge principle? As mentioned before, the spatial integrations over time-components of the currents J_μ and $J_\mu^\dagger$ carrying charges appeared in the weak interaction from the algebra $SU(2)_{\mathrm{L}}$. Thus, we need to introduce the gauge group $SU(2)$. Moreover, we would like to introduce the electromagnetic field as a gauge field in order to describe the weak interaction and the electromagnetic interaction

K. Nishijima, *Quantum Field Theory*,
https://doi.org/10.1007/978-94-024-2190-3_16

simultaneously, so we need to start with the gauge group

$$SU(2) \times U(1) . \tag{16.1}$$

Recalling the formula,

$$Q = e\left(I_3 + \frac{Y}{2}\right) , \tag{16.2}$$

we consider $I_i, i = 1, 2, 3$, and Y in this equation as the generators for the algebra of the group (16.1). Here, I_i is the isospin. Originally, the isospin was considered only for hadrons, but this type of isospin was defined for both hadrons and leptons. This is obvious from (13.188), which expresses the universality of the weak interaction. Hence, the quantities I_i are called *weak isospin*.

The number of generators in (16.1) is $g = 3 + 1$. If we choose the Higgs field φ to be a two-component complex scalar field, then $n = 4$, and in this case there is only one invariant, viz.,

$$\varphi_1^\dagger \varphi_1 + \varphi_2^\dagger \varphi_2 , \tag{16.3}$$

so $\nu = 1$. Therefore, the number of vector particles with finite masses is $r = n - \nu = 4 - 1 = 3$, while $g - r = 1$, so there is one massless gauge boson, i.e., the photon field. We write the gauge fields corresponding to $SU(2)$ and $U(1)$ as $\hat{W}_\mu$ and B_μ, respectively. Therefore, considering that the Higgs field φ has weak isospin and weak hypercharge, the Lagrangian density for this system is

$$\begin{aligned}\mathscr{L} = &-\frac{1}{4}\left(\partial_\mu \hat{W}_\nu - \partial_\nu \hat{W}_\mu + g\hat{W}_\mu \times \hat{W}_\nu\right)^2 - \frac{1}{4}\left(\partial_\mu B_\nu - \partial_\nu B_\mu\right)^2 \\ &-\frac{1}{2}\varphi^\dagger\left(\overleftarrow{\partial}_\mu + \mathrm{i}g\frac{\hat{\tau}}{2}\hat{W}_\mu - \mathrm{i}\frac{g'}{2}B_\mu\right)\left(\overrightarrow{\partial}_\mu + \mathrm{i}g\frac{\hat{\tau}}{2}\hat{W}_\mu - \mathrm{i}\frac{g'}{2}B_\mu\right)\varphi \\ &-U(\varphi) ,\end{aligned} \tag{16.4}$$

where we have assumed that both the transformation property and the electric charge of φ are the same as those of the K-meson doublet (K^+, K^0). Following Kibble's theory, to express the four components of φ using $\theta_i,\ i = 1, 2, 3$, and ρ, we write

$$\varphi = \mathrm{e}^{\mathrm{i}\hat{\theta}\hat{\tau}/2}\begin{pmatrix}0\\ \rho\end{pmatrix} . \tag{16.5}$$

As mentioned earlier, $\hat{\theta}$ is absorbed to give masses to the intermediate gauge particles. However, in the approximation which neglects the effects of ρ, we must make the replacement

$$\varphi \to \begin{pmatrix} 0 \\ \eta \end{pmatrix} . \tag{16.6}$$

Then, setting $\rho = \eta + \rho'$ and rewriting (16.4), in the unitarity gauge,

$$\begin{aligned} \mathscr{L}_{\mathrm{U}} = & -\frac{1}{4}\left(\partial_\mu \hat{W}_\nu - \partial_\nu \hat{W}_\mu + g\hat{W}_\mu \times \hat{W}_\nu\right)^2 - \frac{1}{4}(\partial_\mu B_\nu - \partial_\nu B_\mu)^2 \\ & -\frac{1}{4}g^2\eta^2 W_\mu^\dagger W_\mu - \frac{1}{8}\eta^2(gW_\mu^3 + g'B_\mu)^2 - U(\eta) \\ & + \text{terms containing } \rho' , \end{aligned} \tag{16.7}$$

where W_μ and $W_\mu^\dagger$ denote the charged intermediate gauge fields, and W_μ^3 denotes a neutral intermediate gauge field. Diagonalizing the mass terms, we define the combinations

$$Z_\mu = \frac{gW_\mu^3 + g'B_\mu}{\sqrt{g^2 + g'^2}} , \quad m_{\mathrm{Z}}^2 = \frac{1}{4}\eta^2(g^2 + g'^2) , \tag{16.8}$$

$$W_\mu = \frac{W_\mu^1 - \mathrm{i}W_\mu^2}{\sqrt{2}} , \quad m_{\mathrm{W}}^2 = \frac{1}{4}\eta^2 g^2 . \tag{16.9}$$

Moreover, the combination orthogonal to Z_μ is the massless photon field:

$$A_\mu = \frac{-g'W_\mu^3 + gB_\mu}{\sqrt{g^2 + g'^2}} . \tag{16.10}$$

Thus,

$$W_\mu^3 = \frac{gZ_\mu - g'A_\mu}{\sqrt{g^2 + g'^2}} , \quad B_\mu = \frac{g'Z_\mu + gA_\mu}{\sqrt{g^2 + g'^2}} . \tag{16.11}$$

Using the expressions for Z_μ and A_μ, we rewrite $\mathscr{L}_{\mathrm{U}}$ in the form

$$\begin{aligned} \mathscr{L}_{\mathrm{U}} = & -\frac{1}{2}W_{\mu\nu}^\dagger W_{\mu\nu} - m_{\mathrm{W}}^2 W_\mu^\dagger W_\mu - \frac{1}{4}A_{\mu\nu}A_{\mu\nu} \\ & -\frac{1}{4}Z_{\mu\nu}Z_{\mu\nu} - \frac{1}{2}m_{\mathrm{Z}}^2 Z_\mu Z_\mu \\ & \left.\begin{aligned} & -\mathrm{i}g\sin\theta_{\mathrm{W}} A_\mu (W_{\mu\nu}^\dagger W_\nu - W_{\mu\nu} W_\nu^\dagger - W_{\mu\nu} W_\nu^\dagger) \\ & +g^2\sin^2\theta_{\mathrm{W}} (A_\mu A_\nu W_\mu^\dagger W_\nu - A_\mu A_\nu W_\nu^\dagger W_\nu) \end{aligned}\right\} \quad \text{(Dirac-type EM interaction)} \\ & +\mathrm{i}g\sin\theta_{\mathrm{W}} A_{\mu\nu} W_\mu^\dagger W_\nu \quad \text{(anomalous magnetic moment)} \end{aligned}$$

$$
\begin{aligned}
&+ig\cos\theta_W Z_\mu(W_{\mu\nu}^\dagger W_\nu - W_{\mu\nu}W_\nu^\dagger) \\
&+g^2\cos^2\theta_W(Z_\mu Z_\nu W_\mu^\dagger W_\nu - Z_\mu Z_\mu W_\nu^\dagger W_\nu) \\
&-ig\cos\theta_W Z_{\mu\nu}W_\mu^\dagger W_\nu \\
&+g^2\sin\theta_W\cos\theta_W(2A_\mu Z_\mu W_\nu^\dagger W_\nu - A_\mu Z_\nu W_\mu^\dagger W_\nu - A_\nu Z_\mu W_\mu^\dagger W_\nu) \\
&+\frac{g^2}{2}(W_\mu^\dagger W_\mu^\dagger W_\nu W_\nu - W_\mu^\dagger W_\nu^\dagger W_\mu W_\nu) \\
&-U(\eta) + \text{terms containing } \rho' \,. \qquad (16.12)
\end{aligned}
$$

We have assumed that the first component of φ has an electric charge e, like K^+, and that the second component is neutral. Then it turns out that W has an electric charge e. From the electromagnetic interaction of W in the equation above,

$$e = -g\sin\theta_W \,, \qquad (16.13)$$

where θ_W, defined by

$$\frac{g'}{\sqrt{g^2+g'^2}} = \sin\theta_W \,, \qquad \frac{g}{\sqrt{g^2+g'^2}} = \cos\theta_W \,, \qquad (16.14)$$

is called the *Weinberg angle*. The angle is also called the *weak mixing angle*, introduced by S. Glashow in his 1961 paper [153]. A symbol like $A_{\mu\nu}$ means $A_{\mu\nu} = \partial_\mu A_\nu - \partial_\nu A_\mu$. Combining (16.8), (16.9), and (16.14),

$$m_W = m_Z\cos\theta_W \,. \qquad (16.15)$$

Experimentally,

$$\sin^2\theta_W = 0.22 \,. \qquad (16.16)$$

It turns out that Fermi's weak interaction is produced by relating J_μ to $J_\mu^\dagger$ via W. Like the nuclear force, this is a combination of two Yukawa interactions. At energies much lower than m_W, this reduces to the Fermi interaction. Comparing the gauge theory and the Fermi four-fermion interaction at low energies, we obtain

$$\frac{G_F}{\sqrt{2}} = \left(\frac{g}{2\sqrt{2}}\right)^2 \frac{1}{m_W^2} = \frac{e^2}{8m_W^2\sin^2\theta_W} \,. \qquad (16.17)$$

From this,

$$m_W^2 = \frac{\sqrt{2}}{8}\frac{g^2}{G_F} > \frac{\sqrt{2}}{8}\frac{e^2}{G_F} = (37.3\,\text{GeV})^2 \,. \qquad (16.18)$$

If the Weinberg angle is determined, it turns out that m_W and m_Z are also determined. Indeed, approximate values for these masses were predicted theoretically. In 1983 at CERN, the W and Z bosons were detected and their masses were also measured [154–157]:

$$m_W = 80.8 \pm 2.7\,\text{GeV}\,, \quad m_Z = 92.9 \pm 1.6\,\text{GeV}\,. \tag{16.19}$$

These values were quite close to the predictions.

According to the gauge theory, in the weak interaction, there are neutral and charged intermediate vector bosons, namely, Z^0 and $W^{\pm}$, respectively. Weak interactions with neutral particles involve phenomena that cannot be described by the Fermi interaction, such as neutrino scattering. Such phenomena were confirmed early on using a large bubble chamber at CERN in [158].

16.2 Introducing Fermions

In the last section, we formulated the electroweak unification using only gauge fields and a Higgs field. However, a realistic model cannot be constructed without fermions like quarks and leptons. We thus begin by introducing leptons. According to the (V − A) theory, what appears in the weak interaction are left-handed leptons. Hence, what interacts with the gauge fields are three generations of left-handed leptons:

$$L_1 = \frac{1}{2}(1+\gamma_5)\begin{pmatrix}\nu_e\\ e\end{pmatrix}\,, \quad L_2 = \frac{1}{2}(1+\gamma_5)\begin{pmatrix}\nu_\mu\\ \mu\end{pmatrix}\,, \quad L_3 = \frac{1}{2}(1+\gamma_5)\begin{pmatrix}\nu_\tau\\ \tau\end{pmatrix}\,. \tag{16.20}$$

On the other hand, right-handed leptons do not interact with $SU(2)$ gauge fields. We therefore choose the Lagrangian density for the left-handed leptons to be

$$-\sum_j \bar{L}_j\gamma_\mu\left(\partial_\mu - \mathrm{i}g\frac{\hat{\tau}}{2}\hat{W}_\mu - \mathrm{i}\frac{g'}{2}B_\mu\right)L_j\,. \tag{16.21}$$

Regarding the interaction with $\hat{W}_\mu$, since L_i is a doublet like φ, we have chosen the same form. In contrast, Y given by (16.2) is $+1$ for φ carrying the charge $(+e, 0)$, and -1 for L carrying the charge $(0, -1)$. Thus, as for the interaction with B_μ, we have chosen the sign opposite to that of φ. Regarding the neutrino, there is no right-handed component because the mass vanishes. Therefore, we write the right-handed component as

$$R_1 = \frac{1}{2}(1-\gamma_5)e\,, \quad R_2 = \frac{1}{2}(1-\gamma_5)\mu\,, \quad R_3 = \frac{1}{2}(1-\gamma_5)\tau\,. \tag{16.22}$$

This time, since neutrinos are singlets, they do not interact with $\hat{W}_\mu$. However, since Y becomes -2, a plausible choice is

$$-\sum_j \bar{R}_j \gamma_\mu (\partial_\mu - \mathrm{i}g' B_\mu) R_j \ . \tag{16.23}$$

The sum of (16.21) and (16.23) yields the Lagrangian density for leptons.

We shall now introduce quarks. Note that the quark doublets appearing in the weak interaction are not necessarily mass eigenstates. They are obtained by the rotation (13.192), introduced by Cabibbo, of the doublet of mass eigenstates. Experimentally, the angle θ_C is given by

$$\sin\theta_\mathrm{C} = 0.22 \ . \tag{16.24}$$

This value can be derived using (13.178), (13.150), and (13.180), whence the doublet of $SU(2)_\mathrm{L}$ is

$$q_\mathrm{L} = \frac{1}{2}(1+\gamma_5)\begin{pmatrix} u \\ d' \end{pmatrix} , \tag{16.25}$$

and the others are all singlets of $SU(2)_\mathrm{L}$, i.e., they are u_R, d_R, s_R, s_L and so on. However, it will be shown in the next section that this choice yields a constellation of difficulties.

As already mentioned in Sect. 13.6, Glashow, Iliopoulos, and Maiani proposed what is known as the GIM mechanism [130]. They introduced a fourth quark c, and formed two doublets of $SU(2)_\mathrm{L}$:

$$\frac{1}{2}(1+\gamma_5)\begin{pmatrix} u \\ d' \end{pmatrix} , \quad \frac{1}{2}(1+\gamma_5)\begin{pmatrix} c \\ s' \end{pmatrix} . \tag{16.26}$$

One particular merit of this model will be discussed in the next section. In this model, although the interaction with $\hat{W}_\mu$ is the same as that of the leptons, if we now compute Y from (16.2), we obtain $1/3$. The Lagrangian density including gauge interactions of the left-handed quarks for each generation is then

$$-\bar{q}_\mathrm{L}\gamma_\mu\left(\partial_\mu - \mathrm{i}g\frac{\hat{\tau}}{2}\hat{W}_\mu + \mathrm{i}\frac{g'}{6}B_\mu\right)q_\mathrm{L} \ . \tag{16.27}$$

On the other hand, since right-handed quarks are all singlets of $SU(2)_\mathrm{L}$, $Y/2 = 2/3$ for quarks with charge $2e/3$ and $Y/2 = -1/3$ for quarks with charge $-e/3$. Hence, if

$$q_\mathrm{R} = \frac{1}{2}(1-\gamma_5)q \ , \tag{16.28}$$

then the Lagrangian density is given by

$$-\bar{q}_{\mathrm{R}}\gamma_\mu\left(\partial_\mu + \mathrm{i}\frac{g'}{2}YB_\mu\right)q_{\mathrm{R}}\ . \tag{16.29}$$

The sum of (16.27) and (16.29) gives the Lagrangian for quarks. Hence, the interactions of fermions and gauge fields are given by:

$$\mathscr{L}_{\mathrm{fG}} = g\hat{W}_\mu\hat{j}_\mu - g'B_\mu\left(j_\mu^{\mathrm{em}} - j_\mu^3\right)\,, \tag{16.30}$$

$$j_\mu^{\mathrm{em}} = -\mathrm{i}\bar{e}\gamma_\mu e - \mathrm{i}\bar{\mu}\gamma_\mu\mu - \mathrm{i}\bar{\tau}\gamma_\mu\tau + \mathrm{i}\left(\frac{2}{3}\bar{u}\gamma_\mu u - \frac{1}{3}\bar{d}\gamma_\mu d + \cdots\right)\,, \tag{16.31}$$

$$\hat{j}_\mu = \sum_j \mathrm{i}\bar{L}_j\gamma_\mu\frac{\hat{\tau}}{2}L_j + \sum_q \mathrm{i}\bar{q}_{\mathrm{L}}\gamma_\mu\frac{\hat{\tau}}{2}q_{\mathrm{L}}\ . \tag{16.32}$$

We rewrite these interactions using A_μ and Z_μ:

$$\mathscr{L}_{\mathrm{fG}} = \frac{g}{\sqrt{2}}\left(W_\mu j_\mu^\dagger + W_\mu^\dagger j_\mu\right) - g\sin\theta_{\mathrm{W}}\, j_\mu^{\mathrm{em}}A_\mu + \frac{g}{\cos\theta_{\mathrm{W}}}\left(j_\mu^3 - \sin^2\theta_{\mathrm{W}}\, j_\mu^{\mathrm{em}}\right)Z_\mu\ . \tag{16.33}$$

The explicit forms of the various currents are:

$$j_\mu = \mathrm{i}\bar{e}\gamma_\mu\frac{1+\gamma_5}{2}\nu_{\mathrm{e}} + \cdots + \mathrm{i}\bar{d}'\gamma_\mu\frac{1+\gamma_5}{2}u + \cdots\,, \tag{16.34}$$

$$j_\mu^\dagger = \mathrm{i}\bar{\nu}_{\mathrm{e}}\gamma_\mu\frac{1+\gamma_5}{2}e + \cdots + \mathrm{i}\bar{u}\gamma_\mu\frac{1+\gamma_5}{2}d' + \cdots\,, \tag{16.35}$$

$$\begin{aligned} j_\mu^3 &= \frac{1}{2}\left(\mathrm{i}\bar{\nu}_{\mathrm{e}}\gamma_\mu\frac{1+\gamma_5}{2}\nu_{\mathrm{e}} - \mathrm{i}\bar{e}\gamma_\mu\frac{1+\gamma_5}{2}e\right) + \cdots \\ &\quad + \frac{1}{2}\left(\mathrm{i}\bar{u}\gamma_\mu\frac{1+\gamma_5}{2}u - \mathrm{i}\bar{d}'\gamma_\mu\frac{1+\gamma_5}{2}d'\right) + \cdots\ . \end{aligned} \tag{16.36}$$

Although we have not included the fermion mass terms, these masses are generated via the Higgs mechanism. For this purpose, we introduce Yukawa-type interactions between the Higgs field and fermions. In fact, introducing these interactions contradicts the general idea that all interactions be gauge interactions, so we may have to consider some dynamical symmetry breaking in the future. Considering only the first generation, the interactions invariant under $SU(2)_{\mathrm{L}}$ are

$$\mathscr{L}_{\mathrm{f}\varphi} = -f(\bar{L}_1\varphi)R_1 - f_\uparrow(\bar{q}_{\mathrm{L}}\varphi^{\mathrm{G}})u_{\mathrm{R}} - f_\downarrow(\bar{q}_{\mathrm{L}}\varphi)d'_{\mathrm{R}} + \mathrm{h.c.}\,, \tag{16.37}$$

where q_{L} is given by (16.25) and φ^G is the G-conjugation of φ given by

$$\varphi^G = \begin{pmatrix} \varphi_2^* \\ -\varphi_1^* \end{pmatrix} . \tag{16.38}$$

So, if we make the replacement (16.6) for φ, then

$$\mathscr{L}_{\mathrm{f}\varphi} \to -f\eta\bar{e}e - f_{\uparrow}\eta\bar{u}u - f_{\downarrow}\eta\bar{d}'d' + \cdots . \tag{16.39}$$

From this, we obtain the following formulas for the masses:

$$m_{\mathrm{e}} = f\eta , \quad m_u = f_{\uparrow}\eta , \quad m_{d'} = f_{\downarrow}\eta . \tag{16.40}$$

In this approximation, d' is still a mass eigenstate. A weak point of this method is that, for each fermion, we need to introduce another parameter f. Thus, this part is phenomenological, apart from the gauge principle.

16.3 GIM Mechanism

In the quark model in which only u, d, and s are taken into account, the hadron contribution to j_μ^3 is given by

$$\frac{1}{2}\left(\mathrm{i}\bar{u}\gamma_\mu \frac{1+\gamma_5}{2} u - \bar{d}'\gamma_\mu \frac{1+\gamma_5}{2} d' \right) . \tag{16.41}$$

This contains a strangeness-changing part, i.e., in the second term there appears

$$-\frac{1}{2}\sin\theta_{\mathrm{C}} \left(\mathrm{i}\bar{d}\gamma_\mu \frac{1+\gamma_5}{2} s + \mathrm{i}\bar{s}\gamma_\mu \frac{1+\gamma_5}{2} d \right) . \tag{16.42}$$

Such terms can lead to the following decay pattern, where a neutral current changes the strangeness:

$$\mathrm{K}_{\mathrm{L}}^0 \to \mu^+ + \mu^- . \tag{16.43}$$

Experimentally, the branching ratio is very small, viz., $(9.1 \pm 1.9) \times 10^{-9}$. That means that there must be some mechanism that suppresses the process. The GIM mechanism is based on the assumption that there is a fourth quark c, and each item in (16.26) is an $SU(2)_{\mathrm{L}}$ doublet. The hadron parts of the charged current are then given by

$$j_\mu = \mathrm{i}\bar{d}'\gamma_\mu \frac{1+\gamma_5}{2} u + \mathrm{i}\bar{s}'\gamma_\mu \frac{1+\gamma_5}{2} c + \cdots , \tag{16.44}$$

$$j_\mu^\dagger = \mathrm{i}\bar{u}\gamma_\mu \frac{1+\gamma_5}{2} d' + \mathrm{i}\bar{c}\gamma_\mu \frac{1+\gamma_5}{2} s' + \cdots . \tag{16.45}$$

The hadron part of the neutral current then becomes

$$\begin{aligned} j_\mu^3 &= \frac{1}{2}\left(\mathrm{i}\bar{u}\gamma_\mu \frac{1+\gamma_5}{2} u + \mathrm{i}\bar{c}\gamma_\mu \frac{1+\gamma_5}{2} c - \mathrm{i}\bar{d}'\gamma_\mu \frac{1+\gamma_5}{2} d' - \mathrm{i}\bar{s}'\gamma_\mu \frac{1+\gamma_5}{2} s' \right) \\ &= \frac{1}{2}\left(\mathrm{i}\bar{u}\gamma_\mu \frac{1+\gamma_5}{2} u + \mathrm{i}\bar{c}\gamma_\mu \frac{1+\gamma_5}{2} c - \mathrm{i}\bar{d}\gamma_\mu \frac{1+\gamma_5}{2} d' - \mathrm{i}\bar{s}\gamma_\mu \frac{1+\gamma_5}{2} s \right) , \end{aligned} \tag{16.46}$$

so the flavour is conserved. In the neutral current, strangeness is conserved. In fact, the reason why (16.43) still occurs is that the flavour is not strictly conserved, since the flavour symmetry is broken. The above mechanism is called the *GIM mechanism* [130]. As already mentioned in Sect. 13.7, the fourth quark c was subsequently discovered by Ting and Richter et al. in [132, 133].

16.4 Anomalous Terms and Generation of Fermions

So far we have investigated several kinds of symmetries in classical theories or at the lowest order in perturbation. At first, such symmetries were assumed to hold true in general if higher order corrections were included. However, during the 1960s, through research by Adler et al. [159, 160], it gradually became clearer that a certain kind of symmetry is broken by higher order corrections or quantization of a system. In other words, a certain kind of symmetry is not consistent with quantization. Investigating field equations or Green's functions, it sometimes happens that terms are produced which break originally (classically) existing symmetries. These are called *anomalous terms*. Since the details will be discussed in the last chapter, we shall only outline the issue very briefly here. If a certain kind of symmetry is lost by the emergence of anomalous terms, it may happen that renormalizability is also lost. Hence, the requirement that anomalous terms should not appear is sometimes used as a restriction on a model. We shall give an example in this section.

We begin with Adler's anomalous term. In QED, we define the axial vector A_μ, the pseudoscalar current P, and a pseudoscalar term C by

$$A_\mu = \mathrm{i}\bar{\psi}\gamma_\mu\gamma_5\psi , \quad P = \mathrm{i}\bar{\psi}\gamma_5\psi , \quad C = \boldsymbol{E}\cdot\boldsymbol{H} = \frac{\mathrm{i}}{8}\epsilon_{\alpha\beta\gamma\delta}F_{\alpha\beta}F_{\gamma\delta} , \tag{16.47}$$

and we assume that they have been renormalized multiplicatively. In the Heisenberg representation, we have

$$\partial_\lambda \boldsymbol{A}_\lambda = 2m\boldsymbol{P} - \frac{e^2}{2\pi^2}\boldsymbol{C} \,. \tag{16.48}$$

Since in the classical theory, or at the lowest order in perturbation theory, the second term on the right-hand side does not exist, this term is the anomaly. In QED, the axial vector current does not appear in the interaction, so this anomalous term is harmless, but in theories where the axial vector current appears, as in the Weinberg–Salam model, the emergence of such an anomalous term is problematic. It breaks important symmetries, and blocks renormalizability as well. We must therefore choose models in which the anomalous term does not appear. As will be discussed later, the anomalous term appearing here is produced by the contribution from fermion loops, so let us begin by writing down the interaction terms of fermions and gauge fields.

In the Weinberg–Salam model, the representative terms in the Lagrangian density can be written as

$$\mathrm{i}\bar{L}\gamma_\mu\left(\frac{g}{2}\hat{C}\hat{W}_\mu + \frac{g'}{2}C_0 B_\mu\right)L + \mathrm{i}\bar{R}\gamma_\mu g' D B_\mu R \,. \tag{16.49}$$

For the lepton doublet, (ν_e, e), and the lepton singlet e, we have

$$\hat{C} = \hat{\tau}\,, \quad C_0 = 1\,, \quad D = \frac{1-\tau_3}{2}\,. \tag{16.50}$$

For the quark doublet and right-handed singlet, we have

$$\hat{C} = \hat{\tau}\,, \quad C_0 = -\frac{1}{3}\,, \quad D = -\frac{1}{3} + \frac{1-\tau_3}{2}\,. \tag{16.51}$$

The relations with the electric charge Q, in units of e, are

$$Q = -D = \frac{1}{2}(C_3 - C_0)\,. \tag{16.52}$$

If we write the field operator of a quark or a lepton as ψ and we introduce the notation

$$L = \frac{1}{2}(1+\gamma_5)\psi\,, \quad R = \frac{1}{2}(1-\gamma_5)\psi\,, \tag{16.53}$$

then (16.49) becomes

$$\mathrm{i}\bar{\psi}\gamma_\mu \mathscr{V}_\mu \psi + \mathrm{i}\bar{\psi}\gamma_\mu\gamma_5\mathscr{A}_\mu\psi\,. \tag{16.54}$$

However, corresponding to (16.49),

$$\mathscr{V}_\mu = \frac{g}{4}\hat{C}\hat{W}_\mu - \frac{g'}{4}C_3 B_\mu + \frac{g'}{2}C_0 B_\mu \,,$$
$$\mathscr{A}_\mu = \frac{g}{4}\hat{C}\hat{W}_\mu + \frac{g'}{4}C_3 B_\mu \,. \tag{16.55}$$

According to Bardeen [161], the anomalous terms from the fermion loops are

$$\partial_\lambda(\bar{\psi}\gamma_\lambda\gamma_5 C_j\psi) = \text{normal term} + \frac{1}{4\pi^2}\epsilon_{\mu\nu\rho\sigma}\,\mathrm{Tr}[C_j\{\cdots\}] \,, \tag{16.56}$$

where

$$\begin{aligned}\{\cdots\} = &\frac{1}{4}\mathscr{V}_{\mu\nu}\mathscr{V}_{\rho\sigma} + \frac{1}{12}\mathscr{A}_{\mu\nu}\mathscr{A}_{\rho\sigma}\\ &+\frac{2}{3}\mathrm{i}\big(\mathscr{A}_\mu\mathscr{A}_\nu\mathscr{V}_{\rho\sigma} + \mathscr{A}_\mu\mathscr{V}_{\nu\rho}\mathscr{A}_\sigma + \mathscr{V}_{\mu\nu}\mathscr{A}_\rho\mathscr{A}_\sigma\big)\\ &-\frac{8}{3}\mathscr{A}_\mu\mathscr{A}_\nu\mathscr{A}_\rho\mathscr{A}_\sigma \,,\end{aligned} \tag{16.57}$$

$$\begin{aligned}\mathscr{V}_{\mu\nu} &= \partial_\mu\mathscr{V}_\nu - \partial_\nu\mathscr{V}_\mu - \mathrm{i}[\mathscr{V}_\mu, \mathscr{V}_\nu] - \mathrm{i}[\mathscr{A}_\mu, \mathscr{A}_\nu] \,,\\ \mathscr{A}_{\mu\nu} &= \partial_\mu\mathscr{A}_\nu - \partial_\nu\mathscr{A}_\mu - \mathrm{i}[\mathscr{V}_\mu, \mathscr{A}_\nu] - \mathrm{i}[\mathscr{A}_\mu, \mathscr{V}_\nu] \,.\end{aligned} \tag{16.58}$$

We now consider the condition for such anomalous terms to disappear. This means that the trace should vanish. Expressing the products of the matrices by separating them into commutators and anti-commutators, the trace is eventually described by a coefficient corresponding to f_{abc} or d_{abc} in the case of $SU(3)$. As discussed by Georgi and Glashow in [162], the condition for the anomalous terms to vanish is

$$d_{abc} = 0 \,. \tag{16.59}$$

This is because at least one factor in any trace becomes d. So it is required that, for $C_i,\ i = 1, 2, 3$, and C_0, the above equation holds true. Firstly, putting C_0 aside, for $SU(2)$,

$$\mathrm{Tr}\{C_i, C_j\}C_k = 0 \,. \tag{16.60}$$

Note that, since $\mathscr{A}_\mu$ is not included in C_0, what matters is

$$\mathrm{Tr}\{C_i, C_j\}C_0 = \mathrm{Tr}\{C_i, C_j\}(C_3 - 2Q) \,. \tag{16.61}$$

Since this becomes non-zero only for $i = j$, and $C_i^2 = 1$, the condition for the trace to disappear turns out to be

$$\mathrm{Tr}\, Q = 0\,. \tag{16.62}$$

Hence, the sum of the electric charges of all elementary particles with spin 1/2 should vanish. The existence of the c quark provides a way to satisfy this condition, because the sum of the charges of all the leptons and quarks in each generation is zero. The different generations are:

$$\text{leptons} \quad \begin{pmatrix} \nu_\mathrm{e} \\ e^- \end{pmatrix}, \begin{pmatrix} \nu_\mu \\ \mu^- \end{pmatrix}, \ \ldots,$$

$$\text{quarks} \quad \begin{pmatrix} u \\ d \end{pmatrix}, \begin{pmatrix} c \\ s \end{pmatrix}, \ \ldots.$$

The sum of the electric charges of the leptons in each generation is $0 - 1 = -1$, while the sum of the electric charges of the quarks in each generation is

$$3 \times \left(\frac{2}{3} - \frac{1}{3} \right) = 1\,,$$

assuming that each quark has three possible colours. This would not work without the c quark. This is why the GIM mechanism is useful for eliminating anomalies.

From the standpoint of cancelling anomalies, it is plausible to consider each generation to consist of a neutral neutrino, a lepton with a negative electric charge, three-coloured quarks with an electric charge $2e/3$, and three-coloured quarks with an electric charge $-e/3$. Moreover, since there is a third generation of leptons, viz., (ν_τ, τ), we need to consider a third generation of quarks as well. Hence, the quarks t and b were postulated. For the third generation, the reader is referred to the discussion in Sect. 13.6.

In addition, in the discussion above we see that the colour degrees of freedom play an important role. There have to be three colour states for each quark for this to work. In the following, we will give another argument which supports this conclusion.

As already discussed in Sect. 8.5, the amplitude for the decay process $\pi^0 \to 2\gamma$ is given by (8.103). The effective Hamiltonian density corresponding to this is

$$\mathscr{H}_{\mathrm{eff}} = \frac{\alpha}{\pi} \frac{G}{m} \varphi \boldsymbol{E} \cdot \boldsymbol{H}\,. \tag{16.63}$$

We now derive the effective Hamiltonian density corresponding to (16.63) by using the quark model, and analyze hints of the presence of the colour.

We introduce the electromagnetic interaction to the quark model as a perturbation and check its effects. To do so, we consider the axial vector current

$$A_\lambda^a = \sum_q \mathrm{i}\bar{q}\gamma_\lambda\gamma_5 \frac{\tau_a}{2} q \quad (q \text{ is only for the first generation}) . \tag{16.64}$$

In QCD, referring to (15.44), the following PCAC relation [129] holds true:

$$\partial_\lambda \boldsymbol{A}_a^1 = -f_\pi \mu^2 \boldsymbol{\varphi}_a . \tag{16.65}$$

Note that, introducing the electromagnetic interaction as a perturbation, because of Adler's anomalous terms, the equation above transforms to

$$\partial_\lambda \boldsymbol{A}_\lambda^a = -f_\pi \mu^2 \boldsymbol{\varphi}_a + b\delta_{a3} \boldsymbol{E} \cdot \boldsymbol{H} . \tag{16.66}$$

We will show how to compute the coefficient b in the last chapter. It turns out to be the same as that of $\pi^0 \to 2\gamma$, viz.,

$$b = -\frac{2\alpha}{\pi} N_\mathrm{C} S , \tag{16.67}$$

where N_C is the number of quark colours and S is given by

$$S = \sum_{u,d} \left(\frac{\tau_3}{2}\right)_a Q_a^2 = \frac{1}{2}\left(\frac{2}{3}\right)^2 - \frac{1}{2}\left(-\frac{1}{3}\right)^2 = \frac{1}{6} . \tag{16.68}$$

In the sigma model introduced in Sect. 15.2, the term we need to add to the Lagrangian density to change the PCAC relation from (16.65) to (16.66) is

$$\Delta \mathscr{L} = \frac{b}{f_\pi} \varphi_3 \boldsymbol{E} \cdot \boldsymbol{H} . \tag{16.69}$$

In fact, in this case, the PCAC relation assumes the form

$$\partial_\lambda \boldsymbol{A}_\lambda^a = -f_\pi \mu^2 \boldsymbol{\varphi}_a + b\delta_{a3} \frac{\boldsymbol{\sigma}}{f_\pi} \boldsymbol{E} \cdot \boldsymbol{H} . \tag{16.70}$$

Here, replacing $\boldsymbol{\sigma} \to \eta$ and $f_\pi \to \eta$, we note that (16.70) coincides with (16.66). Then, using the effective Lagrangian density of $\pi^0 \to 2\gamma$ in (16.69), based on the quark model, we can compute this decay width and compare it with the experimental results:

$$\Gamma(\pi^0 \to 2\gamma) = \frac{\mu^3}{64\pi^3} \left(\frac{2\alpha}{f_\pi} N_\mathrm{C} S\right)^2 = (7.95 \pm 0.55)\,\mathrm{eV} . \tag{16.71}$$

This implies that

$$N_C = 3.06 \pm 0.10 \,. \tag{16.72}$$

This provides experimental support for the idea that there are three colour states. Moreover, since π^0 is a bound state of the form $q\bar{q}$, the computation is difficult. To avoid this, we used the sigma model in the computation above, whence the computation itself is phenomenological.

16.5 Grand Unified Theory

The Weinberg–Salam theory succeeded in unifying the electromagnetic and weak interactions by applying the gauge principle, thereby producing the *electroweak interaction*. On the other hand, the strong interaction is obtained by applying the gauge principle to the colour group $SU(3)$. The dynamics of the strong interaction is called *quantum chromodynamics*, abbreviated as QCD. Since the gauge particles are bosons which form hadrons by gluing quarks together, they are called *gluons*. This symmetry is not broken, so the gluons are assumed to be massless and the strong force should be long range.

Therefore, the gauge group governing the strong interaction and the electroweak interaction is

$$SU(3) \times SU(2) \times U(1) \,. \tag{16.73}$$

Here we come up with the idea that, if we start with a larger gauge group G, and consider G to decompose by spontaneous symmetry breaking into (16.73), then we may obtain a unified description of the strong, weak, and electromagnetic interactions. Such a theory is called a *grand unified theory*. If this gauge theory is to be specified by just one coupling constant, a simple group G is preferred, i.e., we assume that G does not have any invariant subgroup except for G itself and the identity. We thus look for a simple group G satisfying the condition

$$G \supset SU(3) \times SU(2) \times U(1) \supset SU(3)_{\text{colour}} \times U(1)_{\text{em}} \,. \tag{16.74}$$

Regarding the group (16.73), due to spontaneous symmetry breaking, the strict symmetries surviving to the end are only the colour $SU(3)$ and the electromagnetic $U(1)$. Since the ranks of $SU(3)$, $SU(2)$, and $U(1)$ are 2, 1, and 1, respectively, the rank of G must be at least 4. Although there have been many proposals for such a group, we shall consider as an example the group $SU(5)$, suggested by Georgi and Glashow in [163].

We note that gravity is not included in this grand unified theory. The energies at which gravity becomes important are

$$E \gtrsim \left(\frac{\hbar c}{G}\right)^{1/2} c^2 \sim 10^{19}\,\text{GeV} , \tag{16.75}$$

where G is the gravitational constant and $(\hbar c/G)^{1/2}$ is the *Planck mass*. Thus, the grand unified theory without gravity, known also as GUT, should be formulated at energies less than 10^{19} GeV.

In general, the conditions that the group G must satisfy are:

1. G is simple and only one coupling constant is introduced.
2. Electric charges of quarks and leptons should satisfy the following inequality, which is established by observation:

$$\left|\frac{Q_\text{e} + Q_\text{p}}{Q_\text{p}}\right| \lesssim \mathcal{O}(10^{-21}) . \tag{16.76}$$

3. The rank of G is greater than or equal to 4:

$$G \supset SU(3) \times SU(2) \times U(1) . \tag{16.77}$$

4. To accommodate the Dirac particle, a complex representation should be included.

In the following, we consider $G = SU(5)$. First, we consider three generations of quarks and leptons:

$$\begin{aligned}
&\text{I} . \quad \begin{pmatrix} \nu_\text{e} \\ e \end{pmatrix}_\text{L} , \quad \begin{pmatrix} u \\ d \end{pmatrix}_\text{L} , \quad \bar{e}_\text{R} , \ u_\text{R} , \ d_\text{R} , \\
&\text{II} . \quad \begin{pmatrix} \nu_\mu \\ \mu \end{pmatrix}_\text{L} , \quad \begin{pmatrix} c \\ s \end{pmatrix}_\text{L} , \quad \bar{\mu}_\text{R} , \ c_\text{R} , \ s_\text{R} , \\
&\text{III} . \quad \begin{pmatrix} \nu_\tau \\ \tau \end{pmatrix}_\text{L} , \quad \begin{pmatrix} t \\ b \end{pmatrix}_\text{L} , \quad \bar{\tau}_\text{R} , \ t_\text{R} , \ b_\text{R} .
\end{aligned}$$

All quarks have three possible colours. The first generation is thus

$$(\nu_\text{e}, e, \bar{e}, u_i, d_i, \bar{u}_i, \bar{d}_i)_\text{L} , \quad i = 1, 2, 3, \tag{16.78}$$

where $\bar{u}_\mathrm{L} = \overline{(u_\mathrm{R})}$. There are 15 components altogether. We use them to express the fundamental representation $\bar{\mathbf{5}}$ and the anti-symmetric part $\mathbf{10}$ of $\mathbf{5} \times \mathbf{5}$:

$$\bar{\mathbf{5}} \oplus \mathbf{10}\,, \tag{16.79}$$

$$\bar{\mathbf{5}} = \begin{pmatrix} \bar{d}_\mathrm{R} \\ \bar{d}_\mathrm{Y} \\ \bar{d}_\mathrm{B} \\ e \\ \nu \end{pmatrix}_\mathrm{L}, \qquad \mathbf{10} = \begin{pmatrix} 0 & \bar{u}_\mathrm{B} & -\bar{u}_\mathrm{Y} & -u_\mathrm{R} & -d_\mathrm{R} \\ -\bar{u}_\mathrm{B} & 0 & \bar{u}_\mathrm{R} & -u_\mathrm{Y} & -d_\mathrm{Y} \\ \bar{u}_\mathrm{Y} & -\bar{u}_\mathrm{R} & 0 & -u_\mathrm{B} & -d_\mathrm{B} \\ u_\mathrm{R} & u_\mathrm{Y} & u_\mathrm{B} & 0 & -\bar{e} \\ d_\mathrm{R} & d_\mathrm{Y} & d_\mathrm{B} & \bar{e} & 0 \end{pmatrix}_\mathrm{L}, \tag{16.80}$$

where R, Y, and B indicate the three colours, corresponding to red, yellow, and blue. Expressing them in terms of the transformation properties under $SU(3)$ and $SU(2)$, for instance,

$$\bar{\mathbf{5}} = (\bar{\mathbf{3}}, \mathbf{1}) + (\mathbf{1}, \mathbf{2}) = \bar{d} + \begin{pmatrix} \nu_\mathrm{e} \\ e \end{pmatrix}_\mathrm{L},$$

$$(\mathbf{3}, \mathbf{2}) = \begin{pmatrix} u \\ d \end{pmatrix}_\mathrm{L}, \qquad (\mathbf{1}, \mathbf{1}) = \bar{e}_\mathrm{L} = \overline{(e_\mathrm{R})}\,.$$

In this theory, since the charge operator Q is one of the generators of $SU(5)$, the trace vanishes. Hence, the sum of the charges of the members of $\bar{\mathbf{5}}$ also vanishes:

$$-3Q_\mathrm{d} + Q_\mathrm{e} = 0\,, \quad \text{i.e.,} \quad Q_\mathrm{d} = \frac{1}{3} Q_\mathrm{e} = -\frac{1}{3}\,. \tag{16.81}$$

Since the trace of Q is zero, there is no anomalous term.

In addition, there are 24 gauge particles in $SU(5)$. Half of them are accounted for by 8 gluons and 4 electroweak gauge particles. The rest play the role of connecting leptons with quarks. Considering this kind of gauge particle as a mediator, a nucleon can decay into leptons. We call this gauge particle X, and the width of the nucleon decay becomes the same as the width of the μ decay:

$$\Gamma(N \to \text{lepton}) \propto a_\mathrm{G}^2 \frac{m_\mathrm{p}^5}{M_\mathrm{X}^4}\,. \tag{16.82}$$

The appearance of the fifth power of the mass of the decaying particle is analogous to what we see in the μ decay. The fourth power of the mass of the intermediate gauge particle comes from the squared propagator of this boson. Using the coupling constant g in the grand unified theory, α_G is given by

$$\alpha_\mathrm{G} = \frac{g^2}{4\pi}\,. \tag{16.83}$$

Then the lower bound for the proton lifetime is

$$\tau_{\mathrm{p}} > 10^{32}\ \mathrm{yr}\,. \tag{16.84}$$

From this, a lower bound for M_{X} is obtained:

$$M_{\mathrm{X}} \gtrsim 10^{15}\,\mathrm{GeV} \ll \text{Planck mass}\,. \tag{16.85}$$

To evaluate M_{X}, we use the renormalization group. Even if the symmetry is broken at low energies, we assume that the breaking becomes gradually less important at high energies, until in the end, at an energy M_{X}, the symmetry is recovered. This is analogous to the fact that, even if the rotational symmetry is broken in a ferromagnet at low energies, a phase transition occurs at high energies and the rotational symmetry is then recovered. In this case, what corresponds to the temperature is a value of the four-momentum transfer. Writing this as Q^2, the effective coupling constant becomes a function of Q^2. This way of thinking is used in the renormalization group method, discussed in Chap. 20. Thus, writing the effective coupling constants at low energies corresponding to the groups $SU(3)$, $SU(2)$, and $U(1)$ as $\alpha_3(Q^2)$, $\alpha_2(Q^2)$, and $\alpha_1(Q^2)$, respectively, if the symmetry is indeed recovered at high energies, we have

$$\alpha_1({M_{\mathrm{X}}}^2) = \alpha_2({M_{\mathrm{X}}}^2) = \alpha_3({M_{\mathrm{X}}}^2) \equiv \alpha_{\mathrm{G}}({M_{\mathrm{X}}}^2)\,. \tag{16.86}$$

However, at low energies, they differ from each other. With increasing Q^2, $\alpha_2(Q^2)$ and $\alpha_3(Q^2)$ decrease, while $\alpha_1(Q^2)$ increases. Using the renormalization group, their Q^2-dependence is

$$SU(3) \qquad \frac{1}{\alpha_3(Q^2)} - \frac{1}{\alpha_3({M_{\mathrm{X}}}^2)} = -\frac{1}{4\pi}\left(11 - \frac{2}{3}n_{\mathrm{f}}\right)\ln\frac{{M_{\mathrm{X}}}^2}{Q^2}\,, \tag{16.87}$$

$$SU(2) \qquad \frac{1}{\alpha_2(Q^2)} - \frac{1}{\alpha_2({M_{\mathrm{X}}}^2)} = -\frac{1}{4\pi}\left(\frac{22}{3} - \frac{2}{3}n_{\mathrm{f}}\right)\ln\frac{{M_{\mathrm{X}}}^2}{Q^2}\,, \tag{16.88}$$

$$U(1) \qquad \frac{1}{\alpha_1(Q^2)} - \frac{1}{\alpha_1({M_{\mathrm{X}}}^2)} = \frac{1}{4\pi}\frac{2}{3}n_{\mathrm{f}}\ln\frac{{M_{\mathrm{X}}}^2}{Q^2}\,, \tag{16.89}$$

where n_{f} is the number of flavours. It turns out that, if α_1, α_2, and α_3 are determined experimentally at low energies, then we can find the value of M_{X}. Let us therefore determine M_{X} from the Weinberg angle.

Using (16.2) in the Weinberg–Salam theory, measuring Q in units of e, and writing I_i as T_i, we have

$$Q = T_3 + \frac{1}{2}Y\,. \tag{16.90}$$

We understand that the coupling constants multiplying $\hat{T}$ and $Y/2$ are g and g', respectively. Here, $\hat{T}$ and Y are connected to $\hat{W}_\mu$ and B_μ, respectively, but in $SU(5)$, normalizing Y again,

$$\mathrm{Tr}(T_i T_j) = \frac{1}{2}\delta_{ij} , \qquad i, j = 1, 2, \ldots, 24 . \tag{16.91}$$

Therefore, considering the representation $\bar{\mathbf{5}}$ and taking

$$Y = cT_Y , \tag{16.92}$$

we can determine c. We thus consider the square of the following equation:

$$Q = T_3 + \frac{c}{2}T_Y . \tag{16.93}$$

Therefore, since there exists no quark but $\overline{(d_R)}$, this does not contribute to T_3, and from (16.91),

$$\sum_{\bar{5}} Q^2 = 3\left(\frac{1}{3}\right)^2 + 1 = \frac{4}{3} , \quad \sum_{\bar{5}} T_3^2 = \sum_{\bar{5}} T_Y^2 = \frac{1}{2} , \quad \sum_{\bar{5}} T_3 T_Y = 0 , \tag{16.94}$$

$$\sum_{\bar{5}} Q^2 = \sum_{\bar{5}} \left(T_3 + \frac{c}{2}T_Y\right)^2 = \frac{1}{2} + \frac{c^2}{8} = \frac{4}{3} . \tag{16.95}$$

Thus, the value of c is determined. We obtain

$$Y = 2\sqrt{\frac{5}{3}}T_Y . \tag{16.96}$$

The coefficient of T_3 is g and that of Y is $g'/2$. Therefore, setting

$$gT_i W_{\mu i} = g_2 T_i W_{\mu i} , \quad \frac{1}{2}g' Y B_\mu = g_1 T_Y B_\mu , \tag{16.97}$$

the coupling constants of $SU(2)$ and $U(1)$ are g_2 and g_1, respectively. In the limit where $SU(5)$ is exact, we should have $g_1 = g_2$. From the equation above, we obtain

$$g' = \sqrt{\frac{3}{5}}g_1 . \tag{16.98}$$

We assume that these coupling constants have been defined at low energies. Hence, the Weinberg angle is given by

$$\sin^2\theta_{\rm W} = \frac{g'^2}{g^2+g'^2} = \frac{\frac{3}{5}{g_1}^2}{{g_2}^2+\frac{3}{5}{g_1}^2} \,. \tag{16.99}$$

Moreover, e^2 is given by

$$e^2 = \frac{g^2 g'^2}{g^2+g'^2} = \frac{\frac{3}{5}{g_1}^2{g_2}^2}{{g_2}^2+\frac{3}{5}{g_1}^2} \,. \tag{16.100}$$

This implies

$$\alpha = \frac{3}{5}\frac{\alpha_1\alpha_2}{\alpha_2+\frac{3}{5}\alpha_1} = \alpha_2\sin^2\theta_{\rm W} \,. \tag{16.101}$$

If we take the difference between (16.88) and (16.89), we obtain

$$\frac{1}{\alpha_2} - \frac{1}{\alpha_1} = -\frac{1}{4\pi}\frac{22}{3}\ln\frac{M_{\rm X}^2}{Q^2} \,. \tag{16.102}$$

Additionally, expressing (16.99) in terms of α_1 and α_2,

$$\sin^2\theta_{\rm W} = \frac{\frac{3}{5}\alpha_1}{\alpha_2+\frac{3}{5}\alpha_1} \,. \tag{16.103}$$

Now, eliminating α_1 and α_2 using (16.101), (16.102), and (16.103), we obtain an equation in which all the constants are known:

$$\sin^2\theta_{\rm W} = \frac{3}{8}\left(1 - \frac{\alpha}{4\pi}\frac{110}{9}\ln\frac{M_{\rm X}^2}{Q^2}\right) , \tag{16.104}$$

where Q^2 is of low energy and is assumed to be of order $({\rm GeV})^2$. Therefore, if we choose

$$M_{\rm X} \sim 10^{16}\,{\rm GeV} \,, \tag{16.105}$$

the experimental value of θ_W is reproduced:

$$\sin^2 \theta_W = 0.21 . \tag{16.106}$$

However, since the lifetime of the proton is about 10^{31} years, it is hard to reach the experimental lower bound of 10^{32} years. Although this model is simple, it does not necessarily agree with experiment. However, it is not until we arrive at the grand unified theory that we recognize that quarks and leptons are originally of the same species. On this point, it seems that a clear understanding has been obtained of the problem of the universality of quarks and leptons discussed in Sect. 13.6.

Chapter 17
Path-Integral Quantization Method

Apart from the canonical quantization method, another way to quantize fields is the path-integral method. In many problems, the two methods give the same results, but for the quantization of gauge fields the latter is more effective. It was after the path-integral method had been well understood that the canonical quantization method came to be used. The path-integral method was invented by Feynman [96], and applied to the gauge theory by Faddeev and Popov in [164].

17.1 Quantization of a Point-Particle System

We begin with the easiest quantization of a point-particle system. In the Heisenberg picture, the operators depend on the time variable, and an operator Q is expressed as

$$Q_{\mathrm{H}}(t) \ . \tag{17.1}$$

Its eigenstate is expressed as $|q, t\rangle_{\mathrm{H}}$, whence

$$Q_{\mathrm{H}}(t)|q, t\rangle_{\mathrm{H}} = q|q, t\rangle_{\mathrm{H}} \ , \tag{17.2}$$

where q is one of the eigenvalues of Q. The relationship with the Schrödinger picture is

$$Q_{\mathrm{H}}(t) = \mathrm{e}^{\mathrm{i}Ht} Q_{\mathrm{S}} \mathrm{e}^{-\mathrm{i}Ht} \ , \tag{17.3}$$

$$|q\rangle = \mathrm{e}^{-\mathrm{i}Ht}|q, t\rangle_{\mathrm{H}} \ , \tag{17.4}$$

$$Q_{\mathrm{S}}|q\rangle = q|q\rangle \ . \tag{17.5}$$

K. Nishijima, *Quantum Field Theory*,
https://doi.org/10.1007/978-94-024-2190-3_17

We introduce a transformation function:

$$F(q',t';q,t) = {}_{\rm H}\langle q',t'|q,t\rangle_{\rm H} = \langle q'|\exp[-\mathrm{i}H(t'-t)]|q\rangle\,. \tag{17.6}$$

So far the discussion has been based on the canonical formalism. With this as the starting point, we shall now introduce the path integral.

We begin by dividing $(t-t')$ into $(n+1)$ equal intervals:

$$t' = t+(n+1)\epsilon\,, \quad t_l = t+l\epsilon\,. \tag{17.7}$$

Therefore, from the complete set of $|q_l,t_l\rangle$, we obtain

$$F(q',t';q,t) = \int \mathrm{d}q_1(t_1)\cdots\int \mathrm{d}q_n(t_n)\langle q',t'|q_n,t_n\rangle\langle q_n,t_n|q_{n-1},t_{n-1}\rangle\ldots\langle q_1,t_1|q,t\rangle\,, \tag{17.8}$$

where we have omitted the indices. If we choose ϵ small enough, then

$$\langle q',\epsilon|q,0\rangle = \langle q'|\mathrm{e}^{-\mathrm{i}\epsilon H}|q\rangle = \delta(q'-q) - \mathrm{i}\epsilon\langle q'|H|q\rangle + \mathcal{O}(\epsilon^2)\,. \tag{17.9}$$

The first term expresses the orthonormality of the eigenstates. We choose the following form for H:

$$H = \frac{1}{2}P^2 + V(Q)\,. \tag{17.10}$$

Therefore, the second term on the right-hand side of (17.9) becomes

$$\begin{aligned}\langle q'|H(P,Q)|q\rangle &= \int \frac{\mathrm{d}p}{2\pi}\exp\left[\mathrm{i}p(q'-q)\right]\left[\frac{1}{2}p^2+V(q)\right]\\ &= \int \frac{\mathrm{d}p}{2\pi}\exp\left[\mathrm{i}p(q'-q)\right]H\left(p,\frac{q+q'}{2}\right)\,.\end{aligned} \tag{17.11}$$

Thus, to first order in ϵ,

$$\langle q_l,t_l|q_{l-1},t_{l-1}\rangle = \int \frac{\mathrm{d}p}{2\pi}\exp\left[\mathrm{i}p(q_l-q_{l-1}) - \mathrm{i}\epsilon H\left(p,\frac{1}{2}(q_l+q_{l-1})\right)\right]\,. \tag{17.12}$$

Setting $q_0 = q$ and $q_{n+1} = q'$ and inserting this into (17.8), we obtain

$$F(q',t';q,t) = \lim_{n\to\infty}\int\prod_{j=1}^{n}\mathrm{d}q_j\int\prod_{j=1}^{n+1}\frac{\mathrm{d}p_j}{2\pi}\exp\left\{\mathrm{i}\sum_{l=1}^{n+1}\left[p_l(q_l+q_{l-1}) - H\left(p_l,\frac{1}{2}(q_l+q_{l-1})\right)(t_l-t_{l-1})\right]\right\}. \tag{17.13}$$

We see from this that the number of integration variables p_j is one greater than the number of q_j. Moreover, the normalization condition is determined by unitarity:

$$\int \mathrm{d}q' F(q',t';q_1,t)F^*(q',t';q_2,t) = \int \mathrm{d}q' \langle q',t'|q_1,t\rangle\langle q',t'|q_2,t\rangle^* = \delta(q_1 - q_2)\ . \tag{17.14}$$

Bearing this in mind, without paying too much attention to the normalization, we write the limit (17.13) as

$$F(q',t';q,t) = \int \mathscr{D}p\mathscr{D}q \exp\left\{\mathrm{i}\int_t^{t'} \mathrm{d}t\big[p\dot{q} - H(p,q)\big]\right\}. \tag{17.15}$$

Applying this formula to H as given above, the p-integral can be carried out:

$$\int_{-\infty}^{\infty} \frac{\mathrm{d}p}{2\pi} \exp\left[\mathrm{i}\left(p\dot{q} - \frac{1}{2}p^2\right)\epsilon\right] = \frac{1}{\sqrt{2\pi\mathrm{i}\epsilon}} \exp\left(\frac{1}{2}\mathrm{i}\dot{q}^2\epsilon\right). \tag{17.16}$$

Therefore,

$$\begin{aligned} F(q',t';q,t) &= \lim_{n\to\infty}\left(\frac{1}{\sqrt{2\pi\mathrm{i}\epsilon}}\right)^{n+1}\int\prod_{j=1}^{n}\mathrm{d}q_j \\ &\qquad \times \exp\left\{\mathrm{i}\sum_{l=1}^{n+1}\epsilon\left[\frac{1}{2}\left(\frac{q_l - q_{l-1}}{\epsilon}\right)^2 - V\left(\frac{q_l + q_{l-1}}{2}\right)\right]\right\} \\ &= \int \mathscr{D}q \exp\left[\mathrm{i}\int_t^{t'} \mathrm{d}t L(q,\dot{q})\right], \end{aligned} \tag{17.17}$$

where

$$L(q,\dot{q}) = \frac{1}{2}\dot{q}^2 - V(q)\ . \tag{17.18}$$

Thus, the argument of the exponential is just the action integral:

$$S = \int_t^{t'} \mathrm{d}t\, L(q,\dot{q})\ . \tag{17.19}$$

As shown above, it turns out that the path integral can be expressed by both the Hamiltonian formalism and the Lagrangian formalism. However, as is clear from the derivation, the Hamiltonian formalism is more fundamental.

In the case of the Lagrangian formalism, the formula above is sometimes deformed to some extent. For example, consider the Lagrangian

$$L = \frac{1}{2}\dot{q}^2 f(q) \ , \tag{17.20}$$

$$p = \frac{\partial L}{\partial \dot{q}} = \dot{q} f(q) \ , \quad H(p,q) = p\dot{q} - L = \frac{1}{2}p^2[f(q)]^{-1} \ . \tag{17.21}$$

In this case, the transformation function is

$$\begin{aligned}
F(q',t';q,t) &= \lim_{n\to\infty}\int\prod_{j=1}^{n}\mathrm{d}q_j\int\prod_{j=1}^{n+1}\mathrm{d}p_j \\
&\qquad \times \exp\left[\mathrm{i}\left\{\sum_{l=1}^{n+1}p_l(q_l-q_{l-1})-\frac{1}{2}\epsilon p_l^2\left[f\left(\frac{q_l+q_{l-1}}{2}\right)\right]^{-1}\right\}\right] \\
&= \lim_{n\to\infty}\int\prod_{j=1}^{n}\mathrm{d}q_j\sqrt{\frac{1}{2\pi\mathrm{i}\epsilon}f\left(\frac{q_j+q_{j-1}}{2}\right)} \\
&\qquad \times \exp\left\{\mathrm{i}\sum_{l=1}^{n+1}\epsilon\left[\frac{1}{2}\left(\frac{q_l-q_{l-1}}{\epsilon}\right)^2 f\left(\frac{q_l+q_{l-1}}{2}\right)\right]\right\} \\
&= \lim_{n\to\infty}\left(\frac{1}{\sqrt{2\pi\mathrm{i}\epsilon}}\right)^{n+1}\int\prod_{j=1}^{n}\mathrm{d}q_j \\
&\qquad \times \exp\left\{\mathrm{i}\sum_{l=1}^{n+1}\epsilon\left[\frac{1}{2}\left(\frac{q_l-q_{l-1}}{\epsilon}\right)^2 f\left(\frac{q_l+q_{l-1}}{2}\right)-\frac{\mathrm{i}}{2\epsilon}\ln f\left(\frac{q_l+q_{l-1}}{2}\right)\right]\right\} \\
&= \int \mathscr{D}q\,\exp(\mathrm{i}S_{\mathrm{eff}}) \ ,
\end{aligned} \tag{17.22}$$

where

$$S_{\mathrm{eff}} = \int_t^{t'}\mathrm{d}t\left[L(q,\dot{q}) - \frac{\mathrm{i}}{2}\delta(0)\ln f(q)\right] . \tag{17.23}$$

To obtain this result, we made the substitution

$$\sum\epsilon \to \int \mathrm{d}t \ , \qquad \frac{1}{\epsilon} \to \mathrm{d}t^{-1} \to \delta(0) \ . \tag{17.24}$$

As we have seen above, this method can be used to obtain the effective Lagrangian in the computation of Feynman's equations. What is important here is that, when we go from the Hamiltonian formalism to the Lagrangian formalism, $\mathrm{d}q$ is replaced by

$$\mathrm{d}q \rightarrow \mathrm{d}q\sqrt{f(q)}\ . \tag{17.25}$$

This form is invariant under the general coordinate transformation. In fact, in Riemannian geometry, when the line element

$$\mathrm{d}s^2 = g_{ij}\mathrm{d}x^i\mathrm{d}x^j \tag{17.26}$$

is given, the volume element which is invariant under the general coordinate transformation is

$$\mathrm{d}V = \sqrt{|g|}\mathrm{d}x^1 \ldots \mathrm{d}x^n\ . \tag{17.27}$$

The substitution (17.25) is a special case of this. Here g is the determinant defined by

$$g = \det(g_{ij})\ . \tag{17.28}$$

The substitution (17.25) corresponds to (17.27) with $n = 1$. This result will be extended to the field theory case later and applied to reproduce the result in Sect. 11.5. Moreover, it is easy to extend the previous result to the system with N degrees of freedom.

Next, we consider the matrix elements of the operators. We impose the condition

$$t' > t_0 > t\ . \tag{17.29}$$

In this case, the matrix element of $Q(t_0)$ is

$$\begin{aligned}\langle q', t'|Q(t_0)|q, t\rangle = \int \mathrm{d}q_1(t_1)\ldots\int \mathrm{d}q_n(t_n)\langle q', t'|q_n, t_n\rangle\langle q_n, t_n|q_{n-1}, t_{n-1}\rangle \\ \ldots\langle q_{i+1}, t_{i+1}|q_i, t_i\rangle\langle q_i, t_i|Q(t_i)|q_{i-1}, t_{i-1}\rangle\ldots\langle q_1, t_1|q, t\rangle,\end{aligned} \tag{17.30}$$

where $t_0 = t_i$. Therefore,

$$\langle q_i, t_i|Q(t_i)|q_{i-1}, t_{i-1}\rangle = q_i\langle q_i, t_i|q_{i-1}, t_{i-1}\rangle\ . \tag{17.31}$$

Thus,

$$\langle q', t'|Q(t_0)|q, t\rangle = \int \mathscr{D}p\mathscr{D}q\ q(t_0)\exp\left\{\mathrm{i}\int_t^{t'}\mathrm{d}t\left[p\dot{q} - H(p, q)\right]\right\}\ . \tag{17.32}$$

We now consider two operators $Q(t_1)$ and $Q(t_2)$, with the condition

$$t' > t_1 > t_2 > t \ . \tag{17.33}$$

Then, $\langle q', t' | Q(t_1) Q(t_2) | q, t \rangle$ can also be written in a form similar to (17.32). In general,

$$\begin{aligned} &\langle q', t' | T^*[Q(t_1) \dots Q(t_N)] | q, t \rangle \\ &\qquad = \int \mathscr{D}p\mathscr{D}q \ q(t_1) \dots q(t_N) \exp\left\{ \mathrm{i} \int_t^{t'} \mathrm{d}t \big[p\dot{q} - H(p, q) \big] \right\}, \end{aligned} \tag{17.34}$$

where $t_1, \dots, t_N$ are between t' and t. In this way, if we compute the right-hand side of (17.34), it naturally becomes a matrix element of the time-ordered operator product.

We now compute the matrix element for the ground state. Assuming

$$T' > t' > t > T \ , \tag{17.35}$$

we include an external field between t' and t, and add a term $J(t)q(t)$:

$$\begin{aligned} \langle Q', T' | Q, T \rangle_J &= \int \mathscr{D}p\mathscr{D}q \exp\left\{ \mathrm{i} \int_T^{T'} \mathrm{d}t \big[p\dot{q} - H(p, q) - Jq \big] \right\} \\ &= \int \mathrm{d}q' \mathrm{d}q \langle Q', T' | q', t' \rangle \langle q', t' | q, t \rangle_J \langle q, t | Q, T \rangle \ . \end{aligned} \tag{17.36}$$

We consider $|n\rangle$ to be an energy eigenstate such that

$$H|n\rangle = E_n |n\rangle \ , \quad \langle q | n \rangle = \varphi_n(q) \ . \tag{17.37}$$

Therefore,

$$\begin{aligned} \langle q, t | Q, T \rangle &= \langle q | \exp[-\mathrm{i}H(t - T)] | Q \rangle \\ &= \sum_n \varphi_n(q) \varphi_n^*(Q) \exp[-\mathrm{i}E_n(t - T)] \ . \end{aligned} \tag{17.38}$$

In order to read off the ground state, identified by $n = 0$, from this sum, when we set $T \to \mathrm{i}\infty$, only $n = 0$ survives, according to the third law of thermodynamics:

$$\lim_{T \to \mathrm{i}\infty} \exp(-\mathrm{i}E_0 T) \langle q, t | Q, T \rangle = \varphi_0(q, t) \varphi_0^*(Q) \ , \tag{17.39}$$

where $\varphi_0(q,t) = \varphi_0(q)\exp(-\mathrm{i}E_0 t)$. Similarly,

$$\lim_{\substack{T' \to -\mathrm{i}\infty \\ T \to \mathrm{i}\infty}} \frac{\langle Q', T' | Q, T\rangle_J}{\exp\big[-\mathrm{i}E_0(T'-T)\big]\varphi_0^*(Q)\varphi_0(Q')} = \int \mathrm{d}q \int \mathrm{d}q' \varphi_0^*(q',t')\big\langle q',t'|q,t\big\rangle_J \varphi_0(q,t)\,. \tag{17.40}$$

The right-hand side of this equation can be written as $\langle 0|0\rangle_J$. Since this is a generating functional, we write

$$\mathscr{T}[J] = \langle 0|0\rangle_J\,. \tag{17.41}$$

Therefore, from the definition,

$$\begin{aligned} \left.\frac{\delta^n \mathscr{T}[J]}{\delta J(t_1)\ldots\delta J(t_n)}\right|_{J=0} &= (-\mathrm{i})^n\big\langle 0\big|T^*[Q(t_1)\ldots Q(t_n)]\big|0\big\rangle \\ &= (-\mathrm{i})^n \int \mathrm{d}q \int \mathrm{d}q' \varphi_0^*(q',t')\varphi_0(q,t) \\ &\quad \times \int \mathscr{D}p\mathscr{D}q\; q(t_1)\ldots q(t_n)\exp\left\{\mathrm{i}\int_t^{t'} \mathrm{d}t\big[p\dot{q} - H(p,q)\big]\right\}. \end{aligned} \tag{17.42}$$

Moreover, apart from a factor that is independent of J,

$$\mathscr{T}[J] \sim \lim_{\substack{T' \to -\mathrm{i}\infty \\ T \to \mathrm{i}\infty}} \big\langle Q', T' | Q, T\big\rangle_J\,. \tag{17.43}$$

Therefore, since the generating functional is normalized by $\mathscr{T}[0] = \langle 0|0\rangle = 1$,

$$\begin{aligned} \mathscr{T}[J] &= \frac{\lim\langle Q', T'|Q, T\rangle_J}{\lim\langle Q', T'|Q, T\rangle_{J=0}} \\ &= \frac{\lim \int \mathscr{D}q \exp\left\{\mathrm{i}\int_T^{T'} \mathrm{d}t\big[L_{\mathrm{eff}}(q,\dot{q}) - J(t)q(t)\big]\right\}}{\lim \int \mathscr{D}q \exp\left[\mathrm{i}\int_T^{T'} \mathrm{d}t L_{\mathrm{eff}}(q,\dot{q})\right]}\,. \end{aligned} \tag{17.44}$$

where lim means $T' \to -\mathrm{i}\infty$ and $T \to \mathrm{i}\infty$ as above. Since taking this limit corresponds to rotating 90 degrees clockwise with respect to the origin, we can also implement the analytic continuation by replacing t by $t(1-\mathrm{i}\epsilon)$ and then taking the limit $\epsilon \to +0$. As an example, we consider the action integral when $V = \omega^2 q^2/2$. Replacing t by $t(1-\mathrm{i}\epsilon)$ in

$$\exp\left(-\frac{\mathrm{i}}{2}\omega^2 \int_T^{T'} \mathrm{d}t\, q^2\right) \tag{17.45}$$

is equivalent to replacing ω^2 as follows:

$$\omega^2 \to \omega^2(1 - \mathrm{i}\epsilon) = \omega^2 - \mathrm{i}\epsilon \ . \tag{17.46}$$

In the limit $\epsilon \to +0$, the generating function then becomes a generating function which gives a correct analytic continuation.

17.2 Quantization of Fields

In field theory, the field quantity $\varphi(\boldsymbol{x})$ at each point in space is an independent dynamical variable. Thus, if we make the replacement

$$q_n(t) \to \varphi(\boldsymbol{x}, t) \ , \tag{17.47}$$

then everything is exactly the same as before. In the limit, the generating functional $\mathscr{T}[J]$ behaves as follows:

$$\mathscr{T}[J] \sim \lim \int \mathscr{D}\varphi \mathscr{D}\pi \exp\left\{\mathrm{i}\int \mathrm{d}^4x\big[\pi(x)\dot{\varphi}(x) - \mathscr{H}\big(\varphi(x), \pi(x)\big) - J(x)\varphi(x)\big]\right\} , \tag{17.48}$$

$$\left.\frac{\delta^n \mathscr{T}[J]}{\delta J(x_1)\dots\delta J(x_n)}\right|_{J=0} = (-\mathrm{i})^n \langle \mathbf{0}|T^*[\boldsymbol{\varphi}(x_1)\dots\boldsymbol{\varphi}(x_n)]|\mathbf{0}\rangle. \tag{17.49}$$

In particular, in a simple linear theory,

$$\mathscr{T}[J] \sim \lim \int \mathscr{D}\varphi \exp\left\{\mathrm{i}\int \mathrm{d}^4x\big[\mathscr{L}\big(\varphi(x)\big) - J(x)\varphi(x)\big]\right\} . \tag{17.50}$$

In order to implement the analytic continuation, instead of taking the limit, as in the case of a point-particle system, we make the replacement $m^2 \to m^2 - \mathrm{i}\epsilon$. As an example, we consider a free scalar field:

$$\mathscr{L} = -\frac{1}{2}\big[(\partial_\lambda \varphi)^2 + m^2\varphi^2\big] . \tag{17.51}$$

We start by evaluating the integral

$$I = \int \prod_j \mathrm{d}x_j \exp\left[-\mathrm{i}\left(\sum_{j,k}\frac{1}{2}A_{jk}x_j x_k + \sum_j x_j y_j\right)\right] , \tag{17.52}$$

where A is a real symmetric matrix. To carry out this integration, we make the change of variables

$$x_i = -(A^{-1})_{ij} y_j + z_i \,. \tag{17.53}$$

Therefore,

$$\frac{1}{2}\sum_{j,k} A_{j,k} x_j x_k + \sum_j x_j y_j = \frac{1}{2}\sum_{j,k} A_{jk} z_j z_k - \frac{1}{2}\sum_{j,k} (A^{-1})_{jk} y_j y_k \,. \tag{17.54}$$

Thus,

$$\begin{aligned} I &= \int \prod_j \mathrm{d}z_j \exp\left[-\mathrm{i}\left(\frac{1}{2}\sum_{j,k} A_{jk} z_j z_k - \frac{1}{2}\sum_{j,k} (A^{-1})_{jk} y_j y_k \right)\right] \\ &\propto \exp\left[\frac{\mathrm{i}}{2}\sum_{j,k} (A^{-1})_{jk} y_j y_k \right] \,. \end{aligned} \tag{17.55}$$

We then make the substitutions

$$x_i \to \varphi(x) \,, \quad y_i \to J(x) \,, \quad \sum \to \int \mathrm{d}^4 x \,, \quad A_{jk} \to A(x-y) \,. \tag{17.56}$$

Hence, for the scalar field described by the Lagrangian (17.51),

$$\begin{aligned} A(x-y) &= -(\Box_x - m^2)\delta^4(x-y) \,, \\ A^{-1}(x-y) &= \frac{1}{(2\pi)^4}\int \mathrm{d}^4 p \frac{\mathrm{e}^{\mathrm{i}p\cdot(x-y)}}{p^2 + m^2 - \mathrm{i}\epsilon} = \mathrm{i}\Delta_{\mathrm{F}}(x-y) \,. \end{aligned} \tag{17.57}$$

Thus,

$$\mathscr{T}[J] = \exp\left[-\frac{1}{2}\int \mathrm{d}^4 x \int \mathrm{d}^4 y \, J(x) \Delta_{\mathrm{F}}(x-y) J(y) \right]. \tag{17.58}$$

In this case, the normalization condition holds true automatically, i.e.,

$$\mathscr{T}[0] = 1 \,. \tag{17.59}$$

If there are interactions, we have to expand as a power series in the coupling constants.

We now derive the effective Lagrangian discussed in Sect. 11.5 using the path-integral method. If C is the inverse matrix of D,

$$\mathscr{L} = -\frac{1}{2} D_{ab}(\varphi) \partial_\lambda \varphi_a \partial_\lambda \varphi_b - V(\varphi) , \tag{17.60}$$

$$\pi_a = D_{ab} \dot{\varphi}_b , \tag{17.61}$$

$$\mathscr{H} = \frac{1}{2} C_{ab} \pi_a \pi_b + \frac{1}{2} D_{ab} \nabla \varphi_a \nabla \varphi_b + V . \tag{17.62}$$

The generating functional $\mathscr{T}[J]$ then has the form

$$\mathscr{T}[J] = \int \mathscr{D}\varphi \mathscr{D}\pi \exp\left\{ \mathrm{i} \int \mathrm{d}^4 x \big[\pi_a(x) \dot{\varphi}_a(x) - \mathscr{H}(x) - J_a(x) \varphi_a(x) \big] \right\} . \tag{17.63}$$

To carry out the integral over π, we make the change of variables

$$\pi_a = \Pi_a + D_{ab} \dot{\varphi}_b , \tag{17.64}$$

$$\pi_a \dot{\varphi}_a - \frac{1}{2} C_{ab} \pi_a \pi_b = D_{ab} \dot{\varphi}_a \dot{\varphi}_b - \frac{1}{2} C_{ab} \Pi_a \Pi_b . \tag{17.65}$$

Therefore,

$$\mathscr{T}[J] = \int \mathscr{D}\varphi \mathscr{D}\Pi \exp\left\{ \mathrm{i} \int \mathrm{d}^4 x \left[\mathscr{L}(x) - \frac{1}{2} C_{ab} \Pi_a(x) \Pi_b(x) - J_a(x) \varphi_a(x) \right] \right\} . \tag{17.66}$$

Integrating over Π, the following factor shows up:

$$\begin{aligned} \prod_x \sqrt{\frac{\det C^{-1}}{2\pi \mathrm{i} \Delta}} &\propto \prod_x \sqrt{\det D} , \quad \Delta = \mathrm{d}^4 x , \\ &= \exp\left[\frac{1}{\Delta} \sum_x \frac{\Delta}{2} \ln(\det D) \right] \\ &= \exp\left[\delta^4(0) \int \mathrm{d}^4 x \frac{1}{2} \ln(\det D) \right] . \end{aligned} \tag{17.67}$$

This corresponds to replacing $\mathscr{L}(x)$ by the effective Lagrangian density

$$\mathscr{L}_{\mathrm{eff}} = \mathscr{L} - \frac{\mathrm{i}}{2}\delta^4(0)\ln(\det D) \ . \tag{17.68}$$

This is what we used in (11.113). While the derivation based on the canonical formalism was complex, the path-integral method has the advantage of being extremely intuitive and simple.

Chapter 18
Quantization of Gauge Fields Using the Path-Integral Method

The quantization of the electromagnetic field, which is the simplest gauge field, was discussed in Chap. 5. The problem arising in that case was that, if we apply the usual quantization discussed in Chap. 4 using the gauge invariant Lagrangian density, then the operator $D(\partial)$ defined in (4.43) does not have an inverse. A related difficulty should thus arise in the path-integral method. In this chapter, we shall see how we can avoid this difficulty.

18.1 Quantization of Gauge Fields

Faddeev and Popov showed for the first time how to quantize gauge fields using the path-integral method [164]. Given a field A, in order to compute its Green's functions, we need to introduce the action integral $S[A]$:

$$S[A] = \int \mathrm{d}^4 x\, \mathscr{L}(x)\,. \tag{18.1}$$

The vacuum expectation value of an arbitrary operator $F[A]$ containing the field A is given by

$$\langle F[A] \rangle = \frac{\displaystyle\int \mathscr{D}A\, F[A] \exp\left\{\mathrm{i}S[A]\right\}}{\displaystyle\int \mathscr{D}A \exp\left\{\mathrm{i}S[A]\right\}}\,. \tag{18.2}$$

However, the denominator and numerator are both divergent. The reason is that, if A is a gauge field, then since all configurations of the gauge field which can be obtained by gauge transformations from any given configuration correspond to exactly the same state physically, the same physical state will appear infinitely many

K. Nishijima, *Quantum Field Theory*,
https://doi.org/10.1007/978-94-024-2190-3_18

times. This divergence corresponds to the fact that the differential operator $D(\partial)$ does not have an inverse. However, if $F[A]$ is a gauge invariant quantity, then such divergences will cancel between the denominator and the numerator.

Hence, if we introduce a gauge function Ω and write the corresponding transformation for the gauge field A as

$$A(x) \to A^{\Omega}(x) , \tag{18.3}$$

we need to divide each path integral in the denominator and in the numerator of (18.2) by

$$\int \mathscr{D}\Omega . \tag{18.4}$$

Faddeev and Popov provided a method to do this. We arrange for one configuration of the gauge field to correspond to one physical state. The condition for picking one configuration is called the *gauge condition*.

18.1.1 A Method to Specify the Gauge Condition

We can specify the gauge condition using a functional $f[A]$ of A and writing

$$f[A] = 0 . \tag{18.5}$$

Alternatively, for arbitrary A, we can choose a suitable Ω and require

$$f\left[A^{\Omega}\right] = 0 . \tag{18.6}$$

We can then define the gauge-invariant functional $\Delta_f[A]$ by

$$\Delta_f[A] \int \mathscr{D}\Omega\delta(f[A^{\Omega}]) = \text{const.} \tag{18.7}$$

Therefore,

$$\frac{\int \mathscr{D}A \exp\left\{\mathrm{i}S[A]\right\}}{\int \mathscr{D}\Omega} \sim \int \mathscr{D}\mathscr{A} \exp\left\{\mathrm{i}S[A]\right\}\Delta_f[A]\delta(f[A]) . \tag{18.8}$$

If we now introduce an external field, the generating functional of the Green's function is

$$\mathscr{T}_f[J] \sim \int \mathscr{D}A \exp\left\{\mathrm{i}\int \mathrm{d}^4x\big[\mathscr{L}(x) - J(x)A(x)\big]\right\} \Delta_f[A]\delta(f[A]) . \qquad (18.9)$$

18.1.2 The Additional Term Method

Adding a term to S and setting

$$\varphi[A]\int \mathscr{D}\Omega \exp\left\{\mathrm{i}\Delta S[A^{\Omega}]\right\} = \text{const.} , \qquad \Delta S = \int \mathrm{d}^4x \Delta\mathscr{L}(x) , \qquad (18.10)$$

as in the case above, the equation corresponding to (18.9) is

$$\mathscr{T}[J] \sim \int \mathscr{D}A \exp\left\{\mathrm{i}\int \mathrm{d}^4x\big[\mathscr{L}(x) + \Delta\mathscr{L}(x) - J(x)A(x)\big]\right\} \varphi[A] . \qquad (18.11)$$

This is the customary way of quantizing gauge fields. We now turn to examples.

18.2 Quantization of the Electromagnetic Field

We apply the above method for quantizing gauge fields to the case of the electromagnetic field, which is the best known Abelian gauge field. The canonical quantization of the electromagnetic field is well understood. We shall now check whether the same result can be obtained using the path-integral method. The Lagrangian is

$$\begin{aligned} \mathscr{L} &= -\bar{\psi}\big[\gamma_\mu(\partial_\mu - \mathrm{i}eA_\mu) + m\big]\psi - \frac{1}{4}F_{\mu\nu}F_{\mu\nu} , \\ &= \mathscr{L}_{\mathrm{f}} + \mathscr{L}_{\mathrm{int}} , \end{aligned} \qquad (18.12)$$

where $F_{\mu\nu} = \partial_\mu A_\nu - \partial_\nu A_\mu$ and

$$\mathscr{L}_{\mathrm{int}} = \mathrm{i}e\bar{\psi}\gamma_\mu\psi A_\mu = j_\mu A_\mu . \qquad (18.13)$$

This Lagrangian density is invariant under the gauge transformations

$$A_\mu(x) \to A_\mu(x) + \partial_\mu\lambda(x) , \quad \psi(x) \to \mathrm{e}^{\mathrm{i}e\lambda(x)}\psi(x) , \quad \bar{\psi}(x) \to \mathrm{e}^{-\mathrm{i}e\lambda(x)}\bar{\psi}(x) . \qquad (18.14)$$

18.2.1 Specifying the Gauge Condition

Here we consider the Lorenz gauge and the Coulomb (or radiation) gauge:

$$f_{\mathrm{L}}[A] = \partial_\mu A_\mu = 0\,, \quad f_{\mathrm{R}}[A] = \operatorname{div} \boldsymbol{A} = 0\,. \tag{18.15}$$

In both cases, if we implement the gauge transformation with the gauge function $-\lambda$, then

$$\Delta_{\mathrm{L}}[A] \int \mathscr{D}\lambda\,\delta(\partial_\mu A_\mu - \Box\lambda) = \text{const.}\,, \tag{18.16}$$

$$\Delta_{\mathrm{R}}[A] \int \mathscr{D}\lambda\,\delta(\operatorname{div} \boldsymbol{A} - \Delta\lambda) = \text{const.}\,, \tag{18.17}$$

noting that Δ does not depend on A in either case. For this reason, QED remains simple. We thus consider the generating functional in the Lorenz gauge:

$$\mathscr{T}[J,\eta,\bar{\eta}] \sim \int \mathscr{D}A_\mu \mathscr{D}\psi \mathscr{D}\bar{\psi}\, \exp\left[\mathrm{i}\int \mathrm{d}^4x(\mathscr{L} - \bar{\eta}\psi - \bar{\psi}\eta - J_\mu A_\mu)\right]\delta(\partial_\mu A_\mu)\,. \tag{18.18}$$

Setting $e = 0$, we carry out the path integral for the free field.

For the fermionic field, we take η and $\bar{\eta}$ to be anti-commuting c-numbers and consider

$$\int \mathscr{D}\psi \mathscr{D}\bar{\psi}\, \exp\left\{-\mathrm{i}\int \mathrm{d}^4x\big[\bar{\psi}(\gamma_\mu\partial_\mu + m)\psi + \bar{\eta}\psi + \bar{\psi}\eta\big]\right\}\,. \tag{18.19}$$

In order to evaluate this integral, we generalize the example in Sect. 17.2. We define an inner product by

$$(x, Ax) = \sum_{j,k} x_j^* A_{jk} x_k\,. \tag{18.20}$$

We then write a generalization of the integral (17.52):

$$I = \int \prod_j \mathrm{d}x_j \prod_k \mathrm{d}x_k^* \exp\Big\{-\mathrm{i}\big[(x, Ax) + (x, y) + (y, x)\big]\Big\}\,. \tag{18.21}$$

Introducing the change of variables

$$x_j = -(A^{-1})_{jk} y_k + z_j\,, \tag{18.22}$$

we can carry out the integral, viz.,

$$I = \int \prod_j \mathrm{d}z_j \prod_k \mathrm{d}z_k^* \exp\{-\mathrm{i}[(z, Az) - (y, A^{-1}y)]\} \propto \exp\left[\mathrm{i}(y, A^{-1}y)\right]. \tag{18.23}$$

Making the same replacement as (17.56),

$$A^{-1}(x-y) = \frac{1}{(2\pi)^4} \int \mathrm{d}^4 p \frac{\mathrm{e}^{\mathrm{i}p\cdot(x-y)}}{\mathrm{i}p\cdot\gamma + m - \mathrm{i}\epsilon} = \mathrm{i}S_\mathrm{F}(x-y) . \tag{18.24}$$

The integral (18.19) thus assumes the form

$$\exp\left[-\int \mathrm{d}^4 x \int \mathrm{d}^4 y \bar{\eta}(x) S_\mathrm{F}(x-y)\eta(y)\right] . \tag{18.25}$$

Although we should in fact take into account the anti-commutativity of the variables ψ, $\bar{\psi}$, η, and $\bar{\eta}$, here we have just given the result by analogy.

Now, for the electromagnetic field, using

$$\delta(\partial_\mu A_\mu) \sim \int \mathscr{D}B \exp\left[\mathrm{i}\int \mathrm{d}^4 x\, B(x)\partial_\mu A_\mu(x)\right] \tag{18.26}$$

and integrating by parts in the exponent above, we obtain

$$\int \mathscr{D}A_\mu \mathscr{D}B \exp\left[-\mathrm{i}\int \mathrm{d}^4 x \left\{\frac{1}{2}\left[A_\mu(\partial_\mu\partial_\nu - \delta_{\mu\nu}\Box)A_\nu + \partial_\mu B A_\mu - B\partial_\mu A_\mu\right] + J_\mu A_\mu\right\}\right] . \tag{18.27}$$

For a pair (A_μ, B), the operator corresponding to A_{ij} is then expressed by the matrix

$$\begin{pmatrix} \partial_\mu\partial_\nu - \delta_{\mu\nu} & \partial_\mu \\ -\partial_\nu & 0 \end{pmatrix} . \tag{18.28}$$

Its inverse matrix appears in the propagator. It can be shown to be

$$\begin{pmatrix} \dfrac{1}{\Box}\left(\dfrac{\partial_\mu\partial_\nu}{\Box} - \delta_{\mu\nu}\right) & -\dfrac{\partial_\mu}{\Box} \\ \dfrac{\partial_\nu}{\Box} & 0 \end{pmatrix} . \tag{18.29}$$

Inserting this, the path integral (18.27) becomes

$$\exp\left[-\frac{1}{2}\int \mathrm{d}^4 x \int \mathrm{d}^4 y J_\mu(x) D_{\mu\nu}(x-y) J_\nu(y)\right] , \tag{18.30}$$

where

$$D_{\mu\nu}(x) = \frac{-\mathrm{i}}{(2\pi)^4} \int \mathrm{d}^4k \frac{1}{k^2 - \mathrm{i}\epsilon} \left(\delta_{\mu\nu} - \frac{k_\mu k_\nu}{k^2 - \mathrm{i}\epsilon} \right) \mathrm{e}^{\mathrm{i}k\cdot x} . \tag{18.31}$$

The propagator appearing here, corresponding to the gauge condition (18.15), is written in the Landau gauge. The introduction of the auxiliary field B has already been discussed in Sect. 15.5.

18.2.2 The Additional Term Method

As an additional term, we choose

$$\Delta S = \int \mathrm{d}^4x \Delta\mathscr{L} , \quad \Delta\mathscr{L} = -\frac{1}{2\alpha}(\partial_\mu A_\mu)^2 . \tag{18.32}$$

Since $\varphi[A]$ does not depend on A, the generating functional has the simple form

$$\mathscr{T}[J, \eta, \bar{\eta}] \sim \int \mathscr{D}A_\mu \mathscr{D}\psi \mathscr{D}\bar{\psi} \exp\left[\mathrm{i} \int \mathrm{d}^4x (\mathscr{L} + \Delta\mathscr{L} - J_\mu A_\mu - \bar{\eta}\psi - \bar{\psi}\eta) \right] . \tag{18.33}$$

The ψ-part is the same as above, but the propagator of the electromagnetic field is

$$D_{\mu\nu} = \frac{-\mathrm{i}}{(2\pi)^4} \int \mathrm{d}^4k \frac{1}{(k^2 - \mathrm{i}\epsilon)^2} (k^2\delta_{\mu\nu} - k_\mu k_\nu + \alpha k_\mu k_\nu) \mathrm{e}^{\mathrm{i}k\cdot x} , \tag{18.34}$$

where α is a gauge parameter. This form coincides with the integral expression already derived in (12.251) with $\sigma = 0$.

18.2.3 Ward–Takahashi Identity

The path-integral method gives the same result as the canonical quantization. We can use this method to derive other properties, such as the Ward–Takahashi identity [118, 119].

The propagator of the electron in the Landau gauge is

$$\langle \boldsymbol{\psi}(x)\bar{\boldsymbol{\psi}}(y) \rangle_{\mathrm{L}} = \frac{1}{N_{\mathrm{L}}} \int \mathscr{D}A_\mu \mathscr{D}\psi \mathscr{D}\bar{\psi} \, \psi(x)\bar{\psi}(y) \mathrm{e}^{\mathrm{i}S[\psi,\bar{\psi},A]} \delta(\partial_\mu A_\mu) , \tag{18.35}$$

where N_{L} is a normalization factor given by

$$N_{\mathrm{L}} = \int \mathscr{D}A_\mu \mathscr{D}\psi \mathscr{D}\bar{\psi}\ \mathrm{e}^{\mathrm{i}S[\psi,\bar{\psi},A_\mu]}\delta(\partial_\mu A_\mu) \ . \tag{18.36}$$

We make the change of variables

$$\psi \to \psi' = \mathrm{e}^{\mathrm{i}e\lambda}\psi \ , \quad \bar{\psi} \to \bar{\psi}' = \mathrm{e}^{-\mathrm{i}e\lambda}\bar{\psi} \ . \tag{18.37}$$

Changing the integration variables to ψ' and $\bar{\psi}'$, and then rewriting them again as ψ and $\bar{\psi}$, the expression (18.35) takes the form

$$\begin{aligned}\langle \boldsymbol{\psi}(x)\bar{\boldsymbol{\psi}}(y)\rangle_{\mathrm{L}} =& \frac{1}{N_{\mathrm{L}}}\int \mathscr{D}A_\mu \mathscr{D}\psi \mathscr{D}\bar{\psi}\ \psi(x)\bar{\psi}(y)\mathrm{e}^{\mathrm{i}S[\psi,\bar{\psi},A]}\delta(\partial_\mu A_\mu) \\ &\times \exp\left\{\mathrm{i}e\big[\lambda(x)-\lambda(y)\big] - \mathrm{i}\int \mathrm{d}^4z j_\mu(z)\partial_\mu\lambda(z)\right\} ,\end{aligned} \tag{18.38}$$

where the last term is originated from the electron part of the Lagrangian density. Carrying out the functional differentiation of this equation with respect to $\lambda(x)$, and then setting that $\lambda = 0$,

$$\mathrm{i}e\big[\delta^4(x-z)-\delta^4(y-z)\big]\langle \boldsymbol{\psi}(x),\bar{\boldsymbol{\psi}}(y)\rangle_{\mathrm{L}} + \mathrm{i}\partial_\mu\langle \boldsymbol{j}_\mu(z),\boldsymbol{\psi}(x),\bar{\boldsymbol{\psi}}(y)\rangle_{\mathrm{L}} = 0 \ . \tag{18.39}$$

Taking the Fourier transform of this equation, we obtain the Ward–Takahashi identity (12.200):

$$-\mathrm{i}(p-q)_\mu S'_{\mathrm{F}}(p)\Gamma_\mu(p,q)S'_{\mathrm{F}}(q) = S'_{\mathrm{F}}(p) - S'_{\mathrm{F}}(q) \ . \tag{18.40}$$

The discussion about the derivation above only refers to the fermionic (electron) part, and not the electromagnetic field, so it turns out that this result holds true for any gauge fields.

Next, we discuss the gauge transformations for Green's functions.

18.2.4 Gauge Transformations for Green's Functions

We ask ourselves what kind of relations exist among Green's functions in different gauges. As an example, we investigate the relation between the Landau gauge and the radiation gauge, viz.,

$$\langle \boldsymbol{\psi}(x)\bar{\boldsymbol{\psi}}(y)\rangle_{\mathrm{R}} = \frac{1}{N_{\mathrm{R}}}\int \mathscr{D}A_\mu \mathscr{D}\psi \mathscr{D}\bar{\psi}\ \psi(x)\bar{\psi}(y)\mathrm{e}^{\mathrm{i}S[\psi,\bar{\psi},A]}\delta(\mathrm{div}\,\boldsymbol{A}) \ , \tag{18.41}$$

$$N_{\mathrm{R}} = \int \mathscr{D}A_\mu \mathscr{D}\psi \mathscr{D}\bar{\psi}\ \mathrm{e}^{\mathrm{i}S[\psi,\bar{\psi},A]}\delta(\mathrm{div}\,\boldsymbol{A}) \ . \tag{18.42}$$

We insert the following factor into the denominator and the numerator:

$$\int \mathscr{D}\lambda\delta(\Box\lambda - \partial_\mu A_\mu) = \text{const.} \tag{18.43}$$

If we now carry out a gauge transformation, then S and $\mathscr{D}A_\mu\mathscr{D}\psi\mathscr{D}\bar{\psi}$ are invariant. In the integral,

$$\delta(\Box\lambda - \partial_\mu A_\mu) \to \delta(\partial_\mu A_\mu) , \quad \delta(\text{div}\, \boldsymbol{A}) \to \delta(\text{div}\, \boldsymbol{A} + \Delta\lambda) , \tag{18.44}$$

and in the numerator, the following factor shows up:

$$\exp\left\{\mathrm{i}e\left[\lambda(x) - \lambda(y)\right]\right\} . \tag{18.45}$$

From (18.44),

$$\text{div}\, \boldsymbol{A} + \Delta\lambda = 0 , \tag{18.46}$$

so by solving this equation, we can find the constraint on λ :

$$\lambda(x) = \int \mathrm{d}^4 z \boldsymbol{\phi}(x - z) \cdot \boldsymbol{A}(z) , \tag{18.47}$$

$$\boldsymbol{\phi}(x) = -\delta(x_0)\nabla\left(\frac{1}{4\pi|\boldsymbol{x}|}\right) . \tag{18.48}$$

When we carry out the functional integration with respect to λ, $\delta(\text{div}\, \boldsymbol{A} + \Delta\lambda)$ disappears, whence

$$\left\langle\boldsymbol{\psi}(x)\bar{\boldsymbol{\psi}}(y)\right\rangle_{\mathrm{R}} = \left\langle\boldsymbol{\psi}(x)\bar{\boldsymbol{\psi}}(y)\exp\left\{\mathrm{i}e\int \mathrm{d}^4 z\left[\boldsymbol{\phi}(x - z) - \boldsymbol{\phi}(y - z)\right]\boldsymbol{A}(z)\right\}\right\rangle_{\mathrm{L}} . \tag{18.49}$$

Thus the propagator in the radiation gauge has been expressed in terms of the propagator in the Landau gauge.

18.3 Quantization of Non-Abelian Gauge Fields

Using the standard path-integral method for quantizing gauge fields, we consider the non-Abelian gauge fields.

18.3.1 A Method to Specify the Gauge Condition

We choose the gauge condition

$$f[A] = \partial_\mu A^a_\mu = 0 . \tag{18.50}$$

Applying an infinitesimal gauge transformation, from (14.23),

$$\delta A^a_\mu = \frac{1}{g}\partial_\mu \lambda^a + f_{abc} A^b_\mu \lambda^c \equiv \frac{1}{g}(\mathscr{D}_\mu \lambda)^a . \tag{18.51}$$

Thus, under an infinitesimal gauge transformation,

$$\left(\partial_\mu A^a_\mu\right)^\Omega = \partial_\mu\left(A^a_\mu + \delta A^a_\mu\right) = \partial_\mu\left[A^a_\mu + \frac{1}{g}(D_\mu\lambda)^a\right] . \tag{18.52}$$

In the usual way, we compute $\Delta_f[A]$. If we use (18.7), we have

$$\begin{aligned}\int \mathscr{D}\Omega\,\delta\left[(\partial_\mu A^a_\mu)^\Omega\right] &= \int \mathscr{D}\lambda\,\delta\left[(\partial_\mu A^a_\mu)^\Omega\right] \\ &= \int \mathscr{D}\left[(\partial_\mu A^a_\mu)^\Omega\right]\left\{\frac{\mathscr{D}\left[(\partial_\mu A^a_\mu)^\Omega\right]}{\mathscr{D}\lambda}\right\}^{-1}\delta\left[(\partial_\mu A^a_\mu)^\Omega\right] \\ &= \left\{\frac{\mathscr{D}\left[(\partial_\mu A^a_\mu)^\Omega\right]}{\mathscr{D}\lambda}\right\}^{-1} . \end{aligned} \tag{18.53}$$

Thus, $\Delta_f[A]$ is the functional Jacobian

$$\Delta_f[A] = \frac{\mathscr{D}\left[(\partial_\mu A^a_\mu)^\Omega\right]}{\mathscr{D}\lambda} = \det\left(-\frac{1}{g}\partial_\mu D_\mu\right) . \tag{18.54}$$

Normalizing this determinant to unity when $A^a_\mu = 0$,

$$\Delta_f[A] = \det\left(\frac{\partial_\mu D_\mu}{\Box}\right) . \tag{18.55}$$

To compute this expression, we use the method due to 't Hooft in 1971 [165]. The generating functional for the Green's functions is

$$\mathscr{T}[J] = \int \mathscr{D}A_\mu \exp\left\{\mathrm{i}S[A, J]\right\}\Delta_f[A]\delta(\partial_\mu A_\mu) , \tag{18.56}$$

where

$$S[A, J] = S[A] - \int \mathrm{d}^4x \, A^a_\mu(x) J^a_\mu(x) \, . \tag{18.57}$$

The δ-function in the integral is given by

$$\delta(\partial_\mu A_\mu) \sim \int \mathscr{D}B \exp\left[\mathrm{i} \int \mathrm{d}^4x \, B(x) \partial_\mu A_\mu(x)\right] , \tag{18.58}$$

but note that we do not discuss the normalization here. We use

$$\int \prod_j \mathrm{d}x_j \mathrm{d}y_j \exp\left[\mathrm{i}(x, Ay)\right] = (2\pi)^n (\det A)^{-1} \, . \tag{18.59}$$

And so we obtain

$$\det\left(\frac{\partial_\mu D_\mu}{\Box}\right)^{-1} \sim \int \mathscr{D}\varphi \mathscr{D}\bar{\varphi} \exp\left[\mathrm{i} \int \mathrm{d}^4x \, \bar{\varphi}(x) \partial_\mu D_\mu \varphi(x)\right] . \tag{18.60}$$

Note also that (18.60) is the inverse of (18.55). Let us therefore consider how to obtain the inverse.

We treat the expression (18.60) as a sum of loop contributions obtained by contractions among the scalar fields φ and $\bar{\varphi}$, while the gauge field A appears as an external line. According to the discussion in Sect. 11.2, this sum is the connected part, so in order to derive the inverse, we need to invert the sign of the connected part. However, the connected part consists of single loops obtained by contracting φ and $\bar{\varphi}$ before A is quantized. Thus, we must reverse the sign of each loop. As mentioned in Sect. 8.4, this reversal happens when φ and $\bar{\varphi}$ are anti-commutative, i.e., when they obey Fermi statistics. For a path integral involving these so-called *Grassmann numbers*, which anti-commute, we need an additional discussion, but for the moment we avoid getting further involved and just write down the result:

$$\Delta_f[A] \sim \int \mathscr{D}\varphi \mathscr{D}\bar{\varphi} \exp\left[\mathrm{i} \int \mathrm{d}^4x \, \bar{\varphi}(x) \partial_\mu D_\mu \varphi(x)\right] . \tag{18.61}$$

Although φ and $\bar{\varphi}$ are scalar fields, they obey Fermi statistics. It turns out that this introduces an indefinite metric. The effective Lagrangian density in this theory is

$$\mathscr{L}[A] + \bar{\varphi} \partial_\mu D_\mu \varphi + B \partial_\mu A_\mu \, . \tag{18.62}$$

Here we have summed indices standing for components, although this has not been written explicitly. This Lagrangian corresponds to the one in the Landau gauge in QED. The scalar fields φ and $\bar{\varphi}$ are called *Faddeev–Popov ghost fields*.

18.3.2 The Additional Term Method

We take $\Delta\mathscr{L}$ to be expressed in terms of $\partial_\mu A_\mu$ and define $\varphi[A]$ by

$$\varphi[A] \int \mathscr{D}\Omega \exp\left\{\mathrm{i}\Delta S\left[(\partial_\mu A_\mu)^\Omega\right]\right\} = \text{const.} \tag{18.63}$$

In order to compute this, we use

$$\exp\left\{\mathrm{i}\Delta S\left[(\partial_\mu A_\mu)^\Omega\right]\right\} \sim \int \mathscr{D}C \, \mathrm{e}^{\mathrm{i}\Delta S[C]} \delta\left[(\partial_\mu A_\mu)^\Omega - C\right] . \tag{18.64}$$

This computation is the same as the example above, so $\varphi[A]$ can be readily derived. As a consequence, the effective Lagrangian density, considering the first term as a gauge-invariant term, is

$$\mathscr{L}[A] + \bar{\varphi}\partial_\mu D_\mu \varphi + \Delta\mathscr{L} . \tag{18.65}$$

We choose the following form for $\Delta\mathscr{L}$:

$$\Delta\mathscr{L} = -\frac{1}{2\alpha}(\partial_\mu A_\mu)^2 . \tag{18.66}$$

Therefore, the total Lagrangian density is

$$\mathscr{L}[A] + \bar{\varphi}\partial_\mu D_\mu \varphi - \frac{1}{2\alpha}(\partial_\mu A_\mu)^2 . \tag{18.67}$$

18.3.3 Hermitization of the Lagrangian Density

In the discussion so far, we used the effective Lagrangian density to compute the S-matrix and Green's functions. In the operator formalism, the Lagrangian density should be Hermitian. The Faddeev–Popov ghost term in (18.67) is not Hermitian. Integrating this term by parts,

$$\mathscr{L}_{\mathrm{FP}} \sim -\partial_\mu \bar{\varphi} D_\mu \varphi = -\partial_\mu \bar{\varphi}^a \left(\partial_\mu \varphi^a + g f_{abc} A_\mu^b \varphi^c\right) . \tag{18.68}$$

If φ and $\bar{\varphi}$ are Hermitian, then (18.67) is obviously not Hermitian. This is because φ and $\bar{\varphi}$ are anti-commutative scalar fields. We thus change the phase of this part:

$$\mathscr{L}_{\mathrm{FP}} \to \mathrm{e}^{\mathrm{i}\alpha} \mathscr{L}_{\mathrm{FP}} . \tag{18.69}$$

Consequently, the phases of the ghost propagator and the coupling constant for the ghost and the gauge field change according to

$$\langle \varphi(x)\bar{\varphi}(y)\rangle = D_F(x-y) \to e^{-i\alpha} D_F(x-y) , \tag{18.70}$$

$$g \to g e^{i\alpha} . \tag{18.71}$$

Note that, when φ and $\bar{\varphi}$ only appear in closed loops, the numbers of D_F and g are the same, so the contributions to the S-matrix or Green's functions are invariant under the above phase transformation. That is, it turns out that the phase α can be freely chosen. If we choose $e^{i\alpha} = -i$, and write c and $\bar{c}$ instead of φ and $\bar{\varphi}$ [see (4.89)], we have

$$c^\dagger = c , \quad \bar{c}^\dagger = \bar{c} , \tag{18.72}$$

$$\mathscr{L}_{FP} = i\partial_\mu \bar{c} D_\mu c . \tag{18.73}$$

This Lagrangian density is then Hermitian.

The Lagrangian density in a general gauge (also called the α-gauge) is

$$\mathscr{L} = \mathscr{L}_{int} + \mathscr{L}_{GF} + \mathscr{L}_{FP} , \tag{18.74}$$

where, dropping indices for the gauge field,

$$\mathscr{L}_{inv} = -\frac{1}{4} F_{\mu\nu} F_{\mu\nu}, \tag{18.75}$$

$$\mathscr{L}_{GF} = (\partial_\mu B) A_\mu + \frac{\alpha}{2} B \cdot B , \quad \text{or} \quad -\frac{1}{2\alpha}(\partial_\mu A_\mu)^2, \tag{18.76}$$

$$\mathscr{L}_{FP} = i\partial_\mu \bar{c} D_\mu c. \tag{18.77}$$

The first term in (18.74) is gauge invariant, the second is a gauge-fixing term, and the third is a ghost term. This form was given by Kugo and Ojima in [166].

18.3.4 Gauge Transformations of Green's Functions

In the last section, we investigated the relations among Green's functions defined using different gauge conditions in QED. Here we discuss the different relations among Green's functions defined by including an additional term. The gauge-invariant term is the same, and we thus treat two theories which are physically equivalent in different gauges. Hence, we introduce two Lagrangian densities and two action integrals:

$$\mathscr{L}_{II} = \mathscr{L}_I + \Delta\mathscr{L} , \quad S_{II} = S_I + \Delta S . \tag{18.78}$$

Then considering the field operators $A, B, C, \ldots$, we introduce the Green's function in the second gauge:

$$\langle \boldsymbol{ABC\ldots} \rangle_{II} = \frac{1}{N_{II}} \int \mathscr{D}A_\mu \ldots ABC\ldots \exp(\mathrm{i}S_{II}) \,, \tag{18.79}$$

$$N_{II} = \int \mathscr{D}A_\mu \ldots \exp(\mathrm{i}S_{II}) \,. \tag{18.80}$$

Then from (18.78), we decompose S_{II}, considering ΔS as a perturbation and treating $\exp(\mathrm{i}\Delta S)$ like $A, B, C, \ldots$. Therefore,

$$\begin{aligned} \langle \boldsymbol{ABC\ldots} \rangle_{II} &= \frac{1}{N_{II}} \int \mathscr{D}A_\mu \ldots ABC\ldots \exp(\mathrm{i}\Delta S)\exp(\mathrm{i}S_I) \\ &= \frac{N_I}{N_{II}}\frac{1}{N_I} \int \mathscr{D}A_\mu \ldots ABC\ldots \exp(\mathrm{i}\Delta S)\exp(\mathrm{i}S_I) \\ &= \frac{N_I}{N_{II}} \big\langle \boldsymbol{ABC\ldots} \exp(\mathrm{i}\Delta \boldsymbol{S})\big\rangle_I \,, \end{aligned} \tag{18.81}$$

$$\frac{N_I}{N_{II}} = \frac{1}{N_I} \int \mathscr{D}A_\mu \ldots \exp(\mathrm{i}\Delta S)\exp(\mathrm{i}S_I) = \big\langle \exp(\mathrm{i}\Delta \boldsymbol{S})\big\rangle_I \,. \tag{18.82}$$

We thus obtain

$$\big\langle \boldsymbol{ABC\ldots} \big\rangle_{II} = \frac{\big\langle \boldsymbol{ABC\ldots} \exp(\mathrm{i}\Delta \boldsymbol{S})\big\rangle_I}{\big\langle \exp(\mathrm{i}\Delta \boldsymbol{S})\big\rangle_I} \,. \tag{18.83}$$

This gives the relation among Green's functions in two different gauges. For example, considering (18.74), we choose the Landau gauge with $\alpha = 0$ and the gauge with $\alpha \neq 0$ for $\mathscr{L}_I$ and $\mathscr{L}_{II}$, respectively, and distinguish the Landau gauge by the index L. Then,

$$\langle \boldsymbol{ABC\ldots} \rangle_\alpha = \frac{\left\langle \boldsymbol{ABC\ldots} \exp\left[\frac{\mathrm{i}\alpha}{2}\int \mathrm{d}^4x\, \boldsymbol{B}(x)\cdot \boldsymbol{B}(x)\right]\right\rangle_{\mathrm{L}}}{\left\langle \exp\left[\frac{\mathrm{i}\alpha}{2}\int \mathrm{d}^4x\, \boldsymbol{B}(x)\cdot \boldsymbol{B}(x)\right]\right\rangle_{\mathrm{L}}} \,. \tag{18.84}$$

This equation shows the α-dependence of an arbitrary Green's function. We may also interpret operators appearing in the discussion above as being *unrenormalized*. The subscript L indicates that these Green's functions should be evaluated in Heisenberg's picture in the Landau gauge, while the subscript α indicates that they should be evaluated in the Heisenberg picture in the gauge $\alpha \neq 0$. This formula provides a basis for the discussion about the gauge invariance of various kinds of Green's functions.

18.4 Axial Gauge

In the last section, we introduced the effective Lagrangian density in the covariant gauge, and thereby understood the need for the Faddeev–Popov ghost. However, if we do not require manifest Lorentz covariance, there is a gauge in which we can quantize without ghosts. This is the axial gauge.

We replace the gauge condition (18.50) discussed in the last section by

$$n_\mu A_\mu^a = 0\,, \tag{18.85}$$

where n_μ is a constant vector. For the infinitesimal gauge transformation (18.51),

$$\delta\left(n_\mu A_\mu^a\right) = \frac{1}{g} n_\mu \partial_\mu \lambda^a + f_{abc}(n_\mu A_\mu^b)\lambda^c\,. \tag{18.86}$$

Under the gauge condition (18.85),

$$\delta\left(n_\mu A_\mu^a\right) = \frac{1}{g} n_\mu \partial_\mu \lambda^a\,. \tag{18.87}$$

This is independent of A_μ. Hence,

$$\int \mathscr{D}\Omega\,\delta[(n_\mu A_\mu^a)^\Omega]$$

does not involve A_μ and $\Delta_f[A]$ is a constant. Therefore, it turns out that the Faddeev–Popov ghost term is not produced here.

Using the additional term method, if we choose

$$\Delta S[A] = -\frac{1}{2\alpha}\int \mathrm{d}^4x \left(n_\mu A_\mu^a\right)^2\,, \tag{18.88}$$

and set

$$\varphi[A] \int \mathscr{D}\Omega \exp\left\{\mathrm{i}S\left[A^\Omega\right]\right\} = \text{const.}\,, \tag{18.89}$$

then once again $\varphi[A]$ does not involve A_μ. Hence, we consider the effective Lagrangian density

$$\mathscr{L} = \mathscr{L}_{\text{inv}} - \frac{1}{2\alpha}\left(n_\mu A_\mu\right)^2\,. \tag{18.90}$$

Now quantizing, leaving only terms in quadratic form and applying the variation principle,

$$D_{\mu\nu}(\partial)A^a_\nu = 0\ , \tag{18.91}$$

where

$$D_{\mu\nu}(\partial) = (\delta_{\mu\nu} - \partial_\mu\partial_\nu) - \frac{1}{\alpha}n_\mu n_\nu\ . \tag{18.92}$$

Making the substitution $\partial_\mu\partial_\nu \to -k_\mu k_\nu$ in momentum space,

$$D_{\mu\nu}(k) = -k^2\delta_{\mu\nu} + k_\mu k_\nu - \frac{1}{\alpha}n_\mu n_\nu\ , \tag{18.93}$$

$$\left[\delta_{\mu\nu} - \frac{n_\mu k_\lambda + n_\lambda k_\mu}{n\cdot k} + \frac{n^2+\alpha k^2}{(n\cdot k)^2}k_\mu k_\nu\right]D_{\lambda\nu}(k) = -k^2\delta_{\mu\nu}\ . \tag{18.94}$$

Thus, it turns out that the propagator is

$$\frac{1}{k^2 - \mathrm{i}\epsilon}\left[\delta_{\mu\nu} - \frac{n_\mu k_\nu + n_\nu k_\mu}{n\cdot k} + \frac{n^2+\alpha k^2}{(n\cdot k)^2}k_\mu k_\nu\right]. \tag{18.95}$$

Problems with this gauge include the question of how to treat the pole $n\cdot k = 0$, and showing that computations of various physical quantities do not depend on the choice of n.

18.5 Feynman Rules in the α-Gauge

We now introduce the Feynman rules for the Lagrangian density (18.74). First, note that

$$F^a_{\mu\nu} = \partial_\mu A^a_\nu - \partial_\nu A^a_\mu + g(A_\mu \times A_\nu)^a\ , \tag{18.96}$$

$$(D_\mu c)^a = \partial_\mu c^a + g(A_\mu \times c)^a\ . \tag{18.97}$$

For the gauge group indices, we use the inner and the outer product symbols:

$$A\cdot B = \sum_a A^a B_a\ , \quad (A\times B)^a = \sum_{b,c} f_{abc}A^b B^c\ . \tag{18.98}$$

We split the Lagrangian density (18.74) into the free part $\mathscr{L}_{\mathrm{f}}$ and the interaction part $\mathscr{L}_{\mathrm{int}}$:

$$\mathscr{L}_{\mathrm{f}} = -\frac{1}{4}(\partial_\mu A_\nu^a - \partial_\nu A_\mu^a)^2 - \frac{1}{2\alpha}(\partial_\mu A_\mu^a)^2 + \mathrm{i}\partial_\mu \bar{c}\partial_\mu c \,, \tag{18.99}$$

$$\mathscr{L}_{\mathrm{int}} = -g\partial_\mu A_\nu(A_\mu \times A_\nu) - \frac{1}{4}g^2(A_\mu \times A_\nu)(A_\mu \times A_\nu) + \mathrm{i}g\partial_\mu \bar{c}(A_\mu \times c) \,. \tag{18.100}$$

Expressing the Feynman rule in the Lagrangian formalism, it turns out that we assign the factors $\mathrm{i}(2\pi)^4$, $-\mathrm{i}/(2\pi)^4$, and (-1) to each vertex, propagator, and closed ghost loop, respectively. Considering the gauge particle as the gluon, the propagators are

$$\text{gluon} \qquad \frac{\delta_{ab}}{k^2 - \mathrm{i}\epsilon}\left[\delta_{\mu\nu} - (1-\alpha)\frac{k_\mu k_\nu}{k^2 - \mathrm{i}\epsilon}\right] , \tag{18.101}$$

$$\text{ghost} \qquad \frac{\mathrm{i}\delta_{ab}}{k^2 - \mathrm{i}\epsilon} \,. \tag{18.102}$$

Moreover, we find the following three types of vertex function:

1. three-gluon vertex, 2. four-gluon vertex, 3. ghost–gluon vertex.

1. Three-gluon vertex (Fig. 18.1). Taking all the momenta of the incoming gluons, the vertex function is

$$-\mathrm{i}gf_{abc}\big[\delta_{\beta\gamma}(r-q)_\alpha + \delta_{\gamma\alpha}(p-r)_\beta + \delta_{\alpha\beta}(q-p)_\alpha\big] \,. \tag{18.103}$$

2. Four-gluon vertex (Fig. 18.2). In this case, the vertex function is

$$\begin{aligned} &- g^2 f_{gac} f_{gbd}(\delta_{\alpha\beta}\delta_{\gamma\delta} - \delta_{\alpha\delta}\delta_{\beta\gamma}) - g^2 f_{gad} f_{gbc}(\delta_{\alpha\beta}\delta_{\gamma\delta} - \delta_{\alpha\gamma}\delta_{\beta\delta}) \\ &- g^2 f_{gab} f_{gcd}(\delta_{\alpha\gamma}\delta_{\delta\delta} - \delta_{\alpha\delta}\delta_{\beta\gamma}) \,. \end{aligned} \tag{18.104}$$

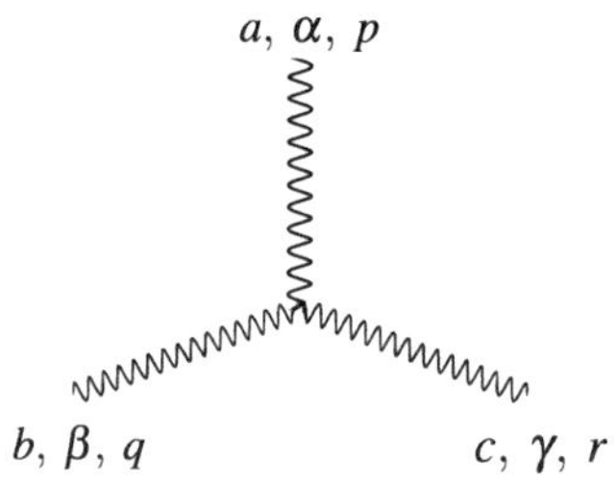

Fig. 18.1 Three-gluon vertex

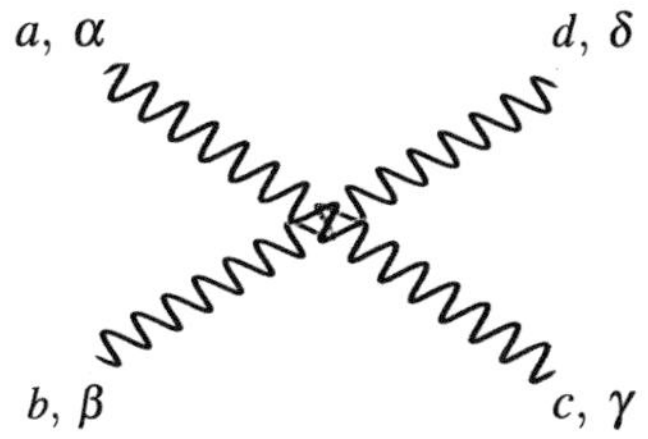

Fig. 18.2 Four-gluon vertex

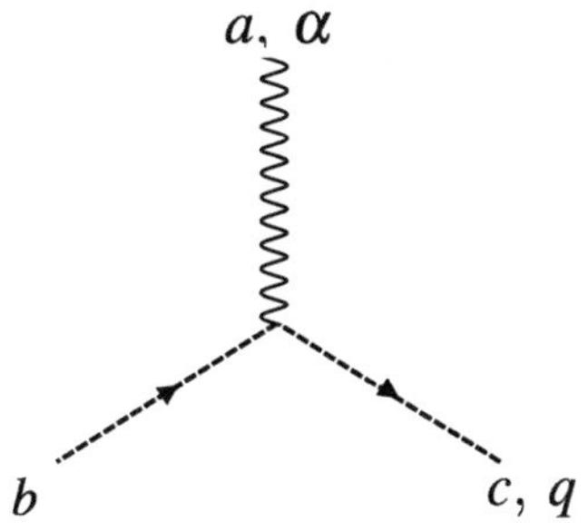

Fig. 18.3 Ghost–gluon vertex

3. Ghost–gluon vertex (Fig. 18.3). Considering the ghost lines to be directed from $\bar{c}$ to c and q to be the outgoing momentum, the vertex function is

$$-\,g f_{abc} q_\alpha \,. \tag{18.105}$$

Combining the above propagators and vertex functions, we can compute the S-matrix elements or Green's functions. Note that the four-momentum is conserved at each vertex. For the total amplitude, we then have conservation of four-momentum, viz., a factor

$$\delta^4(P_f - P_i)\,. \tag{18.106}$$

Moreover, we have to integrate over all the four-momenta k_i in closed loops, which are not affected by the overall conservation of four-momentum, i.e., we introduce the integrals

$$\int \mathrm{d}^4 k_1 \ldots \int \mathrm{d}^4 k_l\,, \tag{18.107}$$

where l is the number of closed loops.

Chapter 19
Becchi–Rouet–Stora Transformations

We consider a system in which a gauge field and fundamental fermions interact with each other. In this case, local gauge transformations are defined for the gauge field and the fermion field. However, in order to quantize these fields, we need to introduce the gauge-fixing term and the Faddeev–Popov ghost term. Consequently, the invariance under the local gauge transformation is broken. However, a new global invariance shows up in its place. This is invariance under the BRS transformation found by Becchi et al. in [167]. In this chapter, we discuss the properties of this transformation.

19.1 BRS Transformations

The Lagrangian density for a system where a gauge field and a fermion field interact is written as

$$\mathscr{L} = \mathscr{L}_{\text{inv}} + \mathscr{L}_{\text{GF}} + \mathscr{L}_{\text{FP}} \,, \tag{19.1}$$

which is the same as the one in (18.74). Regarding the gauge-fixing term and the ghost term, these are the same as given in (18.76) and (18.77), but the gauge-invariant term is

$$\mathscr{L}_{\text{int}} = -\frac{1}{4}F_{\mu\nu}F_{\mu\nu} - \bar{\psi}(\gamma_\mu D_\mu + m)\psi \,, \tag{19.2}$$

where the covariant derivative for the fermion field is

$$D_\mu\psi = (\partial_\mu - igt \cdot A_\mu)\psi = \left(\partial_\mu - ig\sum_a t_a A_\mu^a\right)\psi \,. \tag{19.3}$$

K. Nishijima, *Quantum Field Theory*,
https://doi.org/10.1007/978-94-024-2190-3_19

Local gauge transformations are now defined for the gauge field and the fermion field, which are called *fundamental fields*. For a gauge function $\lambda(x)$, the infinitesimal gauge transformations for the fundamental fields are

$$\delta A^a_\mu(x) = \frac{1}{g}\big(D_\mu\lambda(x)\big)^a\,, \quad \delta\psi(x) = \mathrm{i}\big[\lambda(x)\cdot t\big]\psi(x)\,. \tag{19.4}$$

The transformations obtained by making the following replacement are called the *BRS transformations*:

$$\lambda^a(x) \to gc^a(x)\,. \tag{19.5}$$

In the following, we indicate these transformations by the same symbol δ used in (19.4).

Since $c^a(x)$ are anti-commuting fields, these transformations change the statistics of the field operators. Taking this into account, the BRS transformations for fundamental fields are

$$\delta A_\mu = D_\mu c\,, \quad \delta\psi = \mathrm{i}g(c\cdot t)\psi\,, \quad \delta\bar{\psi} = -\mathrm{i}g\bar{\psi}(c\cdot t)\,. \tag{19.6}$$

From these, the following equations are obtained directly:

$$\delta F_{\mu\nu} = gF_{\mu\nu}\times c\,, \tag{19.7}$$

$$\delta\mathscr{L}_{\text{inv}} = 0\,. \tag{19.8}$$

Note that the local gauge transformations are not defined for the auxiliary fields B, c, and $\bar{c}$, so we cannot define the BRS transformation by the substitution (19.5). We thus introduce the BRS transformations by requiring the total Lagrangian density (19.1) to be invariant under the transformations. To do this, we first write down the field equations derived by the variational principle:

$$D_\mu F_{\mu\nu} + \partial_\nu B - \mathrm{i}g\partial_\nu\bar{c}\times c + \mathrm{i}g\bar{\psi}\gamma_\nu t\psi = 0\,, \tag{19.9}$$

$$\partial_\mu A_\mu = \alpha B\,, \tag{19.10}$$

$$\partial_\mu D_\mu c = D_\mu\partial_\mu\bar{c} = 0\,. \tag{19.11}$$

Assuming that δ and ∂_μ commute, Eqs. (19.10) and (19.11) imply

$$\alpha\delta B = \delta(\partial_\mu A_\mu) = \partial_\mu\delta A_\mu = \partial_\mu D_\mu c = 0\,.$$

Thus,

$$\delta B = 0\,. \tag{19.12}$$

Moreover, if we rewrite (19.11), we have

$$0 = \delta(\partial_\mu D_\mu c) = \partial_\mu \delta(D_\mu c) = \partial_\mu \delta^2 A_\mu \ .$$

We can thus set

$$\delta(D_\mu c) = \delta^2 A_\mu = 0 \ . \tag{19.13}$$

Finally, taking into account (19.8), we require

$$\delta \mathscr{L} = \delta(\mathscr{L}_{\rm GF} + \mathscr{L}_{\rm FP}) = 0 \ . \tag{19.14}$$

Hence, considering (19.12) and (19.13),

$$\begin{aligned} \delta(\mathscr{L}_{\rm GF} + \mathscr{L}_{\rm FP}) &= \delta\big(\partial_\mu B \cdot A_\mu + \mathrm{i}\partial_\mu \bar{c} \cdot D_\mu c\big) \\ &= \partial_\mu B \cdot D_\mu c + \mathrm{i}\partial_\mu(\delta\bar{c}) \cdot D_\mu c \\ &= \partial_\mu(B + \mathrm{i}\delta\bar{c}) \cdot D_\mu c \ . \end{aligned}$$

In order for this to vanish, we must set

$$\delta\bar{c} = \mathrm{i}B \ . \tag{19.15}$$

Finally, in order to determine δc, we consider

$$\begin{aligned} \delta(D_\mu c) &= \delta(\partial_\mu c) + g(D_\mu c) \times c + g A_\mu \times \delta c \\ &= D_\mu\left(\delta c + \frac{1}{2} g c \times c\right) = 0 \ . \end{aligned}$$

From this,

$$\delta c = -\frac{1}{2} g c \times c \ . \tag{19.16}$$

All the BRS transformations for auxiliary fields are thus determined:

$$\delta B = 0 \ , \quad \delta\bar{c} = \mathrm{i}B \ , \quad \delta c = -\frac{1}{2} g c \times c \ , \tag{19.17}$$

and the total Lagrangian density is invariant under the BRS transformation.

19.2 BRS Charge

If a theory is BRS invariant, according to Noether's theorem, there should exist a conserved quantity corresponding to this invariance. This is called the *BRS charge*, denoted Q_{B}. In the following, we shall derive an expression for this quantity. First, we give the equal-time canonical commutation relations necessary for quantization. Using A_j^a, $j = 1, 2, 3$, and B^a as independent variables, the field operators conjugate to each field operator are

$$\pi_j^a = \mathrm{i}F_{4j}^a\,, \quad \pi_{\mathrm{B}}^a = A_0^a = -\mathrm{i}A_4^a\,. \tag{19.18}$$

Therefore, for $x_0 = y_0$,

$$\left[A_4^a(x), B^b(y)\right] = \delta_{ab}\delta^3(x-y)\,, \quad \left[A_j^a(x), F_{4k}^b(y)\right] = \delta_{ab}\delta_{jk}\delta^3(x-y)\,. \tag{19.19}$$

For the ghost fields, if $x_0 = y_0$, we have

$$\left\{c^a(x), \dot{\bar{c}}^b(y)\right\} = -\delta_{ab}\delta^3(x-y)\,, \quad \left\{D_4c^a(x), \bar{c}^b(y)\right\} = -\mathrm{i}\delta_{ab}\delta^3(x-y)\,. \tag{19.20}$$

We can now construct the Noether current:

$$\begin{aligned} J_\mu^{\mathrm{B}} = \sum_\alpha \delta\varphi_\alpha \frac{\partial \mathscr{L}}{\partial \varphi_{\alpha,\mu}} &= -D_\nu c\cdot F_{\mu\nu} - \mathrm{i}g\bar{\psi}\gamma_\mu(c\cdot t)\psi - B\cdot D_\mu c + \frac{\mathrm{i}}{2}g\partial_\mu\bar{c}\cdot(c\times c) \\ &= -\partial_\nu(F_{\mu\nu}\cdot c) - B\cdot D_\mu c + \partial_\mu B\cdot c - \frac{\mathrm{i}}{2}g\partial_\mu\bar{c}\cdot(c\times c)\,. \end{aligned} \tag{19.21}$$

This is a conserved quantity. The spatial integral of its time component is Q_{B}:

$$\partial_\mu J_\mu^{\mathrm{B}} = 0\,, \tag{19.22}$$

$$Q_{\mathrm{B}} = \int \mathrm{d}^3x\, J_0^{\mathrm{B}} = \int \mathrm{d}^3x \left[B\cdot\dot{c} - \dot{B}\cdot c - gB\cdot(A_0\times c) + \frac{\mathrm{i}}{2}g\dot{\bar{c}}\cdot(c\times c)\right]. \tag{19.23}$$

Therefore, the BRS transformation can be written

$$\delta\phi = \mathrm{i}[Q_{\mathrm{B}}, \phi]_\mp\,. \tag{19.24}$$

If ϕ is even-ordered with respect to c and $\bar{c}$, we choose $(-)$, and if it is odd-ordered, we choose $(+)$.

It is also a straightforward matter to show that

$$\delta\left[-B\cdot D_\mu c+\partial_\mu B\cdot c-\frac{\mathrm{i}}{2}g\partial_\mu\bar{c}\cdot(c\times c)\right]=0\,. \tag{19.25}$$

Thus,

$$\delta Q_{\mathrm{B}}=\int \mathrm{d}^3x\,\delta J_0^{\mathrm{B}}=0\,, \tag{19.26}$$

whence

$$\delta Q_{\mathrm{B}}=\mathrm{i}\{Q_{\mathrm{B}},Q_{\mathrm{B}}\}=2\mathrm{i}Q_B^2=0\,.$$

This proves the nilpotency of Q_{B} :

$$Q_{\mathrm{B}}^2=0\,. \tag{19.27}$$

This is the most remarkable property of the BRS charge. The BRS transformation itself has this property, i.e., using (19.27),

$$\delta^2\phi=\begin{cases}-\{Q_{\mathrm{B}},[Q_{\mathrm{B}},\phi]\}=0\,,\ \phi\text{ is even-ordered w.r.t. }c\text{ and }\bar{c}\,,\\ -\{Q_{\mathrm{B}},\{Q_{\mathrm{B}},\phi\}\}=0\,,\ \phi\text{ is odd-ordered w.r.t. }c\text{ and }\bar{c}\,.\end{cases} \tag{19.28}$$

We thus see that the BRS transformation is nilpotent.

Ghost terms appear now only in the ghost term, which is invariant under the scale transformation

$$c\to \mathrm{e}^{\lambda}c\,,\quad \bar{c}\to \mathrm{e}^{-\lambda}\bar{c}\,. \tag{19.29}$$

Since both c and $\bar{c}$ are Hermitian, we cannot introduce a phase transformation. The Noether current corresponding to the above transformation is

$$J_\mu^c=\left(c\frac{\partial\mathscr{L}}{\partial c_{,\mu}}-\bar{c}\frac{\partial\mathscr{L}}{\partial\bar{c}_{,\mu}}\right)=\mathrm{i}(\partial_\mu\bar{c}\cdot c-\bar{c}\cdot D_\mu c)\,, \tag{19.30}$$

$$\partial_\mu J_\mu^c=0\,. \tag{19.31}$$

We thus introduce the conserved quantity Q_c :

$$Q_c=\int \mathrm{d}^3x J_0^c=\mathrm{i}\int \mathrm{d}^3x\left[\bar{c}\cdot\dot{c}-\dot{\bar{c}}\cdot c-g\bar{c}\cdot(A_0\times c)\right]. \tag{19.32}$$

This quantity satisfies the commutation relations

$$\mathrm{i}[Q_c, c(x)] = c(x)\ , \quad \mathrm{i}[Q_c, \bar{c}(x)] = -\bar{c}(x)\ , \tag{19.33}$$

and Q_c commutes with all other fields. Therefore, in general,

$$\mathrm{i}[Q_c, \phi] = N\phi\ , \tag{19.34}$$

where N is the number of $\bar{c}$ minus the number of c included in ϕ as factors. N is called the *ghost number of* ϕ. In addition, both Q_B and Q_c are Hermitian:

$$Q_\mathrm{B}^\dagger = Q_\mathrm{B}\ , \quad Q_c^\dagger = Q_c\ . \tag{19.35}$$

From the structure of Q_B given in (19.23), its ghost number is unity. Therefore, from (19.34),

$$\mathrm{i}[Q_c, Q_\mathrm{B}] = Q_\mathrm{B}\ . \tag{19.36}$$

Combining (19.27) and (19.36) results in the *BRS algebra*. Since commutation relations and anti-commutation relations coexist in this algebra, it is called a *graded Lie algebra*, rather than a Lie algebra.

19.3 Another BRS Transformation

The BRS transformation can be obtained by replacing an infinitesimal gauge function by the ghost field c, as shown in (19.37). So what happens if we then replace an infinitesimal gauge function by $\bar{c}$? This issue was investigated by Curci and Ferrari in [168], Ojima in [169], Bonora and Tonin in [170], and so on. For the fundamental fields, denoting such a transformation by the symbol $\bar{\delta}$,

$$\bar{\delta}A_\mu = D_\mu \bar{c}\ , \quad \bar{\delta}\psi = \mathrm{i}g(\bar{c}\cdot t)\psi\ , \quad \bar{\psi} = -\mathrm{i}g\bar{\psi}(\bar{c}\cdot t)\ . \tag{19.37}$$

This transformation also leaves invariant the gauge invariant term, in exactly the same way as δ. To determine the transformation of the auxiliary field, we note the identity

$$\mathrm{i}\partial_\mu \bar{c}\cdot D_\mu c + \partial_\mu B\cdot A_\mu = \mathrm{i}D_\mu \bar{c}\cdot \partial_\mu c - \partial_\mu \bar{B}\cdot A_\mu\ , \tag{19.38}$$

where $\bar{B}$ is defined by

$$B + \bar{B} - \mathrm{i}gc\times\bar{c} = 0\ . \tag{19.39}$$

It is easily shown that

$$\bar{\delta}\bar{B} = 0\,, \quad \bar{\delta}c = \mathrm{i}\bar{B}\,, \quad \bar{\delta}\bar{c} = -\frac{1}{2}g\bar{c}\times\bar{c}\,. \tag{19.40}$$

Writing the BRS charge corresponding to this transformation as $\bar{Q}_{\mathrm{B}}$, the BRS algebra mentioned in the previous section is extended as follows:

$$Q_{\mathrm{B}}^2 = \bar{Q}_{\mathrm{B}}^2 = Q_{\mathrm{B}}\bar{Q}_{\mathrm{B}} + \bar{Q}_{\mathrm{B}}Q_{\mathrm{B}} = 0\,, \tag{19.41}$$

$$\mathrm{i}[Q_c, Q_{\mathrm{B}}] = Q_{\mathrm{B}}\,, \quad \mathrm{i}[Q_c, \bar{Q}_{\mathrm{B}}] = -\bar{Q}_{\mathrm{B}}\,. \tag{19.42}$$

Moreover, from (19.41) we obtain

$$\delta^2 = \bar{\delta}^2 = \bar{\delta}\delta + \bar{\delta}\delta = 0\,. \tag{19.43}$$

In particular, the last equation is obtained from the identities

$$\begin{aligned} &\big\{Q_{\mathrm{B}}, [\bar{Q}_{\mathrm{B}}, \phi]\big\} + \big\{\bar{Q}_{\mathrm{B}}, [Q_{\mathrm{B}}, \phi]\big\} = \big[\{Q_{\mathrm{B}}, \bar{Q}_{\mathrm{B}}\}, \phi\big] \qquad \text{(even ordered)}\,,\\ &\big\{Q_{\mathrm{B}}, \{\bar{Q}_{\mathrm{B}}, \phi\}\big\} + \big[\bar{Q}_{\mathrm{B}}, \{Q_{\mathrm{B}}, \phi\}\big] = \big[\{Q_{\mathrm{B}}, \bar{Q}_{\mathrm{B}}\}, \phi\big] \qquad \text{(odd ordered)}\,, \end{aligned} \tag{19.44}$$

depending on whether ϕ is even-ordered or odd-ordered with respect to c and $\bar{c}$, respectively. If, in addition, the symmetry under the global gauge transformation, i.e., when the gauge function is constant, is not spontaneously broken, the colour charge Q^a given by

$$J_\mu = -A_\mu \times F_{\mu\nu} - A_\mu \times B + \mathrm{i}\bar{\psi}\gamma_\mu t\psi + \mathrm{i}(\bar{c}\times D_\mu c) - \mathrm{i}(\partial_\mu\bar{c}\times c)\,, \tag{19.45}$$

$$Q^a = \int \mathrm{d}^3x\, J_0^a\,, \tag{19.46}$$

is conserved. This J_μ becomes a source for the gauge field in the sense that

$$\partial_\mu F_{\mu\nu} = -gJ_\nu + \mathrm{i}\delta\bar{\delta}A_\nu\,. \tag{19.47}$$

Moreover, Q^a satisfies the commutation relations

$$[\psi(x), Q^a] = t^a\psi(x)\,, \ldots \tag{19.48}$$

$$[Q^a, Q^b] = \mathrm{i}f_{abc}Q^c\,. \tag{19.49}$$

Since none of Q_{B}, $\bar{Q}_{\mathrm{B}}$, and Q_c have colour, they commute with Q^a, i.e.,

$$[Q^a, Q_{\mathrm{B}}] = [Q^a, \bar{Q}_{\mathrm{B}}] = [Q^a, Q_c] = 0\,. \tag{19.50}$$

Put another way, the colour charge is a BRS-invariant quantity:

$$\delta Q^a = \bar{\delta} Q^a = 0 . \tag{19.51}$$

19.4 BRS Identity and Slavnov–Taylor Identity

For a system including gauge fields, even where the symmetry is spontaneously broken, we can consider that the BRS invariance is not broken, i.e., even if some Q^a is not conserved, the conservation law for Q_{B} may still hold true. An indication of this is the fact that the unitarity of the S-matrix is based on the BRS invariance, as will be shown later. As a consequence, the following equation holds true for the vacuum $|\mathbf{0}\rangle$:

$$Q_{\mathrm{B}}|\mathbf{0}\rangle = 0 . \tag{19.52}$$

Consequently, for arbitrary field operators $\boldsymbol{O}_1(x_1), \ldots, \boldsymbol{O}_n(x_n)$,

$$\big\langle \mathbf{0}\big|\delta T[\boldsymbol{O}_1(x_1)\ldots \boldsymbol{O}_n(x_n)]\big|\mathbf{0}\big\rangle = 0 . \tag{19.53}$$

This is called the *BRS identity*. It will be exemplified in the following. From now on, we shall use the notation

$$\big\langle \mathbf{0}\big|T[\boldsymbol{O}_1(x_1)\ldots \boldsymbol{O}_n(x_n)]\big|\mathbf{0}\big\rangle = \big\langle \boldsymbol{O}_1(x_1)\ldots \boldsymbol{O}_n(x_n)\big\rangle .$$

If we have

$$\delta\boldsymbol{O}_1(x_1) = \ldots = \delta\boldsymbol{O}_n(x_n) = 0 , \quad \boldsymbol{O}(x) = \delta\boldsymbol{P}(x) , \tag{19.54}$$

for a set of operators $\boldsymbol{O}_1(x_1)$, ..., $\boldsymbol{O}_n(x_n)$, and $\boldsymbol{O}(x)$, then from (19.53),

$$\big\langle \boldsymbol{O}(x)\boldsymbol{O}_1(x_1)\ldots \boldsymbol{O}_n(x_n)\big\rangle = 0 . \tag{19.55}$$

We thus consider (18.84). In the unrenormalized Landau gauge, since

$$\delta\boldsymbol{B}(x) = 0 , \quad \boldsymbol{B}(x) = -\mathrm{i}\delta\bar{\boldsymbol{c}}(x) , \tag{19.56}$$

the Green's function including arbitrary numbers of $\boldsymbol{B}$ turns out to vanish, so that

$$\left\langle \exp\left[\frac{\mathrm{i}\alpha}{2}\int \mathrm{d}^4x\, \boldsymbol{B}(x)\ldots \boldsymbol{B}(x)\right]\right\rangle_{\mathrm{L}} = 1 . \tag{19.57}$$

Moreover, if in (18.84) we have

$$\delta \boldsymbol{A} = \delta \boldsymbol{B} = \delta \boldsymbol{C} = \ldots = 0 , \tag{19.58}$$

then for an unrenormalized Green's function,

$$\langle \boldsymbol{ABC}\ldots\rangle_\alpha = \langle \boldsymbol{ABC}\ldots\rangle_{\mathrm{L}} , \tag{19.59}$$

which means that an unrenormalized Green's function does not depend on the unrenormalized gauge parameter α. This result will be further generalized.

As in Sect. 18.3, we consider a class of Lagrangian densities with the same gauge-invariant term. We assume that the difference between two arbitrary Lagrangian densities belonging to this class can be written as the BRS transformation of an operator:

$$\Delta\mathscr{L} = \mathscr{L}_{II} - \mathscr{L}_I = \delta\mathscr{M} , \tag{19.60}$$

$$\Delta S = S_{II} - S_I = \delta \int \mathrm{d}^4 x \mathscr{M} . \tag{19.61}$$

If in (18.83) we now have

$$\big\langle \exp(\mathrm{i}\Delta \boldsymbol{S})\big\rangle_I = 1 , \tag{19.62}$$

and (18.91) are satisfied once again, then

$$\langle \boldsymbol{ABC}\ldots\rangle_{II} = \langle \boldsymbol{ABC}\ldots\rangle_I , \tag{19.63}$$

for an unrenormalized Green's function. This is important. As will be discussed later, when computing observables such as the S-matrix using the LSZ reduction formula, we start with the BRS invariant Green's function. Since its normalization is determined by a renormalization condition like (11.122), it turns out that, within the class satisfying the condition (19.61), the S-matrix is independent of the choice of gauge.

As another application, using

$$-\mathrm{i}\delta T\big[A^a_\mu(x), \bar{c}^b(y)\big] = T\big[A^a_\mu(x), B^b(y)\big] - \mathrm{i}T\big[D_\mu c^a(x), \bar{c}^b(y)\big] , \tag{19.64}$$

we obtain the BRS identity in the form

$$\big\langle \boldsymbol{A}^a_\mu(x), \boldsymbol{B}^b(y)\big\rangle = \mathrm{i}\big\langle D_\mu \boldsymbol{c}^a(x), \bar{\boldsymbol{c}}^b(y)\big\rangle . \tag{19.65}$$

Using the field equation and the canonical commutation relation, we see immediately that the four-divergence of this equation is equal to $-\mathrm{i}\delta_{ab}\delta^4(x-y)$. Repeating the discussion in Sect. 15.1,

$$\left\langle \boldsymbol{A}^a_\mu(x), \boldsymbol{B}^b(y)\right\rangle = -\delta_{ab}\partial_\mu D_F(x-y)\ . \tag{19.66}$$

This is important because this equation assumes the same form in both the renormalized and the unrenormalized formalisms.

A set of relations which hold true among Green's functions can be derived from the BRS identity. This corresponds to the Ward–Takahashi identity in QED, and is called the *Slavnov–Taylor identity* in QCD [171, 172]. We introduce an external field:

$$\begin{aligned}\mathscr{S} = J_\mu\cdot\boldsymbol{A}_\mu + \bar{J}_\alpha\cdot\boldsymbol{\psi}_\alpha + J_\alpha\cdot\bar{\boldsymbol{\psi}}_\alpha + \bar{J}_c\cdot\boldsymbol{c} + J_{\bar{c}}\cdot\bar{\boldsymbol{c}} + J_B\cdot\boldsymbol{B}\\ +K_\mu\cdot\delta\boldsymbol{A}_\mu + \bar{K}_\alpha\cdot\delta\boldsymbol{\psi}_\alpha + K_\alpha\cdot\delta\bar{\boldsymbol{\psi}}_\alpha + K_c\cdot\delta\boldsymbol{c}\ .\end{aligned} \tag{19.67}$$

The fact that $\mathrm{i}\bar{\boldsymbol{c}}$ is equal to $\mathrm{i}\boldsymbol{B}$ has already been included in (19.67). If $\mathscr{S}$ is bosonic, then J_α, $\bar{J}_\alpha$, K_α, and $\bar{K}_\alpha$ are external fields that anti-commute with the fermionic fields, while $\bar{J}_c$, $J_{\bar{c}}$, K_α, and $\bar{K}_\alpha$ are external fields that anti-commute with the ghost fields. These are considered to be anti-commutative external fields. Hence,

$$\exp(-\mathrm{i}W) = \left\langle\boldsymbol{0}\middle|T\exp\left[-\mathrm{i}\int\mathrm{d}^4x\,\mathscr{S}(x)\right]\middle|\boldsymbol{0}\right\rangle . \tag{19.68}$$

The BRS identity in this case is

$$\left\langle\boldsymbol{0}\middle|T\left[\int\mathrm{d}^4x\,\delta\mathscr{S}(x), \exp\left[-\mathrm{i}\int\mathrm{d}^4y\mathscr{S}(y)\right]\right]\middle|\boldsymbol{0}\right\rangle = 0\ . \tag{19.69}$$

We introduce the notation

$$\left\langle O(x)\right\rangle \equiv \frac{\langle\boldsymbol{0}|T\left[\boldsymbol{O}(x), \exp\left[-\mathrm{i}\int\mathrm{d}^4y\mathscr{S}(y)\right]\right]|\boldsymbol{0}\rangle}{\langle\boldsymbol{0}|\exp\left[-\mathrm{i}\int\mathrm{d}^4y\mathscr{S}(y)\right]|\boldsymbol{0}\rangle}\ . \tag{19.70}$$

Using the nilpotency of the BRS transformation and the fact that $\bar{J}_c$ and $J_{\bar{c}}$ anti-commute with Q_B,

$$\langle\delta\mathscr{S}\rangle = J_\mu\langle\delta A_\mu\rangle + \bar{J}_\alpha\langle\delta\psi_\alpha\rangle + J_\alpha\langle\delta\bar{\psi}_\alpha\rangle - \bar{J}_c\langle\delta c\rangle - J_{\bar{c}}\langle\delta\bar{c}\rangle = 0\ . \tag{19.71}$$

Differentiating anti-commuting external fields from the left-hand side,

$$\begin{aligned}&\frac{\delta W}{\delta J_\mu} = \langle A_\mu\rangle\ , \quad \frac{\delta W}{\delta\bar{J}_\alpha} = \langle\psi_\alpha\rangle\ , \quad \frac{\delta W}{\delta J_\alpha} = \langle\bar{\psi}_\alpha\rangle,\\ &\frac{\delta W}{\delta J_B} = \langle B\rangle\ , \quad \frac{\delta W}{\delta\bar{J}_c} = \langle c\rangle\ , \quad \frac{\delta W}{\delta J_{\bar{c}}} = \langle\bar{c}\rangle\ .\end{aligned} \tag{19.72}$$

We now introduce the Legendre transformation in order to have $\langle\varphi\rangle$ and K as independent variables instead of J and K :

$$\Gamma = W - J_\mu \cdot \langle A_\mu \rangle - \bar{J}_\alpha \cdot \langle \psi_\alpha \rangle - J_\alpha \cdot \langle \bar{\psi}_\alpha \rangle - \bar{J}_c \cdot \langle c \rangle - J_{\bar{c}} \cdot \langle \bar{c} \rangle - J_B \cdot \langle B \rangle \ . \tag{19.73}$$

Differentiating this from the left-hand side,

$$\frac{\delta \Gamma}{\delta \langle \phi \rangle} = \begin{cases} -J_\phi \ , \ \phi = A_\mu, B \\ J_\phi \ , \quad \phi = c, \bar{c}, \psi_\alpha, \bar{\psi}_\alpha \ . \end{cases} \tag{19.74}$$

Further, the derivative with respect to K is

$$\frac{\delta \Gamma}{\delta K} = \frac{\delta W}{\delta K} \ . \tag{19.75}$$

From this,

$$\frac{\delta \Gamma}{\delta K_\mu} = \langle \delta A_\mu \rangle \ , \quad \frac{\delta \Gamma}{\delta \bar{K}_\alpha} = \langle \delta \psi_\alpha \rangle \ , \quad \frac{\delta \Gamma}{\delta K_\alpha} = \langle \delta \bar{\psi}_\alpha \rangle \ , \quad \frac{\delta \Gamma}{\delta K_c} = \langle \delta c \rangle \ , \tag{19.76}$$

while $\langle \delta \bar{c} \rangle$ is given not by the derivative with respect to Γ but by

$$\langle \delta \bar{c} \rangle = \mathrm{i} \langle B \rangle \ . \tag{19.77}$$

Expressing (19.71) in terms of derivatives with respect to Γ,

$$\begin{aligned} 0 &= \frac{\delta \Gamma}{\delta \langle A_\mu \rangle} \langle \delta A_\mu \rangle - \frac{\delta \Gamma}{\delta \langle \psi_\alpha \rangle} \langle \delta \psi_\alpha \rangle - \frac{\delta \Gamma}{\delta \langle \bar{\psi}_\alpha \rangle} \langle \delta \bar{\psi}_\alpha \rangle + \frac{\delta \Gamma}{\delta \langle c \rangle} \langle \delta c \rangle + \frac{\delta \Gamma}{\delta \langle \bar{c} \rangle} \langle \delta \bar{c} \rangle \\ &= \langle \delta A_\mu \rangle \frac{\delta \Gamma}{\delta \langle A_\mu \rangle} + \langle \delta \psi_\alpha \rangle \frac{\delta \Gamma}{\delta \langle \psi_\alpha \rangle} + \langle \delta \bar{\psi}_\alpha \rangle \frac{\delta \Gamma}{\delta \langle \bar{\psi}_\alpha \rangle} + \langle \delta c \rangle \frac{\delta \Gamma}{\delta \langle c \rangle} + \langle \delta \bar{c} \rangle \frac{\delta \Gamma}{\delta \langle \bar{c} \rangle} \\ &= \frac{\delta \Gamma}{\delta K_\mu} \frac{\delta \Gamma}{\delta \langle A_\mu \rangle} + \frac{\delta \Gamma}{\delta \bar{K}_\alpha} \frac{\delta \Gamma}{\delta \langle \psi_\alpha \rangle} + \frac{\delta \Gamma}{\delta K_\alpha} \frac{\delta \Gamma}{\delta \langle \bar{\psi}_\alpha \rangle} + \frac{\delta \Gamma}{\delta K_c} \frac{\delta \Gamma}{\delta \langle c \rangle} + \mathrm{i} \langle B \rangle \frac{\delta \Gamma}{\delta \langle \bar{c} \rangle} \ . \end{aligned} \tag{19.78}$$

This is the *Slavnov–Taylor identity* [171, 172].

Furthermore, from the field equations with external fields,

$$-J_B = \partial_\mu \langle A_\mu \rangle - \alpha \langle B \rangle \ , \quad J_{\bar{c}} = \mathrm{i} \partial_\mu \langle D_\mu c \rangle \ . \tag{19.79}$$

These give the following constraints for Γ with external fields:

$$\frac{\delta \Gamma}{\delta \langle B \rangle} = \partial_\mu \langle A_\mu \rangle - \alpha \langle B \rangle \ , \quad \mathrm{i} \partial_\mu \left(\frac{\delta \Gamma}{\delta K_\mu} \right) = \frac{\delta \Gamma}{\delta \langle \bar{c} \rangle} \ . \tag{19.80}$$

If B is not used, $\langle B \rangle$ should be replaced by

$$\langle B \rangle \to \frac{1}{\alpha} \partial_\mu \langle A_\mu \rangle \, . \tag{19.81}$$

This corresponds to setting $J_B = 0$ at the outset. Differentiating (19.78) with respect to the external fields and then setting the external fields equal to zero, we obtain many identities between the Green's functions.

19.5 Representations of the BRS Algebra

To complete the quantization of gauge theories and understand their structure, it is important to know the representation of the BRS algebra. We first investigate the representation of the BRS algebra but excluding $\bar{Q}_{\text{B}}$:

$$\mathrm{i}[Q_c, Q_{\text{B}}] = Q_{\text{B}} \, , \quad Q_{\text{B}}^2 = 0 \, . \tag{19.82}$$

However, we restrict the representations to those used in gauge theories, rather than considering abstract representations. As mentioned before, Q_c and Q_{B} are both Hermitian operators, and this turns out to be a key fact in obtaining representations.

We first note that all operators in gauge theories can be decomposed into a sum of operators satisfying

$$\mathrm{i}[Q_c, \boldsymbol{\phi}] = N\boldsymbol{\phi} \, . \tag{19.83}$$

The state $\boldsymbol{\phi}|\mathbf{0}\rangle$ satisfies

$$\mathrm{i}Q_c\boldsymbol{\phi}|\mathbf{0}\rangle = N\boldsymbol{\phi}|\mathbf{0}\rangle \, , \quad N \text{ integer} , \tag{19.84}$$

where we have assumed that the vacuum $|\mathbf{0}\rangle$ is annihilated by Q_c, i.e.,

$$Q_c|\mathbf{0}\rangle = 0 \, . \tag{19.85}$$

Equation (19.84) indicates that all eigenvalues of the Hermitian operators are pure imaginary. Moreover, the second power of the Hermitian operator Q_{B} is zero. All these facts indicate that we must have an indefinite metric. In the following, we will use the result in Chap. 5, and the reader is referred to the discussion there for the above details.

The important point regarding this algebraic representation is to obtain simultaneously the metric matrix η and the representation matrices of the operators Q_{B} and Q_c. Writing the representation matrices of Q_{B} and $\mathrm{i}Q_c$ as q and n, respectively, the

Hermiticity condition for Q_B and Q_c can be written in the form

$$\eta q = q^\dagger \eta \,, \quad \eta n = -n^\dagger \eta \,, \tag{19.86}$$

where we have used (5.34). The representation of the BRS algebra satisfies

$$[n, q] = q \,, \quad q^2 = 0 \,. \tag{19.87}$$

We thus seek a representation which diagonalizes n. This corresponds to choosing a basis satisfying (19.84) as basis in the state vector space $\mathscr{V}$. Thus,

$$n^\dagger = n \,. \tag{19.88}$$

The diagonal elements of the matrix n are integers, as already mentioned. According to (19.86), n anti-commutes with η. Therefore, we decompose $\mathscr{V}$ in terms of the eigenvalues N of n,

$$\mathscr{V} = \sum_{N=-\infty}^{\infty} \bigoplus \mathscr{V}^{(N)} \,, \tag{19.89}$$

$$\mathscr{V}^{(N)} = \left\{ |x\rangle / (\mathrm{i}Q_c - N)|x\rangle = 0, |x\rangle \in \mathscr{V} \right\} . \tag{19.90}$$

The fact that n anti-commutes with η means that, multiplying a vector by the matrix η, $\mathscr{V}^{(N)}$ turns into $\mathscr{V}^{(-N)}$, i.e.,

$$\eta \mathscr{V}^{(N)} = \mathscr{V}^{(-N)} \,, \tag{19.91}$$

where the manipulation of multiplying a vector by a matrix is defined as follows. If we take $\left\{ |e_j\rangle \right\}$ as a set of basis vectors, an arbitrary vector $|x\rangle$ is expressed as

$$|x\rangle = \sum_j x_j |e_j\rangle \,. \tag{19.92}$$

Multiplying by η is then defined by

$$\eta |x\rangle = \sum_{j,k} \eta_{jk} x_k |e_j\rangle \,. \tag{19.93}$$

Moreover, we choose η to be the standard form mentioned in Chap. 5, whence

$$\eta^2 = \mathbf{1} \,. \tag{19.94}$$

We must now obtain the matrix representing q. To do this, we first define the BRS singlet and the BRS doublet, following Kugo and Ojima [173], then refine them.

1. **BRS singlet.** If a vector $|f\rangle$ cannot be written in the form $|f\rangle = Q_B|g\rangle$ and it satisfies

$$Q_B|f\rangle = 0 , \tag{19.95}$$

then $|f\rangle$ is called a BRS singlet.
2. **BRS doublet.** If there exist vectors $|f\rangle$ and $|g\rangle$ satisfying

$$|f\rangle = Q_B|g\rangle \neq 0 , \quad \text{and hence} , \quad Q_B|f\rangle = 0 , \tag{19.96}$$

we say that $|f\rangle$ and $|g\rangle$ form a BRS doublet. Note also that $|g\rangle$ is called a *parent vector* (parent state) and $|f\rangle$ is called a *daughter vector* (daughter state).

A weak point of the above definition is the lack of uniqueness. If we consider a singlet state and parent and daughter states of a doublet as $|\mathrm{s}\rangle$, $|\mathrm{p}\rangle$, and $|\mathrm{d}\rangle$, respectively, then $|\mathrm{s}\rangle + |\mathrm{d}\rangle$ also satisfies the first condition (1), while $|\mathrm{p}\rangle + |\mathrm{s}\rangle$ also satisfies the condition for the parent in (2). In order to define these uniquely within a given representation, we use the metric matrix η.

We define two subspaces

$$\mathscr{V}_\mathrm{d} = q\mathscr{V} = \left\{ q|x\rangle : |x\rangle \in \mathscr{V} \right\} , \tag{19.97}$$

$$\mathscr{V}_\mathrm{p} = q^\dagger \mathscr{V} = \left\{ q^\dagger |x\rangle : |x\rangle \in \mathscr{V} \right\} . \tag{19.98}$$

Firstly, $Q_B|x\rangle = q|x\rangle$, and since Q_B annihilates the singlet and the daughter state, $\mathscr{V}_\mathrm{d}$ should include every daughter state, so in this sense, it is a set of daughter states. Therefore, considering an arbitrary vector $|y\rangle \in \mathscr{V}_\mathrm{d}$, it can always be written in the form

$$\mathscr{V}_\mathrm{d} \ni |y\rangle = Q_B|x\rangle . \tag{19.99}$$

Setting $|\tilde{y}\rangle = \eta|y\rangle$ and using the definition of η,

$$\langle x|Q_B|\tilde{y}\rangle = \langle y|\tilde{y}\rangle = \sum_{j,k,l} y_j^* \eta_{jk}\eta_{kl} y_l = \sum_j |y_j|^2 \neq 0 , \tag{19.100}$$

where we have used (19.94). This implies that

$$Q_B|\tilde{y}\rangle = q|\tilde{y}\rangle \neq 0 . \tag{19.101}$$

Hence, $\eta|y\rangle$ is a parent vector. In fact,

$$\eta\mathscr{V}_\mathrm{d} = \eta q \mathscr{V} = q^\dagger \eta \mathscr{V} = q^\dagger \mathscr{V} = \mathscr{V}_\mathrm{p} , \tag{19.102}$$

where we have used the fact that η is non-degenerate, viz.,

$$\eta \mathscr{V} = \mathscr{V} . \tag{19.103}$$

We thus see that $\mathscr{V}_{\mathrm{p}}$ is a set of parent states. Alternatively, it can be considered that (19.97) and (19.98) define daughter and parent states, or from (19.102), a parent state is created from a daughter state by multiplying by η.

Furthermore, if η is the standard form, then (16.93) implies

$$\eta \mathscr{V}_{\mathrm{p}} = \mathscr{V}_{\mathrm{d}} . \tag{19.104}$$

Next, we decompose $\mathscr{V}$ into a direct sum of three subspaces. If $\mathscr{V}_{\mathrm{S}}$ is the space of singlets,

$$\mathscr{V} = \mathscr{V}_{\mathrm{S}} \oplus \mathscr{V}_{\mathrm{p}} \oplus \mathscr{V}_{\mathrm{d}} . \tag{19.105}$$

Therefore, from (19.102), (19.103), and (19.104),

$$\eta \mathscr{V}_{\mathrm{S}} = \mathscr{V}_{\mathrm{S}} . \tag{19.106}$$

This means that the singlet state is orthogonal to the doublet state. The structure of η is

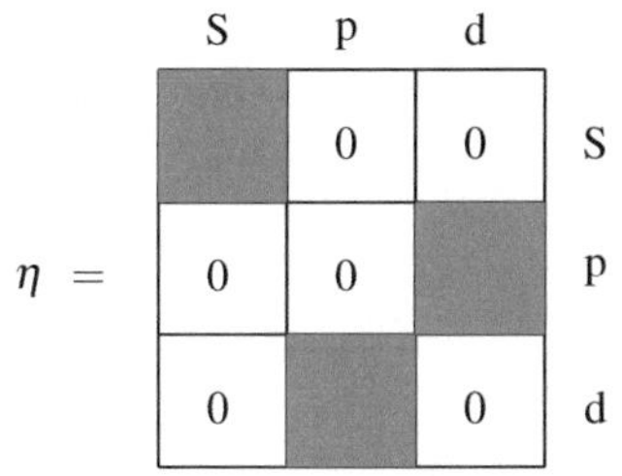

(19.107)

where the shaded regions are non-degenerate matrices. What can be understood from this is that both the parent vectors belonging to $\mathscr{V}_{\mathrm{p}}$ and the daughter vectors belonging to $\mathscr{V}_{\mathrm{d}}$ have zero norm. Non-zero inner products appear only between a parent vector and a daughter vector or between singlets.

Now, Q_{B} annihilates singlets, so if $\{0\}$ is the set consisting only of the zero vector,

$$q \mathscr{V}_{\mathrm{S}} = \{0\} . \tag{19.108}$$

Therefore,

$$q^{\dagger} \mathscr{V}_{\mathrm{S}} = q^{\dagger} \eta \mathscr{V}_{\mathrm{S}} = \eta q \mathscr{V}_{\mathrm{S}} = \{0\} . \tag{19.109}$$

Thus, singlets turn out to be annihilated by both q and $q^\dagger$. Additionally, since the second power of either q or $q^\dagger$ is zero,

$$q\mathscr{V}_{\rm d} = q^\dagger \mathscr{V}_{\rm p} = \{0\}\ . \tag{19.110}$$

Thus,

$$q\mathscr{V}_{\rm p} = q\mathscr{V}_{\rm p} \oplus q\mathscr{V}_{\rm d} \oplus q\mathscr{V}_{\rm S} = q\mathscr{V} = \mathscr{V}_{\rm d}\ , \tag{19.111}$$

$$q^\dagger\mathscr{V}_{\rm d} = q^\dagger\mathscr{V}_{\rm d} \oplus q^\dagger\mathscr{V}_{\rm p} \oplus q^\dagger\mathscr{V}_{\rm S} = q^\dagger\mathscr{V} = \mathscr{V}_{\rm p}\ . \tag{19.112}$$

These relations will be used to determine the matrix q. Introducing the ghost number N, we consider the subspace

$$\mathscr{V}_{\rm p}^{(N)} = \mathscr{V}_{\rm p} \cap \mathscr{V}^{(N)}\ . \tag{19.113}$$

From the commutation relation between n and q,

$$q\mathscr{V}_{\rm p}^{(N)} = \mathscr{V}_{\rm d}^{(N+1)}\ . \tag{19.114}$$

For η, we have

$$\eta\mathscr{V}_{\rm p}^{(N)} = \mathscr{V}_{\rm d}^{(-N)}\ , \quad \eta\mathscr{V}_{\rm d}^{(N+1)} = \mathscr{V}_{\rm p}^{(-N-1)}\ . \tag{19.115}$$

Taking into account these relations and the condition (19.86) on q, in the representation where n is diagonal,

$$q = \begin{array}{cccc|l} (N+1,{\rm d}) & (N,{\rm p}) & (-N,{\rm d}) & (-N-1,{\rm p}) & \\ \hline 0 & h & 0 & 0 & (N+1,{\rm d}) \\ 0 & 0 & 0 & 0 & (N,{\rm p}) \\ 0 & 0 & 0 & h^\dagger & (-N,{\rm d}) \\ 0 & 0 & 0 & 0 & (-N-1,{\rm p}) \end{array} \tag{19.116}$$

where h and $h^\dagger$ are non-degenerate matrices. If we choose the basis in each subspace properly, then at the end of the day the irreducible representation in the doublet space has the form

$$\eta = \left(\begin{array}{cc|cc} & & 0 & 1 \\ & & 1 & 0 \\ \hline 0 & 1 & & \\ 1 & 0 & & \end{array}\right), \quad n = \left(\begin{array}{cc|cc} N+1 & 0 & & \\ 0 & N & & \\ \hline & & -N & 0 \\ & & 0 & -N-1 \end{array}\right),$$

$$q = \left(\begin{array}{cc|cc} 0 & a & & \\ 0 & 0 & & \\ \hline & & 0 & a \\ & & 0 & 0 \end{array}\right), \tag{19.117}$$

where a is a real number which cannot be determined by (19.87) alone, and the blanks are zero matrices. The four states in the basis of this representation form the quartet found by Kugo and Ojima.

What we have discussed above concerns the representation for the BRS doublet. For the singlet, two kinds of representation can be obtained:

- $N \neq 0$

$$n = \begin{pmatrix} N & \\ & -N \end{pmatrix}, \quad \eta = \begin{pmatrix} 0 & 1 \\ 1 & 0 \end{pmatrix}, \quad q = \begin{pmatrix} 0 & 0 \\ 0 & 0 \end{pmatrix}, \tag{19.118}$$

- $N = 0$

$$n = (0), \quad \eta = (\pm 1), \quad q = (0). \tag{19.119}$$

In order to formulate gauge theories, we introduce the following requirement:

A subspace of the BRS singlet $\mathscr{V}_S$ belongs to the positive-definite metric.

With this assumption, the BRS singlet representation in a gauge theory is restricted to

$$n = (0), \quad \eta = (1), \quad q = (0). \tag{19.120}$$

19.6 Unitarity of the S-Matrix

Since gauge theories have many fields belonging to the indefinite metric, there remains the problem of how to eliminate these from the S-matrix. We discuss this issue using the method suggested by Kugo and Ojima. Kugo and Ojima defined the physical state space $\mathscr{V}_{\text{phys}}$ by [174]

$$\mathscr{V}_{\text{phys}} = \Big\{ |x\rangle : Q_{\text{B}}|x\rangle = 0\,, |x\rangle = 0\,,\ |x\rangle \in \mathscr{V} \Big\}\,. \tag{19.121}$$

It is therefore clear from the discussion in the last section that

$$\mathscr{V}_{\text{phys}} = \mathscr{V}_{\text{S}} \oplus \mathscr{V}_{\text{d}}\,. \tag{19.122}$$

Hence, given the structure of η, if $|f\rangle$ and $|g\rangle$ belong to $\mathscr{V}_{\text{phys}}$, we have

$$\langle f|g\rangle = \langle f|P(\mathscr{V}_{\text{S}})|g\rangle\,, \tag{19.123}$$

where $P(\mathscr{V}_{\text{S}})$ is the projection operator onto the BRS singlet subspace $\mathscr{V}_{\text{S}}$.

Since Q_{B} is a conserved quantity, it commutes with the S-matrix, i.e.,

$$[Q_{\text{B}}, S] = 0\,. \tag{19.124}$$

Hence, if both $|f\rangle$ and $|g\rangle$ belong to $\mathscr{V}_{\text{S}}$, we have

$$Q_{\text{B}}S|f\rangle = 0\,, \quad Q_{\text{B}}S|g\rangle = 0\,. \tag{19.125}$$

Thus,

$$S|f\rangle,\ S|g\rangle \in \mathscr{V}_{\text{phys}}\,. \tag{19.126}$$

So, changing the unitarity condition by using (19.123),

$$\langle f|g\rangle = \big\langle f\big|S^{\dagger}S\big|g\big\rangle = \big\langle f\big|S^{\dagger}P(\mathscr{V}_{\text{S}})S\big|g\big\rangle\,. \tag{19.127}$$

It thus turns out that the unitarity condition for the S-matrix is expressed only by states in $\mathscr{V}_{\text{phys}}$, which belongs to the positive-definite metric, and the BRS doublets are completely eliminated.

The discussion above can be applied to QED. We use the Lagrangian density in the form (19.1). In this case, since the gauge group is $U(1)$, the structure constant f is zero. The ghost part is

$$\mathscr{L}_{\text{FP}} = \mathrm{i}\partial_{\mu}\bar{c}\partial_{\mu}c\,. \tag{19.128}$$

The ghost fields are then free fields:

$$\Box c = \Box\bar{c} = 0\,. \tag{19.129}$$

To eliminate the ghost fields, which do not exist in standard QED, we restrict the physical states by

$$c^{(+)}(x)|\text{phys}\rangle = \bar{c}^{(+)}|\text{phys}\rangle = 0\,. \tag{19.130}$$

Further, we define them using Q_{B}:

$$Q_{\text{B}}|\text{phys}\rangle = 0\,. \tag{19.131}$$

This means that consistency between (19.130) and (19.131) is an issue. In QED,

$$\{Q_{\text{B}}, c^{(+)}(x)\} = 0\,, \quad \{Q_{\text{B}}, \bar{c}^{(+)}(x)\} = B^{(+)}(x)\,, \tag{19.132}$$

so it turns out that (19.131) requires the condition

$$B^{(+)}(x)|\text{phys}\rangle = 0\,, \tag{19.133}$$

which is nothing but the Lorenz condition introduced in Sect. 5.4. Since Q_{B} commutes with $B^{(+)}(x)$, it does not give any further condition. Moreover, in this case Q_{B} is given by

$$Q_{\text{B}} = \int \mathrm{d}^3x (B\dot{c} - \dot{B}c)\,, \tag{19.134}$$

so if (19.130) and (19.133) are satisfied, then (19.131) is satisfied automatically. Hence, in QED, it turns out that the Kugo–Ojima condition coincides with the Lorenz condition for the states with no ghosts.

19.7 Representations of the Extended BRS Algebra

In Sect. 19.5, we investigated representations of the BRS algebra without $\bar{Q}_{\text{B}}$. In this section, we add comments on representations of the BRS algebra including $\bar{Q}_{\text{B}}$, as defined by (19.41) and (19.42).

We first decompose the state space into the singlet and the doublet:

$$\mathscr{V} = \mathscr{V}_{\text{S}} \oplus \mathscr{V}_{\text{D}}\,. \tag{19.135}$$

This decomposition is common to both Q_{B} and $\bar{Q}_{\text{B}}$. The doublet space can be decomposed as

$$\mathscr{V}_{\text{D}} = \mathscr{V}_{\text{p}} \oplus \mathscr{V}_{\text{d}} = \mathscr{V}_{\bar{\text{p}}} \oplus \mathscr{V}_{\bar{\text{d}}}\,. \tag{19.136}$$

However, considering $\bar{q}$ as a representation of $\bar{Q}_B$, new subspaces are defined by

$$\mathscr{V}_{\bar{p}} = \bar{q}^\dagger \mathscr{V} = \bar{q}^\dagger \mathscr{V}_D , \quad \mathscr{V}_{\bar{d}} = \bar{q} \mathscr{V} = \bar{q} \mathscr{V}_D . \tag{19.137}$$

For each of Q_B and $\bar{Q}_B$, we introduce decompositions into p and d. A subspace

$$Q_B \bar{Q}_B \mathscr{V} = \bar{Q}_B Q_B \mathscr{V} = \mathscr{V}^{(d,d)} \tag{19.138}$$

is a set of daughter states for both Q_B and $\bar{Q}_B$. Thus,

$$\mathscr{V}^{(d,d)} = \mathscr{V}_d \cap \mathscr{V}_{\bar{d}} . \tag{19.139}$$

Multiplying this subspace by η, it becomes a set of parent states for both Q_B and $\bar{Q}_B$, i.e.,

$$\eta \mathscr{V}^{(d,d)} = \mathscr{V}^{(p,p)} = \mathscr{V}_p \cap \mathscr{V}_{\bar{p}} . \tag{19.140}$$

Starting with this subspace, we introduce the two subspaces

$$Q_B \mathscr{V}^{(p,p)} = \mathscr{V}^{(d,p)} , \quad \bar{Q}_B \mathscr{V}^{(p,p)} = \mathscr{V}^{(p,d)} , \tag{19.141}$$

which are related by

$$\eta \mathscr{V}^{(d,p)} = \mathscr{V}^{(p,d)} . \tag{19.142}$$

Therefore, the doublet space decomposes as follows:

$$\mathscr{V}_D = \mathscr{V}^{(p,p)} \oplus \mathscr{V}^{(p,d)} \oplus \mathscr{V}^{(d,p)} \oplus \mathscr{V}^{(d,d)} . \tag{19.143}$$

In contrast, the singlet space has the following properties:

$$Q_B \mathscr{V}_S = \bar{Q}_B \mathscr{V}_S = \{0\} , \quad \eta \mathscr{V}_S = \mathscr{V}_S . \tag{19.144}$$

The discussion in this section is descriptive and results are given without proof, but we shall give the details in the remainder of this chapter. Note that the decomposition (19.143) shows that there is a one-to-one correspondence between the four subspaces. In the following, we shall express the elements in $\mathscr{V}_D$ in the form (p,d), using the transformation properties for Q_B and $\bar{Q}_B$. We express relations such as $Q_B|a\rangle \sim |b\rangle$, $\bar{Q}_B|c\rangle \sim |d\rangle$, where $\sim$ stands for equality up to a numerical coefficient, in the following graphical way:

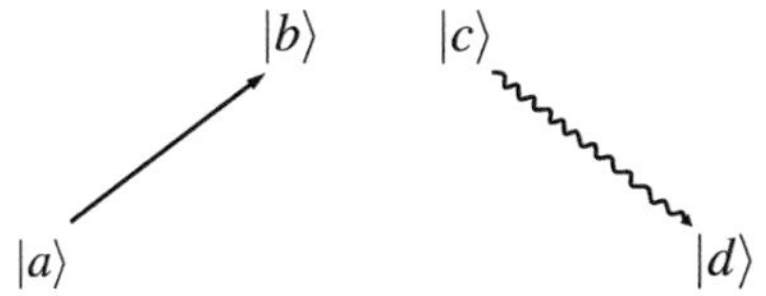

(19.145)

Then the graphical representation of the one-to-one correspondence between the elements of the four subspaces in (19.143) is

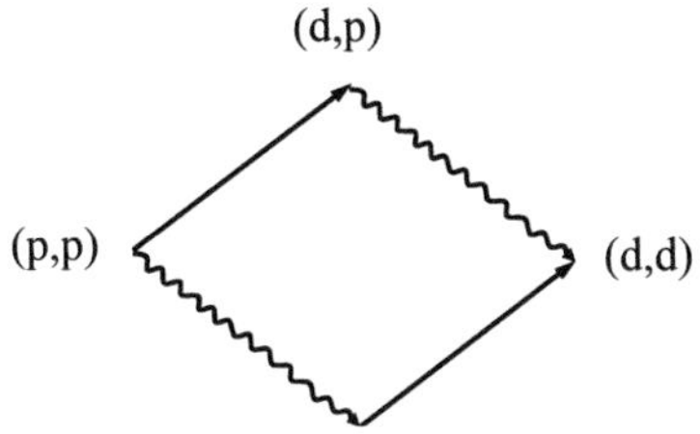

(19.146)

These four elements form a quartet.

19.8 Representations of BRS Transformations for Auxiliary Fields

We have considered BRS singlets and BRS doublets for state vectors, and this can be extended to operators, i.e., operators can be classified using states obtained by multiplying the vacuum by operators. However, since Heisenberg operators are not irreducible, it is simpler to consider asymptotic fields. As already shown in (19.65) and (19.66), multiplying the vacuum by A_μ, B, $D_\mu c$, and $\bar{c}$ produces massless scalar states. Since the operators which create these particles should be included, we consider them as asymptotic fields and write them as follows:

$$\boldsymbol{A} \to \partial_\mu \chi \;, \quad \boldsymbol{B} \to \beta \;, \quad \boldsymbol{D}_\mu c \to \partial_\mu \gamma \;, \quad \bar{\boldsymbol{c}} \to \bar{\gamma} \;. \tag{19.147}$$

Similarly, we introduce the following asymptotic fields:

$$\bar{\boldsymbol{B}} \to \bar{\beta} \;, \quad D_\mu \bar{\boldsymbol{c}} \to \partial_\mu \bar{\Gamma} \;, \quad \boldsymbol{c} \to \Gamma \;. \tag{19.148}$$

It is easily checked in the Landau gauge that these create massless particles. Therefore, from the BRS transformations for Heisenberg operators,

$$\delta\chi = \gamma\,, \quad \delta\bar{\gamma} = \mathrm{i}\beta\,, \quad \delta\gamma = 0\,, \quad \delta\beta = 0\,, \tag{19.149}$$

$$\bar{\delta}\chi = \bar{\Gamma}\,, \quad \bar{\delta}\Gamma = \mathrm{i}\bar{\beta}\,, \quad \bar{\delta}\bar{\Gamma} = 0\,, \quad \bar{\delta}\bar{\beta} = 0\,. \tag{19.150}$$

From these asymptotic fields, choosing the Landau gauge, we obtain

$$\begin{aligned}\langle\chi^a(x), \beta^b(y)\rangle &= -\langle\chi^a(x), \bar{\beta}^b(y)\rangle = \mathrm{i}\langle\gamma^a(x), \bar{\gamma}^b(y)\rangle = \mathrm{i}\langle\Gamma^a(x), \bar{\Gamma}^b(y)\rangle \\ &= -\delta_{ab} D_{\mathrm{F}}(x-y)\,.\end{aligned} \tag{19.151}$$

In the following, we consider two cases:

1. **Quartet representation.** In perturbation theories, the following conditions hold true:

$$\gamma = \Gamma\,, \quad \bar{\Gamma} = \bar{\gamma}\,. \tag{19.152}$$

In this case,

$$\bar{\delta}\delta\chi = \bar{\delta}\gamma = \bar{\delta}\Gamma = \mathrm{i}\bar{\beta}\,, \qquad \delta\bar{\delta}\chi = \delta\bar{\Gamma} = \delta\bar{\gamma} = \mathrm{i}\beta\,.$$

Thus, taking into account the anti-commutativity of δ and $\bar{\delta}$,

$$\bar{\beta} = -\beta\,. \tag{19.153}$$

In this case, the auxiliary fields form a quartet expressed by the following graph:

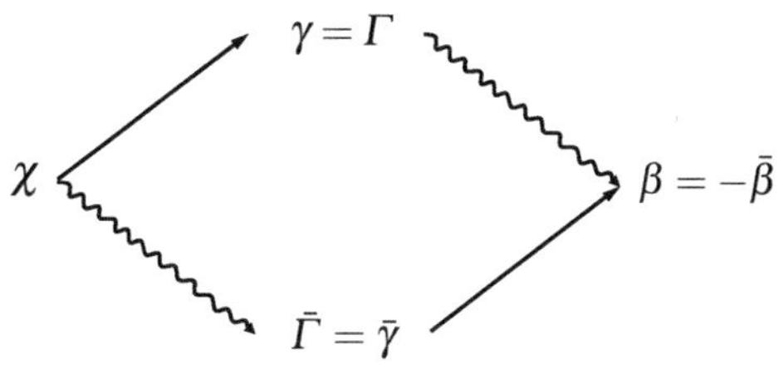

(19.154)

This quartet was introduced by Kugo and Ojima.

2. **Octet representation.** Next, we consider what happens if we take γ and $\bar{\Gamma}$ to be different from Γ and $\bar{\gamma}$:

$$\gamma \neq \Gamma\,, \quad \bar{\Gamma} \neq \bar{\gamma}\,. \tag{19.155}$$

In this case, if we include β, and taking into account that there are at least five auxiliary fields, it will be impossible to accommodate all the auxiliary fields unless we introduce at least two quartets. In this case too, we have to write down equations for the BRS transformations twice, as in (19.149) and (19.150). Hence, in addition to the four asymptotic fields in (19.155), we introduce two new asymptotic fields β, $\bar{\beta}$ and d, $\bar{d}$:

$$\begin{aligned} \delta\Gamma &= \mathrm{i}d\ , \quad \bar{\delta}\Gamma = \mathrm{i}\beta\ , \quad \delta\bar{\beta} = -\gamma\ , \quad \bar{\delta}d = \gamma\ , \\ \delta\bar{\gamma} &= \mathrm{i}\beta\ , \quad \bar{\delta}\bar{\gamma} = -\mathrm{i}\bar{d}\ , \quad \delta\bar{d} = \Gamma\ , \quad \bar{\delta}\beta = \Gamma\ . \end{aligned} \tag{19.156}$$

These equations include (19.149) and (19.150). In addition, although χ has vanished, if we interpret

$$\chi = \beta - \bar{\beta} + (\text{singlet})\ , \tag{19.157}$$

then it does not after all contradict (19.149) and (19.150). Now, d and $\bar{d}$ become massless asymptotic fields for $\boldsymbol{c} \times \boldsymbol{c}$ and $\bar{\boldsymbol{c}} \times \bar{\boldsymbol{c}}$, as will be clear from the discussion in the next section.

Additionally, for d and $\bar{d}$, it can be shown immediately that

$$\left\langle \bar{d}^a(x), d^b(y) \right\rangle = \delta_{ab} D_F(x-y)\ . \tag{19.158}$$

We use this to make a graphical representation for an octet corresponding to (19.156):

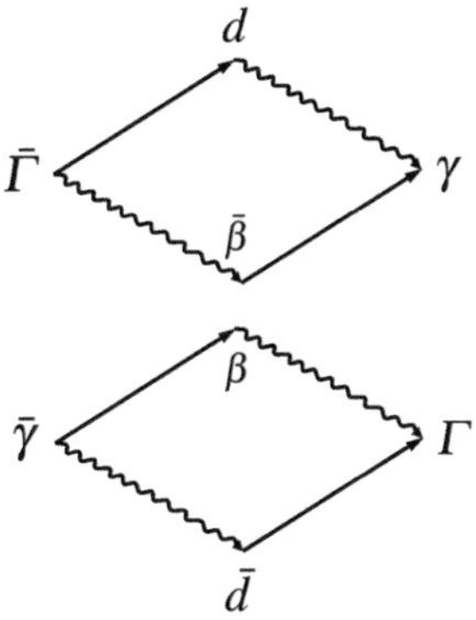

(19.159)

This octet was considered by Kugo.

Now that representations like the quartet and the octet have been obtained, we can investigate what kinds of representations exist more generally. We will discuss this in the next section.

19.9 Representations of BRSNO Algebras

We have discussed the fact that, in general gauge theories, an extended BRS invariance holds true and the conserved quantities form the BRS algebras. It was pointed out by Nakanishi and Ojima in [175] that two more conserved charges exist in the Landau gauge. In this gauge, the sum of the gauge-fixing term and the ghost term is

$$\mathscr{L}_{\mathrm{GF}} + \mathscr{L}_{\mathrm{FP}} = \partial_\mu B\, A_\mu + \mathrm{i}\partial_\mu \bar{c}\, D_\mu c\ . \tag{19.160}$$

This Lagrangian density is invariant under the following transformations among auxiliary fields:

$$\boldsymbol{B} \to \boldsymbol{B} - \mathrm{i}g(\boldsymbol{c} \times \boldsymbol{c})\lambda\ , \quad \bar{\boldsymbol{c}} \to \bar{\boldsymbol{c}} - 2\boldsymbol{c}\lambda\ , \quad \boldsymbol{c} \to \boldsymbol{c}\ , \tag{19.9.2a}$$

where λ is a transformation parameter. Another transformation is

$$\boldsymbol{B} \to \boldsymbol{B} + \mathrm{i}g(\bar{\boldsymbol{c}} \times \bar{\boldsymbol{c}})\lambda\ , \quad \boldsymbol{c} \to \boldsymbol{c} + 2\bar{\boldsymbol{c}}\lambda\ , \quad \bar{\boldsymbol{c}} \to \bar{\boldsymbol{c}}\ . \tag{19.9.2b}$$

Writing the Hermitian generators of these transformations as $Q(c,c)$ and $Q(\bar{c},\bar{c})$, respectively,

$$\mathrm{i}\big[Q(c,c), \boldsymbol{B}\big] = -\mathrm{i}g\boldsymbol{c} \times \boldsymbol{c}\ , \quad \mathrm{i}\big[Q(c,c), \bar{\boldsymbol{c}}\big] = -2\boldsymbol{c}\ , \quad \mathrm{i}\big[Q(c,c), \boldsymbol{c}\big] = 0\ , \tag{19.9.3a}$$

$$\mathrm{i}\big[Q(\bar{c},\bar{c}), \boldsymbol{B}\big] = \mathrm{i}g\bar{\boldsymbol{c}} \times \bar{\boldsymbol{c}}\ , \quad \mathrm{i}\big[Q(\bar{c},\bar{c}), \boldsymbol{c}\big] = 2\bar{c}\ , \quad \mathrm{i}\big[Q(\bar{c},\bar{c}), \bar{\boldsymbol{c}}\big] = 0\ . \tag{19.9.3b}$$

In contrast, the fundamental fields are invariant under these transformations. Hence, adding $Q(c,c)$ and $Q(\bar{c},\bar{c})$ to the extended BRS algebra (19.41) and (19.42), we obtain the following relations:

$$\begin{aligned}
&\mathrm{i}\big[Q_c, Q(c,c)\big] = 2Q(c,c)\ , \quad \mathrm{i}\big[Q_c, Q(\bar{c},\bar{c})\big] = -2Q(\bar{c},\bar{c})\ ,\\
&\big[Q(c,c), \bar{Q}_{\mathrm{B}}\big] = 2\mathrm{i}Q_{\mathrm{B}}\ , \quad \big[Q(\bar{c},\bar{c}), Q_{\mathrm{B}}\big] = -2\mathrm{i}Q_{\mathrm{B}}\ ,\\
&\big[Q(c,c), Q_{\mathrm{B}}\big] = 0\ , \quad \big[Q(\bar{c},\bar{c}), \bar{Q}_{\mathrm{B}}\big] = 0\ ,\\
&\big[Q(c,c), Q(\bar{c},\bar{c})\big] = 4\mathrm{i}Q_c\ .
\end{aligned} \tag{19.9.4}$$

Although it is possible to give explicit forms of $Q(c,c)$ and $Q(\bar{c},\bar{c})$, we omit them here because such expressions will not be necessary. In the Landau gage, both $Q(c,c)$ and $Q(\bar{c},\bar{c})$ can be conserved quantities, but in other gauges they are not conserved. However, the commutation relations given above still hold true. The algebra that extends the extended BRS algebra by including (19.9.4) is called the *BRSNO algebra*, which we denote by $\mathscr{A}$. Moreover, excluding the anti-commuting operators Q_{B} and $\bar{Q}_{\mathrm{B}}$, the subspace consisting only of Q_c, $Q(c, c)$, and $Q(\bar{c}, \bar{c})$ is denoted by $\mathscr{B}$. This can be considered as the bosonic part of the algebra $\mathscr{A}$.

To obtain representations of the algebra $\mathscr{A}$, we introduce the following notation:

$$\mathrm{i}Q_c \to 2J_x\ , \quad Q(c, c) \to 2J_+\ , \quad Q(\bar{c}, \bar{c}) \to 2J_-\ ,$$
$$\mathrm{i}Q_{\mathrm{B}} \to \alpha\ , \quad \bar{Q}_{\mathrm{B}} \to \beta\ . \tag{19.9.5}$$

In addition, in the following discussion, we restrict the representations of the algebra $\mathscr{A}$ to those which can be realized in gauge theories, rather than considering completely abstract representations. First, we obtain a representation of the Lie algebra $\mathscr{B}$. We write down the commutation relations which characterize $\mathscr{B}$ using the symbols above:

$$\big[J_z, J_+\big] = J_+\ , \quad \big[J_z, J_-\big] = -J_-\ , \quad \big[J_+, J_-\big] = 2J_z\ . \tag{19.9.6}$$

These coincide with the commutation relations among the three components of the angular momentum. Hence, in gauge theories and in the Landau gauge, we will prove that the representation of $\mathscr{B}$ coincides with that of the angular momentum. To do so, we begin with the following requirements:

1. The vacuum state $|\mathbf{0}\rangle$ must satisfy the condition

$$J_+|\mathbf{0}\rangle = J_-|\mathbf{0}\rangle = J_z|\mathbf{0}\rangle\ . \tag{19.9.7}$$

2. The vacuum state must be cyclic with respect to the algebra formed by fundamental fields and auxiliary fields, i.e., combining states constructed by multiplying the vacuum by products of field operators, we should be able to generate the entire state vector space $\mathscr{V}$.

Since the fundamental fields and $B - \bar{B}$ commute with the three elements of $\mathscr{B}$, these belong to the one-dimensional representation of β. Therefore, to make other representations, we must use $\boldsymbol{c}$, $\bar{\boldsymbol{c}}$, and $\boldsymbol{B} + \bar{\boldsymbol{B}}$ or $\boldsymbol{c} \times \bar{\boldsymbol{c}}$. We have the following commutation relations:

$$\begin{aligned}
&[J_z, \boldsymbol{c}] = \frac{1}{2}\boldsymbol{c}\ , \quad [J_z, \bar{\boldsymbol{c}}] = -\frac{1}{2}\bar{\boldsymbol{c}}\ ,\\
&\mathrm{i}[J_+, \bar{\boldsymbol{c}}] = -\boldsymbol{c}\ , \quad \mathrm{i}[J_+, \boldsymbol{c}] = 0\ ,\\
&\mathrm{i}[J_-, \boldsymbol{c}] = \bar{\boldsymbol{c}}\ , \quad \mathrm{i}[J_-, \bar{\boldsymbol{c}}] = 0\ .
\end{aligned} \tag{19.9.8}$$

We then define a doublet and a triplet by

$$\phi\,(1/2, 1/2) = \mathrm{i}\boldsymbol{c}\ , \quad \phi\,(1/2, -1/2) = \bar{\boldsymbol{c}}\ , \tag{19.9.9}$$

$$\phi(1,1) = \frac{\mathrm{i}}{\sqrt{2}}(\boldsymbol{c}\times\boldsymbol{c})\ , \quad \phi(1,0) = \boldsymbol{c}\times\bar{\boldsymbol{c}}\ , \quad \phi(1,-1) = -\frac{\mathrm{i}}{\sqrt{2}}(\bar{\boldsymbol{c}}\times\bar{\boldsymbol{c}})\ . \tag{19.9.10}$$

We immediately obtain

$$\begin{aligned} \big[J_z, \phi(j,m)\big] &= m\phi(j,m)\ , \\ \big[J_+, \phi(j,m)\big] &= \sqrt{(j-m)(j+m+1)}\phi(j,m+1)\ , \\ \big[J_-, \phi(j,m)\big] &= \sqrt{(j+m)(j-m+1)}\phi(j,m-1)\ . \end{aligned} \tag{19.9.11}$$

Since these relations are the same as those for angular momentum, starting with the doublet and the triplet and using the Clebsch–Gordan coefficients for the angular momentum, we can make a product representation:

$$\phi(J,M,c) = \sum_{m_1+m_2=M} \phi(j_1,m_1,a)\phi(j_2,m_2,b)\big\langle j_1,j_2,m_1,m_2 | j_1,j_2,J,M\big\rangle\ , \tag{19.9.12}$$

where a, b, and c represent other quantum numbers. It is clear that such a product representation again satisfies (19.9.11). Hence, if we take

$$\boldsymbol{J}^2 = \frac{1}{2}(J_+J_- + J_-J_+) + J_z^2\ , \tag{19.9.13}$$

then clearly, from (19.9.7) and (19.9.11), for the polynomial of field operators $\phi(j,m,a)$,

$$\boldsymbol{J}^2\phi(j,m,a)|\boldsymbol{0}\rangle = j(j+1)\phi(j,m,a)|\boldsymbol{0}\rangle\ , \quad J_z\phi(j,m,a)|\boldsymbol{0}\rangle = m\phi(j,m,a)|\boldsymbol{0}\rangle\ . \tag{19.9.14}$$

We thus see that the representation of $\mathscr{B}$ is exactly the same as that of the angular momentum. Next, we introduce α and β, with commutation relations

$$[J_z,\alpha] = \frac{1}{2}\alpha\ , \quad [J_z,\beta] = -\frac{1}{2}\beta\ , \quad [J_+,\beta] = \alpha\ , \quad [J_-,\alpha] = \beta\ . \tag{19.9.15}$$

We consider the product $\alpha\beta$:

$$[J_+,\alpha\beta] = [J_-,\alpha\beta] = [J_z,\alpha\beta] = 0\ . \tag{19.9.16}$$

The subspace $\mathscr{V}^{(\mathrm{d,d})}$ can now be written as $\alpha\beta\mathscr{V}$, and from (19.9.16),

$$\boldsymbol{J}\mathscr{V}^{(\mathrm{d,d})} \subset \mathscr{V}^{(\mathrm{d,d})} . \tag{19.9.17}$$

This means that irreducible representations of $\mathscr{B}$ can be constructed within $\mathscr{V}^{(\mathrm{d,d})}$. Next, introducing the metric matrix η and expressing the representation matrices of J_+, J_- and J_z using the same letters, we find that

$$\eta J_+ \eta = J_+^\dagger , \quad \eta J_- \eta = J_-^\dagger , \quad \eta J_z \eta = -J_z^\dagger . \tag{19.9.18}$$

We have chosen η to have the standard form, so

$$\eta\beta\eta = \beta^\dagger , \quad \eta\alpha\eta = -\alpha^\dagger . \tag{19.9.19}$$

In the angular momentum representation, we have

$$J_z^\dagger = J_z , \quad J_+^\dagger = J_- , \quad J_-^\dagger = J_+ , \tag{19.9.20}$$

so, combining (19.9.18) with (19.9.20), we find

$$\eta J_z \eta = -J_z , \quad \eta J_+ \eta = J_- , \quad \eta J_- \eta = J_+ . \tag{19.9.21}$$

Thus,

$$\eta \boldsymbol{J}^2 \eta = \boldsymbol{J}^2 , \tag{19.9.22}$$

so we see that the metric matrix connects states having the same value of the quantum number j. It has been shown in (19.140) that η sends $\mathscr{V}^{(\mathrm{d,d})}$ to $\mathscr{V}^{(\mathrm{p,p})}$, so from (19.9.17),

$$\boldsymbol{J}\mathscr{V}^{(\mathrm{p,p})} \subset \mathscr{V}^{(\mathrm{p,p})} . \tag{19.9.23}$$

Therefore, it turns out that irreducible representations of $\mathscr{B}$ can also be constructed within $\mathscr{V}^{(\mathrm{p,p})}$. We thus construct within $\mathscr{V}^{(\mathrm{p,p})}$ an irreducible representation of $\mathscr{B}$ corresponding to one value of j, writing its basis as

$$|j, m, a\rangle \in \mathscr{V}^{(\mathrm{p,p})} , \quad m = -j, \ldots, j . \tag{19.9.24}$$

Therefore,

$$\alpha|j, m, a\rangle \in \mathscr{V}^{(\mathrm{d,p})} , \quad \beta|j, m, a\rangle \in \mathscr{V}^{(\mathrm{p,d})}. \tag{18.9.25}$$

Moreover,

$$\alpha\beta|j, m, a\rangle \in \mathscr{V}^{(\mathrm{d,d})} . \tag{19.9.26}$$

We thus obtain $4(2j+1)$ basis vectors for a representation of the algebra $\mathscr{A}$. This set of basis vectors is closed under operations by elements of $\mathscr{A}$, i.e.,

$$|j,m,a\rangle\,,\quad \alpha|j,m,a\rangle\,,\quad \beta|j,m,a\rangle\,,\quad \alpha\beta|j,m,a\rangle\,. \tag{19.9.27}$$

If we fix m in the above set, they become a BRS quartet. In addition, if we change m, $\{|j,m,a\rangle\}$ and $\{\alpha\beta|j,m,a\rangle\}$ belong to a representation of the three-dimensional rotation group D_j. On the other hand, $\{\alpha|j,m,a\rangle\}$ and $\{\beta|j,m,a\rangle\}$ become linear combinations of $D_{j+1/2}$ and $D_{j-1/2}$:

$$\begin{aligned}\left|j+\frac{1}{2},m+\frac{1}{2},a\right\rangle &= \sqrt{\frac{j+m+1}{2j+1}}\alpha|j,m,a\rangle + \sqrt{\frac{j-m}{2j+1}}\beta|j,m,a\rangle\,,\\ \left|j-\frac{1}{2},m-\frac{1}{2},a\right\rangle &= \sqrt{\frac{j-m}{2j+1}}|j,m,a\rangle + \sqrt{\frac{j+m+1}{2j+1}}\beta|j,m,a\rangle\,.\end{aligned} \tag{19.9.28}$$

We write the resulting representation of $\mathscr{A}$ as $\mathscr{D}_j$. All representations of $\mathscr{A}$ except for singlets turn out to be designated by one quantum number j. We can express the representation $\mathscr{D}_j$ graphically (see Fig. 19.1). In this case, since $J_z = \mathrm{i}Q_c/2$,

$$m = \frac{1}{2}N\,, \tag{19.9.29}$$

whence m is one-half of the ghost number.

We now see that the quartet and the octet introduced in the last section belong to the representations $\mathscr{D}_0$ and $\mathscr{D}_1$, respectively. In particular, considering auxiliary

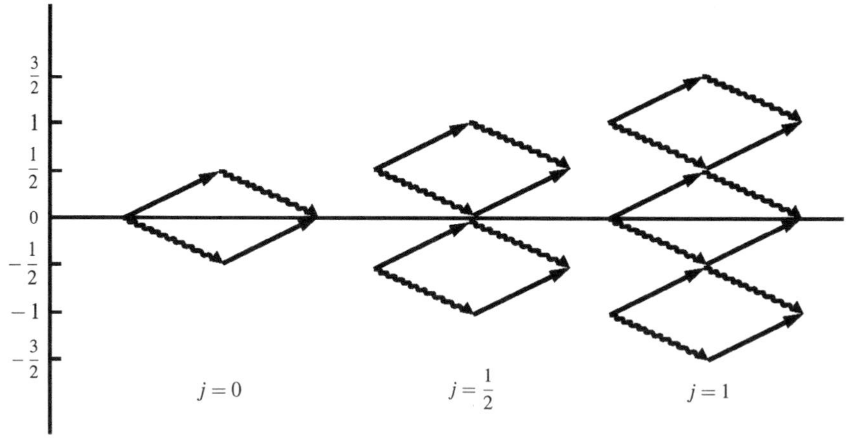

Fig. 19.1 Irreducible representations $\mathscr{D}_j$ of the algebra $\mathscr{A}$

fields, depending on whether j is an integer or a half integer, we find the following difference:

$$j = \text{integer} \qquad \beta + \bar{\beta} = 0\,, \tag{19.9.30}$$

$$j = \text{half integer} \qquad \beta + \bar{\beta} \neq 0\,. \tag{19.9.31}$$

In the latter case, $c \times c$ and $\bar{c} \times \bar{c}$ turn out to possess massless asymptotic fields from (19.9.3a) and (19.9.3b). Similarly, from (19.39), $c \times \bar{c}$ possesses a massless asymptotic field as well. From (19.9.9), it belongs to the triplet of $\mathscr{B}$. According to the discussion in Chap. 21, colour confinement occurs when j is a half integer.

We should also determine the representation of the matrix η of the indefinite metric. Note that the representation $\mathscr{D}_j$ is not necessarily closed under η. One representation $\mathscr{D}_j(a)$ may sometimes be transferred to another representation $\mathscr{D}_j(b)$. In this case, we need the following two irreducible representations:

$$\eta\mathscr{D}_j(a) = \mathscr{D}_j(b)\,, \quad \eta\mathscr{D}_j(b) = \mathscr{D}_j(a)\,. \tag{19.9.32}$$

We define a phase by

$$\eta|j, m, a\rangle = \alpha\beta|j, -m, b\rangle\,. \tag{19.9.33}$$

The following relations are then obtained automatically:

$$\begin{aligned} \eta(\alpha|j, m, a\rangle) &= -\beta|j, -m, b\rangle\,, \\ \eta(\beta|j, m, a\rangle) &= -\alpha|j, -m, b\rangle\,, \\ \eta(\alpha\beta|j, m, a\rangle) &= |j, -m, b\rangle\,. \end{aligned} \tag{19.9.34}$$

In this way, we can show that (19.9.20) is satisfied.

For the irreducible representation $\mathscr{D}_j$, going back to the notation in Sect. 19.5 and writing

$$\alpha = iq\,, \quad \beta = \bar{q}\,, \tag{19.9.35}$$

it can be shown that

$$\begin{aligned} &q^2 = \bar{q}^2 = (q^\dagger)^2 = (\bar{q}^\dagger)^2 = 0\,, \\ &\{q, \bar{q}\} = \{q^\dagger, \bar{q}^\dagger\} = \{q, \bar{q}^\dagger\} = \{q^\dagger, \bar{q}\} = 0\,. \end{aligned} \tag{18.9.36}$$

If we choose (19.9.27) as a basis, then for this normalization, we have

$$\Delta = qq^\dagger + q^\dagger q = \bar{q}\bar{q}^\dagger + \bar{q}^\dagger\bar{q} = 1\,. \tag{19.9.37}$$

This Δ is in fact a Casimir operator commuting with all elements in $\mathscr{A}$. However, since Δ vanishes for singlets,

$$\Delta|f\rangle = \begin{cases} |f\rangle\,, & |f\rangle \in \mathscr{V}_{\mathrm{D}}\,, \\ 0\,, & |f\rangle \in \mathscr{V}_{\mathrm{S}}\,. \end{cases} \tag{19.9.38}$$

Moreover, it is easily shown that Δ also commutes with η.

Chapter 20
Renormalization Group

The simple term "renormalization group" has a variety of meanings. This is because various kinds of transformation go by this name. Moreover, it is used in slightly different ways in several different areas of physics. Indeed, different kinds of quantity are renormalized under the renormalization transformations: for instance, free energies in condensed matter physics and Green's functions in elementary particle physics. If we ask what feature of the renormalization group is the greatest common factor, then we can say that it is a dilation transformation for lengths, momenta, and so on. Note also that the renormalization group really is a group in the mathematical sense. The qualifier "renormalization" refers to the fact that this group is related to manipulations, or renormalizations. In elementary particle theories, two main kinds of renormalization group are considered.

The first kind concerns transformations occurring within one given world. In order to solve the same equations of motion, we define a fundamental quantity, say a Green's function, under different boundary conditions. The boundary conditions are distinguished by certain parameters. We assume that every Green's function essentially includes information about solutions to the equations of motion, and any observable can be expressed using any Green's function. Therefore, observables turn out to be independent of the parameters distinguishing boundary conditions. They are invariant quantities under variations of such parameters. On the other hand, Green's functions with different values of the parameters are related to one another. We can investigate correspondences between Green's functions under variations of the parameters. The set of these parameter transformations forms the first kind of renormalization group. This group of transformations is closed in any given world. Observables are invariant quantities under the renormalization group, but quantities which vary under these transformations are not dynamical parameters, so this does not correspond to any conservation law.

The second kind of renormalization group differs from the first kind because it is not restricted to one given world. An example in classical mechanics is Reynolds' similarity law in fluid dynamics. Even if a viscosity, a density, and other quantities

K. Nishijima, *Quantum Field Theory*,
https://doi.org/10.1007/978-94-024-2190-3_20

which characterize a fluid are different, if the Reynolds' number is the same, and if we know the behaviour of one fluid, even a different kind of fluid, then using a suitable dictionary, the behaviours of other fluids can be understood as well. In the realm of elementary particles, if we consider changing masses or coupling constants, this is not restricted to one specific world.

20.1 Renormalization Group for QED

A renormalization group of the first kind described above was introduced for QED, first by Stueckelberg and Petermann in [176], and then by Gell-Mann and Low in [177]. There is a pedagogical description in Chap. 8 of the textbook by Bogoliubov and Shirkov, entitled *Introduction to the Theory of Quantized Fields* [178].

In the following, we label unrenormalized field operators and physical quantities with a subscript 0 to distinguish them from renormalized quantities. The Lagrangian density for QED is

$$\mathscr{L} = -\frac{1}{4} F^{(0)}_{\mu\nu} F^{(0)}_{\mu\nu} - \frac{1}{2\alpha_0} \left(\partial_\mu A^{(0)}_\mu\right)^2 - \bar{\psi}^{(0)} (\gamma_\mu D_\mu + m_0) \psi^{(0)} . \tag{20.1}$$

The renormalization is defined by the following equations:

$$A^{(0)}_\mu(x) = Z_3^{1/2} A_\mu(x) , \quad \psi^{(0)}(x) = Z_2^{1/2} \psi(x). \tag{20.2}$$

The renormalized parameters are

$$e_0^2 = Z_3^{-1} e^2 , \quad \alpha_0 = Z_3 \alpha . \tag{20.3}$$

Here we have used Ward's identity, viz.,

$$Z_1 = Z_2 . \tag{20.4}$$

Labeling quantities quantized on the mass shell with a subscript C, we introduce the Green's function

$$\langle \mathbf{0} | T[\boldsymbol{A}_\mu(x) \boldsymbol{A}_\nu(y)] | \mathbf{0} \rangle = \frac{-\mathrm{i}}{(2\pi)^4} \int \mathrm{d}^4 k \, \mathrm{e}^{\mathrm{i}k\cdot(x-y)} D_{\mathrm{FC}}(k)_{\mu\nu} , \tag{20.5}$$

$$D_{\mathrm{FC}}(k)_{\mu\nu} = \frac{1}{k^2 - \mathrm{i}\epsilon} \left[\left(\delta_{\mu\nu} - \frac{k_\mu k_\nu}{k^2 - \mathrm{i}\epsilon} d_{\mathrm{C}}(k^2) + \alpha \frac{k_\mu k_\nu}{k^2 - \mathrm{i}\epsilon} \right) \right] . \tag{20.6}$$

where d_{C} has the form

$$d_{\mathrm{C}}(k^2) = 1 + k^2 \int \mathrm{d}\kappa^2 \frac{C(\kappa^2)}{k^2 + \kappa^2 - \mathrm{i}\epsilon} . \tag{20.7}$$

From (12.249), to lowest order, $C(\kappa^2)$ has the form

$$C(\kappa^2) = \frac{e^2}{12\pi^2}\frac{1}{\kappa^2}\left(1+\frac{2m^2}{\kappa^2}\right)\sqrt{1-\frac{4m^2}{\kappa^2}}\theta(\kappa^2-4m^2)\ . \tag{20.8}$$

When we compute a Green's function, we express contributions corresponding to the Feynman diagrams in terms of combinations of propagators, vertex functions, and coupling constants. When we modify the normalizations of S_{F}, D_{F}, Γ_μ, and so on, this leads to variations in the coupling constants. However, regarding the Green's function as a whole, these variations cannot be absorbed completely, and the overall normalization turns out to be changed. A special case is the renormalization

$$S'_{\mathrm{F}}(p, e_0^2) = Z_2 S_{\mathrm{FC}}(p, e^2)\ , \tag{20.9}$$

$$D'_{\mathrm{F}}(k, e_0^2) = Z_3 D_{\mathrm{FC}}(k, e^2)\ , \tag{20.10}$$

$$\Gamma'_\mu(p, q, k, e_0^2) = Z_1^{-1}\Gamma_{\mu C}(p, q, k, e^2)\ , \tag{20.11}$$

where S_{FC} and D_{FC} have been normalized at $p\cdot\gamma = \mathrm{i}m$ and $k^2 = 0$, respectively. We thus take the normalization points to be those for space-like momenta:

$$p\cdot\gamma = \lambda'\ ,\quad k^2 = \lambda^2\ . \tag{20.12}$$

That is, introducing $s(\lambda, \lambda', p)$ and $d(\lambda, \lambda', k)$ by

$$(\mathrm{i}p\cdot\gamma + m)S_{\mathrm{F}}(\lambda, \lambda', p) \equiv s(\lambda, \lambda', p)\ , \tag{20.13}$$

$$(\delta_{\mu\nu}k^2 - k_\mu k_\nu)D_{\mathrm{F}}(\lambda, \lambda', k)_{\nu\sigma} \equiv \left(\delta_{\mu\sigma} - \frac{k_\mu k_\nu}{k^2}\right)d(\lambda, \lambda', k)\ , \tag{20.14}$$

we assume that the following normalization has been taken for (20.12):

$$s(\lambda, \lambda', p) = 1\ ,\quad d(\lambda, \lambda', k) = 1\ . \tag{20.15}$$

S_{F} and D_{F} introduced in (20.13) and (20.14) differ from S_{FC} and D_{FC} only by normalizations. Using (20.9) and (20.10), we can now write the equations connecting the unrenormalized Green's functions with the above Green's functions:

$$S'_{\mathrm{F}}(p, e_0^2) = Z_2(\lambda, \lambda'; e_\lambda^2)S_{\mathrm{F}}(\lambda, \lambda'; p, e_\lambda^2)\ , \tag{20.16}$$

$$D'_{\mathrm{F}}(p, e_0^2) = Z_3(\lambda, \lambda'; k, e_\lambda^2)\ , \tag{20.17}$$

$$e_\lambda^2 = Z_3(\lambda, \lambda'; e_\lambda^2)e_0^2\ , \tag{20.18}$$

where e_λ^2 is the parameter for perturbative expansion. The mass shell is specified by

$$\lambda' = im \ , \quad \lambda = 0 \ . \tag{20.19}$$

Setting $\lambda' = im$ in (20.16) and $\lambda = 0$ in (20.17),

$$S'_F(p, e_0^2) = Z_2(\lambda, im, e_\lambda^2) S_{FC}(p, e^2) \ , \tag{20.20}$$

$$D'_F(k, e_0^2) = Z_3(0, \lambda', e^2) D_{FC}(k, e^2) \ . \tag{20.21}$$

We then introduce the ratios

$$z_2(\lambda, \lambda', e^2) = Z_2(\lambda, \lambda'; e_\lambda^2)/Z_2(\lambda, im; e_\lambda^2) \ , \tag{20.22}$$

$$z_3(\lambda, \lambda', e^2) = Z_3(\lambda, \lambda'; e_\lambda^2)/Z_3(0, \lambda'; e^2) \ . \tag{20.23}$$

Therefore, for propagators with pole factors removed,

$$z_2(\lambda, \lambda'; e^2) s(\lambda, \lambda', p, e_\lambda^2) = s_C(p, e^2) \ , \tag{20.24}$$

$$z_3(\lambda, \lambda'; e^2) d(\lambda, \lambda', k, e_\lambda^2) = d_C(k, e^2) \ , \tag{20.25}$$

where z_2 and z_3 are finite factors. Now, from (20.15),

$$z_2(\lambda, \lambda'; e^2) = s_C(\lambda', e^2) \ , \quad z_3(\lambda, \lambda'; e^2) = d_C(\lambda, e^2) \ . \tag{20.26}$$

and

$$e_\lambda^2 = e^2 d_C(\lambda, e^2) \ . \tag{20.27}$$

All the basic renormalization equations are expressed by (20.24)–(20.27).

20.2 Approximate Equations for the Renormalization Group

In the renormalization group equations introduced in the previous section, we can obtain s and d when m^2 is assumed to be small.

20.2.1 Approximation Neglecting Vacuum Polarization

First, an approximation neglecting vacuum polarization is

$$d_C = 1 \ . \tag{20.28}$$

In this case, from (20.24), (20.26), and (20.28), the equation for s is

$$s(\lambda', p, e^2) = \frac{s_C(p, e^2)}{s_C(\lambda', e^2)} . \tag{20.29}$$

We thus set

$$s_C(p, e^2) = s_C\left(\frac{p^2}{m^2}, e^2\right) , \quad s_C(\lambda', e^2) = s_C\left(\frac{\lambda'^2}{m^2}, e^2\right) . \tag{20.30}$$

If we assume that $s(\lambda', p, e^2)$ does not depend on m^2,

$$s(\lambda', p, e^2) = s\left(\frac{p^2}{\lambda'^2}, e^2\right) . \tag{20.31}$$

This becomes an approximation when λ' and p are far larger than m :

$$s\left(\frac{p^2}{\lambda'^2}, e^2\right) = \frac{s_C\left(\frac{p^2}{m^2}, e^2\right)}{s_C\left(\frac{\lambda'^2}{m^2}, e^2\right)} . \tag{20.32}$$

Solutions to this are

$$\begin{aligned} s_C\left(\frac{p^2}{m^2}, e^2\right) &= C(e^2)\left(\frac{p^2}{m^2}\right)^{\gamma(e^2)} , \\ s_C\left(\frac{\lambda'^2}{m^2}, e^2\right) &= C(e^2)\left(\frac{\lambda'^2}{m^2}\right)^{\gamma(e^2)} , \end{aligned} \tag{20.33}$$

where, in perturbation theory, $C(e^2)$ includes infrared divergences. However, this factor is eliminated from

$$s\left(\frac{p^2}{\lambda'^2}, e^2\right) = \left(\frac{p^2}{\lambda'^2}\right)^{\gamma(e^2)} . \tag{20.34}$$

In fact, the exponent $\gamma(e^2)$ depends on how we choose the gauge. Therefore, even if the mass is zero, if we normalize at a point off the mass shell, then infrared divergences do not appear. In addition, the reader is referred to (20.102).

20.2.2 *Approximation Taking into Account Vacuum Polarization*

Here we also assume that s does not depend on m^2. This is another high-energy approximation. In this case, we write s in the form

$$s(\lambda, \lambda', p, e_\lambda^2) \approx s\left(\frac{p^2}{\lambda^2}, \frac{p^2}{\lambda'^2}, e_\lambda^2\right) . \tag{20.35}$$

Therefore, the renormalization group equation under this approximation is

$$s\left(\frac{p^2}{\lambda^2}, \frac{p^2}{\lambda'^2}, e_\lambda^2\right) = \frac{s_C\left(\frac{p^2}{m^2}, e^2\right)}{s_C\left(\frac{\lambda'^2}{m^2}, e^2\right)} . \tag{20.36}$$

Moreover, assuming that d does not depend on m^2, we introduce the approximation

$$d(\lambda, k, e_\lambda^2) = d\left(\frac{k^2}{\lambda^2}, \frac{m^2}{\lambda^2}, e_\lambda^2\right) \approx d\left(\frac{k^2}{\lambda^2}, e_\lambda^2\right) . \tag{20.37}$$

Thus, an approximate equation for d is

$$d\left(\frac{k^2}{\lambda^2}, e_\lambda^2\right) = \frac{d_C\left(\frac{k^2}{m^2}, e^2\right)}{d_C\left(\frac{\lambda^2}{m^2}, e^2\right)} , \tag{20.38}$$

$$e_\lambda^2 = e^2 d_C\left(\frac{\lambda^2}{m^2}, e^2\right) . \tag{20.39}$$

Since no gauge parameter has been included here, we can assume that this is in the Landau gauge. In a general gauge, the following parameter is included:

$$\alpha_\lambda e_\lambda^2 = \alpha e^2 . \tag{20.40}$$

Combining (20.38) with (20.39),

$$e^2 d_C\left(\frac{k^2}{m^2}, e^2\right) = e^2 d_C\left(\frac{\lambda^2}{m^2}, e^2\right) d\left(\frac{k^2}{\lambda^2}, e^2 d_C\left(\frac{\lambda^2}{m^2}, e^2\right)\right) . \tag{20.41}$$

In order to modify this equation, we set

$$e^2 d_{\mathrm{C}}\left(\frac{k^2}{m^2}, e^2\right) = g\left(\frac{k^2}{m^2}, e^2\right) , \tag{20.42}$$

and the above equation takes the form

$$g\left(\frac{k^2}{m^2}, e^2\right) = Q\left(\frac{k^2}{\lambda^2}, g\left(\frac{\lambda^2}{m^2}, e^2\right)\right) . \tag{20.43}$$

This functional equation can be solved as we shall now see. We first rewrite (20.43) in the form

$$g(x, e^2) = Q\left(\frac{x}{y}, g(y, e^2)\right) . \tag{20.44}$$

Then, using $g = g(x, e^2)$ and $g' = g(y, e^2)$, we invert to express x and y in terms of g and g':

$$x = h(g, e^2) , \quad y = h(g', e^2) . \tag{20.45}$$

Inserting this into (20.44),

$$g = Q\left(\frac{h(g, e^2)}{h(g', e^2)}, g'\right) . \tag{20.46}$$

Since this means that the ratio of $h(g, e^2)$ and $h(g', e^2)$ does not depend on e^2, h can be written as a product of a function of g and a function of e^2, viz.,

$$h(g, e^2) = \frac{G(g)}{\phi\left(e^2\right)} , \quad h(g', e^2) = \frac{G(g')}{\phi\left(e^2\right)} . \tag{20.47}$$

This means that

$$G(e_\lambda^2) = \frac{\lambda^2}{m^2}\phi\left(e^2\right) . \tag{20.48}$$

If F is the inverse function of G, then

$$e_\lambda^2 = e^2 d_{\mathrm{C}}\left(\frac{\lambda^2}{m^2}, e^2\right) = F\left(\frac{\lambda^2}{m^2}\phi(e^2)\right) , \tag{20.49}$$

or

$$e^2 d_C\left(\frac{\lambda^2}{m^2}, e^2\right) = g\left(\frac{k^2}{m^2}, e^2\right) = F\left(\frac{k^2}{m^2}\phi(e^2)\right) = F\left(\frac{k^2}{\lambda^2}\frac{\lambda^2}{m^2}\phi(e^2)\right). \tag{20.50}$$

Combining (20.48) with (20.50),

$$e^2 d_C\left(\frac{k^2}{m^2}, e^2\right) = F\left(\frac{k^2}{\lambda^2}G\left(e_\lambda^2\right)\right). \tag{20.51}$$

Note that, from (20.39),

$$e^2 d_C\left(\frac{\lambda^2}{m^2}, e^2\right) d\left(\frac{k^2}{\lambda^2}, e^2 d_C\left(\frac{\lambda^2}{m^2}, e^2\right)\right) = e_\lambda^2 d\left(\frac{k^2}{\lambda^2}, e_\lambda^2\right). \tag{20.52}$$

Comparing this with (20.41) and (20.51),

$$F\left(\frac{k^2}{\lambda^2}G\left(e_\lambda^2\right)\right) = e_\lambda^2 d\left(\frac{k^2}{\lambda^2}, e_\lambda^2\right), \tag{20.53}$$

$$\frac{k^2}{\lambda^2}G\left(e_\lambda^2\right) = G\left(e_\lambda^2 d\left(\frac{k^2}{\lambda^2}, e_\lambda^2\right)\right). \tag{20.54}$$

Thus,

$$\ln\frac{k^2}{\lambda^2} = \ln G\left(e_\lambda^2 d\left(\frac{k^2}{\lambda^2}, e_\lambda^2\right)\right) - \ln G\left(e_\lambda^2\right). \tag{20.55}$$

We now define the function $\psi(x)$ introduced by Gell-Mann and Low:

$$\frac{1}{\psi(x)} = \frac{\mathrm{d}}{\mathrm{d}x}\ln G(x). \tag{20.56}$$

Therefore,

$$\ln\frac{k^2}{\lambda^2} = \int_{e_\lambda^2}^{e_\lambda^2 d\left(\frac{k^2}{\lambda^2}, e_\lambda^2\right)} \frac{\mathrm{d}x}{\psi(x)}. \tag{20.57}$$

Similarly, an equation derived from (20.50), viz.,

$$\frac{k^2}{m^2}\phi(e^2) = G\left(e^2 d_C\left(\frac{k^2}{m^2}, e^2\right)\right), \tag{20.58}$$

can also be written in the form

$$\ln\frac{k^2}{m^2} = \int_{F(\phi(e^2))}^{e^2 d_C\left(\frac{k^2}{m^2}, e^2\right)} \frac{dx}{\psi(x)} . \tag{20.59}$$

In perturbation theory,

$$F\left(\phi(e^2)\right) = e^2 d_C\left(1, e^2\right) = e^2 - \frac{5}{36\pi^2}e^4 + \cdots \equiv e'^2 . \tag{20.60}$$

It follows immediately from (20.50) that

$$\left[m\frac{\partial}{\partial m} + \beta(e)\frac{\partial}{\partial e}\right] e^2 d_C\left(\frac{k^2}{m^2}, e^2\right) = 0 , \tag{20.61}$$

where

$$e\beta(e) = \frac{\phi(e^2)}{\phi'(e^2)} . \tag{20.62}$$

Additionally, from (20.59),

$$\left[m\frac{\partial}{\partial m} + \frac{\psi\left(e'^2\right)}{e'}\frac{\partial}{\partial e'}\right] e^2 d_C\left(\frac{k^2}{m^2}, e^2\right) = 0 . \tag{20.63}$$

This implies that

$$\frac{\psi\left(e'^2\right)}{e'}\frac{\partial}{\partial e'} = \beta(e)\frac{\partial}{\partial e} . \tag{20.64}$$

Equation (20.61) is the Callan–Symanzik equation [179–181], but without a mass term.

If we compute ψ and β in perturbation theory, we find

$$\psi(x) = \frac{1}{12\pi^2}\left(x^2 + \frac{3}{16\pi^2}x^3 + \cdots\right) , \tag{20.65}$$

$$\beta(e) \approx \frac{\psi(e^2)}{e} \approx \frac{e^3}{12\pi^2} . \tag{20.66}$$

In fact, since the function d does not depend on the gauge parameter α in QED, the above discussion holds true in arbitrary gauges. However, the function s does depend on α.

After the preparation above, we come back to the equation for s. We rewrite (20.35) using the result for d :

$$s\left(\frac{p^2}{\lambda^2}, \frac{p^2}{\lambda'^2}, e_\lambda^2\right) = s\left(\frac{p^2}{\lambda^2}, \frac{p^2}{\lambda'^2}, F\left(\frac{\lambda^2}{m^2}\phi(e^2)\right)\right) . \tag{20.67}$$

However, according to (20.36), this function should not depend on λ, so the right-hand side of this equation should only be a function of

$$\frac{p^2}{m^2}\phi(e^2) \quad \text{and} \quad \frac{p^2}{\lambda'^2} . \tag{20.68}$$

Therefore,

$$s\left(\frac{p^2}{\lambda^2}, \frac{p^2}{\lambda'^2}, e_\lambda^2\right) = R\left(\frac{p^2}{\lambda'^2}, \frac{p^2}{m^2}\phi(e^2)\right) . \tag{20.69}$$

In every gauge except for the Landau gauge, this depends on the variable αe^2 as well. The reason is that, since this equation does not depend on λ, α should always be included in the combination $\alpha_\lambda e_\lambda^2 = \alpha e^2$. Setting

$$s_{\mathrm{C}}\left(\frac{p^2}{m^2}, e^2, \alpha e^2\right) = A(e^2, \alpha e^2) H\left(\frac{p^2}{\lambda'^2}, \alpha e^2, \frac{p^2}{m^2}\phi(e^2)\right) , \tag{20.70}$$

we see that (20.69) holds true. This H satisfies

$$\mathscr{D}H \equiv \left[m\frac{\partial}{\partial m} + \beta(e)\frac{\partial}{\partial e} + \delta(e,\alpha)\frac{\partial}{\partial \alpha}\right] H = 0 , \tag{20.71}$$

where

$$\delta(e,\alpha) = -\frac{2\alpha}{e}\beta(e) . \tag{20.72}$$

We now define γ_{el} by

$$\mathscr{D} \ln A = -2\gamma_{\mathrm{el}}(e,\alpha) . \tag{20.73}$$

Thus,

$$(\mathscr{D} + 2\gamma_{\mathrm{el}})\, s_{\mathrm{C}}\left(\frac{p^2}{m^2}, e^2, \alpha e^2\right) = 0 . \tag{20.74}$$

This equation is also called the Callan–Symanzik equation without the mass term.

In this section, we have treated renormalization groups of the first kind, assuming that the mass m^2 is smaller than the mass corresponding to the four-momentum or the renormalization point. As a result, a derivative with respect to the mass has been included. However, if we think of this as a derivative with respect to the four-momentum, then the story can be considered to be closed in a world with a fixed mass. In this section we have treated an approximation which holds true only if m^2 is small, but what would the renormalization group equations look like if we did not make this approximation? We will discuss this issue in the next section.

20.3 Ovsianikov's Equation

We begin with the equation for d in which m is not neglected:

$$d\left(\frac{k^2}{\lambda^2}, \frac{m^2}{\lambda^2}, e_\lambda^2\right) = \frac{d_C\left(\frac{k^2}{m^2}, e^2\right)}{d_C\left(\frac{\lambda^2}{m^2}, e^2\right)} . \tag{20.75}$$

Eliminating d_C from this equation,

$$d\left(\frac{k^2}{\lambda_2^2}, \frac{m^2}{\lambda_2^2}, e_2^2\right) = \frac{d\left(\frac{k^2}{\lambda_1^2}, \frac{m^2}{\lambda_1^2}, e_1^2\right)}{d\left(\frac{\lambda_2^2}{\lambda_1^2}, \frac{m^2}{\lambda_1^2}, e_1^2\right)} , \tag{20.76}$$

$$e_2^2 = e_1^2 d\left(\frac{\lambda_2^2}{\lambda_1^2}, \frac{m^2}{\lambda_1^2}, e_1^2\right) , \tag{20.77}$$

where we have used

$$d\left(1, \frac{m^2}{\lambda^2}, e_1^2\right) = 1 , \tag{20.78}$$

considering that e_1 and e_2 express e_{λ_1} and e_{λ_2}, respectively. In addition, the combination

$$e_\lambda^2 d\left(\frac{k^2}{\lambda^2}, \frac{m^2}{\lambda^2}, e_\lambda^2\right) = e^2 d_C\left(\frac{k^2}{m^2}, e^2\right) \tag{20.79}$$

is called an *invariant charge* because it does not depend on λ. We now introduce the variables

$$t = \frac{\lambda_2^2}{\lambda_1^2}\,, \quad x = \frac{k^2}{\lambda_1^2}\,, \quad y = \frac{m^2}{\lambda_1^2}\,, \quad e^2 = e_1^2\,. \tag{20.80}$$

Hence, (20.76) becomes a rigorous equation with the form

$$e^2 d(x, y, e^2) = e^2 d(t, y, e^2) d\left(\frac{x}{t}, \frac{y}{t}, e^2 d\left(t, y, e^2\right)\right)\,. \tag{20.81}$$

Similar equations can be found for the other functions. In the expression

$$S_{\mathrm{F}}(\lambda, \lambda', p, e_\lambda^2) = \frac{1}{p^2 + m^2 - \mathrm{i}\epsilon}\Big[-\mathrm{i}(p\cdot\gamma)a(\lambda, \lambda', p, e_\lambda^2) + mb(\lambda, \lambda', p, e_\lambda^2)\Big]\,, \tag{20.82}$$

we write both a and b as

$$s\left(x, y, a, e^2\right)\,, \tag{20.83}$$

where

$$x = \frac{p^2}{\lambda^2}\,, \quad y = \frac{m^2}{\lambda^2}\,, \quad a = \frac{m^2}{\lambda'^2}\,, \quad e^2 = e_\lambda^2. \tag{20.84}$$

In the following, we do not write a and αe^2 explicitly. Therefore, the equation for s is

$$\frac{s(x, y, e^2)}{s(t, y, e^2)} = \frac{s\left(\frac{x}{t}, \frac{y}{t}, e^2 d\left(t, y, e^2\right)\right)}{s\left(1, \frac{y}{t}, e^2 d\left(t, y, e^2\right)\right)}\,. \tag{20.85}$$

A similar equation holds true for the vertex function as well. Expanding the vertex function in a covariant polynomial,

$$\Gamma_\mu(p, q, k) = \sum_A c_\mu^A(p, q, \gamma)\Gamma_A\left(p^2, q^2, k^2\right)\,, \tag{20.86}$$

and writing the coefficients as $\Gamma_A(x, y, z, u, e^2)$, we find

$$\frac{\Gamma_A\left(x, y, z, u, e^2\right)}{\Gamma_A\left(t, y, z, u, e^2\right)} = \frac{\Gamma_A\left(\frac{x}{t}, \frac{y}{t}, \frac{z}{t}, \frac{u}{t}, e^2 d\left(t, u, e^2\right)\right)}{\Gamma_A\left(1, \frac{y}{t}, \frac{z}{t}, \frac{u}{t}, e^2 d\left(t, u, e^2\right)\right)}\,, \tag{20.87}$$

where the variables are

$$x = \frac{p^2}{\lambda^2}\,, \quad y = \frac{q^2}{\lambda^2}\,, \quad z = \frac{k^2}{\lambda^2}\,, \quad u = \frac{m^2}{\lambda^2}\,, \quad e^2 = e_\lambda^2\,. \tag{20.88}$$

Since a condition like (20.78) cannot be used in these equations, we can not eliminate expressions like $s(1,\ldots)$ and $\Gamma_A(1,\ldots)$. We therefore introduce the following functions:

$$\Phi\left(y, e^2\right) = \frac{\partial}{\partial x} d\left(x, y, e^2\right)\Big|_{x=1}\,, \tag{20.89}$$

$$\Psi\left(y, e^2\right) = \frac{\partial}{\partial x} \ln s\left(x, y, e^2\right)\Big|_{x=1}\,. \tag{20.90}$$

Now, differentiating the functional equations (20.81) and (20.85) with respect to x and setting $t = x$, we obtain

$$\frac{\partial}{\partial x} e^2 d\left(x, y, e^2\right) = \frac{e^2 d\left(x, y, e^2\right)}{x} \Phi\left(\frac{y}{x}, e^2 d\left(x, y, e^2\right)\right)\,, \tag{20.91}$$

$$\frac{\partial}{\partial x} s\left(x, y, e^2\right) = \frac{s\left(x, y, e^2\right)}{x} \Psi\left(\frac{y}{x}, e^2 d\left(x, y, e^2\right)\right)\,. \tag{20.92}$$

These are *Ovsianikov's equations*. Both are highly non-linear, but Φ and Ψ determine the initial conditions for d and s at $x = 1$ and they also determine their differential equations.

If we assume that $k^2 \gg m^2$, i.e., $x \gg y$, and we neglect any y-dependence, then they coincide with the results in the previous section.

1. Neglecting *y*

We now have

$$\frac{\partial}{\partial x} e^2 d\left(x, e^2\right) = \frac{e^2 d\left(x, e^2\right)}{x} \Phi\left(e^2 d\left(x, e^2\right)\right)\,, \tag{20.93}$$

$$\frac{\partial}{\partial x} s\left(x, e^2\right) = \frac{s\left(x, e^2\right)}{x} \Psi\left(e^2 d\left(x, e^2\right)\right)\,. \tag{20.94}$$

Therefore,

$$\ln x = \int_{e^2}^{e^2 d(x,e^2)} \frac{\mathrm{d}z}{z\Phi(z)}\,. \tag{20.95}$$

If we set $\psi(z) = z\Phi(z)$, this result coincides with (20.57). Similarly, if we solve the equation for s, then

$$\ln \frac{s\left(x, e^2\right)}{s\left(x_0, e^2\right)} = \int_{e^2 d(x_0, e^2)}^{e^2 d(x, e^2)} dz \frac{\Psi(z)}{z\Phi(z)} . \tag{20.96}$$

In perturbation theory,

$$d\left(x, e^2\right) = 1 + \frac{e^2}{12\pi^2} \ln x + \frac{e^4}{16\pi^2}\left(\frac{1}{4\pi^2} \ln x + \frac{1}{16\pi} \ln^2 x\right) + \cdots , \tag{20.97}$$

$$\Phi(z) = \frac{\partial}{\partial x} d(x, z)\bigg|_{x=1} = \frac{z}{12\pi^2}\left(1 + \frac{3z}{16\pi^2} + \cdots\right) \quad (z \ll 1) . \tag{20.98}$$

2. Solution Around the Mass Shell

Implementing a perturbative expansion of the function s in the vicinity of the mass shell, we obtain an infrared divergence. We thus derive a solution that is not based on perturbation theory, using the renormalization group equations. In the vicinity of the mass shell, we set

$$x \approx 1 , \quad y \approx -1 , \tag{20.99}$$

then differentiate (20.85) with respect to x and set $t = x'$, whence

$$\ln \frac{s\left(x, -1, e^2\right)}{s\left(x_0, -1, e^2\right)} = \int_{x_0}^{x} \frac{dx'}{x'} \frac{\partial}{\partial \xi} \ln s\left(\xi, -\frac{1}{x'}, e^2 d\left(x', -1, e^2\right)\right)\bigg|_{\xi=1} .$$

To lowest order in both a and b,

$$\ln \frac{s\left(x, -1, e^2\right)}{s\left(x_0, -1, e^2\right)} \approx -\frac{3e^2}{8\pi^2} \ln \frac{x-1}{x_0 - 1} . \tag{20.100}$$

Therefore,

$$s\left(x, -1, e^2\right) \propto (x-1)^{-3e^2/8\pi^2} , \tag{20.101}$$

or

$$s\left(\frac{p^2}{m^2}, e^2\right) \propto \left(1 + \frac{p^2}{m^2}\right)^{-3e^2/8\pi^2} . \tag{20.102}$$

In a general gauge, we replace the factor of 3 in the exponent by $(3 - \alpha)$. Each term of the perturbative expansion in this equation diverges on the mass shell.

Ovsianikov's equation is exact but non-linear. However, we can derive a linear equation. Although in the above we differentiated the renormalization group functional equation with respect to x and then set $t = x$, this time we differentiate it with respect to t and set $t = 1$:

$$\left[\Phi\left(y, e^2\right) - \left(x\frac{\partial}{\partial x} + y\frac{\partial}{\partial y}\right) + e^2\Phi\left(y, e^2\right)\frac{\partial}{\partial e^2}\right] d\left(x, y, e^2\right) = 0 . \quad (20.103)$$

This is linear. Up to now $x = k^2/\lambda^2$ and $y = m^2/\lambda^2$, but from now on we shall write λ as μ. Therefore,

$$-2\left(x\frac{\partial}{\partial x} + y\frac{\partial}{\partial y}\right) = \mu\frac{\partial}{\partial \mu} . \quad (20.104)$$

This yields an equation of the form

$$\left[\mu\frac{\partial}{\partial \mu} + \beta(e, y)\frac{\partial}{\partial e} + 2\gamma_{\text{ph}}(e, y)\right] d\left(x, y, e^2\right) = 0 , \quad (20.105)$$

$$\beta(e, y) = e\Phi\left(y, e^2\right) , \quad \gamma_{\text{ph}} = \Phi\left(y, e^2\right) . \quad (20.106)$$

20.4 Linear Equations for the Renormalization Group

We consider a generalization of the linear equation given at the end of the last section. To do this, we start with the renormalization group equation. As an example, we choose the ϕ^4-theory:

$$\mathscr{L} = -\frac{1}{2}\left[(\partial_\lambda \varphi)^2 + m^2\varphi^2\right] - g\frac{\varphi^4}{4!} . \quad (20.107)$$

Indicating the unrenormalized Green's functions with a subscript 0, the renormalized Green's function is given by

$$G_0^{(n)}(p_1, \ldots, p_n; g_0) = Z_3^{n/2}(\mu) G^{(n)}\big(p_1, \ldots, p_n; g(\mu)\big) , \quad (20.108)$$

where the renormalized coupling constant and four-point vertex function are given by

$$g_0 = Z_1(\mu) Z_3^{-2}(\mu) g(\mu) , \quad (20.109)$$

$$\Gamma\big(p_i; g(\mu)\big) = Z_1(\mu)\Gamma_0(p_i; g_0) , \quad i = 1, 2, 3, 4 , \quad (20.110)$$

and Z_1, Z_3, and so on have been determined by the renormalization conditions

$$(p^2+m^2)G^{(2)}\big(p;g(\mu)\big)\big|_{p^2=\mu^2}=-1\,, \tag{20.111}$$

$$\Gamma\big(p_i;g(\mu)\big)=1\,,\quad \text{with}\quad p_i\cdot p_j=\frac{\mu^2}{3}(4\delta_{ij}-1)\,. \tag{20.112}$$

The four-point vertex function Γ_0 is defined by the unrenormalized four-point Green's function without the propagators and the coupling constant. For the definition of the n-point Green's function, the reader is referred to Sect. 15.3.

Since the unrenormalized Green's function is not related to the renormalization point μ,

$$\mu\frac{\partial}{\partial\mu}G_0^{(n)}(p_1,\ldots,p_n;g_0)=0\,. \tag{20.113}$$

Inserting (20.108) into this equation, we obtain

$$\left(\mu\frac{\partial}{\partial\mu}+\beta\frac{\partial}{\partial g}+n\gamma\right)G^{(n)}\big(p_1,\cdots,p_n;g(\mu)\big)=0\,, \tag{20.114}$$

where

$$\beta=\mu\frac{\partial}{\partial\mu}g(\mu)=\left(2\mu\frac{\partial}{\partial\mu}\ln Z_3-\mu\frac{\partial}{\partial\mu}\ln Z_1\right)g(\mu)\,, \tag{20.115}$$

$$\gamma=\frac{1}{2}\mu\frac{\partial}{\partial\mu}\ln Z_3\,. \tag{20.116}$$

Both of these are functions of m^2/μ^2 and g. From now on, we shall write $g(\mu)$ simply as g. Therefore, writing (20.114) in more detail, we have

$$\left[\mu\frac{\partial}{\partial\mu}+\beta\left(g,\frac{m^2}{\mu^2}\right)\frac{\partial}{\partial g}+n\gamma\left(g,\frac{m^2}{\mu^2}\right)\right]G^{(n)}(p_1,\ldots,p_n;\mu,g)=0\,. \tag{20.117}$$

In this equation, g and μ are treated as independent variables.

Similarly, we can write down a linear renormalization equation in QED. If the numbers of photon legs and electron legs in the Green's function are n and l, respectively, we have

$$\begin{aligned}&G_0^{(n,l)}(k_1,\ldots,k_n;p_1,\ldots,p_l;e_0)\\&\qquad=Z_3^{n/2}(\mu)Z_2^{l/2}(\mu)G^{(n,l)}\big(k_1,\ldots,k_n;p_1,\ldots,p_l;e(\mu)\big)\,,\end{aligned} \tag{20.118}$$

where

$$e_0^2 = Z_3^{-1}(\mu)e^2(\mu) . \tag{20.119}$$

This implies

$$\left[\mu\frac{\partial}{\partial\mu} + \beta\left(e, \frac{m^2}{\mu^2}\right)\frac{\partial}{\partial e} + n\gamma_{\text{ph}}\left(e, \frac{m^2}{\mu^2}\right) + l\gamma_{\text{ph}}\left(e, \frac{m^2}{\mu^2}\right)\right]G^{(n,l)}(\ldots;\mu, e) = 0. \tag{20.120}$$

This equation is in the Landau gauge and a generalization of (20.93). In a general gauge α,

$$\left[\mu\frac{\partial}{\partial\mu} + \beta\left(e, \frac{m^2}{\mu^2}\right)\frac{\partial}{\partial e} + \delta\left(e, \frac{m^2}{\mu^2}, \alpha\right)\frac{\partial}{\partial\alpha} + \gamma^{(n,l)}\right]G^{(n,l)}(\ldots;\mu, e, \alpha) = 0 , \tag{20.121}$$

where

$$\gamma^{(n,l)} = n\gamma_{\text{ph}}\left(e, \frac{m^2}{\mu^2}\right) + l\gamma_{\text{el}}\left(e, \frac{m^2}{\mu^2}, \alpha\right) . \tag{20.122}$$

As stated in the last section,

$$\gamma_{\text{ph}}\left(e, \frac{m^2}{\mu^2}\right) = \Phi\left(\frac{m^2}{\mu^2}, e^2\right) , \tag{20.123}$$

$$\beta\left(e, \frac{m^2}{\mu^2}\right) = e\gamma_{\text{ph}}\left(e, \frac{m^2}{\mu^2}\right) , \quad \delta\left(e, \frac{m^2}{\mu^2}, \alpha\right) = -2\alpha\gamma_{\text{ph}}\left(e, \frac{m^2}{\mu^2}\right) . \tag{20.124}$$

The last expression was derived from (12.98). The renormalization equations derived in this section are linear and homogeneous, but the coefficients are functions of two variables, so they are hard to deal with. In contrast, the coefficients in the Callan–Symanzik equation introduced in the next section are single-variable functions, while the equation itself becomes inhomogeneous.

20.5 Callan–Symanzik Equation

So far we have considered a renormalization group of the first kind. In this section, we move on to the second kind, i.e., we introduce the idea of changing the particle mass.

We first define a Fourier transform of the Green's function. Using the notation $(\mathrm{d}p) = \mathrm{d}^4 p_1 \dots \mathrm{d}^4 p_n$,

$$\begin{aligned}\langle \mathbf{0} | T[\varphi(x_1)\dots\varphi(x_n)] | \mathbf{0} \rangle_{\text{conn}} \qquad\qquad (20.125) \\ = \left[\frac{\mathrm{i}}{(2\pi)^4}\right]^{n-1} \int (\mathrm{d}p) \mathrm{e}^{\mathrm{i}(p_1\cdot x_1 + \dots + p_n \cdot x_n)} \delta^4(p_1 + \dots + p_n) G^{(n)}(p_1, \dots, p_n) ,\end{aligned}$$

$$\begin{aligned}\langle \mathbf{0} | T[\boldsymbol{A}(x)\varphi(x_1)\dots\varphi(x_n)] | \mathbf{0} \rangle_{\text{conn}} \qquad\qquad (20.126) \\ = \left[\frac{\mathrm{i}}{(2\pi)^4}\right]^{n} \int \mathrm{d}^4 q (\mathrm{d}p) \mathrm{e}^{\mathrm{i}(q\cdot x + p_1\cdot x_1 + \dots + p_n \cdot x_n)} \delta^4(q + p_1 + \dots + p_n) A^{(n)}(q; p_1, \dots, p_n) .\end{aligned}$$

We start with an unrenormalized theory, then introduce the quantity

$$T^{(0)}(x) = m_0^2 \big[\varphi^{(0)}(x)\big]^2 . \qquad (20.127)$$

To lowest order,

$$G_0^{(2)}(p) = -\frac{1}{p^2 + m^2 - \mathrm{i}\epsilon} . \qquad (20.128)$$

Therefore,

$$m_0 \frac{\partial}{\partial m_0} \left(-\frac{1}{p^2 + m_0^2 - \mathrm{i}\epsilon} \right) = \frac{1}{p^2 + m_0^2 - \mathrm{i}\epsilon} 2m_0^2 \frac{1}{p^2 + m_0^2 - \mathrm{i}\epsilon} . \qquad (20.129)$$

This equation tells us that a derivative with respect to m_0 is equivalent to inserting $T^{(0)}$ in an unrenormalized form, i.e.,

$$m_0 \frac{\partial}{\partial m_0} G_0^{(n)}(p_1, \dots, p_n) = T_0^{(n)}(0; p_1, \dots, p_n) \equiv \Delta G_0^{(n)}(p_1, \dots, p_n) . \qquad (20.130)$$

Next, we consider a renormalization on the mass shell:

$$G_0^{(n)}(p_1, \dots, p_n) = Z_3^{n/2} G^{(n)}(p_1, \dots, p_n; g) . \qquad (20.131)$$

From now on, we consider all renormalized quantities like m and g to be functions of m_0, g_0, and a cut-off momentum Λ. We think about the derivative with respect to m_0 in the case where g_0 and Λ are fixed. First, we define Z by

$$Z m_0 \frac{\partial m}{\partial m_0} = m . \qquad (20.132)$$

Therefore, for functions of g and m,

$$Zm_0\frac{\partial}{\partial m_0}f = \left(Zm_0\frac{\partial m}{\partial m_0}\frac{\partial}{\partial m} + Zm_0\frac{\partial g}{\partial m_0}\frac{\partial}{\partial g}\right)f \equiv \mathscr{D}f\ , \tag{20.133}$$

where

$$\mathscr{D} = m\frac{\partial}{\partial m} + \beta(g)\frac{\partial}{\partial g}\ , \quad \beta(g) = Zm_0\frac{\partial g}{\partial m_0}\ . \tag{20.134}$$

Thus, we redefine T by

$$T(x) = Zm_0^2\left[\varphi^{(0)}(x)\right]^2\ . \tag{20.135}$$

This yields the Callan–Symanzik equation [179–181]:

$$\left[m\frac{\partial}{\partial m} + \beta(g)\frac{\partial}{\partial g} + n\gamma_\varphi(g)\right]G^{(n)}(p_1, \ldots, p_n) = T^{(n)}(0; p_1, \ldots, p_n)\ , \tag{20.136}$$

where γ_φ is called the *anomalous dimension of the field* φ, defined by

$$\gamma_\varphi(g) = \frac{1}{2}Zm_0\frac{\partial}{\partial m_0}\ln Z_3\ . \tag{20.137}$$

Denoting the multiplicative renormalization of φ^2 by S, we rewrite T as

$$T(x) = m^2 S(x)\ . \tag{20.138}$$

Focusing on the pole at $p^2 + m^2 = 0$ when $n = 2$, multiplying (20.136) corresponding to $n = 2$ by $(p^2 + m^2)^2$, and taking the limit $p^2 + m^2 \to 0$, the left-hand side becomes $2m^2$. Therefore, the right-hand side should be $2m^2$ as well. Using the LSZ reduction formula, S turns out to satisfy the following renormalization condition:

$$\sqrt{2p_0}\langle \boldsymbol{p}|S(x)|\boldsymbol{p}\rangle\sqrt{2p_0} = 1\ . \tag{20.139}$$

We now define a multiple mass insertion term. Inserting T into $G^{(n)}$ a total of k times and writing it as $\delta^k G^{(n)}$,

$$\left\{\mathscr{D} + n\gamma_\varphi(g) - k\left[2 - \gamma_S(g)\right]\right\}\Delta^k G^{(n)} = \Delta^{k+1}G^{(n)}\ , \tag{20.140}$$

where

$$\gamma_S(g) = Zm_0 \frac{\partial}{\partial m_0} \ln Z_S \ , \quad \left[\varphi^{(0)}(x)\right]^2 = Z_S S(x) \ . \tag{20.141}$$

On the other hand, an unrenormalized multiple mass insertion term is

$$\Delta^k G_k^{(n)} = (2m_0^2)^k \left(\frac{\partial}{\partial m_0^2}\right)^k G_0^{(n)} \ , \tag{20.142}$$

and its renormalization is

$$\Delta^k G^{(n)} = Z^k Z_3^{-n/2} \Delta^k G_0^{(n)} \ . \tag{20.143}$$

Inserting this into the Callan–Symanzik equation,

$$\gamma_S(g) = 2(1 - Z) - m_0 \frac{\partial}{\partial m_0} Z \ . \tag{20.144}$$

According to perturbation theory,

$$\beta(g) = \frac{3g^2}{16\pi^2} \ , \quad \gamma_\varphi(g) = \frac{g^2}{12(16\pi^2)^2} \ , \quad \gamma_S(g) = -\frac{g}{16\pi^2} \ . \tag{20.145}$$

An advantage compared to the renormalization group of the first kind is that β and γ_φ become functions of g alone. A disadvantage is that the equation becomes inhomogeneous. However, if $p_i^2 \gg m^2$, since the momentum dependence of $\Delta G^{(n)}$ decreases by a factor of momentum squared compared to $G^{(n)}$, we can neglect $\Delta G^{(n)}$ in comparison to $G^{(n)}$. We then obtain a homogeneous equation:

$$\left[\mathscr{D} + n\gamma_\varphi(g)\right] G^{(n)} \approx 0 \ . \tag{20.146}$$

Similarly, if $\mu^2 \gg m^2$, even in the renormalization group of the first kind, the m^2/μ^2-dependence of the coefficient function is dropped, and it becomes a function of g alone, so the equations coincide.

We now consider QED. The renormalizations are

$$G_0^{(n,l)} = Z_3^{n/2} Z_2^{l/2} G^{(n,l)} \ , \tag{20.147}$$

$$\Delta^k G_0^{(n,l)} = m_0^k \left(\frac{\partial}{\partial m_0}\right)^k G_0^{(n,l)} = Z^{-k} Z_3^{n/2} Z_2^{l/2} \Delta^k G^{(n,l)} \ , \tag{20.148}$$

$$Zm_0 \frac{\partial}{\partial m_0} m = m \ . \tag{20.149}$$

When we differentiate with respect to m_0, we fix e_0, α_0, and Λ, and we define the following quantities:

$$\begin{aligned}
\beta(e) &= Zm_0 \frac{\partial e}{\partial m_0} , \\
\gamma_{\mathrm{el}}(e, \alpha) &= \frac{1}{2} Zm_0 \frac{\partial}{\partial m_0} \ln Z_2 , \\
\gamma_{\mathrm{ph}}(e) &= \frac{1}{2} Zm_0 \frac{\partial}{\partial m_0} \ln Z_3 , \\
\delta(e, \alpha) &= Zm_0 \frac{\partial}{\partial m_0} \alpha .
\end{aligned} \tag{20.150}$$

Combining these with the Ward identities,

$$e^2 = Z_3 e_0^2 , \quad e^2 \alpha = e_0^2 \alpha_0 , \tag{20.151}$$

we obtain the following equations, which correspond to (20.124):

$$\beta(e) = e \gamma_{\mathrm{ph}}(e) , \quad \delta(e, \alpha) = -2\alpha \gamma_{\mathrm{ph}}(e) . \tag{20.152}$$

We thus introduce the differential operator

$$\mathscr{D} = m \frac{\partial}{\partial m} + \beta(e) \frac{\partial}{\partial e} - 2\alpha \gamma_{\mathrm{ph}}(e) \frac{\partial}{\partial \alpha} . \tag{20.153}$$

Therefore, the Callan–Symanzik equation in QED becomes

$$\left[\mathscr{D} + \gamma^{(n,l)} - k(1 - \gamma_S)\right] \Delta^k G^{(n,l)} = \Delta^{k+1} G^{(n,l)} . \tag{20.154}$$

In this case, it turns out that $m(\bar{\psi}\psi)_{\mathrm{R}}$ is inserted as T. $(\bar{\psi}\psi)_{\mathrm{R}}$ is the quantity obtained when $\bar{\psi}\psi$ has been properly renormalized in a multiplicative way. The renormalization condition is

$$\langle \boldsymbol{p} | [\bar{\boldsymbol{\psi}}(x) \boldsymbol{\psi}(x)]_{\mathrm{R}} | \boldsymbol{p} \rangle = \bar{u}(p) u(p) , \tag{20.155}$$

where u has been normalized so that $u^\dagger(p) u(p) = 1$.

In general, the renormalization may be not only multiplicative, but also a linear transformation:

$$O_i^{(0)} = \sum_j Z_{ij} O_j . \tag{20.156}$$

For the Green's function,

$$O_{i(0)}^{(n,l)} = \sum_j Z_3^{n/2} Z_2^{l/2} Z_{ij} O_j^{(n,l)} . \tag{20.157}$$

Therefore,

$$\left[\mathscr{D} + \gamma^{(n,l)} - k(1-\gamma_S)\right]\Delta^k O_i^{(n,l)} + \sum_j \gamma_{ij} \Delta^k O_j^{(n,l)} = \Delta^{k+1} O_i^{(n,l)} , \tag{20.158}$$

where

$$\gamma_S = 1 - Z - m_0 \frac{\partial Z}{\partial m_0} , \quad \gamma_{ij} = \sum_k Z_{ik}^{-1} \left(Z m_0 \frac{\partial}{\partial m_0} Z_{kj} \right) . \tag{20.159}$$

To lowest order in QED,

$$\beta = \frac{e^3}{12\pi^2} , \quad \gamma_{\text{ph}} = \frac{e^2}{12\pi^2} , \quad \gamma_{\text{el}} = \frac{e^2}{16\pi^2}\alpha , \quad \gamma_S = -\frac{3e^2}{8\pi^2} . \tag{20.160}$$

Next, let us consider renormalizing the electron mass multiplicatively. This is possible only if the divergence of the self-energy becomes logarithmic:

$$Z_m = \frac{m}{m_0} , \quad \gamma_m = Z m_0 \frac{\partial}{\partial m_0} \ln Z_m . \tag{20.161}$$

Thus, immediately from the definition,

$$\gamma_m = 1 - Z . \tag{20.162}$$

Thus, taking γ_m as a function of e alone,

$$\begin{aligned} \gamma_S &= 1 - Z - Z m_0 \frac{\partial}{\partial m_0} \ln Z \\ &= 1 - Z - Z m_0 \frac{\partial}{\partial m_0} \ln(1 - \gamma_m) \\ &= \gamma_m - \mathscr{D} \ln(1 - \gamma_m) \\ &= \gamma_m - \beta(e) \frac{\partial}{\partial e} \ln(1 - \gamma_m) , \end{aligned} \tag{20.163}$$

or

$$\beta \frac{\mathrm{d}\gamma_m}{\mathrm{d}e} = (1 - \gamma_m)(\gamma_S - \gamma_m) . \tag{20.164}$$

Therefore, the equation for Z is

$$\left[\beta\frac{\mathrm{d}}{\mathrm{d}e}+(1-\gamma_S)\right]Z^{-1}(e)=1\,,\quad Z(0)=1\,. \tag{20.165}$$

From this, we see that Z is a finite function of e alone.

20.6 Homogeneous Callan–Symanzik Equation

The Callan–Symanzik equation introduced in the previous section is inhomogeneous. It is in fact a system of equations for an infinite number of functions $G^{(n)}$, $\Delta G^{(n)}$, and so on. We shall now rewrite them in a homogeneous and closed form. This is actually possible, but the price to pay for this is that the number of independent variables increases by one.

In order to write this infinite number of functions collectively, we introduce a generating function:

$$G^{(n)}(p_1,\ldots,p_n;m^2,g,K)=\sum_{j=0}^{\infty}\frac{K^j}{j!}\Delta^j G^{(n)}(p_1,\ldots,p_n;m^2,g)\,. \tag{20.166}$$

$$\tilde{\mathscr{D}}=\mathscr{D}-\left\{1+\left[2-\gamma_S(g)\right]K\right\}\frac{\partial}{\partial K}\,. \tag{20.167}$$

Therefore, from (20.140),

$$\left[\tilde{\mathscr{D}}+n\gamma_\varphi(g)\right]G^{(n)}(p_1,\cdots,p_n;m,g,K)=0\,. \tag{20.168}$$

As for the renormalization group of the first kind, we have thus obtained a homogeneous form. Note that the number of independent variables also coincides. In the case of the first kind, those variables are μ, ν, and m, while in the case of the second kind, they are m, g, and K. Therefore, by a suitable variable transformation, each can be transformed into the other. This was proven by Higashijima and the author in [182].

Similarly in QED, if we take

$$G^{(n,l)}(p_i;m,e,K)=\sum_{j=0}^{\infty}\frac{K^j}{j!}\Delta^j G^{(n,l)}(p_j;m,e)\,, \tag{20.169}$$

$$\tilde{\mathscr{D}}=\mathscr{D}-\left\{1+\left[1-\gamma_S(e)\right]K\right\}\frac{\partial}{\partial K}\,, \tag{20.170}$$

then

$$\big[\tilde{\mathscr{D}} + \gamma^{(n,l)}\big]G^{(n,l)}(p_i; m, e, K) = 0 . \tag{20.171}$$

Now, an infinitesimal renormalization group transformation varies a function Q of m, g, and K via an infinitesimal parameter $\delta\rho$ (scalar theory):

$$\delta Q = (\tilde{\mathscr{D}} Q)\delta\rho , \tag{20.172}$$

where

$$\delta m = m\delta\rho , \quad \delta g = \beta(g)\delta\rho , \quad \delta K = -\Big\{1 + \big[2 - \gamma_S(g)\big]K\Big\}\delta\rho . \tag{20.173}$$

Clearly, m is not fixed, so this is a property of the second kind. From (20.168), the response of the Green's function is

$$\delta G^{(n)} = -n\gamma_\varphi(g)G^{(n)}\delta\rho . \tag{20.174}$$

The operators yielding finite transformations are then

$$R_\rho = \exp(\rho\tilde{\mathscr{D}}) . \tag{20.175}$$

These operators form a group in the sense that

$$R_\rho R_{\rho'} = R_{\rho+\rho'} . \tag{20.176}$$

We thus set

$$R_\rho[g] = \bar{g}(\rho) , \quad R_\rho[K] = \bar{K}(\rho) , \quad R_\rho[m] = m\mathrm{e}^\rho . \tag{20.177}$$

This tells us, for instance, that $\bar{g}(\rho)$ is the solution of (20.173) for the initial condition $\bar{g}(0) = g$. Therefore, from (20.173),

$$G^{(n)}(p_i; m, g, K) = \exp\left[n\int_0^\rho \mathrm{d}\rho'\gamma_\varphi\big(\bar{g}(\rho')\big)\right]G^{(n)}\big(p_i; m\mathrm{e}^\rho, \bar{g}(\rho), \bar{K}(\rho)\big) . \tag{20.178}$$

Note that, from a dimensional analysis,

$$G^{(n)}\big(p_i; m\mathrm{e}^\rho, \bar{g}(\rho), \bar{K}(\rho)\big) = \mathrm{e}^{(4-3n)\rho}G^{(n)}\big(p_i\mathrm{e}^{-\rho}, m, \bar{g}(\rho), \bar{K}(\rho)\big) . \tag{20.179}$$

Combining these two equations, we obtain the response of the Green's function for a large momentum.

Since m is not an invariant quantity under the renormalization group, it is not a physical mass appearing as a pole in the Green's function. In particular, K is a parameter which shifts the mass. We thus eliminate K by a suitable renormalization group transformation. Hence, we use

$$\bar{K}(\rho) = 0 \tag{20.180}$$

to define ρ. For this ρ, $\bar{m}(\rho) = m\mathrm{e}^{\rho}$ becomes a physical mass. In this case, (20.178) becomes

$$G^{(n)}(p_i; m, g, k) = Z^n G^{(n)}(p_i; \bar{m}, \bar{g}, 0) \ . \tag{20.181}$$

Multiplying the left-hand side by $(\tilde{\mathscr{D}} + n\gamma_\varphi)$, we obtain zero, so the right-hand side should also vanish under this operation:

$$\left[(\tilde{\mathscr{D}}\bar{m})\frac{\partial}{\partial \bar{m}} + (\tilde{\mathscr{D}}\bar{g})\frac{\partial}{\partial \bar{g}} + n\bar{\gamma}_\varphi\right] G^{(n)}(p_i; \bar{m}, \bar{g}, 0) = 0 \ , \tag{20.182}$$

$$\bar{\gamma}_\varphi = \gamma_\varphi + \tilde{\mathscr{D}}(\ln Z) \ . \tag{20.183}$$

However, since the homogeneous equation cannot be expressed in terms of just two variables $\bar{m}$ and $\bar{g}$, we deduce that

$$\tilde{\mathscr{D}}\bar{m} = 0 \ , \quad \tilde{\mathscr{D}}\bar{g} = 0 \ , \quad \bar{\gamma}_\varphi = 0 \ . \tag{20.184}$$

Hence, in the physical quantity, $\bar{m}$ becomes invariant under the renormalization group transformation, and if we take

$$\bar{m}^2 = m^2 F(g, K) \ , \quad F(g, 0) = 1 \ , \tag{20.185}$$

then

$$\left[\beta(g)\frac{\partial}{\partial g} - \left\{1 + [2 - \gamma_S(g)]K\right\}\frac{\partial}{\partial K} + 2\right] F(g, K) = 0 \ . \tag{20.186}$$

This equation can be used to determine the dependence of the physical mass on g and K.

So far we have given the renormalization group equation for G, but it can sometimes be given for the one-particle irreducible Γ introduced in Sect. 15.3. We shall thus obtain an equation for Γ. We first introduce a generating functional including a mass-insertion term:

$$\mathscr{T}[J] = \left\langle \mathbf{0} \middle| T \exp\left\{-\mathrm{i}\int \mathrm{d}^4x \left[J(x)\boldsymbol{\varphi}(x) + Km^2\mathbf{S}(x)\right]\right\} \middle| \mathbf{0} \right\rangle \ , \tag{20.187}$$

$$\mathscr{T}[J] = \exp\big(-\mathrm{i}W[J]\big) \ , \quad \Gamma = W - \int \mathrm{d}^4x J(x)\langle\boldsymbol{\varphi}(x)\rangle \ . \tag{20.188}$$

Since $W[J]$ is a generating functional for G, (20.168) can be written in the form

$$\left[\tilde{\mathscr{D}} + \gamma_\varphi(g) \int \mathrm{d}^4x\, J(x) \frac{\delta}{\delta J(x)}\right] W[J] = 0\,. \tag{20.189}$$

The variation of W is now

$$\delta W = \left.\frac{\partial W}{\partial m}\right|_J \delta m + \left.\frac{\partial W}{\partial g}\right|_J \delta g + \left.\frac{\partial W}{\partial K}\right|_J \delta K + \int \mathrm{d}^4x \frac{\delta W}{\delta J(x)} \delta J(x)\,. \tag{20.190}$$

As we have seen in Sect. 15.3,

$$\frac{\delta W}{\delta J(x)} = \langle \boldsymbol{\varphi}(x) \rangle\,, \qquad \frac{\delta \Gamma}{\delta \langle \boldsymbol{\varphi}(x) \rangle} = -J(x)\,. \tag{20.191}$$

Thus, using (20.188), we compute $\delta \Gamma$:

$$\delta \Gamma = \left.\frac{\partial W}{\partial m}\right|_J \delta m + \left.\frac{\partial W}{\partial g}\right|_J \delta g + \left.\frac{\partial W}{\partial K}\right|_J \delta K + \int \mathrm{d}^4x\, \delta \langle \boldsymbol{\varphi}(x) \rangle \frac{\delta \Gamma}{\delta \langle \boldsymbol{\varphi}(x) \rangle}\,. \tag{20.192}$$

Therefore,

$$\left.\frac{\partial \Gamma}{\partial m}\right|_{\langle \varphi \rangle} = \left.\frac{\partial W}{\partial m}\right|_J\,, \quad \left.\frac{\partial \Gamma}{\partial g}\right|_{\langle \varphi \rangle} = \left.\frac{\partial W}{\partial g}\right|_J\,, \quad \left.\frac{\partial \Gamma}{\partial K}\right|_{\langle \varphi \rangle} = \left.\frac{\partial W}{\partial K}\right|_J\,. \tag{20.193}$$

Additionally,

$$\int \mathrm{d}^4x\, J(x) \frac{\delta W}{\delta J(x)} = -\int \mathrm{d}^4x \langle \boldsymbol{\varphi}(x) \rangle \frac{\delta \Gamma}{\delta \langle \boldsymbol{\varphi}(x) \rangle}\,. \tag{20.194}$$

If we rewrite (20.189) using (20.193) and (20.194), then

$$\left[\tilde{\mathscr{D}} - \gamma_\varphi(g) \int \mathrm{d}^4x \langle \boldsymbol{\varphi}(x) \rangle \frac{\delta}{\delta \langle \boldsymbol{\varphi}(x) \rangle}\right] \Gamma\big[\langle \boldsymbol{\varphi}(x) \rangle\big] = 0\,. \tag{20.195}$$

A renormalization group equation for each function Γ is

$$\big[\tilde{\mathscr{D}} - n\gamma_\varphi(g)\big] \Gamma^{(n)}(p_1, \ldots, p_n; m, g, K) = 0\,. \tag{20.196}$$

Comparing this equation with (20.168), the coefficient of γ_φ has the opposite sign.

So far we have investigated renormalization groups of the first and second kind for the scalar theory and QED. However, it is for non-Abelian gauge theories that the renormalization group shows its real worth. The first reason is this: as mentioned in Chap. 13, it is believed that strong interactions among quarks can be described by the non-Abelian gauge theory with the gauge group $SU(3)$. A second reason is that non-Abelian gauge theories have a crucially different property which other interactions do not possess: at high energies, effective interactions become weak, whence perturbation theory can be applied. This property is called *asymptotic freedom*. Its renormalization group equation facilitates comparison with experiments.

It will turn out that a distinguishing feature of quarks mentioned in Chap. 13 can be explained using the renormalization group equation.

20.7 Renormalization Group for Non-Abelian Gauge Theories

Combining (19.10) and (19.66),

$$\frac{\partial}{\partial x_\mu}\langle \mathbf{0}|T[\boldsymbol{A}^a_\mu(x), \boldsymbol{A}^b_\nu(y)]|\mathbf{0}\rangle = a\delta_{ab}\frac{\partial}{\partial x_\nu}D_{\mathrm{F}}(x-y)\ . \tag{20.197}$$

This equation has the same form for both renormalized quantities and unrenormalized quantities. In particular, the longitudinal part of the gauge field is not affected by the renormalization. We write the propagators for the transverse and ghost parts in unrenormalized forms:

$$D_0^{\mathrm{tr}}(k)_{\mu\nu} = \frac{1}{k^2-\mathrm{i}\epsilon}\left(\delta_{\mu\nu} - \frac{k_\mu k_\nu}{k^2-\mathrm{i}\epsilon}\right)d_0(k^2)\ , \tag{20.198}$$

$$\langle \mathbf{0}|T[\boldsymbol{c}^a(x), \bar{\boldsymbol{c}}^b(y)]|\mathbf{0}\rangle = \delta_{ab}\frac{1}{(2\pi)^4}\int \mathrm{d}^4k\, \mathrm{e}^{\mathrm{i}k\cdot(x-y)}G_0(k^2)\ , \tag{20.199}$$

and we introduce the renormalization constants

$$Z_3 = d_0(\mu^2)\ ,\quad \tilde{Z}_3 = G_0(\mu^2)\ . \tag{20.200}$$

We also introduce the renormalization constants for vertex functions:

gauge field $$\Gamma_0^{abc}(p,q,r)_{\lambda\mu\nu}\big|_{p^2=q^2=r^2=\mu^2} = Z_1^{-1}\Gamma_{\mathrm{bare}}^{abc}(p,q,r)_{\lambda\mu\nu}\big|_{p^2=q^2=r^2=\mu^2}\ , \tag{20.201}$$

ghost field $$\Gamma_0^{abc}(p,q,r)_\lambda|_{p^2=q^2=r^2=\mu^2} = \tilde{Z}_1^{-1}\Gamma_{\mathrm{bare}}^{abc}(p;q,r)_\lambda|_{p^2=q^2=r^2=\mu^2}\ . \tag{20.202}$$

The subscript "bare" indicates the vertex function in the Born approximation, where g has been subtracted from the one given in Sect. 18.4. Taking functional derivatives with respect to $\langle c\rangle$, $\langle A\rangle$, and $\langle B\rangle$ of the Slavnov–Taylor identity [171, 172] introduced in Sect. 19.5 and setting the external fields equal to zero, we can show that

$$Z_3/Z_1 = \tilde{Z}_3/\tilde{Z}_1\ . \tag{20.203}$$

We skip the details of the proof. This equation is called the *Slavnov–Taylor identity*. The renormalizations of the gauge parameter and the coupling constant are

$$\alpha_0 = Z_3\alpha\ , \quad g_0 = Z_1 Z_3^{-3/2} g\ . \tag{20.204}$$

If we now consider an n-point Green's function for the gauge field, its renormalization is

$$G_0^{(n)}(p_1,\ldots,p_n) = Z_3^{n/2} G^{(n)}(p_1,\ldots,p_n)\ , \tag{20.205}$$

where the vector indices have been omitted. The left-hand side is a function of g_0, α_0, and Λ, and the right-hand side is a function of g, α, and μ. Since the left-hand side does not depend on the renormalization point μ, we have

$$\mu\frac{\partial}{\partial\mu}G_0^{(n)}(p_1,\ldots,p_n) = 0\ . \tag{20.206}$$

Using this and referring to the discussion in Sect. 20.4,

$$\left[\mu\frac{\partial}{\partial\mu}+\beta(g,\alpha)\frac{\partial}{\partial g}+\delta(g,\alpha)\frac{\partial}{\partial\alpha}+n\gamma_V(g,\alpha)\right]G^{(n)}(p_1,\ldots,p_n) = 0\ , \tag{20.207}$$

$$\begin{aligned}
\beta(g,\alpha) = \mu\frac{\partial g}{\partial\mu}\bigg|_{g_0,\alpha_0,\Lambda}\ , \quad \gamma_V(g,\alpha) = \mu\frac{\partial}{\partial\mu}\ln Z_3^{1/2}\bigg|_{g_0,\alpha_0,\Lambda}\ , \\
\delta(g,\alpha) = \mu\frac{\partial\alpha}{\partial\mu}\bigg|_{g_0,\alpha_0,\Lambda} = -2\alpha\gamma_V(g,\alpha)\ .
\end{aligned} \tag{20.208}$$

Among the parameters appearing above, μ and Λ are the only ones with physical dimensions. Thus, g, Z_3, α, and so on must be functions of Λ/μ, whence

$$\mu\frac{\partial}{\partial\mu}f(\Lambda/\mu) = -\frac{\partial}{\partial\ln\Lambda}f(\Lambda/\mu)\ .$$

Since the Z factors become unity to lowest order,

$$\beta = -g\frac{\partial}{\partial\ln\Lambda}\left(\frac{Z_3^{3/2}}{Z_1}\right)\ , \quad \gamma_V = -\frac{1}{2}\frac{\partial}{\partial\ln\Lambda}Z_3\ . \tag{20.209}$$

Therefore, we need to know the coefficients of $\ln\Lambda$ in Z_1 and Z_3.

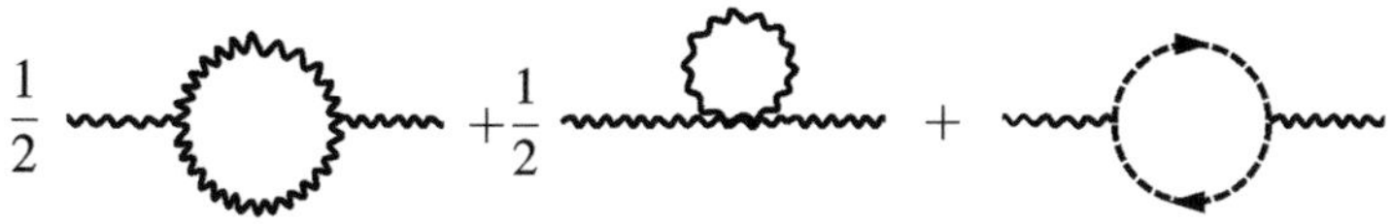

Fig. 20.1 Feynman diagram for calculating Z_3. Gluons are represented by *wavy lines* and ghosts by *dashed lines*

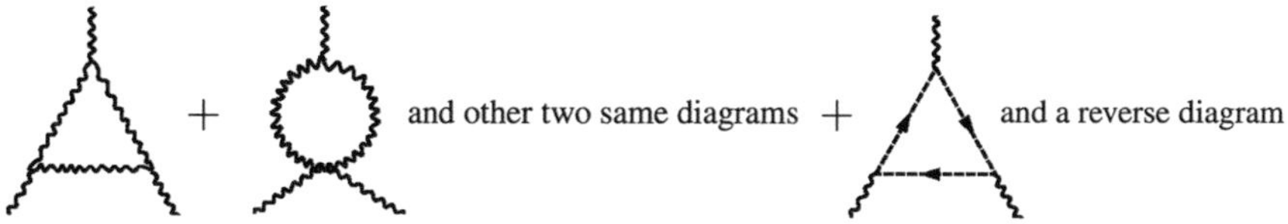

Fig. 20.2 Feynman diagram for calculating Z_1

To compute Z_3, we consider the Feynman diagram in Fig. 20.1. Note that the wavy line stands for the gluon and the dashed line with the arrow stands for the ghost. The result is

$$Z_3 = 1 + \frac{g^2}{16\pi^2}\left(\frac{13}{3} - \alpha\right) C_2(G) \ln \Lambda , \tag{20.210}$$

where

$$\sum_{c,d} f_{acd} f_{bcd} = \delta_{ab} C_2(G) . \tag{20.211}$$

In particular, for $SU(N)$, $C_2(G) = N$. In order to compute Z_1, we consider the Feynman diagrams in Fig. 20.2. The result is

$$Z_1 = 1 + \frac{g^2}{16\pi^2}\left(\frac{17}{6} - \frac{3}{2}\alpha\right) C_2(G) \ln \Lambda . \tag{20.212}$$

Similarly, the Z factors for the ghost are

$$\tilde{Z}_3 = 1 + \frac{g^2}{16\pi^2}\left(\frac{3}{2} - \frac{1}{2}\alpha\right) C_2(G) \ln \Lambda , \tag{20.213}$$

$$\tilde{Z}_1 = 1 + \frac{g^2}{16\pi^2}(-\alpha) C_2(G) \ln \Lambda . \tag{20.214}$$

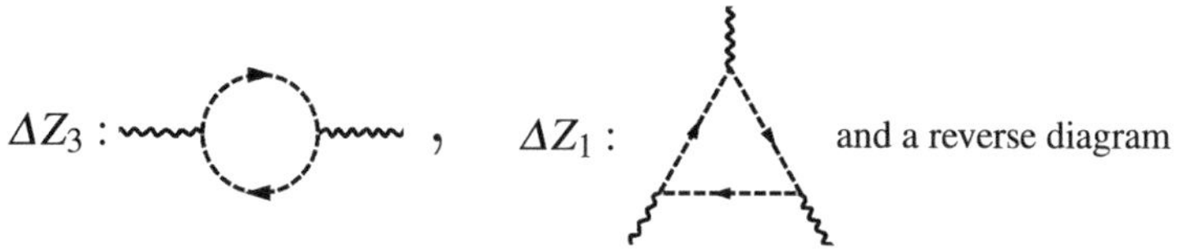

Fig. 20.3 Feynman diagrams for ΔZ_1 and ΔZ_3

Thus,

$$\beta(g,\alpha) = -\frac{g^3}{16\pi^2}\left(\frac{11}{3}\right)C_2(G)\,, \tag{20.215}$$

$$\gamma_V(g,\alpha) = -\frac{g^2}{32\pi^2}\left(\frac{13}{3}-\alpha\right)C_2(G)\,. \tag{20.216}$$

If there are fermions, their contributions to Z_1 and Z_3 are

$$\Delta Z_1 = \Delta Z_3 = -\frac{g^2}{8\pi^2}\frac{4}{3}T(R)\ln\Lambda\,, \tag{20.217}$$

where

$$\mathrm{Tr}(t^a t^b) = \delta_{ab}T(R)\,, \quad rT(R) = d(R)C_2(R)\,, \tag{20.218}$$

with r the number of creation operators, $d(R)$ the dimension of the representation, and $C_2(R)$ as defined in Sect. 13.2. The corresponding Feynman diagrams are shown in Fig. 20.3. For the N-dimensional fundamental representation of $SU(N)$,

$$T(R) = \frac{1}{2}\,, \quad C_2(R) = \frac{N^2-1}{2N}\,. \tag{20.219}$$

Therefore, assuming that all fermions obey the fundamental representation and denoting the number of their species, i.e., flavours, by n_{f},

$$\beta = -\frac{g^3}{16\pi^2}\left[\frac{11}{3}C_2(G) - \frac{4}{3}n_{\mathrm{f}}T(R)\right]\,, \tag{20.220}$$

$$\gamma_V = -\frac{g^2}{32\pi^2}\left[\left(\frac{13}{3}-\alpha\right)C_2(G) - \frac{8}{3}n_{\mathrm{f}}T(R)\right]\,. \tag{20.221}$$

This result was obtained in 1973 by Gross and Wilczek [183] and by Politzer [184].

20.8 Asymptotic Freedom

In Sect. 13.8, we discussed the parton model, concluding that interactions among quarks become weak at short range, i.e., the quarks in a hadron are almost free. In this section, we will prove that QCD based on the colour $SU(3)$ group has this property.

For both the first kind and the second kind, elements of the renormalization group can be written as R_ρ. If ρ is taken as a transformation parameter, infinitesimal transformations can be expressed as follows:

$$\delta\mu = \mu\delta\rho\ , \quad \delta g = \beta(g)\delta\rho\ . \tag{20.222}$$

In the limits $\rho \to \pm\infty$, we then ask whether g approaches a constant limiting value $g_{\pm\infty}$. If it approaches constant values, then g_∞ and $g_{-\infty}$ are called the *ultraviolet fixed point* and the *infrared fixed point*, respectively. In this case,

$$\beta(g_\infty) = 0\ , \quad \text{or} \quad \beta(g_{-\infty}) = 0\ . \tag{20.223}$$

In particular, if $\beta(g) = -bg^3$ $(b > 0)$ when g is small, then $g = 0$ in the limit $\rho \to \infty$. Such a theory is said to be *asymptotically free*. Examples are non-Abelian gauge theories, the $g\phi^4$ theory for $g < 0$, the six-dimensional ϕ^3 theory, and so on.

In a non-Abelian gauge theory, we take $g = g_0$ when $\rho = 0$ and solve the equation for small g:

$$\delta g = -bg^3\delta\rho\ , \quad b > 0\ , \tag{20.224}$$

then

$$\frac{1}{\bar{g}^2(\rho)} = \frac{1}{g_0^2} + 2b\rho\ . \tag{20.225}$$

We can now investigate what can be expected if QCD is asymptotically free.

20.8.1 *Electron–Positron Collision*

An electron–positron collision provides an opportunity to investigate asymptotic freedom. We take j_μ as the four-momentum current density for hadrons. Then, based on the computation given in Sect. 9.4, we can use this j_μ to define $\Pi(s)$ from (20.123) and (20.124). If we consider the lowest order for e and assume that s is large, we obtain (20.124).

When Π includes leptons, the function d appearing in the renormalization group for the electromagnetic field can be written as

$$d = 1 - \Pi \ . \tag{20.226}$$

We note that, in QCD + QED,

$$(\mathscr{D} + 2\gamma_{\text{ph}})d = 0 \ . \tag{20.227}$$

Applying the Landau gauge for both QCD and QED,

$$\mathscr{D} = \mu\frac{\partial}{\partial\mu} + \beta_e\frac{\partial}{\partial e} + \beta_g\frac{\partial}{\partial g} \ . \tag{20.228}$$

Note that both Π and γ_{ph} are of order e^2. If we drop terms of order e^4, then $\gamma_{\text{ph}}\Pi$ and $\beta\partial/\partial e$ can be omitted. Then, applying this approximation to (20.227) and using (20.226),

$$\left[\mu\frac{\partial}{\partial\mu} + \beta(g)\frac{\partial}{\partial g}\right]\Pi = 2\gamma_{\text{ph}} \ . \tag{20.229}$$

In particular, taking the imaginary part, we obtain

$$\left[\mu\frac{\partial}{\partial\mu} + \beta(g)\frac{\partial}{\partial g}\right]\text{Im}\,\Pi\left(\frac{s}{\mu^2}, \frac{m^2}{\mu^2}, g\right) = 0 \ . \tag{20.230}$$

To lowest order in QED,

$$\begin{aligned}\text{Im}\,\Pi &= \frac{1}{3}\sum_i \frac{Q_i^2}{4\pi}\left(1 + \frac{2m_i^2}{s}\right)\sqrt{1 - \frac{4m_i^2}{s}} \\ &\to \frac{1}{3}\sum_i \frac{Q_i^2}{4\pi} \quad (\text{for} s \to \infty) \ .\end{aligned} \tag{20.231}$$

We introduce the following quantity, called the *Drell ratio*:

$$R(s) = \frac{\sigma(\text{e}^+ + \text{e}^- \to \text{hadrons})}{\sigma(\text{e}^+ + \text{e}^- \to \mu^+ + \mu^-)} \ .$$

To lowest order again, for $s \to \infty$, the following scaling law is obtained:

$$R \to \sum_i Q_i^2/e^2 \ . \tag{20.232}$$

Now, at high energies, we set $m^2/\mu^2 \to 0$ for $s \gg \mu^2 \gg m^2$. Therefore, since Π is dimensionless,

$$\left[\mu\frac{\partial}{\partial\mu} + \beta(g)\frac{\partial}{\partial g}\right] \operatorname{Im}\Pi\left(\frac{s}{\mu^2}, g\right) = 0 \,. \tag{20.233}$$

This implies

$$\operatorname{Im}\Pi\left(\frac{s}{\mu^2}, g\right) = \operatorname{Im}\Pi\left(\frac{s_0}{\mu^2}, \bar{g}(\rho)\right) , \quad s = s_0 e^{2\rho} \,. \tag{20.234}$$

If ρ is large, then in QCD $\bar{g}\rho \to 0$, and

$$\operatorname{Im}\Pi\left(\frac{s_0}{\mu^2}, \bar{g}(\rho)\right) = \operatorname{Im}\Pi\left(\frac{s_0}{\mu^2}, 0\right)\left[1 + B\bar{g}^2(\rho) + \cdots\right] . \tag{20.235}$$

We write a solution to (20.225) in the form

$$\bar{g}^2(\rho) = \left(b \ln\frac{s}{\Lambda^2}\right)^{-1} , \quad \Lambda^2 = s_0 \exp\left(-\frac{1}{b g_0^2}\right) . \tag{20.236}$$

Moreover, in $SU(3)$,

$$C_2(G) = 3 \,, \quad T(R) = \frac{1}{2} \,, \quad C_2(R) = \frac{4}{3} \,, \tag{20.237}$$

such that

$$b = \frac{1}{16\pi^2}(11 - 2n_{\mathrm{f}}) \,, \quad B = \frac{3}{16\pi^2}C_2(R) = \frac{1}{4\pi^2} \,. \tag{20.238}$$

If we assume that $s_0/\mu^2 \gg 1$, then at high energies,

$$\operatorname{Im}\Pi\left(\frac{s}{\mu^2}, g\right) = \frac{1}{12\pi}\sum_i Q_i^2\left[1 + \frac{C_i}{\ln(s/\Lambda^2)} + \cdots\right] , \tag{20.239}$$

where

$$C_i = \begin{cases} \dfrac{B}{b} = \dfrac{12}{33 - 2n_{\mathrm{f}}} & (\text{quark}) \,, \\ 0 & (\text{lepton}) \,. \end{cases} \tag{20.240}$$

Thus, assuming that δ_{iq} is 1 for quarks and 0 for leptons,

$$R(s) = \sum_i \frac{Q_i^2}{e^2} \left[1 + \frac{12}{33 - 2n_{\mathrm{f}}} \frac{\delta_{iq}}{\ln(s/\Lambda^2)} + \cdots \right] . \tag{20.241}$$

Hence, for $s \to \infty$, R gets closer to a constant. This is called a *scaling relation*. In addition, the total cross-section to hadrons turns out to be inversely proportional to s. We rewrite the sum over i to take into account the fact that there are three colours for each flavour:

$$\sum_i Q_i^2 = 3 \sum_f Q_f^2 + \sum_l Q_l^2 , \tag{20.242}$$

where indices f and l range over flavours and leptons, respectively, and we assume that the correction term in (20.241) applies only to quarks. This scaling relation agrees well with experiment, supporting the idea that there are three colours.

The branching ratio for τ^- decay is quite similar. Taking into account the large mass of τ, viz., 1784 MeV, the scaling law is considered to apply, so it can decay into each of the following with the same probability:

$$\tau^- \to \mathrm{e}^- + \bar{\nu}_{\mathrm{e}} + \nu_\tau , \quad \tau^- \to \mu^- + \bar{\nu}_\mu + \nu_\tau , \quad \tau^- \to \mathrm{d}' + \bar{\mathrm{u}} + \nu_\tau \quad \text{(3colours)}. \tag{20.243}$$

Thus, the branching ratio for τ to decay into hadrons is expected to be $3/5$. The experimental value is 0.64.

20.8.2 Bjorken Scaling Law

We shall now interpret the Bjorken scaling law [105] discussed in Sect. 13.8 in relation to the question of the asymptotic freedom. However, a rigorous argument requires a complicated procedure involving Wilson's operator expansion, so here we shall limit ourselves to a sketch of the basic scenario.

Around $x^2 = 0$, current densities can be expanded in the following way:

$$\boldsymbol{j}_\mu \left(\frac{x}{2}\right) \boldsymbol{j}_\nu \left(-\frac{x}{2}\right) \sim \sum_n C^{(n)}(x^2, g) x_{\mu_1} \dots x_{\mu_n} \boldsymbol{O}^{(n)}_{\mu\nu:\mu_1\dots\mu_n} . \tag{20.244}$$

This is *Wilson's operator expansion*. Taking the expectation value of this equation in a one-nucleon state and carrying out the Fourier expansion, we obtain a structure function $F_i(x, q^2)$. Here, $\boldsymbol{O}^{(n)}$ is a local operator that is symmetric with respect to $\mu_1, \dots, \mu_n$, and contractions over arbitrary indices have been chosen so that they vanish. Writing q^2 instead of Q^2, we have $x = q^2/2M\nu$.

If we choose a large enough value for q^2, the following equation holds:

$$\int_0^1 \mathrm{d}x\, F_i(x, q^2)x^{n-2} \sim \bar{C}_i^{(n)}(q^2, g)\langle P|\boldsymbol{O}^{(n)}|P\rangle\,, \tag{20.245}$$

where $\bar{C}^{(n)}$ is the Fourier transform of $C^{(n)}$.

If we consider that, in a commutation relation such as (12.180), e is a fixed parameter, the anomalous dimension of j_μ becomes zero. Hence, taking the anomalous dimension of $\boldsymbol{O}^{(n)}$ to γ_n, the anomalous dimension of $\bar{C}^{(n)}$ becomes $-\gamma_n$:

$$[\mathscr{D} - \gamma_n(g)]\bar{C}_i^{(n)}\left(\frac{q^2}{\mu^2}, g\right) = 0\,. \tag{20.246}$$

In order to solve this equation, we set $q^2/\mu^2 = \exp(2\rho)$:

$$\bar{C}_i^{(n)}\left(\frac{q^2}{\mu^2}, g\right) = \bar{C}_i^{(n)}(1, \bar{g}(\rho))\exp\left[-\int_0^\rho \mathrm{d}\rho\gamma_n(\bar{g}(\rho))\right]. \tag{20.247}$$

In general, γ_n can be written in the form

$$\gamma_n(g) = \gamma_n g^2 + \mathscr{O}(g^4)\,, \tag{20.248}$$

whence the above exponential function can be determined:

$$\exp\left[-\int_0^\rho d\rho\gamma_n(\bar{g}(\rho))\right] = \left(1 + bg_0^2 \ln\frac{q^2}{\mu^2}\right)^{-\gamma_n/2b}\,. \tag{20.249}$$

Thus, when q^2 is large,

$$\bar{C}_i^{(n)}\left(\frac{q^2}{\mu^2}, g\right) \approx \text{const.}\left(\ln\frac{q^2}{\mu^2}\right)^{-\gamma_n/2b}\,, \tag{20.250}$$

where we have set $\bar{g}(\rho)$ to zero in the first factor on the right-hand side of (20.247). Hence, although the q^2-dependence remains, and this therefore differs from the original Bjorken scaling law, the dependence is at most logarithmic. This is an interpretation of Bjorken's scaling law based on the asymptotically free QCD. As expected, this result gives support for QCD.

20.9 Gauge Dependence of Green's Functions

In gauge theories, there are many ways to choose the gauge-fixing term and the ghost term, and it is important to show that no physical consequence depends on these choices. For instance, a charge e renormalized in the standard way does not depend on the way we choose the Lagrangian. This is clear from the fact that there are ways to measure e experimentally. Since QED is an infrared-stable theory, we have the so-called *low-energy limit theorem*. At low energies there are rigorous ways to express the S-matrix element in terms of e, and this is why e can be determined experimentally. In contrast, in non-Abelian gauge theories, there is no preferred renormalization prescription like the one in QED. Therefore, it is unclear whether the coupling constant g is defined independently of the gauge; and depending on the renormalization prescription, it may depend on how the gauge is chosen.

In this section, we consider ways to distinguish parameters depending on the gauge choice from those independent of it. First, we discuss the situation where renormalization is not involved. We thus consider an unrenormalized Green's function:

$$G_0^{(n)} = \left\langle \boldsymbol{\varphi}^{(0)}(x_1), \ldots, \boldsymbol{\varphi}^{(0)}(x_n) \right\rangle , \tag{20.251}$$

where we have assumed that $\boldsymbol{\varphi}$ expresses not only a scalar field but also a general field. Now, considering a gauge theory, we choose the Lagrangian (18.74) in Chap. 18. In this case, what distinguishes the Lagrangians is an unrenormalized gauge parameter α_0. For the Green's function (20.251), we can write down two equations. The first is an equation which has appeared many times up to now and the second is the equation derived from (18.84) in Chap. 18:

$$\mu \frac{\partial}{\partial \mu} \left\langle \boldsymbol{\varphi}^{(0)}(x_1) \ldots \boldsymbol{\varphi}^{(0)}(x_n) \right\rangle_{\alpha_0} = 0 , \tag{20.252}$$

$$\begin{aligned} &\alpha_0 \frac{\partial}{\partial \alpha_0} \left\langle \boldsymbol{\varphi}^{(0)}(x_1) \ldots \boldsymbol{\varphi}^{(0)}(x_n) \right\rangle_{\alpha_0} \\ &\quad = \frac{\mathrm{i}\alpha_0}{2} \left\langle \boldsymbol{\varphi}^{(0)}(x_1) \ldots \boldsymbol{\varphi}^{(0)}(x_n) \int \mathrm{d}^4 x \, \boldsymbol{B}^{(0)}(x) \cdot \boldsymbol{B}^{(0)}(x) \right\rangle_{\alpha_0} . \end{aligned} \tag{20.253}$$

In the second equation, the term which should be called a gauge insertion term or something similar to the mass-insertion term in the Callan–Symanzik equation appears on the right-hand side. One property of this term is its invariance under the renormalization, i.e., referring to (19.66), we have

$$\alpha_0 = Z_3 \alpha , \quad \boldsymbol{B}^{(0)}(x) = Z_3^{-1/2} \boldsymbol{B}(x) , \tag{20.254}$$

whence

$$\alpha_0 \boldsymbol{B}^{(0)}(x) \cdot \boldsymbol{B}^{(0)}(x) = \alpha \boldsymbol{B}(x) \cdot \boldsymbol{B}(x) . \tag{20.255}$$

We shall imitate the Callan–Symanzik equation. Setting

$$\boldsymbol{F}(x) = \boldsymbol{B}(x) \cdot \boldsymbol{B}(x) , \tag{20.256}$$

and referring to (20.125) and (20.126), we carry out the Fourier transforms

$$(\mathscr{D} + \gamma_\mathrm{G}) G^{(n)}(p_1, \ldots, p_n) = 0 , \tag{20.257}$$

$$(\bar{\mathscr{D}} + \bar{\gamma}_\mathrm{G}) G^{(n)}(p_1, \ldots, p_n) = -\frac{\alpha}{2} F^{(n)}(0; p_1, \ldots, p_n) , \tag{20.258}$$

where $G^{(n)}$ and $F^{(n)}$ are the renormalized Green's functions:

$$G_0^{(n)} = Z_\mathrm{G} G^{(n)} , \quad \alpha_0 F_0^{(n)} = Z_\mathrm{G} \alpha F^{(n)} . \tag{20.259}$$

Furthermore,

$$\mathscr{D} = \mu \frac{\partial}{\partial \mu} + \beta \frac{\partial}{\partial g} + \delta \frac{\partial}{\partial \alpha} , \quad \bar{\mathscr{D}} = \bar{\beta} \frac{\partial}{\partial g} + \bar{\delta} \frac{\partial}{\partial \alpha} , \tag{20.260}$$

$$\beta = \mu \frac{\partial}{\partial \mu} g \bigg|_{g_0, \alpha_0} , \quad \bar{\beta} = \alpha_0 \frac{\partial}{\partial \alpha_0} g \bigg|_{g_0, \mu} , \tag{20.261}$$

$$\delta = \mu \frac{\partial}{\partial \mu} \alpha \bigg|_{g_0, \alpha_0} , \quad \bar{\delta} = \alpha_0 \frac{\partial}{\partial \alpha_0} \alpha \bigg|_{g_0, \mu} , \tag{20.262}$$

$$\gamma_\mathrm{G} = \mu \frac{\partial}{\partial \mu} \ln Z_\mathrm{G} \bigg|_{g_0, \alpha_0} , \quad \bar{\gamma}_\mathrm{G} = \alpha_0 \frac{\partial}{\partial \alpha_0} \ln Z_\mathrm{G} \bigg|_{g_0, \mu} . \tag{20.263}$$

We introduce the integrability condition

$$\left(\mu \frac{\partial}{\partial \mu} \right) \left(\alpha_0 \frac{\partial}{\partial \alpha_0} \right) = \left(\alpha_0 \frac{\partial}{\partial \alpha_0} \right) \left(\mu \frac{\partial}{\partial \mu} \right) . \tag{20.264}$$

Expressing this in terms of renormalized parameters,

$$\mathscr{D} \bar{\mathscr{D}} = \bar{\mathscr{D}} \mathscr{D} . \tag{20.265}$$

In this theory, there are three renormalized parameters m/μ, g, and α. Introducing three independent functions ρ, σ, and τ, we now implement a variable transformation, giving the following theorem without proof:

Theorem 20.1 *If the commutation condition (20.265) holds true, then we can choose functions ρ, σ, and τ of the three variables m/μ, g, and α satisfying the conditions*

$$\mathscr{D}\rho = 1\,, \quad \bar{\mathscr{D}}\rho = 0\,, \quad \mathscr{D}\sigma = 0\,, \quad \bar{\mathscr{D}}\sigma = 1\,, \quad \mathscr{D}\tau = 0\,, \quad \bar{\mathscr{D}}\tau = 0\,. \tag{20.266}$$

Thus, if m/μ, g, and α can be rewritten in terms of ρ, σ and τ, then (20.257) and (20.258) become

$$\left(\frac{\partial}{\partial\rho} + \gamma_{\mathrm{G}}\right) G^{(n)} = 0\,, \quad \left(\frac{\partial}{\partial\sigma} + \bar{\gamma}_{\mathrm{G}}\right) G^{(n)} = -\frac{\alpha}{2} F^{(n)}\,. \tag{20.267}$$

The integrability condition (20.264) then implies

$$\frac{\partial}{\partial\sigma}\gamma_{\mathrm{G}} = \frac{\partial}{\partial\rho}\bar{\gamma}_{\mathrm{G}}\,. \tag{20.268}$$

Therefore, there is a function Φ satisfying

$$\gamma_{\mathrm{G}} = \frac{\partial}{\partial\rho}\Phi\,, \quad \bar{\gamma}_{\mathrm{G}} = \frac{\partial}{\partial\sigma}\Phi\,. \tag{20.269}$$

Hence, setting

$$G^{(n)} = \mathrm{e}^{-\Phi} I^{(n)}\,, \quad F^{(n)} = \mathrm{e}^{-\Phi} J^{(n)}\,, \tag{20.270}$$

we have

$$\frac{\partial}{\partial\rho} I^{(n)} = 0\,, \quad \frac{\partial}{\partial\sigma} I^{(n)} = -\frac{\alpha}{2} J^{(n)}\,. \tag{20.271}$$

We are especially interested in the case where $\boldsymbol{\varphi}(x_1), \ldots, \boldsymbol{\varphi}(x_n)$ are all BRS-invariant. In this case, $F^{(n)}$ vanishes, and so therefore does $J^{(n)}$, whence I^n becomes a function of τ alone.

It is clear from the previous discussion that ρ is related to a scale transformation, and σ is related to a transformation which changes the gauge. Therefore, *we may say that quantities independent of the unrenormalized gauge parameter α_0 do not depend on σ in a renormalized theory.*

In QED, from the early discussion in this section and (20.40),

$$\bar{\mathscr{D}}e = 0\,, \quad \mathscr{D}(\alpha e^2) = 0\,. \tag{20.272}$$

Therefore, in QED, ρ is expected to become a function of e, and σ a function of αe^2. In QCD, things are more complicated. This is related to the fact that, because the β-function generally depends on the renormalization prescription, it may be a function, not only of g, but also of α. If there is a renormalization prescription satisfying

$$\bar{\mathscr{D}} g = 0 , \tag{20.273}$$

then it may turn out that g has a significance unrelated to the gauge. One way is to renormalize all quantities using (18.84) in the Landau gauge. Hence, assuming that (18.84) also holds true for renormalized quantities, we define Green's functions in the general gauge. Therefore, both β and γ_V turn out not to depend on α. However, ρ can in general be interpreted as playing a role.

Chapter 21
Theory of Confinement

So far we have given indirect evidence that all hadrons consist of quarks. Moreover, the long series of successes of this model make other possibilities quite hard to consider. At the same time, searches for fractional charges have been made, but none has ever been detected. This has led to the conclusion that isolated quarks cannot be observed for some reason, an idea known as the *quark confinement hypothesis*. This hypothesis has in fact been generalized to the idea that no coloured particle can ever be observed. This is the *colour confinement hypothesis*.

Many proposals have been put forward to explain this hypothesis. One idea endorsed by many is that the quark–anti-quark potential may be proportional to distance at long ranges. This idea is closely related to the lattice gauge theories introduced by Wilson [185]. He formulated a condition for quark confinement in such a way that a quantity, which we call a *Wilson loop*, satisfies the area law. He has shown that this condition is satisfied in a strong coupling approximation. However, within the framework of lattice gauge theories, it is hard to formulate a confinement condition for gluons. So in order to discuss confinement for general coloured particles, the author has considered a mechanism for confinement based on an analogy with QED [186].

Firstly, if we ask whether there are ever particles which are created as a consequence of quantization but never observed, we immediately recall the photons corresponding to the longitudinal and scalar waves in QED. If, when we write down the condition for unitarity of the S-matrix, both the initial and final states satisfy the Lorenz condition, the contributions of these photons to intermediate states completely annihilate each other, and so never show up. That is, they cannot be observed. This was discussed in Chap. 5 and Sect. 6.4. This is confinement based on the annihilation mechanism for indefinite metrics, the Lorenz condition being used to eliminate negative norm states.

If we apply this kind of mechanism to confinement for particles with colour, it turns out to be inevitable to introduce indefinite metrics and additional conditions. For non-Abelian gauge fields, all the tools are to hand. That is, in non-Abelian

K. Nishijima, *Quantum Field Theory*,
https://doi.org/10.1007/978-94-024-2190-3_21

gauge fields, indefinite metrics are introduced via auxiliary fields, and additional conditions are defined using the BRS charges. Consequently, if both the initial and final states belong to the BRS singlet, then contributions from the BRS doublets completely annihilate each other in the unitarity condition, which was shown in Sect. 19.6. It thus turns out that colour confinement holds true if all particle states with colours belong to BRS doublets. We thus investigate under what kind of conditions such a thing could be realized.

21.1 Gauge Independence of the Confinement Condition

In the following discussion, we shall treat the question of confinement in a special gauge, and in particular, in the Landau gauge. We shall thus show that the concept of confinement does not depend on the choice of gauge. So if it holds true in one gauge, it will hold true in other gauges as well. For this purpose, we first recall a generalization of the LSZ reduction formula presented in Sect. 11.7. We write (11.161′) in the form

$$\int \mathrm{d}^4x\, \Delta(x-y)K_y ST[\boldsymbol{\varphi}(x_1)\dots\boldsymbol{\varphi}(x_n)\boldsymbol{\varphi}(y)] = \left[\varphi^{\mathrm{in}}(x), ST[\boldsymbol{\varphi}(x_1)\dots\boldsymbol{\varphi}(x_n)]\right] . \tag{21.1}$$

For each local operator $\boldsymbol{\Phi}(x)$, we assume that the following conditions are satisfied:

(1) $\boldsymbol{\Phi}(x)$ is a polynomial of field operators and for an arbitrary elementary particle field $\boldsymbol{\varphi}(x)$,

$$[\boldsymbol{\Phi}(x), \boldsymbol{\varphi}(y)] = 0\,, \quad (x-y)^2 > 0\,. \tag{21.2}$$

(2) $\boldsymbol{\Phi}(x)$ has the same quantum number as $\boldsymbol{\varphi}(x)$ and for a one-particle state $|\boldsymbol{p}\rangle$ of the field $\boldsymbol{\varphi}$,

$$\langle \mathbf{0}|\boldsymbol{\Phi}(x)|\boldsymbol{p}\rangle = \langle \mathbf{0}|\boldsymbol{\varphi}(x)|\boldsymbol{p}\rangle\,. \tag{21.3}$$

In this case, (21.1) extends to

$$\int \mathrm{d}^4y\, \Delta(x-y)K_y ST[\boldsymbol{\varphi}(x_1)\dots\boldsymbol{\varphi}(x_n)\boldsymbol{\Phi}(y)] = \left[\varphi^{\mathrm{in}}(x), ST[\boldsymbol{\varphi}(x_1)\dots\boldsymbol{\varphi}(x_n)]\right] . \tag{21.4}$$

Here we have considered scalar fields, but this can be extended to arbitrary fields, noting that, for fields obeying Fermi statistics, we have to replace the commutator in (21.2) by an anti-commutator.

Since we do not have fundamental fields for composite particles, we change the second condition as follows:

(2′) For a one-particle state of a composite particle,

$$\langle \mathbf{0}|\boldsymbol{\Phi}(x)|\boldsymbol{p}\rangle = \frac{1}{\sqrt{2p_0V}}\mathrm{e}^{\mathrm{i}p\cdot x}\ . \tag{21.5}$$

However, if the spin is not 0, we multiply by a suitable spin function.

In this case, introducing asymptotic fields Φ^{in} and Φ^{out} for a composite particle,

$$\int \mathrm{d}^4y\,\Delta(x-y)K_yST[\ldots\boldsymbol{\Phi}(y)] = \left[\Phi^{\mathrm{in}}(x), ST[\ldots]\right]. \tag{21.6}$$

When we compute the S-matrix element using the LSZ asymptotic condition, field operators appearing in the Green's function are not unique but arbitrary, if the condition above is satisfied. This formulation derives from treatments of composite particles by Haag [187], Zimmermann [188], and the author [189]. For details, the reader is referred to my book *Fields and Particles* [2]. We now return to the question of confinement.

If all particles with colour are confined, then it turns out that all quanta of fundamental fields are confined. Thus, observable quanta will be particles without colours. In QCD, this concerns only hadrons as composite particles. Such hadrons are described by local composite fields. Moreover, if we consider *unrenormalized composite fields* $\boldsymbol{A}, \boldsymbol{B}, \boldsymbol{C}, \ldots$, then as discussed in Sect. 19.4, such fields should be invariant under the BRS transformation. Hence, (19.58) should hold true. In this case, (19.63) also turns out to hold true. Note that, in order to use the LSZ reduction formula for these Green's functions, we should carry out the renormalization in such a way that a normalization condition like (21.5) is satisfied. Note also that, in this case, from (19.63), we have

$$\langle \boldsymbol{A}(x)\boldsymbol{A}^\dagger(y)\rangle_{II} = \langle \mathbf{0}|\boldsymbol{A}(x)\boldsymbol{A}^\dagger(y)|\boldsymbol{p}\rangle_I\ , \tag{21.7}$$

so that the residues at poles corresponding to one-particle states for these two-point functions are equal. Therefore, for the unrenormalized $\boldsymbol{A}$,

$$\langle \mathbf{0}|\boldsymbol{A}(x)|\boldsymbol{p}\rangle_I = \langle \mathbf{0}|\boldsymbol{A}(x)|\boldsymbol{p}\rangle_{II}\ , \tag{21.8}$$

where $|\boldsymbol{p}\rangle$ stands for a one-particle state of a hadron with four-momentum p. Therefore,

$$\frac{\langle \boldsymbol{A}(x)\boldsymbol{B}(y)\ldots\rangle_I}{\sqrt{2p_0V}\langle \boldsymbol{0}|\boldsymbol{A}(0)|\boldsymbol{p}\rangle_I\sqrt{2q_0V}\langle \boldsymbol{0}|\boldsymbol{B}(0)|\boldsymbol{q}\rangle_I\ldots}$$
$$= \frac{\langle \boldsymbol{A}(x)\boldsymbol{B}(y)\ldots\rangle_{II}}{\sqrt{2p_0V}\langle \boldsymbol{0}|\boldsymbol{A}(0)|\boldsymbol{p}\rangle_{II}\sqrt{2q_0V}\langle \boldsymbol{0}|\boldsymbol{B}(0)|\boldsymbol{q}\rangle_{II}\ldots}\,. \tag{21.9}$$

This is the renormalized Green's function which does not depend on gauge choices. It turns out that the S-matrix element obtained by applying the LSZ reduction formula to such a Green's function has meaning independently of how the gauge is chosen. Hence, if the S-matrix corresponding to a reaction among hadrons is unitary in one gauge, it will be unitary in all other gauges. As a consequence, all particles with colour turn out to be confined in any gauge. Thus, in order to discuss the confinement of particles with colour, we only need to show this in one specific gauge.

21.2 Sufficient Condition for Colour Confinement

In Sect. 19.6, we discussed the fact that the unitarity condition for the S-matrix can be expressed in terms of BRS singlet states alone. We thus assume the asymptotic completeness discussed in Chap. 11 only for BRS singlet states. The reason is that it is unclear, due to so-called infrared divergences, whether we can assume the asymptotic condition for BRS doublet states as well, and for singlet states, the existence of the unitary S-matrix forms the basis of the theory.

We begin with the Lagrangian density (19.1). Using the field equations and equal-time commutation relations, we can write down the following Ward–Takahashi identities:

$$\partial_\lambda\langle \delta\bar{\delta}\boldsymbol{A}^a_\lambda(x),\, \boldsymbol{\psi}^\alpha(y),\, \bar{\boldsymbol{\psi}}^\beta(z)\rangle$$
$$= \mathrm{i}gt^a_{\alpha\beta}\Big[\delta^4(x-y)S_{\mathrm{F}}(y-z,\beta) - \delta^4(x-z)S_{\mathrm{F}}(y-z,\alpha)\Big]\,, \tag{21.10}$$

$$\partial_\lambda\langle \delta\bar{\delta}\boldsymbol{A}^a_\lambda(x),\, \boldsymbol{A}^b_\mu(y),\, \boldsymbol{A}^c_\nu(z)\rangle$$
$$= \mathrm{i}gM^a_{bc}\Big[\delta^4(x-y)S_{\mathrm{F}\mu\nu}(y-z,c) - \delta^4(x-z)D_{\mathrm{F}\mu\nu}(y-z,b)\Big]\,, \tag{21.11}$$

where $M^a_{bc} = -\mathrm{i}f_{abc}$ corresponds to the adjoint representation, and S_{F} and D_{F} are propagators of a fermion and a gauge field, respectively. It is only in the presence of

BRS invariance that these identities hold true, even if the symmetry corresponding to the gauge group is spontaneously broken.

So considering (21.10), we begin by defining the vertex function V_λ^a by

$$\langle \delta\bar{\delta}\boldsymbol{A}_\lambda^a(x), \boldsymbol{\psi}^\alpha(y), \bar{\boldsymbol{\psi}}^\beta(z)\rangle = \int \mathrm{d}^4y' \int \mathrm{d}^4z' S_\mathrm{F}(y-y',\alpha) V_\lambda^a(y',z';x)_{\alpha\beta} S_\mathrm{F}(z'-z,\beta)\,. \tag{21.12}$$

We introduce the Fourier transform

$$S_\mathrm{F}(x) = \frac{-\mathrm{i}}{(2\pi)^4} \int \mathrm{d}^4p\, \mathrm{e}^{\mathrm{i}p\cdot x} S_\mathrm{F}(p)\,,$$

$$V_\lambda^a(yz;x)_{\alpha\beta} = \frac{1}{(2\pi)^8} \int \mathrm{d}^4p \int \mathrm{d}^4q\, \mathrm{e}^{\mathrm{i}p\cdot(y-x)+\mathrm{i}q\cdot(x-z)} V_\lambda^a(p,q)_{\alpha\beta}\,. \tag{21.13}$$

Then (21.10) can be written in the form

$$S_\mathrm{F}(p,\alpha)(p-q)_\lambda V_\lambda^a(p,q) S_\mathrm{F}(q,\beta) = \mathrm{i}gt^a_{\alpha\beta}\big[S_\mathrm{F}(p,\alpha) - S_\mathrm{F}(q,\beta)\big]\,. \tag{21.14}$$

What is important here is the identity

$$\partial_\lambda(\delta\bar{\delta}\boldsymbol{A}_\lambda) = -\bar{\delta}(\partial_\lambda D_\lambda \boldsymbol{c}) = 0\,. \tag{21.15}$$

We now investigate whether V_λ has a pole with respect to the variable $(p-q)^2$. First, we assume that x is connected to y' and z' via one-particle lines. This particle can be elementary or composite. In this case, there exists a pole with respect to $(p-q)^2$. For instance, if a particle is created by a scalar field $\boldsymbol{\phi}^b$, such a pole will appear in the Green's function

$$\langle \delta\bar{\delta}\boldsymbol{A}_\lambda^a(x), \boldsymbol{\phi}^b(w)\rangle\,. \tag{21.16}$$

Applying ∂_λ to this expression, since from (21.15) an equal-time commutator appears, the result is proportional to $\delta^4(x-w)$. Thus, its Fourier transformation becomes a constant, which can be written in the form

$$\mathrm{FT}\,\langle \delta\bar{\delta}\boldsymbol{A}_\lambda^a(x), \boldsymbol{\phi}^b(w)\rangle = \frac{k_\lambda}{k^2}\delta_{ab}C_0\,. \tag{21.17}$$

In contrast, if this one-particle state is created by a vector field ϕ_μ^b, we have

$$\mathrm{FT}\,\langle \delta\bar{\delta}\boldsymbol{A}_\lambda^a(x), \boldsymbol{\phi}_\mu^b(w)\rangle = \left(\delta_{\lambda\mu} - \frac{k_\lambda k_\mu}{k^2}\right)\delta_{ab} f(k^2) + \frac{k_\lambda k_\mu}{k^2}\delta_{ab}C_1\,. \tag{21.18}$$

However, the first term on the right-hand side does not appear in (21.14). This is because the left-hand side of (21.14) is multiplied by $k_\lambda = (p-q)_\lambda$. We can thus replace V_λ in (21.14) by

$$\bar{V}_\lambda(p,q) = \frac{(p-q)_\lambda (p-q)_\mu}{(p-q)^2} V_\mu(p,q) , \tag{21.19}$$

which is a projection for a spin 0. With this replacement, the first term on the right-hand side of (21.18) vanishes. We can then rewrite (21.14) in terms of $\bar{V}_\lambda$:

$$S_\mathrm{F}(p,\alpha)(p-q)_\lambda \bar{V}^a_\lambda(p,q) S_\mathrm{F}(q,\beta) = \mathrm{i} g t^a_{\alpha\beta}\big[S_\mathrm{F}(p,\alpha) - S_\mathrm{F}(q,\beta)\big] . \tag{21.20}$$

In $\bar{V}^a_\lambda$, there can still be poles of spin 0. In fact, it is clear from (19.65) and (19.66) that $\delta A_\lambda = D_\lambda c$ creates a pole of spin 0. If so, we may ask whether $\delta\bar{\delta} A_\lambda$ creates a pole of spin 0.

According to (19.147), $D_\lambda c$ includes an asymptotic field $\partial_\lambda \gamma$ of the spin 0 and massless particle. Hence, the condition that V_λ should not include a pole of spin 0 is

$$\bar{\delta}\gamma^a = 0 . \tag{21.21}$$

This condition cannot hold true when a symmetry corresponding to the a th generator of the gauge group is spontaneously broken. So when $\bar{V}_\lambda$ does not include a pole of spin 0, we set

$$p - q = \epsilon P , \quad P^2 \neq 0 , \tag{21.22}$$

and consider the limit $\epsilon \to 0$. By assumption, $\bar{V}_\lambda$ does not include poles of either spin 0 or spin 1, and the projection operator is also finite in this limit:

$$\frac{(p-q)_\lambda (p-q)_\mu}{(p-q)^2} = \frac{P_\lambda P_\mu}{P^2} . \tag{21.23}$$

Thus, we have the following finite limit:

$$\lim_{\epsilon\to 0} \bar{V}_\lambda(p,q) = \bar{V}_\lambda(p,p;P) . \tag{21.24}$$

From (21.20), we obtain

$$S_\mathrm{F}(p,\alpha) P_\lambda \bar{V}_\lambda(p,p;P) S_\mathrm{F}(p,\beta) = \mathrm{i} g t^a_{\alpha\beta} \lim_{\epsilon\to 0} \frac{1}{\epsilon}\big[S_\mathrm{F}(p,\alpha) - S_\mathrm{F}(q,\beta)\big] . \tag{21.25}$$

Although the left-hand side is finite, if the symmetry is broken, we generally have

$$S_\mathrm{F}(p,\alpha) - S_\mathrm{F}(p,\beta) \neq 0 . \tag{21.26}$$

Therefore, the right-hand side diverges and we encounter a contradiction. Thus, it turns out that the condition (21.21) can hold true only if the symmetry is not broken. In this case, if we write

$$S_F(p, \alpha) = S_F(p, \beta) \equiv S_F(p) , \tag{21.27}$$

for components α and β connected by t^a, the right-hand side of (21.25) assumes the form

$$\mathrm{i}gt^a_{\alpha\beta} P_\lambda \frac{\partial}{\partial p_\lambda} S_F(p) . \tag{21.28}$$

If we now assume that quarks belong to BRS singlet states, then (21.25) yields a contradiction. In this case, if we apply the LSZ reduction formula to quarks, the left-hand side implies

$$P_\lambda \langle \boldsymbol{p}, \boldsymbol{\alpha} | \delta\bar{\delta}\bar{\boldsymbol{A}}_\lambda(0) | \boldsymbol{p}, \boldsymbol{\beta} \rangle , \tag{21.29}$$

where $|\boldsymbol{p}, \boldsymbol{\alpha}\rangle$ and $|\boldsymbol{p}, \boldsymbol{\beta}\rangle$ are quark states with four-momentum p. However, $\delta\bar{\delta}\bar{A}_\lambda$ is a (d, d)-type operator. If the quarks belong to BRS singlet states, the matrix element (21.29) should vanish. On the other hand, from the right-hand side, we obtain

$$\mathrm{i}gt^a_{\alpha\beta}\bar{u}(p) \left[\frac{\partial}{\partial p_\lambda} S_F^{-1}(p) \right] u(p) \propto \mathrm{i}gt^a_{\alpha\beta}\bar{u}(p)\mathrm{i}\gamma_\lambda u(p) , \tag{21.30}$$

up to a trivial numerical coefficient. This is finite and never vanishes, which is a contradiction.

This means that, if the condition (21.21) is satisfied, S_F cannot possess a pole corresponding to a BRS singlet state. This in turn means that quarks are confined. Similarly, if we start with (21.11) and assume (21.21), then it can be shown that gluons are also confined. Generalizing this result further, it can be understood that (21.21) is a sufficient condition for all particles with colour to be confined.

Since the condition for confinement has been expressed in terms of the asymptotic field of the auxiliary field, we now investigate the connection with the discussion of the representation in Sect. 19.8. From (19.149),

$$\delta\chi = \gamma . \tag{21.31}$$

Therefore, (21.21) can be expressed in the form

$$\bar{\delta}\delta\chi^a = 0 . \tag{21.32}$$

Hence, using (19.152) which is valid in perturbation theory,

$$\bar{\delta}\delta\chi^a = -\mathrm{i}\beta^a \,, \tag{21.33}$$

so it is clear that confinement does not hold true in perturbation theory. Hence, the quartet representation in Sect. 19.8 does not satisfy the condition for confinement. In contrast, since (21.21) holds in the octet representation given by (19.156), the condition for confinement is satisfied. In this case, as mentioned before, $\boldsymbol{c} \times \boldsymbol{c}$ and $\bar{\boldsymbol{c}} \times \bar{\boldsymbol{c}}$ turn out to have asymptotic fields. Alternatively, since $\delta\boldsymbol{c}$ and $\delta\bar{\boldsymbol{c}}$ have massless spin-0 asymptotic fields, the following quantity turns out to have a massless spin-0 pole:

$$\left\langle \delta\boldsymbol{c}^a(x), \bar{\delta}\bar{\boldsymbol{c}}^b(y)\right\rangle = \left\langle \bar{\delta}\boldsymbol{c}^a(x), \delta\bar{\boldsymbol{c}}^b(y)\right\rangle = -\left\langle \bar{\boldsymbol{B}}^a(x), \boldsymbol{B}^b(y)\right\rangle . \tag{21.34}$$

Referring to the representation theory for the asymptotic field of the auxiliary field in the Landau gauge, as explained in Sect. 19.9, if we take the representation $\mathscr{D}_j$, with j an integer, the condition for confinement is not satisfied, while if j is a half integer, the condition is in fact satisfied.

We now apply the above discussion to the grand unified theory (GUT) discussed in Sect. 16.5. The gauge group G of the grand unified theory, after spontaneous symmetry breaking, is

$$G \to SU(3)_{\text{colour}} \times U(1)_{\text{em}} \,. \tag{21.35}$$

As already discussed, confinement is caused by a *strict symmetry*. Since only singlets for $SU(3)$ can be observed, this means that, among the quanta of gauge fields, only the colourless γ, W, and Z are observed, and if hadrons are colourless bound states of quarks, then they can be observed.

21.3 Colour Confinement and Asymptotic Freedom

In the last section, we gave (21.32) as the condition for confinement. This section has been added to describe subsequent progress [190, 191]. To begin with, according to (19.66) and (19.147), $\boldsymbol{A}_\mu$ and $\boldsymbol{B}$ turn out to have the following asymptotic fields:

$$A^a_\mu(x)^{\text{in}} = \alpha^a_\mu(x) + \partial_\mu\chi^a(x) \,, \quad B^b(y)^{\text{in}} = \beta^b(y) \,, \tag{21.36}$$

where α_μ stands for the asymptotic field for a gluon, and χ and β are those for massless spin-0 fields, which satisfy (19.151). That is,

$$\left\langle \chi^a(x), \beta^b(y)\right\rangle = -\delta_{ab} D_{\text{F}}(x-y) \,. \tag{21.37}$$

Next, we check whether the confinement condition (21.32) is satisfied. Since it is strictly true in QED that

$$\mathrm{i}\delta\bar{\delta}\chi = -\beta\ , \tag{21.38}$$

electrical charge can never be confined.

In QCD, since the computation of $\mathrm{i}\delta\bar{\delta}\chi^a$ is complicated, we consider the lowest order in perturbation theory:

$$\mathrm{i}\delta\bar{\delta}\chi^a \approx -\beta^a\ . \tag{21.39}$$

Higher order corrections alter the coefficient of β on the right-hand side. If this coefficient vanishes in an exact theory, then (21.37) implies that a sufficient condition for confinement is

$$\big\langle \mathrm{i}\delta\bar{\delta}\chi^a(x),\, \chi^b(y)\big\rangle = 0\ . \tag{21.40}$$

We now express the condition including this asymptotic field in terms of the Heisenberg operator. We thus consider a two-point function:

$$\big\langle \mathrm{i}\delta\bar{\delta}\boldsymbol{A}^a_\mu(x),\, \boldsymbol{A}^b_\nu(y)\big\rangle\ . \tag{21.41}$$

The first operator includes $\partial_\mu(\mathrm{i}\delta\bar{\delta}\chi^a)$, the second includes $\partial_\nu\chi^b$, and the left-hand side of (21.40) gives a massless spin-0 pole. We can write the Fourier expression of (21.41), using (21.15):

$$\left(\delta_{\mu\nu} - \frac{k_\mu k_\nu}{k^2 - \mathrm{i}\epsilon}\right)\int \mathrm{d}m^2 \frac{\sigma(m^2)}{k^2 + m^2 - \mathrm{i}\epsilon} + C\frac{k_\mu k_\nu}{k^2 - \mathrm{i}\epsilon}\ . \tag{21.42}$$

The second term corresponds to (21.40). The coefficient C appears in

$$\partial_\mu\big\langle \mathrm{i}\delta\bar{\delta}\boldsymbol{A}^a_\mu(x),\, \boldsymbol{A}^b_\nu(y)\big\rangle = \mathrm{i}\delta_{ab}C\partial_\nu\delta^4(x-y)\ . \tag{21.43}$$

The condition for confinement argued here only concerns the second term in (21.42). Using only the constant C, the sufficient condition (21.40) for confinement reduces to

$$C = 0\ . \tag{21.44}$$

Note also that, given (21.15), Eq. (21.43) yields the following equal-time commutation relation:

$$\delta(x_0 - y_0)\big\langle 0\big|\mathrm{i}\delta\bar{\delta}\boldsymbol{A}^a_0(x),\, \boldsymbol{A}^b_j(y)\big|0\big\rangle = \mathrm{i}\delta_{ab}C\partial_j\delta^4(x-y)\ , \tag{21.45}$$

for $j = 1, 2, 3$. The left-hand side should be evaluated using renormalized operators. Fortunately, both operators are multiplicatively renormalized:

$$\boldsymbol{A}_{\mu}^{(0)} = Z_3^{1/2}\boldsymbol{A}_{\mu}\,, \quad \mathrm{i}\delta\bar{\delta}\boldsymbol{A}_{\mu}^{(0)} = \tilde{Z}_3^{-1}Z_3^{-1/2}(\mathrm{i}\delta\bar{\delta}\boldsymbol{A}_{\mu})\,, \tag{21.46}$$

where we have used the notation in Sect. 20.7. Therefore, the two-point function (21.41) and the constant C satisfy the renormalization group equation in the form

$$(\mathscr{D} - 2\gamma_{\mathrm{FP}})C = 0\,, \tag{21.47}$$

where γ_{FP} is the anomalous dimension of the ghost field and $\mathscr{D}$ is the differential operator appearing in (20.207).

Now recall (19.47). The Ward–Takahashi identities (21.10) and (21.11) used in the last section were in fact derived using this equation. Writing this in the unrenormalized form

$$\mathrm{i}\delta\bar{\delta}\boldsymbol{A}_{\nu}^{(0)} = \partial_{\mu}\boldsymbol{A}_{\mu\nu}^{(0)} + g\partial_{\mu}\big(\boldsymbol{A}_{\mu}^{(0)} \times \boldsymbol{A}_{\nu}^{(0)}\big) + g\boldsymbol{J}_{\nu}^{(0)} \tag{21.48}$$

and inserting in (21.45), we see that the only term contributing to the equal-time commutation relation is the first term involving $\boldsymbol{A}_{\mu\nu}^{(0)} = \partial_{\mu}\boldsymbol{A}_{\nu}^{(0)} - \partial_{\nu}\boldsymbol{A}_{\mu}^{(0)}$. However, if we use renormalized operators, then it turns out that the contribution from the Goto–Imamura–Schwinger term [124, 125] appearing in Sect. 12.8 cannot be neglected. Anyway, inserting the renormalized $\partial_{\mu}\boldsymbol{A}_{\mu\nu}$, we define a constant a by

$$\delta(x_0 - y_0)\langle 0|[\partial_{\mu}\boldsymbol{A}_{\mu 0}^{a}(x), \boldsymbol{A}_{j}^{b}(y)]|0\rangle = \mathrm{i}\delta_{ab}a\partial_j\delta^4(x-y)\,. \tag{21.49}$$

As is clear from the derivation of (12.223), this a is in fact equal to Z_3^{-1}. Additionally, since both sides of (21.49) satisfy the renormalization group equation, which is the same as that of the gluon propagator, we have

$$(\mathscr{D} + 2\gamma_{\mathrm{V}})a = 0\,. \tag{21.50}$$

Comparing this with (21.47), we see that C cannot be equal to a, whence $C - a$ survives as the Goto–Imamura–Schwinger term.

In addition, we introduce the gluon propagator in Fourier representation:

$$D_{F_{\mu\nu}}(k) = \left(\delta_{\mu\nu} - \frac{k_{\mu}k_{\nu}}{k^2 - i\epsilon}\right)D(k^2) + \alpha\frac{k_{\mu}k_{\nu}}{(k^2 - i\epsilon)^2}\,, \tag{21.51}$$

$$D(k^2) = \int \mathrm{d}m^2 \frac{\rho(m^2)}{k^2 + m^2 - \mathrm{i}\epsilon}\,. \tag{21.52}$$

We thus obtain the following relation corresponding to (12.223):

$$a = Z_3^{-1} = \int \mathrm{d}m^2 \rho(m^2) \; . \tag{21.53}$$

Next, we investigate how we can obtain C by solving (21.47). In Sect. 20.6, we used the integral method for the homogeneous Callan–Symanzik equation. Here, we can apply exactly the same method to the renormalization group equation above. Corresponding to (20.175), the renormalization group operator is given by

$$R_\rho = \exp(\rho \mathscr{D}) \; . \tag{21.54}$$

Taking Q as a function of g, α, and μ, we define an application of the renormalization group to Q by

$$\bar{Q}(\rho) = \exp(\rho^{\mathscr{D}}) Q \; , \quad \bar{Q}(0) = Q \; . \tag{21.55}$$

In particular, for an infinitesimal transformation,

$$\delta \bar{Q}(\rho) = \delta\rho \mathscr{D} \bar{Q}(\rho) \; . \tag{21.56}$$

In particular, for g, α, and μ, we obtain

$$\delta\bar{g} = \beta(\bar{g}, \bar{\mu})\delta\rho \; , \quad \delta\bar{\alpha} = -2\bar{\alpha}\gamma_V(\bar{g}, \bar{\alpha}, \bar{\mu})\delta\rho \; , \quad \delta\bar{\mu} = \bar{\mu}\delta\rho \; , \tag{21.57}$$

where the variable (ρ) has been omitted. Their asymptotic values are defined by

$$\bar{g}(\infty) = g_\infty \; , \quad \bar{\alpha}(\infty) = \alpha_\infty \; , \quad \bar{\mu}(\infty) = \infty \; , \tag{21.58}$$

and in particular asymptotic freedom is characterized by $g_\infty = 0$.

For a function $Q(g, \alpha, \mu)$, we clearly have

$$\bar{Q}(\rho) = \exp(\rho \mathscr{D}) \, Q(g, \alpha, \mu) = Q\big(\bar{g}(\rho), \bar{\alpha}(\rho), \bar{\mu}(\rho)\big) \; . \tag{21.59}$$

In particular, choosing a Green's function $G(p_i; g, \alpha, \mu)$ as Q and taking its anomalous dimension to be γ, we have

$$(\mathscr{D} + \gamma)G = 0 \; . \tag{21.60}$$

In order to integrate this equation, we consider

$$\bar{G}(\rho) = \exp(\rho \mathscr{D})G \; , \quad \bar{G}(0) = G \; , \tag{21.61}$$

whence

$$\frac{\partial}{\partial \rho}\bar{G}(\rho) = \exp(\rho\mathscr{D})\mathscr{D}G = -\exp(\rho\mathscr{D})\gamma G = -\bar{\gamma}(\rho)\bar{G}(\rho) . \tag{21.62}$$

Integrating this,

$$G(p_i; g, \alpha, \mu) = \exp\left[\int_0^\rho d\rho\bar{\gamma}(\rho)\right] G\big(p_i; \bar{g}(\rho), \bar{\alpha}(\rho), \bar{\mu}(\rho)\big) . \tag{21.63}$$

This corresponds to (20.178) in the case of the Callan–Symanzik equation, so if we introduce

$$R(k^2) = k^2 D(k^2) , \tag{21.64}$$

this function satisfies the renormalization condition

$$R(\mu^2) = 1 . \tag{21.65}$$

Since the anomalous dimension of R is $2\gamma_V$, we can use (21.63) to obtain

$$R(k^2; g, \alpha, \mu) = \exp\left[2\int_0^\rho \mathrm{d}\rho\bar{\gamma}_V(\rho)\right] R\big(k^2; \bar{g}(\rho), \bar{\alpha}(\rho), \bar{\mu}(\rho)\big) , \tag{21.66}$$

and if we take $k^2 = \bar{\mu}^2(\rho)$ in this equation, then (21.65) implies

$$R\big(\bar{\mu}^2(\rho); g, \alpha, \mu\big) = \exp\left[2\int_0^\rho d\rho\bar{\gamma}_V(\rho)\right] . \tag{21.67}$$

If we now take the limits $\rho \to \infty$ and $\bar{\mu}^2(\rho) \to \infty$, then (21.52) and (21.64) yield

$$\int \mathrm{d}m^2\rho(m^2) = \exp\left[2\int_0^\infty \mathrm{d}\rho\bar{\gamma}_V(\rho)\right] . \tag{21.68}$$

From (21.53), this turns out to be equal to a. We can thus evaluate a using this equation. Once a is known, the next problem is to evaluate C. Although C satisfies (21.47), we should give its boundary condition. Note that, in a theory in which a cut-off is introduced, taking a renormalization point at a high energy by considering $\rho \to \infty$ and $\bar{\mu}(\rho) \to \infty$, we find that $\bar{G}(\rho)$ approaches the Green's function for a free field. We may thus consider that the Goto–Imamura–Schwinger term vanishes in this limit. Hence, we introduce the following boundary condition for C :

$$\lim_{\rho\to\infty}\left[\bar{C}(\rho) - \bar{a}(\rho)\right] = 0 . \tag{21.69}$$

Then, given (21.47) and (21.50) written in the form (21.62), and omitting the variable (ρ), we obtain

$$\frac{\partial}{\partial\rho}\bar{C} = 2\bar{\gamma}_{\mathrm{FP}}\bar{C} , \quad \frac{\partial}{\partial\rho}\bar{a} = -2\bar{\gamma}_{\mathrm{V}}\bar{a} . \tag{21.70}$$

These lead immediately to

$$\left(\frac{\partial}{\partial\rho} - 2\bar{\gamma}_{\mathrm{FP}}\right)(\bar{C} - \bar{a}) = 2(\bar{\gamma}_{\mathrm{V}} + \bar{\gamma}_{\mathrm{FP}})\bar{a} . \tag{21.71}$$

If we solve this equation under the boundary condition (21.69), we find

$$\bar{C}(\rho) = \bar{a}(\rho) - \int_{\rho}^{\infty} \mathrm{d}\rho' 2\big[\bar{\gamma}_{\mathrm{V}}(\rho') + \bar{\gamma}_{\mathrm{FP}}(\rho')\big]\bar{a}(\rho') \exp\left[-2\int_{\rho}^{\rho'} \mathrm{d}\rho'' \bar{\gamma}_{\mathrm{FP}}(\rho'')\right] . \tag{21.72}$$

In particular, if we set $\rho = 0$, then

$$C = a - \int_{0}^{\infty} \mathrm{d}\rho' 2\big[\bar{\gamma}_{\mathrm{V}}(\rho') + \bar{\gamma}_{\mathrm{FP}}(\rho')\big]\bar{a}(\rho') \exp\left[-2\int_{0}^{\rho'} \mathrm{d}\rho'' \bar{\gamma}_{\mathrm{FP}}(\rho'')\right] . \tag{21.73}$$

Since the most important assumption required to obtain this result is (21.69), let us see whether this assumption actually holds true in QED.

Verification in QED

An easy computation in QED yields the value of C to be exactly given by

$$C = 1 . \tag{21.74}$$

Let us therefore check whether this result can be realized using (21.73), obtained by assuming the Goto–Imamura–Schwinger term (21.69). In QED, we can use the simplification

$$\gamma_{\mathrm{FP}} = 0 , \quad \beta(e) = e\gamma_{\mathrm{V}}(e) . \tag{21.75}$$

Therefore,

$$\int_{0}^{\infty} \mathrm{d}\rho\bar{\gamma}_{\mathrm{V}}(\rho) = \int_{e}^{e_{\infty}} \frac{\mathrm{d}\bar{e}}{\beta(\bar{e})}\frac{\beta(\bar{e})}{e} = \ln\frac{e_{\infty}}{e} . \tag{21.76}$$

Then $a(e)$ can be obtained immediately from (21.68):

$$a(e) = (e_\infty/e)^2 . \tag{21.77}$$

Inserting this result into (21.73), we see that (21.74) is indeed realized, so from now on we can also apply (21.69) with confidence in QCD. That is,

$$C = \left(\frac{e_\infty}{e}\right)^2 - 2\int_e^{e_\infty} \frac{\mathrm{d}\bar{e}}{\bar{e}} \left(\frac{e_\infty}{\bar{e}}\right)^2 = 1 . \tag{21.78}$$

For two BRS-invariant Lagrangian densities $\mathscr{L}_1$ and $\mathscr{L}_2$, if there is $\mathscr{M}$ such that

$$\delta\mathscr{L}_1 = \delta\mathscr{L}_2 = 0 , \quad \mathscr{L}_1 - \mathscr{L}_2 = \delta\mathscr{M} , \tag{21.79}$$

then we say that $\mathscr{L}_1$ and $\mathscr{L}_2$ belong to the same gauge class. In this case, as discussed in Sect. 19.4, it is obvious from (19.63) that Green's functions including only BRS-invariant operators will be equal for both Lagrangians. For instance, since for two Lagrangians like those given in (19.1) with different values of α,

$$\mathscr{L}_2 - \mathscr{L}_1 = \delta\mathscr{M} , \quad \mathscr{M} = -\frac{\mathrm{i}}{2}(\alpha_2 - \alpha_1)\bar{c} \cdot B , \tag{21.80}$$

such Lagrangians will belong to the same gauge class. Hence, Green's functions such as those in (19.63) turn out to be independent of α. If we consider the renormalization group equation (21.60) for such Green's functions, then γ does not of course depend on α. Thus, β does not depend on α either, and becomes constant in the same gauge class. This means that the notion of asymptotic freedom becomes a common notion over the same gauge class, i.e., it turns out that there exists a renormalization method such that β does not depend on α. Now integrating (21.57),

$$\ln \frac{\alpha}{\alpha_\infty} = 2\int_0^\infty \mathrm{d}\rho \bar{\gamma}_V(\rho) . \tag{21.81}$$

Alternatively, from (21.53), (21.31), and so on,

$$a = \int \mathrm{d}m^2 \rho(m^2) = \frac{\alpha}{\alpha_\infty} . \tag{21.82}$$

Thus, the problem of obtaining α reduces to the problem of obtaining α_∞. We need therefore to solve (21.57), but γ_V and β are known only in perturbation theory, i.e., as power series expansions of g:

$$\beta(g) = g^3(\beta_0 + \beta_1 g^2 + \cdots) , \quad \gamma_V(g, \alpha) = g^2(\gamma_0 + \gamma_1 g^2 + \cdots) , \tag{21.83}$$

where

$$\gamma_0 = \gamma_{00} + \gamma_{01}\alpha \,, \quad \gamma_1 = \gamma_{10} + \gamma_{11}\alpha + \gamma_{12}\alpha^2 \,, \quad \dots \,. \tag{21.84}$$

In the following, we assume $\beta_0 < 0$ as the condition for asymptotic freedom, and as a consequence, $g_\infty = 0$. Moreover, we assume that $\gamma_{01} > 0$, which always holds true for QCD.

If $g^2 \ll 1$, then for sufficiently large ρ,

$$\bar{g}^2(\rho) = g^2(\mathbf{1} + bg^2\rho)^{-1} \,, \quad b = -2\beta_0 > 0 \,. \tag{21.85}$$

We now consider the integral (21.81):

1. **Convergent case.** If the integral (21.81) converges, since the left-hand side is finite, α_∞ is not 0, but finite. In this case, for large ρ, the asymptotic form of the integrand on the right-hand side is

$$\begin{aligned}\bar{\gamma}_V(\rho) &= \bar{g}^2(\rho)(\gamma_{00} + \gamma_{01}\alpha_\infty) + \mathcal{O}\left(\frac{\mathbf{1}}{\rho^2}\right) \\ &= \frac{1}{b\rho}(\gamma_{00} + \gamma_{01}\alpha_\infty) + \mathcal{O}\left(\frac{\mathbf{1}}{\rho^2}\right) \,.\end{aligned} \tag{21.86}$$

   In order for this integral to converge for large ρ, we must have

$$\gamma_{00} + \gamma_{01}\alpha_\infty = 0 \,, \quad \text{or} \quad \alpha_\infty = -\gamma_{00}/\gamma_{01} \equiv \alpha_0 \,. \tag{21.87}$$

2. **Divergent case.** If the integral (21.81) diverges, then the left-hand side is divergent as well, so the possible values of α_∞ are

$$\alpha_\infty = 0 \quad \text{or} \quad \infty \,. \tag{21.88}$$

As a consequence, there are only three possible values of α_∞, viz., α_0, 0, or ∞.

We now investigate what happens to $\big(\bar{\alpha}(\rho), \bar{g}(\rho)\big)$ for large ρ if we start out with the initial values (α, g). Since $g_\infty = 0$, for sufficiently large ρ, we have $\bar{g}^2(\rho) \ll 1$. We thus follow this flow by starting with a sufficiently small $\bar{g}^2(\rho)$. Apart from the case where $\alpha_\infty = \infty$, the lowest order equation for $\bar{\alpha}$ can be used, and in this case,

$$\frac{\mathrm{d}}{\mathrm{d}\rho} \approx -2\bar{\alpha}(\gamma_{00} + \gamma_{01}\bar{\alpha})\bar{g}^2 = -2\gamma_{01}\bar{g}^2\bar{\alpha}(\bar{\alpha} - \alpha_0) \,, \quad \gamma_{01}\bar{g}^2 > 0 \,. \tag{21.89}$$

Considering this equation in the limit $\rho \to \infty$, when $\bar{\alpha}$ is close to 0 or α_0, it will turn out that α_∞ is 0, α_0, or neither of them. If it is neither of them, then it turns out that $\alpha_\infty = \infty$ or $-\infty$. Since this problem is straightforward, we give only the results:

1. $\gamma_{00} < 0 \quad (\alpha_0 > 0)$

$$\alpha_\infty = \begin{cases} \alpha_0 \,, & \alpha > 0 \,, \\ 0 \,, & \alpha = 0 \,, \\ -\infty \,, & \alpha < 0 \,. \end{cases} \tag{21.90}$$

2. $\gamma_{00} > 0 \quad (\alpha_0 < 0)$

$$\alpha_\infty = \begin{cases} 0 \,, & \alpha > \alpha_0 \,, \\ \alpha_0 \,, & \alpha = \alpha_0 \,, \\ -\infty \,, & \alpha < \alpha_0 \,. \end{cases} \tag{21.91}$$

Here, we began by assuming that g^2 is sufficiently small at the outset. There exists a region where $\alpha_\infty = -\infty$ regardless of the sign of γ_{00}. According to (21.82), in this region, $\alpha = 0$, whence (21.73) implies $C = 0$ in the same region. Since $C = 0$ is a sufficient condition for confinement, if g and α belong to this region, then there is confinement. In this case, there are other regions where $C \neq 0$, but since $C = 0$ is not a necessary condition but a sufficient condition, then there is no contradiction if there is confinement. After all, going back to the origin of this proof, it is concluded that colour confinement occurs if there is asymptotic freedom and a spontaneously unbroken non-Abelian gauge symmetry.

Chapter 22
Anomalous Terms and Dispersion Theory

Field theory has suffered from the problem of divergences since the day of its birth. At first glance, renormalization theory gives a method for removing these divergences, but in this method we have to compute quantities of the type $\infty - \infty$, so some indefiniteness is often left in the result. This is typically exemplified by the self-energy of the photon. Depending on how we compute this quantity, it can be zero or infinity. This is the problem of indefiniteness: depending on how we compute, something that should be gauge invariant turns out not to be invariant. A classic example of this is the problem of vacuum polarization. In Chap. 12, we obtained (12.127), and in order to reach a gauge invariant result, we had to set

$$\int_0^1 \mathrm{d}x \int \mathrm{d}^4 p \frac{\frac{1}{2}p^2 + m^2 + x(1-x)k^2}{\left[p^2 + m^2 + x(1-x)k^2 - \mathrm{i}\epsilon\right]^2} = 0 \,. \tag{22.1}$$

Fukuda and Miyamoto discovered similar things when they computed the two-photon decay of the neutral meson including the vacuum polarization [192]. Steinberger also carried out a similar computation in [193].

In fact, it was not noticed that there were in fact two kinds of singularities, which had become a matter of course around that time: one was the indefiniteness mentioned above, and the other was the anomalous term. In 1951, although Schwinger had proposed a computational method that did not spoil the gauge invariance [194], the problem of anomalous terms had gone unnoticed. So what exactly is this problem? Since a product of operators located at the same space-time point is a highly singular quantity in field theory, it is difficult to define it correctly.

The equation for interacting fields is non-linear, so there is no way to get around this issue. In a classical theory, it is easy to define a product of c-numbers, but in a quantum theory, as a consequence of the quantization, a new kind of singularity is generated. Thus, in a quantum theory, a result derived from a classical field equation does not necessarily hold. Such a term arising from the difference with a classical

K. Nishijima, *Quantum Field Theory*,
https://doi.org/10.1007/978-94-024-2190-3_22

equation is called an *anomalous term*. The issue of anomalous terms is even now a central problem in field theory. It is being investigated in detail using modern mathematical methods such as differential geometry and topology. In this chapter, we will treat this issue from a more elementary point of view.

It was around the end of the 1960s that the difference between indefiniteness and anomalous terms was at last clearly acknowledged. This was a consequence of research by Adler et al. [159, 160]. In this chapter, we will discuss this issue from the point of view of dispersion theory.

As discussed in Sect. 1.4, in dispersion theory, the computation of the S-matrix amplitude is separated into two steps. In the first step, we compute the absorption part of the amplitude from the unitarity condition, and in the second step, we use the scattering formula to compute the dispersion part from the absorption part. In the computations of the first step, the integration is always carried out in a finite phase space, so it converges and there is neither indefiniteness nor divergence. Hence, if divergence arises, it will appear in the second step. In this case, when appropriate, we introduce the dispersion formula with a subtraction. This subtraction corresponds to a renormalization prescription, but since this computation is not of the form $\infty - \infty$, hence different from the Feynman–Dyson theory, no indefiniteness is left at all. However, a weak point in the dispersion theory of the S-matrix is that a complete system for the dispersion formula is not known. However, what we can learn from the above warning is that divergence and indefiniteness in field theory always arise from the dispersion part alone. Computations of the absorption part can be determined uniquely no matter what method we adopt.

In this chapter, we thus introduce the dispersion theory, not for the S-matrix, but for the Green's functions. The advantage is that a complete system is known for the dispersion formula. Using this method, we investigate the mechanism leading to the appearance of anomalous terms.

22.1 Examples of Indefiniteness and Anomalous Terms

As a first example of indefiniteness, we discuss the issue of vacuum polarization.

22.1.1 *Vacuum Polarization*

The integral representation of the propagator in Sect. 12.8 is an example of dispersion relation. If we consider a scalar field, then the absorption part, or the imaginary part, is given by

$$\frac{1}{2}\langle \mathbf{0}|\{\varphi(x), \varphi(y)\}|\mathbf{0}\rangle = \frac{1}{(2\pi)^4}\int \mathrm{d}^4 p\, \mathrm{e}^{\mathrm{i}p\cdot(x-y)} \mathrm{Im}\, \Delta'_{\mathrm{F}}(-p^2) \,. \tag{22.2}$$

Similarly,

$$\frac{1}{2}\langle \mathbf{0}|\{\boldsymbol{j}_\mu(x), \boldsymbol{j}_\nu(y)\}|\mathbf{0}\rangle = \frac{1}{(2\pi)^4}\int \mathrm{d}^4k\, \mathrm{e}^{\mathrm{i}k\cdot(x-y)}\mathrm{Im}\,\Pi_{\mu\nu}(k)\ . \tag{22.3}$$

Inserting an electron–positron pair state into the intermediate state on the left-hand side, we compute to lowest order in perturbation theory:

$$\langle \mathbf{0}|\{\boldsymbol{j}_\mu(x), \boldsymbol{j}_\nu(y)\}|\mathbf{0}\rangle = e^2\mathrm{Tr}\big[\gamma_\mu S^{(+)}(x-y)\gamma_\nu \tilde{S}^{(+)}(x-y)^{\mathrm{T}}\big] + (x \rightleftarrows y, \mu \rightleftarrows \nu)\ , \tag{22.4}$$

where

$$S^{(+)}(x) = (\gamma\cdot\partial - m)\Delta^{(+)}(x)\ , \quad \tilde{S}^{(+)}(x) = (\gamma^{\mathrm{T}}\cdot\partial + m)\Delta^{(+)}(x)\ . \tag{22.5}$$

Inserting this and taking the Fourier transform,

$$\begin{aligned}\mathrm{Tr}[\ldots] = -\frac{1}{(2\pi)^6}\frac{\pi}{3}\int_{4m^2}^{\infty}\mathrm{d}\kappa^2\sqrt{1-\frac{4m^2}{\kappa^2}}\left(2+\frac{4m^2}{\kappa^2}\right)\\ \times\int \mathrm{d}^4k\, \mathrm{e}^{\mathrm{i}k\cdot(x-y)}(\delta_{\mu\nu}k^2 - k_\mu k_\nu)\theta(k_0)\delta(k^2+\kappa^2)\ .\end{aligned} \tag{22.6}$$

Therefore,

$$\mathrm{Im}\,\Pi_{\mu\nu}(k) = \frac{e^2}{3(2\pi)^2}(k_\mu k_\nu - \delta_{\mu\nu}k^2)\int_{4m^2}^{\infty}\mathrm{d}\kappa^2\sqrt{1-\frac{4m^2}{\kappa^2}}\left(1+\frac{2m^2}{\kappa^2}\right)\pi\delta(k^2+\kappa^2)\ . \tag{22.7}$$

If we set

$$\Pi_{\mu\nu}(k) = (k_\mu k_\nu - \delta_{\mu\nu}k^2)\Pi(-k^2)\ , \quad \mathrm{Im}\,\Pi(-k^2) = \pi\sigma(-k^2)\ , \tag{22.8}$$

then

$$\sigma(\kappa^2) = \frac{e^2}{12\pi^2}\sqrt{1-\frac{4m^2}{\kappa^2}}\left(1+\frac{2m^2}{\kappa^2}\right)\theta(\kappa^2 - 4m^2)\ . \tag{22.9}$$

We now compute $\Pi(-k^2)$ using the dispersion relation. Using the renormalization condition $\Pi(0) = 0$ and the dispersion relation in which one subtraction has been carried out,

$$\Pi(-k^2) - \Pi(0) = \int \mathrm{d}\kappa^2\left(\frac{1}{k^2+\kappa^2} - \frac{1}{\kappa^2}\right)\sigma(\kappa^2)\ , \tag{22.10}$$

or

$$\Pi(-k^2) = -k^2 \int \frac{d\kappa^2}{\kappa^2} \frac{\sigma(\kappa^2)}{k^2 + \kappa^2 - i\epsilon} . \tag{22.11}$$

As a result, the vacuum polarization has been computed without any indefiniteness. Inserting (22.11) into (22.8), one obtains the Fourier transform of the following two-point function:

$$\langle \mathbf{0} | T^*[\boldsymbol{j}_\mu(x), \boldsymbol{j}_\nu(y)] | \mathbf{0} \rangle . \tag{22.12}$$

We can consider the covariant product T^* to be defined by the dispersion formula. In the following dispersion theory, we use the T-product in this sense. Then if we compute (22.1) following Feynman and Dyson, we obtain zero for the absorption part. If we differentiate (22.1) twice with respect to k^2, it vanishes. The expression in (22.1) does not have branch points for k^2, so its absorption part vanishes. Although the indefiniteness or the divergence appears only in the dispersion part, the dispersion part can be determined uniquely from the absorption part if we compute using the dispersion relation (22.11).

22.1.2 Goto–Imamura–Schwinger Term

The Goto–Imamura–Schwinger term [124, 125] discussed in Sect. 12.8 is an example of something that does not arise when we use equal-time commutation relations. It is also an example of a singularity of field operators evaluated at a common space-time point. The Schwinger term also plays an important role in string theory.

We now go back to the two-photon decay of the neutral meson mentioned in the introduction to this chapter.

22.1.3 Triangle Anomaly Term

We begin with the Dirac equations in QED:

$$(\gamma \cdot \partial + m)\boldsymbol{\psi} = ie\gamma_\lambda \boldsymbol{A}_\lambda \boldsymbol{\psi} , \quad \bar{\boldsymbol{\psi}}(\gamma \cdot \overleftarrow{\partial} - m) = -ie\bar{\boldsymbol{\psi}}\gamma_\lambda \boldsymbol{A}_\lambda . \tag{22.13}$$

Combining these two equations, we obtain

$$\partial_\lambda \boldsymbol{j}^5_\lambda = 2m\boldsymbol{j}^5 , \tag{22.14}$$

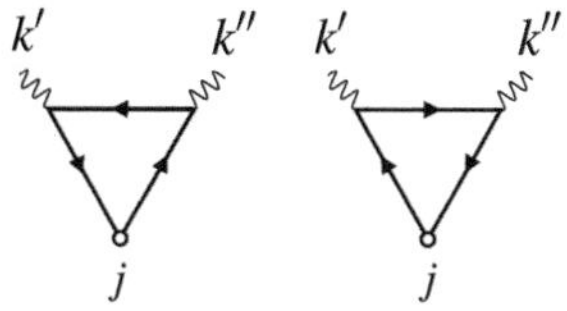

Fig. 22.1 Feynman diagrams for the calculation of the triangle anomaly term

where

$$j^5_\lambda = \mathrm{i}\bar{\psi}\gamma_\lambda\gamma_5\psi\ , \quad j^5 = \mathrm{i}\bar{\psi}\gamma_5\psi\ . \tag{22.15}$$

This equation is obtained using the classical equations of motion. In order to find out whether it is also true in quantum theory, we choose a particular matrix element and compare both sides. Hence, sandwiching this equation between the vacuum $|\mathbf{0}\rangle$ and the two-photon state $|\boldsymbol{k}', \boldsymbol{k}''\rangle$, we compare both sides. Figure 22.1 shows the two triangle diagrams that arise at the lowest order.

We define the form factors $F(s)$ and $G(s)$ by

$$\langle \boldsymbol{k}', \boldsymbol{k}''\mathrm{out}|j^5(0)|\mathbf{0}\rangle = \frac{1}{\sqrt{2k'_0V}\sqrt{2k''_0V}}\mathrm{i}\epsilon_{\alpha\beta\gamma\delta}e'_\alpha e''_\beta k'_\gamma k''_\delta F(s),$$

$$\langle \boldsymbol{k}', \boldsymbol{k}''\mathrm{out}|j^5_\lambda(0)|\mathbf{0}\rangle = \frac{1}{\sqrt{2k'_0V}\sqrt{2k''_0V}}\mathrm{i}\epsilon_{\alpha\beta\gamma\delta}e'_\alpha e''_\beta k'_\gamma k''_\delta(-\mathrm{i})(k'_\lambda + k''_\lambda)G(s)\ , \tag{22.16}$$

where s is the invariant mass squared in the two-photon system, defined by

$$s = -(k' + k'')^2\ , \tag{22.17}$$

and e' and e'' are the polarization vectors of each photon.

The absorption or imaginary parts of these form factors can be computed uniquely if we use either the unitarity condition or Feynman and Dyson's method. The controversial relation (22.14) is expressed using the form factors:

$$sG(s) = 2mF(s)\ . \tag{22.18}$$

We first compute the imaginary parts. According to Kikukawa, taking $\alpha = e^2/4\pi$,

$$\mathrm{Im}\,F(s) = \frac{2\alpha m}{s}\ln\left|\frac{1+\sqrt{1-4m^2/s}}{1-\sqrt{1-4m^2/s}}\right|\theta(s-4m^2)\ , \tag{22.19}$$

$$\mathrm{Im}\,G(s) = \frac{4\alpha m}{s}\ln\left|\frac{1+\sqrt{1-4m^2/s}}{1-\sqrt{1-4m^2/s}}\right|\theta(s-4m^2)\ . \tag{22.20}$$

This implies

$$s \operatorname{Im} G(s) = 2m \operatorname{Im} F(s) \ , \tag{22.21}$$

and (22.18) certainly holds true for the imaginary parts. Regarding the dispersion parts, i.e., the real parts, we compute using the dispersion relations

$$\operatorname{Re} F(s) = \frac{\mathrm{P}}{\pi}\int_{4m^2}^{\infty} \mathrm{d}s' \frac{\operatorname{Im} F(s')}{s'-s} \ , \qquad \operatorname{Re} G(s) = \frac{\mathrm{P}}{\pi}\int_{4m^2}^{\infty} \mathrm{d}s' \frac{\operatorname{Im} G(s')}{s'-s} \ , \tag{22.22}$$

where P stands for the Cauchy principal value. Going back to (22.18), we have

$$\begin{aligned} s \operatorname{Re} G(s) &= \frac{s}{\pi}\int \mathrm{d}s' \frac{\operatorname{Im} G(s')}{s'-s} \\ &= \frac{1}{\pi}\int \mathrm{d}s' \frac{s' \operatorname{Im} G(s')}{s'-s} - \frac{1}{\pi}\int \mathrm{d}s' \operatorname{Im} G(s') \\ &= 2m \operatorname{Re} F(s) - \frac{1}{\pi}\int \mathrm{d}s' \operatorname{Im} G(s') \ . \end{aligned} \tag{22.23}$$

If (22.18) is to hold true, the integral in the second term on the right-hand side must vanish. In fact, the computation tells us that

$$\int_{4m^2}^{\infty} \mathrm{d}s' \operatorname{Im} G(s') = 2\alpha \ . \tag{22.24}$$

This is the anomalous term. If we compute the real parts from (22.22), we find

$$\operatorname{Re} F(s) = \frac{\alpha}{\pi}\frac{m}{s}\left[\pi^2 - \left(\ln\left|\frac{1+\sqrt{1-4m^2/s}}{1-\sqrt{1-4m^2/s}}\right|\right)^2\right] , \tag{22.25}$$

$$\operatorname{Re} G(s) = \frac{2\alpha}{\pi}\frac{m^2}{s}\left\{-\frac{1}{m^2} + \frac{1}{s}\left[\pi^2 - \left(\ln\left|\frac{1+\sqrt{1-4m^2/s}}{1-\sqrt{1-4m^2/s}}\right|\right)^2\right]\right\} . \tag{22.26}$$

Introducing $B(x) = \boldsymbol{E}(x)\cdot\boldsymbol{H}(x)$, to lowest order,

$$\langle \boldsymbol{k}', \boldsymbol{k}'' \mathrm{out} | \boldsymbol{B}(0) | \boldsymbol{0} \rangle = \frac{1}{\sqrt{2k_0' V}\sqrt{2k_0'' V}} \mathrm{i}\epsilon_{\alpha\beta\gamma\delta} e_\alpha' e_\beta'' k_\gamma' k_\delta'' \ . \tag{22.27}$$

Then (22.14) is modified as follows:

$$\partial_\lambda \boldsymbol{j}_\lambda^5 = 2m\boldsymbol{j}^5 - \frac{2\alpha}{\pi}\boldsymbol{E}\cdot\boldsymbol{H}\ . \tag{22.28}$$

22.1.4 Trace Anomaly Term

As we shall see, the derivation of the Callan–Symanzik equation is deeply related to the trace anomaly term. To investigate this anomalous term, we shall consider a neutral scalar theory with Lagrangian density

$$\mathscr{L} = -\frac{1}{2}\left[(\partial_\lambda\varphi)^2 + m^2\varphi\right] - \frac{g}{4!}\varphi^4\ . \tag{22.29}$$

The energy–momentum tensor is

$$T_{\mu\nu} = -\frac{\partial\mathscr{L}}{\partial\varphi_{,\mu}}\varphi_{,\nu} + \delta_{\mu\nu}\mathscr{L}\ , \tag{22.30}$$

with trace

$$\begin{aligned}\boldsymbol{T}_{\mu\mu} &= -\frac{\partial\mathscr{L}}{\partial\boldsymbol{\varphi}_{,\mu}}\boldsymbol{\varphi}_{,\mu} + 4\mathscr{L}\\ &= -(\partial_\mu\boldsymbol{\varphi})^2 - 2m^2\boldsymbol{\varphi}^2 - \frac{g}{6}\boldsymbol{\varphi}^4\\ &= -\partial_\mu(\boldsymbol{\varphi}\partial_\mu\boldsymbol{\varphi}) - m^2\boldsymbol{\varphi}^2\ ,\end{aligned} \tag{22.31}$$

where we have used the field equation. Taking $T = m^2\varphi^2$,

$$\begin{aligned}T\left[\boldsymbol{T}_{\mu\mu}(x)\boldsymbol{\varphi}(x_1)\ldots\boldsymbol{\varphi}(x_n)\right] =& -\partial_\mu T\left[\boldsymbol{\varphi}(x)\partial_\mu\boldsymbol{\varphi}(x),\boldsymbol{\varphi}(x_1)\ldots\boldsymbol{\varphi}(x_n)\right]\\ &- T\left[\boldsymbol{T}(x)\boldsymbol{\varphi}(x_1)\ldots\boldsymbol{\varphi}(x_n)\right]\\ &+ \mathrm{i}\sum_{j=1}^n\delta^4(x-x_j)T\left[\boldsymbol{\varphi}(x_1)\ldots\boldsymbol{\varphi}(x_n)\right]\ .\end{aligned} \tag{22.32}$$

We now take the Fourier transform:

$$\langle \mathbf{0} | T[\boldsymbol{\varphi}(x_1)\dots\boldsymbol{\varphi}(x_n)] | \mathbf{0} \rangle \tag{22.33}$$
$$= \frac{\mathrm{i}^{n-1}}{(2\pi)^{4(n-1)}} \int (\mathrm{d}p) \mathrm{e}^{\mathrm{i}(p_1\cdot x_1+\dots+p_n\cdot x_n)} \delta^4(p_1+\dots+p_n) G^{(n)}(p_1,\dots,p_n) ,$$

$$\langle \mathbf{0} | T[\boldsymbol{T}_{\mu\mu}(x)\boldsymbol{\varphi}(x_1)\cdots\boldsymbol{\varphi}(x_n)] | \mathbf{0} \rangle \tag{22.34}$$
$$= \frac{\mathrm{i}^{n}}{(2\pi)^{4n}} \int (\mathrm{d}p)\mathrm{d}^4 q\, \mathrm{e}^{\mathrm{i}(q\cdot x+p_1\cdot x_1+\dots+p_n\cdot x_n)} \delta^4(q+p_1+\dots+p_n) T^{(n)}_{\mu\mu}(q;p_1,\dots,p_n) ,$$

and so on. Therefore, in the limit $q \to \infty$, Eq. (22.32) becomes

$$T^{(n)}_{\mu\mu}(0;p_1,\dots,p_n) = -T^{(n)}(0;p_1,\dots,p_n) + nG^{(n)}(p_1,\dots,p_n) . \tag{22.35}$$

This is called a *trace identity*, but in fact *this equation is not true*. The reason is that it includes a product of operators at the same space-time point.

We now write down the Ward–Takahashi identity for the energy–momentum tensor. This equation is always true:

$$\frac{\partial}{\partial x_\mu} T\left[\boldsymbol{T}_{\mu\nu}(x)\boldsymbol{\varphi}(x_1)\dots\boldsymbol{\varphi}(x_n)\right] = \mathrm{i}\sum_{j=1}^{n} \delta^4(x-x_j)\frac{\partial}{\partial x_{j\nu}} T\left[\boldsymbol{\varphi}(x_1)\dots\boldsymbol{\varphi}(x_n)\right] . \tag{22.36}$$

Taking the Fourier transform of the vacuum expectation value, we have

$$q_\mu T^{(n)}_{\mu\nu}(q;p_1,\dots,p_n) = \sum_{j=1}^{n}(p_j+q)_\nu G^{(n)}(p_1,\dots,p_j+q,\dots,p_n) . \tag{22.37}$$

Differentiating this equation with respect to q_μ and taking the limit $q \to 0$,

$$T^{(n)}_{\mu\nu}(0;p_1,\dots,p_n) = \left[(n-1) + \sum_{j=1}^{n} p_{j\nu}\frac{\partial}{\partial p_{j\mu}}\right] G^{(n)}(p_1,\dots,p_n) . \tag{22.38}$$

Taking the trace, we obtain

$$T^{(n)}_{\mu\mu}(0;p_1,\dots,p_n) = \left[4(n-1) + \sum_{j=1}^{n} p_j\cdot\frac{\partial}{\partial p_j}\right] G^{(n)}(p_1,\dots,p_n) . \tag{22.39}$$

Since $G^{(n)}$ has the dimension m^{4-3n} and is a homogeneous function of p_j and m,

$$\left(\sum_{j=1}^{n} p_j \cdot \frac{\partial}{\partial p_j} + m\frac{\partial}{\partial m} + 3n - 4\right) G^{(n)}(p_1, \ldots, p_n) = 0\,. \tag{22.40}$$

Combining (22.39) with (22.40), this yields

$$T^{(n)}_{\mu\mu}(0; p_1, \ldots, p_n) = \left(n - m\frac{\partial}{\partial m}\right) G^{(n)}(p_1, \ldots, p_n)\,. \tag{22.41}$$

This equation will be used to derive the Callan–Symanzik equation later.

22.2 Dispersion Theory for Green's Functions

To explain the dispersion theory for Green's functions, we consider a neutral scalar theory. If we introduce a set of Green's functions $\{\bar{\tau}\}$ by (11.176), this set turns out to satisfy a generalized unitarity condition like (11.177). We introduce the dispersion relation by picking out the part corresponding to connected diagrams in $\bar{\tau}$:

$$\bar{\tau}(x_1, \ldots, x_n)_{\text{conn}} = \bar{\rho}(x_1, \ldots, x_n). \tag{22.42}$$

Therefore, a recursion equation similar to (11.29) holds between the sets $\{\bar{\tau}\}$ and $\{\bar{\rho}\}$:

$$\bar{\tau}(x, x_1, \ldots, x_n) = \bar{\rho}(x, x_1, \ldots, x_n) + \sum_{k\neq n} \bar{\rho}(x, x'_1, \ldots, x'_k)\bar{\tau}(x'_{k+1}, \ldots, x'_n)\,, \tag{22.43}$$

where the sum is taken over all ways to divide $(x_1, \ldots, x_n)$ into two sets, $(x'_1, \ldots, x'_k)$ and $(x'_{k+1}, \ldots, x'_n)$. We now take the Fourier transform of $\bar{\rho}$:

$$\bar{\rho}(x_1, \ldots, x_n) = \frac{-\mathrm{i}}{(2\pi)^{4(n-1)}} \int (\mathrm{d}p)\delta^4(p_1 + \cdots + p_n)\mathscr{G}(p_1, \ldots, p_n)\mathrm{e}^{\mathrm{i}(p_1\cdot x_1+\cdots+p_n\cdot x_n)}\,. \tag{22.44}$$

Since $\mathscr{G}$ is a function only of the scalar product $p_\alpha \cdot p_\beta$, we write it as $\mathscr{G}(p_\alpha \cdot p_\beta)$. Using the generalized unitarity condition, if $\mathscr{G}$ is known up to order g^n, then we can compute $\mathrm{Im}\,\mathscr{G}$ up to order g^{n+1}. So if there exists a way to find $\mathrm{Re}\,\mathscr{G}$ for a given $\mathrm{Im}\,\mathscr{G}$, then higher orders of $\mathscr{G}$ can be determined recursively. Moreover, when we compute $\mathrm{Im}\,\mathscr{G}$ from the generalized unitarity condition, since the phase space volume is finite, there are no ultraviolet divergences. To compute $\mathrm{Re}\,\mathscr{G}$ from $\mathrm{Im}\,\mathscr{G}$, we use the following parameter dispersion relation:

$$\mathrm{Re}\,\mathscr{G}(p_\alpha \cdot p_\beta \xi) = \frac{\mathrm{P}}{\pi}\int_{-\infty}^{\infty} \frac{\mathrm{d}\xi'}{\xi' - \xi}\epsilon(\xi')\mathrm{Im}\,\mathscr{G}(p_\alpha \cdot p_\beta \xi') , \tag{22.45}$$

or

$$\mathscr{G}(p_\alpha \cdot p_\beta \xi) = \int_{-\infty}^{\infty} \frac{\mathrm{d}\xi'}{\xi'(1 - \mathrm{i}\epsilon) - \xi}\mathrm{Im}\,\mathscr{G}(p_\alpha \cdot p_\beta \xi') , \tag{22.46}$$

where $p_\alpha \cdot p_\beta \xi$ means that all scalar products between four-momenta should be multiplied by ξ. Since this dispersion relation holds true for all the $\mathscr{G}$, it turns out that we have a complete set for the dispersion relation.

In addition, in order to be able to write the dispersion relation when $\xi = 0$ for all the $\mathscr{G}$, we must have

$$\mathrm{Re}\,\mathscr{G}(0) = \mathrm{const.} , \quad \mathrm{Im}\,\mathscr{G}(0) = 0 . \tag{22.47}$$

On the other hand, a complete set for the dispersion relation for the S-matrix element is not known. The proof of the dispersion relation (22.45) is straightforward. Renormalized Green's functions can be written, at all orders of perturbation theory, in the form

$$\mathscr{G}(p_\alpha \cdot p_\beta) = \int \frac{\sigma(z_j)\mathrm{d}z_1 \ldots \mathrm{d}z_k}{\left[\sum c_{\alpha\beta}(z_j) p_\alpha \cdot p_\beta + M^2(z_j) - \mathrm{i}\epsilon\right]^N} , \tag{22.48}$$

where σ, c, and M are all real numbers and z_j is a suitable Feynman parameter. Using $M^2 \geq 0$ and setting $N = 1$ without loss of generality,

$$\int_0^{\infty} \frac{\mathrm{d}\xi'}{\xi' - \xi}\delta(c\xi' + M^2) = \theta(-c)\frac{1}{c\xi + M^2} , \tag{22.49a}$$

$$\int_{-\infty}^{0} \frac{\mathrm{d}\xi'}{\xi' - \xi}\delta(c\xi' + M^2) = -\theta(c)\frac{1}{c\xi + M^2} . \tag{22.49b}$$

Taking the difference between them, the following dispersion relation is obtained:

$$\int_{-\infty}^{\infty} \frac{\mathrm{d}\xi'}{\xi' - \xi}\varepsilon(\xi')\delta(c\xi' + M^2) = \frac{1}{c\xi + M^2} . \tag{22.50}$$

However, this dispersion integral sometimes diverges. This corresponds to divergences in the field theory. In that case, we must implement the subtraction which corresponds to renormalization. Furthermore, through this subtraction, interactions are introduced in perturbation theory. Note also that the dispersion formula presented in Sect. 22.1 can be proven using a method similar to the one above. We will discuss subtractions in the next section.

22.3 Subtractions in Dispersion Relations

There are dispersion relations with no subtractions, with one subtraction, and with two subtractions:

$$\mathrm{Re}\, f(x) = \frac{1}{\pi}\int \frac{\mathrm{d}x'}{x'-x}\mathrm{Im}\, f(x') \quad \text{(zero)}\,, \tag{22.51}$$

$$\mathrm{Re}\, f(x) = f(a) + \frac{x-a}{\pi}\int \frac{\mathrm{d}x'}{(x'-a)(x'-x)}\mathrm{Im}\, f(x') \quad \text{(one)}\,, \tag{22.52}$$

$$\mathrm{Re}\, f(x) = f(a) + (x-a)f'(a) + \frac{(x-a)^2}{\pi}\int \frac{\mathrm{d}x'}{(x'-a)^2(x'-x)}\mathrm{Im}\, f(x') \quad \text{(two)}\,, \tag{22.53}$$

where the integrals are Cauchy principal values and a is chosen so that $\mathrm{Im}\, f(a) = 0$. If subtractions are necessary, they should be introduced in such a way as to reproduce the results in standard perturbation theory. We should choose subtraction constants so as to determine the normalizations for Green's functions.

In what follows, we start with a two-point function in a neutral scalar theory. Writing $\mathcal{G}^{(2)}(p) = \mathcal{G}(p,-p)$, from the integral representation of the two-point function, we have

$$\mathcal{G}^{(2)}(p) = -(p^2+m^2)\left[1 + (p^2+m^2)\int \mathrm{d}\kappa^2 \frac{\sigma(\kappa^2)}{p^2+\kappa^2-\mathrm{i}\epsilon}\right]. \tag{22.54}$$

If we now replace p^2 by $p^2\xi$, the dispersion relation for ξ requires two subtractions. Note that the subtraction conditions can be obtained immediately from the above integral representation:

$$\mathcal{G}^{(2)} = 0\,, \qquad \frac{\partial \mathcal{G}^{(2)}}{\partial p^2} = -1\,, \quad \text{where} \quad p^2+m^2 = 0\,. \tag{22.55}$$

In fact, these two subtractions correspond to the renormalizations for the mass and the factor Z.

We now consider a four-point function. At the lowest order, corresponding to (20.107),

$$\mathcal{G}^{(4)} = g\,. \tag{22.56}$$

Since $\mathrm{Im}\,\mathscr{G}^{(4)} = 0$ at this order, we see that $\mathscr{G}^{(4)}$ requires one subtraction:

$$\mathrm{Re}\,\mathscr{G}^{(4)}(p_\alpha \cdot p_\beta \xi) = \mathscr{G}^{(4)}(0) + \frac{\xi}{\pi} \int \frac{\mathrm{d}\xi'}{\xi'(\xi' - \xi)} \epsilon(\xi') \mathrm{Im}\,\mathscr{G}^{(4)}(p_\alpha \cdot p_\beta \xi') . \tag{22.57}$$

As the subtraction condition, we equate $\mathscr{G}^{(4)}$ at a subtraction point (s.p.) with the renormalized coupling constant g :

$$\mathscr{G}^{(4)}(\mathrm{s.p.}) = g , \tag{22.58}$$

using (20.112) to choose the s.p. such that

$$p_\alpha \cdot p_\beta = \frac{m^2}{3}(4\delta_{\alpha\beta} - 1) . \tag{22.59}$$

This is the same as the subtraction condition in the S-matrix theory.

We now see that a two-point function needs two subtractions regardless of the details of the interactions. If no further subtractions are required for the other Green's functions, we can show that the fields become free, at least in perturbation theory. To do this, we write the part of $\mathscr{G}^{(n)}$ linear in the coupling constant as $\mathscr{G}^{(1)}$. From the unitarity condition, we can show that

$$\mathrm{Im}\,\mathscr{G}_1^{(n)} = 0 . \tag{22.60}$$

According to the unitarity condition, $\mathrm{Im}\,\mathscr{G}_1^{(n)}$ is higher than second order with respect to $\mathscr{G}$ and $\mathscr{G}^\dagger$, so at least one Green's function becomes a free Green's function $\mathscr{G}_0$. The only Green's function of free fields that does not vanish is a two-point function. However, the two-point function only appears in the unitarity condition in the form

$$\int \mathrm{d}^4 v\, \Delta^{(+)}(u - v)\bar{\rho}(v, x) = - \int \mathrm{d}^4 v\, \Delta^{(+)}(u - v) K_v K_x \Delta'_\mathrm{F}(v - x) . \tag{22.61}$$

The Fourier transform of this expression is

$$\delta(p^2 + m^2)(p^2 + m^2)\left[1 + (p^2 + m^2) \int \mathrm{d}\kappa^2 \frac{\sigma(\kappa^2)}{p^2 + \kappa^2 - \epsilon}\right] = 0 . \tag{22.62}$$

We thus see that a two-point function never appears in the non-linear contributions to the unitarity condition. Hence, if (22.60) holds true and there is no subtraction, we have

$$\mathrm{Re}\,\mathscr{G}_1^{(n)} = 0 , \quad n > 2 , \tag{22.63}$$

and this reduces to a free field.

In the Feynman–Dyson theory, propagators play a fundamental role, but in the dispersion theory, the two-point function never appears in the middle of computations. Other Green's functions are determined independently of the two-point function. However, if other Green's functions are determined, then from the unitarity condition, a two-point function turns out to be determined through the spectral function $\sigma(\kappa^2)$.

22.4 Heisenberg Operators

Here we consider a Heisenberg operator $\boldsymbol{A}(x)$ and, in particular, the problems that occur when $\boldsymbol{A}(x)$ is a product of operators at a common space-time point. We start with the Green's function

$$\bar{\tau}_A(x; x_1, \ldots, x_n) = (-\mathrm{i})^{n+1} K_{x_1} \ldots K_{x_n} \langle \boldsymbol{0} | T[\boldsymbol{A}(x)\boldsymbol{\varphi}(x_1) \ldots \boldsymbol{\varphi}(x_n)] | \boldsymbol{0} \rangle . \tag{22.64}$$

An arbitrary matrix element of $\boldsymbol{A}(x)$ is obtained using the LSZ reduction formula. We set the n four-momenta on the mass shell and take the Fourier transform of $\bar{\tau}_A$, which yields a matrix element of the form

$$\langle \boldsymbol{\beta}, \text{out} | \boldsymbol{A}(x) | \boldsymbol{\alpha}, \text{in} \rangle . \tag{22.65}$$

If the set $\{\bar{\tau}_A\}$ is known, it turns out that the field operator $\boldsymbol{A}(x)$ is determined.

The unitarity condition for $\bar{\tau}_A$ is obtained along the same lines as the $\bar{\tau}$ case:

$$\begin{aligned} 0 = {} & \bar{\tau}_A(x; x_1, \ldots, x_n) + \bar{\tau}_A^*(x; x_1, \ldots, x_n) \\ & + \sum_{\text{comb}} \sum_{l=0}^{\infty} \frac{\mathrm{i}^l}{l!} \int (\mathrm{d}u)(\mathrm{d}v) \bar{\tau}_A(x; x_1, \ldots, x_n) \\ & \qquad \times \Delta^{(+)}(u_1 - v_1) \ldots \Delta^{(+)}(u_l - v_l) \bar{\tau}^*(x'_{k+1}, \ldots, x'_n) \\ & + (\bar{\tau}_A \to \bar{\tau},\ \bar{\tau}^* \to \bar{\tau}_A^*) . \end{aligned} \tag{22.66}$$

Since this unitarity condition is linear in $\bar{\tau}_A$ and $\bar{\tau}_A^*$, we call it a *linear unitarity condition*. We now write the connected part of $\bar{\tau}_A$ as $\bar{\rho}_A$, as in (22.43). This yields the recursion formula

$$\bar{\tau}_A(x; x_1, \ldots, x_n) = \bar{\rho}_A(x; x_1, \ldots, x_n) + \sum_{\substack{\text{comb} \\ k \neq n}} \bar{\rho}_A(x; x'_1, \ldots, x'_k) \bar{\tau}(x'_{k+1}, \ldots, x'_n) . \tag{22.67}$$

Taking the Fourier transform of $\bar{\rho}_A$,

$$\bar{\rho}_A(x; x_1, \ldots, x_n) = \frac{-\mathrm{i}}{(2\pi)^4} \int (\mathrm{d}p)\mathrm{d}^4 q \delta(p_1 + \cdots + p_n + q) \\ \times \mathrm{e}^{\mathrm{i}(q\cdot x + p_1 \cdot x_1 + \cdots + p_n \cdot x_n)} \mathscr{A}(q; p_1, \ldots, p_n). \tag{22.68}$$

The functional system $\{\mathscr{A}\}$ satisfies a dispersion relation similar to $\{\mathscr{G}\}$:

$$\mathrm{Re}\, \mathscr{A}(p_\alpha \cdot p_\beta \xi) = \frac{\mathrm{P}}{\pi} \int_{-\infty}^{\infty} \frac{\mathrm{d}\xi'}{\xi' - \xi} \epsilon(\xi') \mathrm{Im}\, \mathscr{A}(p_\alpha \cdot p_\beta \xi) \ . \tag{22.69}$$

It is obvious that, in perturbation theory, all the $\mathscr{A}$ will vanish unless we introduce a subtraction. In this case, if $\{\mathscr{G}\}$ is known, combining the linear unitarity condition with the dispersion relation, $\{\mathscr{A}\}$ can be determined recursively. In the above discussion, we started with the operator $\boldsymbol{A}$ and set out to determine the set $\{\mathscr{A}\}$, but it sometimes happens that, if we start with two conditions and determine the functional system, that does not necessarily have a corresponding operator realization. Such a functional system turns up when we consider the various Ward–Takahashi identities.

22.5 Subtraction Condition

When we evaluate $\{\mathscr{G}\}$ or $\{\mathscr{A}\}$, we find that the absorption part, i.e., the imaginary part, does not include divergences, due to the unitarity condition. Thus, if divergences appear in field theory, this happens in the dispersion part, i.e., the real part. We compute higher orders recursively by carrying out subtractions for an n-point function $\mathscr{G}^{(n)}$, the number of these subtractions being determined in advance. If a dispersion integral always converges at any higher order due to a previously determined number of subtractions, then that theory is renormalizable. Otherwise, if the higher the order, the more subtractions must be made, then it is a non-renormalizable theory.

In the following, we determine the number of subtractions ensuring renormalizability within the framework of perturbation theory. For this purpose, we make the following assumption:

Assumption For a large value of $|\xi|$ and for almost all $\{p_j\}$, we have

$$\mathscr{G}^{(n)}(p_\alpha \cdot p_\beta \xi) \sim \xi^{c(n)/2} \ , \tag{22.70}$$

where $c(n)$ is an index independent of $\{p_j\}$.

This is true in the Feynman–Dyson theory within the perturbative approach. The meaning of the above equation for an arbitrary positive ϵ is

$$\lim_{\xi\to\infty}\frac{\mathscr{G}^{(n)}(p_\alpha\cdot p_\beta\xi)}{\xi^{c(n)/2+\epsilon}}=0\,,\quad \lim_{t\to\infty}\frac{\mathscr{G}^{(n)}(p_\alpha\cdot p_\beta\xi)}{\xi^{c(n)/2-\epsilon}}\,. \tag{22.71}$$

Furthermore, (22.70) is sometimes written formally as

$$\mathscr{G}^{(n)}(p_\alpha\cdot p_\beta)\sim p^{c(n)}\,. \tag{22.72}$$

To determine $c(n)$, we use the unitarity condition:

$$\begin{aligned}0&=\bar\rho(x_1,\ldots,x_n)+\bar\rho^*(x_1,\ldots,x_n)\\&\quad+\sum_{\text{comb}}{}'\sum_{l=1}^{\infty}\frac{\mathrm{i}^l}{l!}\int(\mathrm{d}u)(\mathrm{d}v)\bar\rho(x'_1,\ldots,x'_k,u_1,\ldots,u_l)\\&\qquad\times\Delta^{(+)}(u_1-v_1)\ldots\Delta^{(+)}(u_l-v_l)\bar\rho^*(x'_{k+1},\ldots,x'_k,v_1,\ldots,v_l)\\&\quad+\text{terms}\geq\text{third order in }\rho\text{ and }\rho^*\,.\end{aligned} \tag{22.73}$$

An exponent of $\operatorname{Im}\mathscr{G}^{(n)}$ never exceeds the order itself. Exponents of the first and second terms are given by the exponent of $\operatorname{Im}\mathscr{G}^{(n)}$, while exponents of other terms are expressed in terms of those of $\mathscr{G}$ and $\mathscr{G}^*$. Therefore,

$$c(n)+4(n-1)\geq\max\big[c(k+l)+c(n-k+l)+4(n+2l-2)-6l\big]\,. \tag{22.74}$$

Inequalities generated by terms higher than the third order hold true automatically if the above inequality does. Furthermore, terms other than $c(n)$ are generated from kinematic factors, and since there is no two-point function,

$$k+l>2\,,\quad n-k+l>2\,. \tag{22.75}$$

We introduce the notation

$$d(n)=c(n)+n-4\,. \tag{22.76}$$

Then, the above inequality can be written in the form

$$d(n)\geq\max\big[d(k+l)+d(n-k+l)\big]\,. \tag{22.77}$$

This immediately implies the necessary condition

$$d(n)\leq 0\,. \tag{22.78}$$

Checking the case with non-negative $c(n)$, in the φ^4-theory, from the lowest order result,

$$c(2) \geq 2\,, \quad c(4) \geq 0\,, \tag{22.79}$$

$$d(2) \geq 0\,, \quad d(4) \geq 0\,. \tag{22.80}$$

Combining this with the necessary condition (22.78), for even n,

$$d(n) = 0\,, \quad \text{or} \quad c(n) = 4 - n\,. \tag{22.81}$$

This corresponds to a renormalizable theory. In contrast, in the case of non-renormalizable theories such as the φ^6-theory, even at the lowest order, it turns out that

$$c(2) \geq 2\,, \quad c(6) \geq 0\,, \quad \text{or} \quad d(2) \geq 0\,, \quad d(6) \geq 2\,, \tag{22.82}$$

so there is no solution consistent with (22.78).

On the other hand, for super-renormalizable theories such as the φ^3-theory,

$$d(2) \geq 0\,, \quad d(3) \geq -1\,. \tag{22.83}$$

Looking more closely, we find that $d(n) = 0$ does not hold true, but

$$d(n) = 2 - n\,. \tag{22.84}$$

In the discussion above, we have considered scalar theories, but extensions are easy. For instance, in the pion–nucleon system, the function $\bar{\tau}$ is given by

$$(-\mathrm{i})^n K_{x_1} \ldots K_{x_n} D_{y_1} \ldots D_{y_l} \tilde{D}_{z_1} \ldots \tilde{D}_{z_l} \times \langle \mathbf{0} | T[\boldsymbol{\varphi}(x_1) \ldots \boldsymbol{\varphi}(x_n) \boldsymbol{\psi}(y_1) \ldots \boldsymbol{\psi}(y_l) \bar{\boldsymbol{\psi}}(z_1) \ldots \bar{\boldsymbol{\psi}}(z_l)] | \mathbf{0} \rangle, \tag{22.85}$$

where

$$D = \gamma \cdot \partial + M\,, \quad \tilde{D} = \gamma^{\mathrm{T}} \cdot \partial - M\,. \tag{22.86}$$

The contraction functions for the nucleon are

$$\langle 0 | \psi^{\mathrm{in}}(x) \bar{\psi}^{\mathrm{in}}(y) | 0 \rangle = -\mathrm{i}S^{(+)}(x-y) = -\mathrm{i}(\gamma \cdot \partial - M)\Delta^{(+)}(x-y)\,, \tag{22.87}$$

$$\langle 0 | \bar{\psi}^{\mathrm{in}}(x) \psi^{\mathrm{in}}(y) | 0 \rangle = -\mathrm{i}\tilde{S}^{(+)}(x-y) = -\mathrm{i}(\gamma^{\mathrm{T}} \cdot \partial + M)\Delta^{(+)}(x-y)\,. \tag{22.88}$$

Setting $m = 2l$, we introduce the exponent of $\mathscr{G}^{(n,m)}$:

$$\mathscr{G}^{(n,m)} \sim p^{c(n,m)} . \tag{22.89}$$

Using the so-called *regular dimensions* 1 and 3/2 of φ and ψ or $\bar{\psi}$, we define

$$d(n,m) = c(n,m) + n + \frac{3}{2}m - 4 . \tag{22.90}$$

The inequality corresponding to (22.77) is then

$$d(n,m) \geq \max\bigl[d(k+l, k'+l') + d(n-k+l, m-k'+l')\bigr] . \tag{22.91}$$

From this inequality, we obtain

$$d(n,m) \leq 0 , \quad \text{or} \quad c(n,m) \leq 4 - n - \frac{3}{2}m . \tag{22.92}$$

We consider the following two kinds of renormalizable interaction known in the pion–nucleon system:

$$\mathrm{i}\bar{\psi}\gamma_5\psi \cdot \varphi , \quad \varphi^4 , \tag{22.93}$$

where the isospin has been omitted. In addition to the conditions arising from the two-point function, viz.,

$$c(2,0) \geq 2 , \quad c(0,2) \geq 1 , \tag{22.94}$$

we obtain the conditions

$$c(1,2) \geq 0 , \quad c(4,0) \geq 0 . \tag{22.95}$$

Again, the unique solution is given by

$$d(n,m) = 0 . \tag{22.96}$$

As we see above, it can be understood that in typical renormalizable theories, we always obtain $d = 0$. For a proof of renormalizability, in addition to this, we have to prove that the exponents of the real and imaginary parts of $\mathscr{G}^{(n)}$ are equal. This requires detailed analyses which will be omitted here.

In the above, we considered the set $\{\mathscr{G}\}$ of typical Green's functions. We now consider $\{\mathscr{A}\}$:

$$\mathscr{A}^{(n,m)}(p_\alpha \cdot p_\beta \xi) \sim \xi^{a(n,m)/2} . \tag{22.97}$$

Using the linear unitarity condition (22.66) corresponding to (22.91), we obtain

$$b(n,m) \geq \max\left[b(k+l, k'+l') + d(n-k+l, m-k'+l')\right] , \tag{22.98}$$

where

$$b(n,m) = a(n,m) + n + \frac{3}{2}m - 4 . \tag{22.99}$$

Therefore, in a renormalizable theory characterized by $d = 0$, the solution (22.98) is

$$b(n,m) = b . \tag{22.100}$$

This b does not depend on the values of n and m. It is called an *index of the set* $\{\mathscr{A}\}$, and is determined by perturbation theory.

For instance, in a scalar theory, if we consider

$$\mathbf{A}(x) = \frac{1}{2}\left[\varphi(x)\right]^2 , \tag{22.101}$$

then at the lowest order,

$$\mathscr{A}^{(2)} = 1 . \tag{22.102}$$

Hence, it turns out that $a(2) = 0$ and $b = -2$. For a polynomial of field operators, it is given by

$$b = (\text{dimension of} A) - 4 . \tag{22.103}$$

Although the dispersion theory is formally elegant, the Feynman–Dyson theory is more useful for practical purposes. However, since there exist neither divergences nor indefiniteness, in a delicate issue such as derivation of the anomalous term, it gives a unique solution, and that is eminently useful. We thus give a theorem which provides a basis for this idea in the φ^4-theory:

Theorem 22.1 *We assume that four sets of functions* $\{\mathscr{A}\}$, $\{\mathscr{B}\}$, $\{\mathscr{C}\}$, *and* $\{\mathscr{D}\}$ *in* φ^4*-theory all satisfy the linear unitarity condition, that the index is* 0 *or* -2, *and that they have a scalar transformation property. Then there exists a linear dependence relation among these four sets of functions.*

To prove this theorem, we introduce the following $\mathscr{S}^{(n)}$:

$$\mathscr{S}^{(n)} = a\mathscr{A}^{(n)} + b\mathscr{B}^{(n)} + c\mathscr{C}^{(n)} + d\mathscr{D}^{(n)} . \tag{22.104}$$

In general, the index of $\{\mathscr{A}\}$ is 0. It satisfies the linear unitarity relation. If we add the following three subtraction conditions, then this set is uniquely determined:

$$\mathscr{S}^{(2)}(0;\, p, -p)\;, \quad \frac{\partial}{\partial p^2}\mathscr{S}^{(2)}(0;\, p, -p) \quad \text{for} p^2 + m^2 = 0\;,$$

$$\mathscr{S}^{(4)}(0;\, p_1, \dots p_4) \quad \text{for } p_\alpha \cdot p_\beta = \frac{m^2}{3}(1 - 4\delta_{\alpha\beta})\;. \tag{22.105}$$

Since these subtraction constants are linear in a, b, and c, if we choose their ratio properly, all three of the above constants can be set to zero. However, the values of $\mathscr{A}, \dots$ at the above subtraction point are assumed to be already known. If all the subtraction constants are zero, from the linear unitarity condition and the dispersion formula, all members of $\{\mathscr{S}\}$ must vanish, i.e., if we determine the ratios of a, b, c, and d, according to the above argument, we must have

$$\mathscr{S}^{(n)} = a\mathscr{A}^{(n)} + b\mathscr{B}^{(n)} + c\mathscr{C}^{(n)} + d\mathscr{D}^{(n)} = 0\;, \tag{22.106}$$

which proves the existence of the linear dependence relation.

22.6 Anomalous Trace Identity

We have already presented the Ward–Takahashi identity for the energy–momentum tensor (22.37) in Sect. 22.1. This can be considered as a defining identity for $T_{\mu\nu}$. We now rewrite it for Green's functions in which the legs of propagators are removed:

$$\begin{aligned} &q_\mu \mathscr{T}^{(n)}_{\mu\nu}(q;\, p_1, \dots, p_n) \qquad (22.107) \\ &\quad = \sum_{j=1}^{n} \frac{{p_j}^2 + m^2}{(p_j + q)^2 + m^2 - \mathrm{i}\epsilon}(p_j + q)_\nu \mathscr{G}^{(n)}(p_1, \dots, p_j + q, \dots, p_n)\;, \end{aligned}$$

where $q + p_1 + \dots + p_n = 0$. We may take this to define $\{\mathscr{T}_{\mu\nu}\}$. If all the p_j are on the mass shell, then for almost all values of q, i.e., when the denominator on the right-hand side does not vanish, the right-hand side does vanish:

$$\langle \boldsymbol{\beta}, \text{out} | \partial_\mu \boldsymbol{T}_{\mu\nu}(x) | \boldsymbol{\alpha}, \text{in} \rangle = 0\;. \tag{22.108}$$

We thus obtain the conservation law

$$\partial_\mu \boldsymbol{T}_{\mu\nu} = 0\;. \tag{22.109}$$

Moreover, Eq. (22.107) implies that the index of $\{\mathscr{T}_{\mu\nu}\}$ is zero, so

$$a(n) = c(n) = 4 - n \ , \tag{22.110}$$

and for $n = 2$, the quantity $\mathscr{T}^{(2)}_{\mu\nu}$ can be determined by solving (22.107):

$$\begin{aligned}\mathscr{T}^{(2)}_{\mu\nu}(q; p_1, p_2) = \delta_{\mu\nu}(p_1 \cdot p_2 - m^2) - (p_{1\mu} \cdot p_{1\nu} + p_{2\mu} p_{1\nu}) + \frac{1}{2}({p_1}^2 + m^2)({p_2}^2 + m^2) \\ \times \left[\delta_{\mu\nu} \int \mathrm{d}\kappa^2 \sigma(\kappa^2) \left(\frac{1}{{p_1}^2 + \kappa^2 - \mathrm{i}\epsilon} + \frac{1}{{p_2}^2 + \kappa^2 - \mathrm{i}\epsilon}\right)\right. \\ \left. + k_\mu k_\nu \int \mathrm{d}\kappa^2 \frac{\sigma(\kappa^2)}{({p_1}^2 + \kappa^2 - \mathrm{i}\epsilon)({p_2}^2 + \kappa^2 - \mathrm{i}\epsilon)}\right] + S^{(1)}_{\mu\nu} F + S^{(2)}_{\mu\nu} G \ ,\end{aligned} \tag{22.111}$$

where $q = -(p_1 + p_2)$ and $k = p_1 - p_2$, while $S^{(1)}_{\mu\nu}$ and $S^{(2)}_{\mu\nu}$ are defined by

$$\begin{aligned} S^{(1)}_{\mu\nu} &= q^2 \delta_{\mu\nu} - q_\mu q_\nu \ , \\ S^{(2)}_{\mu\nu} &= (q \cdot k)^2 q_\mu q_\nu - (q^2)^2 k_\mu k_\nu - q^2 (q \cdot k)(q_\mu k_\nu + q_\nu k_\mu) \ , \end{aligned} \tag{22.112}$$

and satisfy $q_\mu S^{(i)}_{\mu\nu} = 0$. Therefore, for a large value of ξ,

$$F(\xi) \sim \xi^0 \ , \quad G(\xi) \sim \xi^{-2} \ . \tag{22.113}$$

Thus, in order to determine $\{\mathscr{T}_{\mu\nu}\}$ uniquely, we must give a subtraction condition for F. However, this arbitrariness does not remain for $q = 0$.

Since we have $\mathscr{T}^{(4)}_{\mu\nu} \sim \xi^0$ for $n = 4$, even if a tensor such as S_ν appears, its coefficient does not require subtraction. Now, taking $F \neq 0$, this arbitrariness is transmitted to a multi-point function via the linear unitarity condition. So in order to investigate the properties of this term, we consider the operator $\boldsymbol{T} = m^2 \boldsymbol{\varphi}^2$. Considering the corresponding set of functions

$$\mathscr{T}^{(n)}(q; p_1, \ldots, p_n) \ , \tag{22.114}$$

its index is -2, so the unique subtraction is completely determined by

$$\mathscr{T}^{(2)}(0; p_1, \ldots, p_n) = 2m^2 \ , \quad \text{where} \quad p^2 + m^2 = 0 \ . \tag{22.115}$$

Thus, the set starting with $\mathscr{T}^{(2)}$ should be proportional to the set starting with F:

$$F^{(n)}(q; p_1, \ldots, p_n) = a \mathscr{T}^{(n)}(q; p_1, \ldots, p_n) \ . \tag{22.116}$$

Therefore, assuming that $\{\mathscr{T}_{\mu\nu}\}$ is one of the solutions satisfying the Ward–Takahashi identity and the linear unitarity condition, we obtain the following solution with the same property:

$$\mathscr{T}^{(n)}_{\mu\nu}(q; p_1, \ldots, p_n) + a(q^2\delta_{\mu\nu} - q_\mu q_\nu)\mathscr{T}^{(n)}(q; p_1, \ldots, p_n) . \tag{22.117}$$

If we express this in terms of operators, it yields the generalization

$$\boldsymbol{T}_{\mu\nu}(x) \to \boldsymbol{T}_{\mu\nu}(x) + \lambda(\delta_{\mu\nu}\Box - \partial_\mu\partial_\nu)\boldsymbol{\varphi}^2(x) . \tag{22.118}$$

This arbitrariness has been used by Callan et al. to introduce the improved energy–momentum tensor [195].

However, this arbitrariness disappears for $q = 0$, and (22.38) holds. We have already introduced two sets of Green's function systems, viz., $\{\mathscr{T}_{\mu\nu}\}$ and $\mathscr{T}$. We now introduce a third set:

$$\mathscr{W}^{(n)}(q; p_1, \ldots, p_n) = \sum_f \frac{{p_j}^2 + m^2}{(p_j + q)^2 + m^2 - \mathrm{i}\epsilon}\mathscr{G}^{(n)}(p_1, \ldots, p_j + q, \ldots, p_n) . \tag{22.119}$$

This set is the Fourier transform of the Green's function

$$(-\mathrm{i})^n K_{x_1} \ldots K_{x_n} \sum_j \delta^4(x - x_j)\langle\boldsymbol{0}|T[\boldsymbol{\varphi}(x_1) \ldots \boldsymbol{\varphi}(x_n)]|\boldsymbol{0}\rangle . \tag{22.120}$$

It is straightforward to check that this set satisfies the linear unitarity condition. However, assuming that all the q_j are on the mass shell for $q \neq 0$, $\mathscr{W}^{(n)}$ vanishes and hence does not correspond to any operator. This was discussed at the end of Sect. 22.4. This is a set of functions which appear through the connection with the Ward–Takahashi identity.

We now investigate what happens if $q \to 0$. Equation (22.119) is a highly singular function, since the denominator and the numerator cancel. In this case,

$$\mathscr{W}^{(n)}(0; p_1, \ldots, p_n) = n\mathscr{G}^{(n)}(p_1, \ldots, p_n) . \tag{22.121}$$

This function neither vanishes on the mass shell nor satisfies the linear unitarity condition. Moreover, in this case, similar difficulties arise for other sets. This is caused by the pole

$$\left[(p_j + q)^2 + m^2 - \mathrm{i}\epsilon\right]^{-1} . \tag{22.122}$$

In the case of $\{\mathscr{T}_{\mu\nu}\}$ and $\{\mathscr{T}\}$, this pole appears in the diagram when a two-point function is connected to another part by a propagator, as shown in Fig. 22.2. If ${p_j}^2 + m^2 = 0$ for $q = 0$, then (22.122) diverges. This is the reason why in field theory a

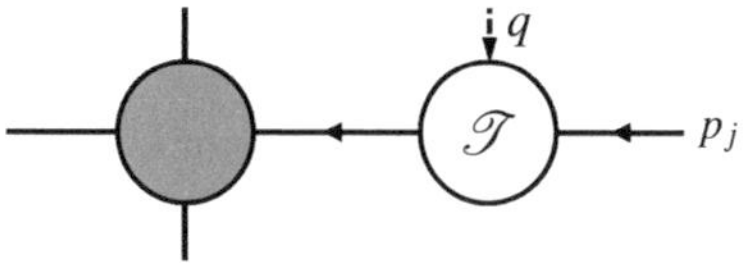

Fig. 22.2 Diagram involving the pole (22.122), when a two-point function is connected to another part by a propagator

one-particle irreducible function is often introduced instead of a connected Green's function. To ensure that there is no pole of the form (22.122) for $q = 0$ even on the mass shell, it turns out that we must take $\mathscr{S}^{(n)}$ to be a suitable linear combination of $\mathscr{T}^{(n)}_{\mu\nu}$, $\mathscr{T}^{(n)}$, and $\mathscr{W}^{(n)}$, and in the vicinity of $p^2 + m^2 = 0$, impose the relation

$$\mathscr{S}^{(2)}(0; p, -p) = \mathscr{O}\big((p^2 + m^2)^2\big) . \tag{22.123}$$

We thus take $\mathscr{S}^{(n)}$ to have the form

$$\mathscr{S}^{(n)} = \mathscr{T}^{(n)}_{\mu\mu} + \mathscr{T}^{(n)} - d_\varphi(g)\mathscr{W}^{(n)} . \tag{22.124}$$

We now determine the coefficient $d_\varphi(g)$ which ensures that the condition (22.123) is satisfied. In addition, we assume that the above $\mathscr{S}^{(n)}$ is defined only for $q = 0$. We thus expand $\mathscr{T}^{(2)}_{\mu\mu}$, $\mathscr{T}^{(2)}$, and $\mathscr{W}^{(2)}$ as power series in $(p^2 + m^2)$ when $q = 0$. If we use (22.111) for $\mathscr{T}^{(2)}_{\mu\mu}$, then

$$\begin{aligned}
\mathscr{T}^{(2)}_{\mu\mu}(0; p, -p) &= -2m^2 - 2(p^2 + m^2) + \mathscr{O}\big((p^2 + m^2)^2\big) , \\
\mathscr{G}^{(2)}(p, -p) &= -(p^2 + m^2) + \mathscr{O}\big((p^2 + m^2)^2\big) , \\
\mathscr{T}^{(2)}(0; p, -p) &= 2m^2 - 2\gamma_\varphi(g)(p^2 + m^2) + \mathscr{O}\big((p^2 + m^2)^2\big) ,
\end{aligned} \tag{22.125}$$

where $\gamma_\varphi(g)$ must be determined by computing higher order terms in perturbation theory. Therefore, in order to satisfy (22.123), we should choose the coefficient $d_\varphi(g)$ as follows:

$$d_\varphi(g) = 1 + \gamma_\varphi(g) . \tag{22.126}$$

Here, $\gamma_\varphi(g)$ and 1 are called the *anomalous dimension of* φ and the *holomorphic dimension*, respectively. For a general n,

$$\mathscr{S}^{(n)} = \mathscr{T}^{(n)}_{\mu\mu}(0; p_1, \dots p_n) + \mathscr{T}^{(n)}(0; p_1, \dots, p_n) - nd_\varphi(g)\mathscr{G}^{(n)}(p_1, \dots, p_n) . \tag{22.127}$$

It turns out that this set satisfies the linear unitarity condition. Moreover, for $n = 2$, the subtraction condition is

$$\mathscr{S}^{(2)}(p, -p) = 0 , \quad \frac{\partial}{\partial p^2}\mathscr{S}^{(2)}(p, -p) = 0 , \quad \text{for } p^2 + m^2 = 0 . \tag{22.128}$$

For $n = 4$, using the subtraction point (22.59), we have to use the result of the computation

$$\mathscr{T}^{(4)}_{\mu\mu}(0; \text{s.p.}) + \mathscr{T}^{(4)}(0; \text{s.p.}) - 4d_\varphi(g)\mathscr{G}^{(4)}(\text{s.p.}) \equiv \beta(g) \,. \tag{22.129}$$

This determines the subtraction condition.

To obtain the linear dependence relation, we introduce another set of functions:

$$\left\{ \frac{\partial}{\partial g} \mathscr{G}^{(n)}(p_1, \ldots, p_n) \right\} . \tag{22.130}$$

Differentiating the non-linear unitarity condition for $\mathscr{G}$ with respect to g, it can be shown that the above set satisfies the linear unitarity condition as well. Furthermore, for $p^2 + m^2$, the subtraction condition is

$$\frac{\partial}{\partial g}\mathscr{G}^{(2)}(p, -p) = 0 \,, \qquad \frac{\partial}{\partial p^2}\left[\frac{\partial}{\partial g}\mathscr{G}^{(2)}(p, -p)\right] = 0 \,, \tag{22.131}$$

$$\frac{\partial}{\partial g}\mathscr{G}^{(4)}(\text{s.p.}) = 1 \,. \tag{22.132}$$

Comparing $\{\mathscr{S}\}$ with $\{\partial\mathscr{G}/\partial g\}$, we see that they are proportional to each other:

$$\mathscr{S}^{(2)}(p_1, \ldots, p_n) = \beta(g)\frac{\partial}{\partial g}\mathscr{G}^{(n)}(p_1, \ldots, p_n) \,. \tag{22.133}$$

This is an example of the linear dependence relation mentioned in the previous section. The four sets and their indices are

$$\begin{array}{cccc} \{\mathscr{T}_{\mu\nu}\} \,, & \{\mathscr{T}\} \,, & \{\mathscr{W}\} \,, & \left\{ \dfrac{\partial}{\partial g}\mathscr{G} \right\} \\ 0 & -2 & 0 & 0 \end{array} \tag{22.134}$$

If we express the relation obtained above by a Green's function without removing the legs of the propagators, we obtain

$$\begin{aligned} T^{(n)}_{\mu\mu}(0; p_1, \ldots, p_n) + T^{(n)}(0; p_1, \ldots, p_n) - nd_\varphi(g)G^{(n)}(p_1, \ldots, p_n) \\ = \beta(g)\frac{\partial}{\partial g}G^{(n)}(p_1, \ldots, p_n) \,. \end{aligned} \tag{22.135}$$

This is called an *anomalous trace identity*. Comparing this with the incorrect (22.35), it should be clear what kinds of anomalous terms have been generated.

Furthermore, combining with (22.41),

$$\left[m\frac{\partial}{\partial m}+\beta(g)\frac{\partial}{\partial g}+n\gamma_\varphi(g)\right]G^{(n)}(p_1,\ldots,p_n)=T^{(n)}(0;p_1,\ldots,p_n)\,. \tag{22.136}$$

This is nothing but the Callan–Symanzik equation given by (20.136). We thus see that, although this equation has been derived from unrenormalized divergent quantities, it can be derived solely from finite fixed quantities. Moreover, in this method, anomalous terms are generated without any ambiguities, so it is a good approach when discussing more delicate issues.

22.7 Triangle Anomaly Terms

In the last section, we investigated the anomalous trace identity. Another typical anomalous term is the triangle anomaly term. We shall now investigate this anomalous term.

We first introduce the *unrenormalized operators*

$$A_\lambda^{(0)}=\mathrm{i}\bar\psi^{(0)}\gamma_\lambda\gamma_5\psi^{(0)}\,,\quad P^{(0)}=\mathrm{i}\bar\psi^{(0)}\gamma_5\psi^{(0)}\,,\quad S^{(0)}=\bar\psi^{(0)}\psi^{(0)}\,, \tag{22.137}$$

$$C_\lambda^{(0)}=\frac{\mathrm{i}}{4}\epsilon_{\lambda\alpha\beta\gamma}\phi_\alpha^{(0)}F_{\beta\gamma}^{(0)}\,,\quad B^{(0)}=\partial_\lambda C_\lambda^{(0)}=\boldsymbol{E}^{(0)}\cdot\boldsymbol{H}^{(0)}\,,$$

where we have used $\phi_\lambda^{(0)}$ for the potential of the electromagnetic field since $A_\lambda^{(0)}$ stands for an axial vector current. The triangle anomaly terms appear in the *renormalized equation*

$$\partial_\lambda(\boldsymbol{A}_\lambda-a\boldsymbol{C}_\lambda)=2m\boldsymbol{P}\,, \tag{22.138}$$

$$a=-\frac{e^2}{2\pi^2}\,, \tag{22.139}$$

where (22.139) is the lowest order result. However, assuming that these operators are renormalized properly, this result becomes strict. This is the *Adler–Bardeen theorem* [196]. We shall also prove this theorem.

We first express (22.138) using Green's functions:

$$\partial_\lambda\big[A_\lambda^{(n,m)}-aC_\lambda^{(n,m)}\big]=2mP^{(n,m)}+W^{(n,m)}\,, \tag{22.140}$$

where these Green's functions are defined by

$$A_\lambda^{(n,m)}(w; x \dots y \dots z \dots) = \langle \mathbf{0} | T[\boldsymbol{A}_\lambda(w)\boldsymbol{\phi}(x) \dots \boldsymbol{\psi}(y) \dots \bar{\boldsymbol{\psi}}(z) \dots] | \mathbf{0} \rangle . \tag{22.141}$$

Here n is the number of fields ϕ and m is the number of fields ψ and $\bar{\psi}$. Additionally,

$$W^{(n,m)}(w; x \dots y \dots z \dots) = -\sum_j \delta^4(w - y_j)(\gamma_5)_j G^{(n,m)}(x \dots y \dots z \dots)$$
$$- \sum_j \delta^4(w - z_j) G^{(n,m)}(x \dots y \dots z \dots)(\gamma_5)_j , \tag{22.142}$$

with

$$G^{(n,m)}(x \dots y \dots z \dots) = \langle \mathbf{0} | T[\boldsymbol{\phi}(x) \dots \boldsymbol{\psi}(y) \dots \bar{\boldsymbol{\psi}}(z) \dots] | \mathbf{0} \rangle . \tag{22.143}$$

Moreover, in order to chop the legs off the propagators, we multiply K, D, $\tilde{D}$, and so on. Here, D and $\tilde{D}$ are given by (22.86) and K is given by

$$K_{\mu\nu}(\partial) = (\delta_{\mu\nu}\Box - \partial_\mu\partial_\nu) + \frac{1}{\alpha}\partial_\mu\partial_\nu . \tag{22.144}$$

This defines Green's functions like the ones in (22.68). For instance,

$$\begin{aligned}(-\mathrm{i})^n K_x \dots D_y \dots \tilde{D}_z \dots \langle \mathbf{0} | T[\boldsymbol{P}(w)\boldsymbol{\phi}(x) \dots \boldsymbol{\psi}(y) \dots \bar{\boldsymbol{\psi}}(z) \dots] | \mathbf{0} \rangle \\ = \frac{1}{(2\pi)^{4(n+m)}} \int (\mathrm{d}k)(\mathrm{d}p)(\mathrm{d}\bar{p})\delta^4(q + k + \dots + p + \dots + \bar{p} + \dots) \\ \times \exp\left[\mathrm{i}(q \cdot w + k \cdot x + \dots + p \cdot y + \dots + \bar{p} \cdot z + \dots)\right] \\ \times \mathscr{P}^{(n,m)}(q; k \dots p \dots \bar{p} \dots) .\end{aligned} \tag{22.145}$$

Equation (22.140) then becomes

$$\mathrm{i}q_\lambda\left[\mathscr{A}_\lambda^{(n,m)} - a\mathscr{C}_\lambda^{(n,m)}\right] = 2m\mathscr{P}^{(n,m)} + \mathscr{W}^{(n,m)} . \tag{22.146}$$

22.7.1 Renormalization Condition

We now investigate several kinds of renormalization condition, i.e., normalization, for Green's functions.

The Set $\{\mathscr{P}\}$

In this case, it can be shown using perturbation theory that the order of divergence is given by

$$d(\mathscr{P}) = 3 - n - \frac{3}{2}m \,. \tag{22.147}$$

From the invariance under charge conjugation, we see that $d(\mathscr{P})$ is non-negative only for the channels (0, 2) and (2, 0). We thus expand $\mathscr{P}^{(0,2)}$ as follows:

$$\begin{aligned}\mathscr{P}^{(0,2)}(q;p,\bar{p}) = (\mathrm{i}\gamma_5)\mathscr{P}_1^{(0,2)} \\ +(\mathrm{i}p\cdot\gamma+m)(\mathrm{i}\gamma_5)\mathscr{P}_2^{(0,2)} + (\mathrm{i}\gamma_5)(-\mathrm{i}\bar{p}\cdot\gamma+m)\tilde{\mathscr{P}}_2^{(0,2)} \\ +\mathrm{i}(p\cdot\gamma+m)(\mathrm{i}\gamma_5)(-\mathrm{i}\bar{p}\cdot\gamma+m)\mathscr{P}_3^{(0,2)} \,. \end{aligned} \tag{22.148}$$

Then $\mathscr{P}_1$ and $\mathscr{P}_3$ are invariant under charge conjugation $p \rightleftarrows \bar{p}$, while $\mathscr{P}_2$ and $\tilde{\mathscr{P}}_2$ are interchanged by this operation. Since $\mathscr{P}^{(0,2)}$ diverges logarithmically overall, only $\mathscr{P}_1$ requires renormalization. We choose the subtraction point such that

$$q = 0 \,, \quad p^2 + m^2 = \bar{p}^2 + m^2 = 0 \,, \tag{22.149}$$

and we choose the subtraction condition

$$\mathscr{P}_1^{(0,2)}(\text{s.p.}) + 2m\mathscr{P}_2^{(0,2)}(\text{s.p.}) = 1 \,. \tag{22.150}$$

Usually, we choose $\mathscr{P}_1^{(0,2)}(\text{s.p.}) = 1$, but in that case the Adler–Bardeen theorem does not hold. Renormalizing as in (22.150), in the vicinity of the mass shell,

$$\mathscr{P}^{(0,2)}(0;p,-p) = \mathrm{i}\gamma_5\left[1 + \mathscr{O}(p^2+m^2)\right] . \tag{22.151}$$

This coincides with the renormalization condition given by Adler and Bardeen.

In the channel (2, 0), a linear divergence is expected. We expand as

$$\mathscr{P}_{\rho\sigma}^{(2,0)}(q;k_1,k_2) = \mathrm{i}\epsilon_{\rho\sigma\alpha\beta}(k_1)_\alpha(k_2)_\beta\mathscr{P}_1^{(2,0)} \,. \tag{22.152}$$

Here ρ and σ are vector indices. Written like this, since $\mathscr{P}_1^{(2,0)}$ goes as p^{-1}, there is no divergence and no renormalization is necessary.

Therefore, only the channel $(0,2)_1$ requires subtraction for $\mathscr{P}$, and since this subtraction condition determines its normalization in this case, it becomes a multiplicative renormalization.

The Set $\{\mathscr{W}\}$

This set is typical of the Ward–Takahashi identity and has the property that it vanishes if all the p_j except q are on the mass shell. In this case, for other Green's functions, for instance for $\mathscr{P}$, this becomes a matrix element of the operator $\boldsymbol{P}$, but vanishes for $\mathscr{W}$. The definition immediately implies the following:

$$\mathscr{W}_1^{(0,2)}(\text{s.p.}) = 0 \,, \tag{22.153}$$

$$\mathscr{W}_2^{(0,2)}(\text{s.p.}) = \tilde{W}_2^{(0,2)}(\text{s.p.}) = -1 \,. \tag{22.154}$$

These give the subtraction condition. In the channel (2, 0),

$$\mathscr{W}^{(2,0)} = 0 \,. \tag{22.155}$$

The Set $\{\mathscr{A}_\lambda\}$

$\mathscr{A}_\lambda$ also has the same index as $\mathscr{P}$. In the channel (0, 2),

$$\begin{aligned}
\mathscr{A}_\lambda^{(0,2)}(q; p, \bar{p}) = {} & (\mathrm{i}\gamma_\lambda\gamma_5)\mathscr{A}_1^{(0,2)} \\
& +(\mathrm{i}p\cdot\gamma + m)(\mathrm{i}\gamma_\lambda\gamma_5)\mathscr{A}_2^{(0,2)} + (\mathrm{i}\gamma_\lambda\gamma_5)(-\bar{p}\cdot\gamma + m)\tilde{\mathscr{A}}_2^{(0,2)} \\
& +(\mathrm{i}p\cdot\gamma + m)(\mathrm{i}\gamma_\lambda\gamma_5)(-\mathrm{i}\bar{p}\cdot\gamma + m)\mathscr{A}_3^{(0,2)} + q_\lambda\gamma_5\mathscr{A}_4^{(0,2)} \\
& +(\mathrm{i}p\cdot\gamma + m)(q_\lambda\gamma_5)\mathscr{A}_5^{(0,2)} + (q_\lambda\gamma_5)(-\mathrm{i}\bar{p}\cdot\gamma + m)\tilde{A}_5^{(0,2)} \\
& +(\mathrm{i}p\cdot\gamma + m)(q_\lambda\gamma_5)(-\mathrm{i}\bar{p}\cdot\gamma + m)\mathscr{A}_6^{(0,2)}.
\end{aligned} \tag{22.156}$$

Only $\{\mathscr{A}_1^{(0,2)}\}$ requires subtractions. The subtraction condition is

$$\mathscr{A}_1^{(0,2)}(\text{s.p.}) = \mathscr{P}_1^{(0,2)}(\text{s.p.}) \,. \tag{22.157}$$

Taking into account the bosonic symmetry, the expansion in the channel (2, 0) is

$$\begin{aligned}
\mathscr{A}_{\lambda\rho\sigma}^{(2,0)}(q; k_1, k_2) = {} & \epsilon_{\rho\sigma\alpha\beta}(k_1)_\alpha(k_2)_\beta\big[(k_1)_\lambda\mathscr{A}_1^{(2,0)} + (k_2)_\lambda\tilde{\mathscr{A}}_1^{(2,0)}\big] \\
& +\big[\epsilon_{\lambda\rho\alpha\beta}(k_1)_\sigma(k_1)_\alpha(k_2)_\beta + (k_1\cdot k_2)\epsilon_{\lambda\rho\sigma\alpha}(k_1)_\alpha\big]\mathscr{A}_2^{(2,0)} \\
& -\big[\epsilon_{\lambda\sigma\alpha\beta}(k_2)_\rho(k_1)_\alpha(k_2)_\beta + (k_1\cdot k_2)\epsilon_{\lambda\rho\sigma\alpha}(k_2)_\alpha\big]\tilde{\mathscr{A}}_2^{(2,0)} \\
& +\big[\epsilon_{\lambda\rho\alpha\beta}(k_2)_\sigma(k_1)_\alpha(k_2)_\beta + {k_2}^2\epsilon_{\lambda\rho\sigma\alpha}(k_1)_\alpha\big]\mathscr{A}_3^{(2,0)} \\
& -\big[\epsilon_{\lambda\sigma\alpha\beta}(k_1)_\rho(k_1)_\alpha(k_2)_\beta + {k_1}^2\epsilon_{\lambda\rho\sigma\alpha}(k_2)_\alpha\big]\tilde{\mathscr{A}}_3^{(2,0)} \\
& +({k_1}^2 - {k_2}^2)\epsilon_{\lambda\rho\sigma\alpha}(k_1 + k_2)_\alpha\mathscr{A}_4^{(2,0)} \\
& +\epsilon_{\lambda\rho\sigma\alpha}(k_1 - k_2)_\alpha\mathscr{A}_5^{(2,0)}.
\end{aligned} \tag{22.158}$$

Note that $\mathscr{A}_j$ and $\tilde{\mathscr{A}}_j$ are mapped to one another for $j = 1, 2, 3$, under the exchange $k_1 \rightleftarrows k_2$. Here, $\mathscr{A}^{(2,0)}$ is expected to diverge linearly. Looking at the total coefficients, we see that only $\mathscr{A}_5$ requires subtractions. For the gauge invariant $\boldsymbol{A}_\lambda$, we have

$$(k_1)_\rho \mathscr{A}^{(2,0)}_{\lambda\rho\sigma} = (k_2)_\sigma \mathscr{A}^{(2,0)}_{\lambda\rho\sigma} = 0 \,. \tag{22.159}$$

From this, we obtain

$$\mathscr{A}^{(2,0)}_4 = \mathscr{A}^{(2,0)}_5 = 0 \,. \tag{22.160}$$

Hence, in this channel, subtractions are not required. The only subtraction condition appears in $(2, 0)_1$, whence it finally becomes a multiplicative renormalization.

The Set $\{\mathscr{C}_\lambda\}$

Since $\{\mathscr{C}_\lambda\}$ is a completely new operator, to determine the renormalization condition, we choose a perturbation theory at the lowest order as a guide. Because the index is the same as that of $\{\mathscr{A}_\lambda\}$, if we first consider the channel $(0, 2)$, it vanishes at the lowest order, so we choose the renormalization condition

$$\mathscr{C}^{(2,0)}_1(\text{s.p.}) = 0 \,. \tag{22.161}$$

Next, in the channel $(2, 0)$, the lowest order computation is

$$\mathscr{C}^{(2,0)}_{\lambda\rho\sigma} = \frac{1}{2}\epsilon_{\lambda\rho\sigma\alpha}(k_1 - k_2)_\alpha \,. \tag{22.162}$$

We thus choose (22.149) as the subtraction point and the renormalization condition

$$\mathscr{C}^{(0,2)}_5(\text{s.p.}) = \frac{1}{2} \,. \tag{22.163}$$

The component above does not disappear, because $\boldsymbol{C}_\lambda$ is not gauge invariant. After all, there are two renormalization conditions for $\boldsymbol{C}_\lambda$, viz., (22.161) and (22.163), and the renormalization is not multiplicative. Instead, renormalizations by linear transformations among operators are required. As given in (20.156), they have the form

$$\boldsymbol{A}^{(0)}_\lambda = Z_A \boldsymbol{A}_\lambda \,, \quad \boldsymbol{C}^{(0)}_\lambda = Z_C \boldsymbol{C}_\lambda + z \boldsymbol{A}_\lambda \,. \tag{22.164}$$

The Set $\{\mathscr{D}\}$

In the Ward–Takahashi identity (22.146) for triangle anomaly terms, $\mathscr{A}_\lambda$ always appears in the form

$$\mathscr{D}^{(n,m)} = \mathrm{i}q_\lambda \mathscr{A}_\lambda^{(n,m)} \ . \tag{22.165}$$

Its power is

$$d(\mathscr{D}) = 4 - n - \frac{3}{2}m \ . \tag{22.166}$$

Thus, (2, 0) yields linear divergences, while $(0, 2)_1$ and $(0, 2)_2$ require subtractions. From the subtraction condition (22.157) for $\{\mathscr{A}_\lambda\}$,

$$\mathscr{D}_1^{(0,2)}(\text{s.p.}) = 2m\mathscr{P}_1^{(0,2)}(\text{s.p.}) \ , \tag{22.167}$$

$$\mathscr{D}_2^{(0,2)}(\text{s.p.}) = \tilde{D}_2^{(0,2)}(\text{s.p.}) = -\mathscr{P}_1^{(0,2)}(\text{s.p.}) \ . \tag{22.168}$$

Similarly,

$$\mathscr{D}_1^{(2,0)}(\text{s.p.}) = 0 \ . \tag{22.169}$$

The Set $\{\mathscr{B}\}$

In this case, $\mathscr{C}_\lambda$ also appears in (22.146) in the form

$$\mathscr{B}^{(n,m)} = \mathrm{i}q_\lambda \mathscr{C}_\lambda^{(n,m)} \ . \tag{22.170}$$

The subtraction conditions are obtained from (22.161) and (22.163) as

$$\mathscr{B}_1^{(0,2)}(\text{s.p.}) = 0 \ , \quad \mathscr{B}_2^{(0,2)}(\text{s.p.}) = \tilde{\mathscr{B}}_2^{(0,2)}(\text{s.p.}) = 0 \ , \tag{22.171}$$

$$\mathscr{B}_1^{(2,0)}(\text{s.p.}) = 0 \ . \tag{22.172}$$

The Set $\{\mathscr{S}\}$

Although the scalar $\boldsymbol{S}$ does not appear in the Ward–Takahashi identity, it appears in the Callan–Symanzik equation, so we give its subtraction conditions. In the channel (0, 2),

$$\begin{aligned}\mathscr{S}^{(0,2)}(q;p,\bar{p}) = \mathscr{S}_1^{(0,2)} + (\mathrm{i}p\cdot\gamma + m)\mathscr{S}_2^{(0,2)} + (-\mathrm{i}\bar{p}\cdot\gamma + m)\tilde{\mathscr{S}}_2^{(0,2)} \\ +(\mathrm{i}p\cdot\gamma + m)(-\mathrm{i}\bar{p}\cdot\gamma + m)\mathscr{S}_3^{(0,2)} \ .\end{aligned} \tag{22.173}$$

Thus, taking (22.149) as a subtraction point, the subtraction condition is

$$\mathscr{S}_1^{(0,2)}(\text{s.p.}) = 1 \ . \tag{22.174}$$

In the channel (2, 0),

$$\mathscr{S}_{\rho\sigma}^{(2,0)}(q; k_1, k_2) = \big[(k_1 \cdot k_2)\delta_{\rho\sigma} - (k_1)_\rho (k_2)_\sigma\big]\mathscr{S}_1^{(2,0)} \ , \tag{22.175}$$

and $\mathscr{S}_1^{(2,0)}$ does not require subtractions.

Since the operators appearing in (22.146) have dimension at most four, their indices are at most zero. The channels which are expected to require subtractions are

$$(0,2)_1 \ , \quad (0,2)_2 \ , \quad (2,0)_1 \ , \quad (4,0) \ , \quad (1,2) \ .$$

The first three channels have already been analyzed. Moreover, we see that subtractions are not in fact necessary in (4, 0) because of the gauge invariance. In the channel (1, 2), if we introduce the set

$$\mathscr{V}^{(n,m)} = \big[2m\mathscr{P}^{(n,m)} + \mathscr{W}^{(n,m)}\big]_{q=0} \ , \tag{22.176}$$

although we have not discussed it here, the *Kroll–Ruderman theorem* [197] implies that

$$\bar{u}(p)\mathscr{V}_\lambda^{(1,2)}(k, p, \bar{p})u(p) = 0 \ , \quad k = 0 \ , \quad p^2 + m^2 = 0 \ , \tag{22.177}$$

where λ is an index of the electromagnetic potential. From this, $\mathscr{V}^{(n,m)}$ can be expanded in terms of quantities which vanish for $q = 0$. The right-hand side of (22.146) is of course proportional to q_λ. If we examine its coefficient, we see that subtractions are not necessary.

We can therefore apply the theorem given at the end of Sect. 22.5 in this case as well. We investigate the value at each subtraction point of the following linear combination:

$$\mathscr{T}^{(n,m)} = \mathscr{D}^{(n,m)} - a\mathscr{B}^{(n,m)} - 2m\mathscr{P}^{(n,m)} - \mathscr{W}^{(n,m)} \ . \tag{22.178}$$

Hence,

$$\mathscr{T}_1^{(0,2)}(\text{s.p.}) = 2m\mathscr{P}_1^{(0,2)}(\text{s.p.}) - 0 - 2m\mathscr{P}_1^{(0,2)}(\text{s.p.}) - 0 = 0 \ , \tag{22.179}$$

$$\mathscr{T}_2^{(0,2)}(\text{s.p.}) = -\mathscr{P}_1^{(0,2)}(\text{s.p.}) - 0 - 2m\mathscr{P}_2^{(0,2)}(\text{s.p.}) - (-1) = 0 \ , \tag{22.180}$$

$$\mathscr{T}_1^{(2,0)}(\text{s.p.}) = 0 - a - 2m\mathscr{P}_1^{(2,0)}(\text{s.p.}) - 0 = 0 \ , \tag{22.181}$$

where (22.181) effectively defines the coefficient a :

$$a = -2m\mathscr{P}_1^{(2,0)}(\text{s.p.}) \ . \tag{22.182}$$

In this way, all the subtraction constants vanish, and from the theorem in Sect. 22.5,

$$\mathscr{T}^{(n,m)} = 0 \ . \tag{22.183}$$

This is the proof of (22.146). Thus, it turns out that (22.138) also holds.

22.7.2 Ward–Takahashi Identity for C_λ

From (22.164), we obtain

$$\boldsymbol{C}_\lambda = -zZ_C^{-1}\boldsymbol{A}_\lambda + \frac{\mathrm{i}}{4}Z_C^{-1}Z_3\epsilon_{\lambda\alpha\beta\gamma}\boldsymbol{\phi}_\alpha \boldsymbol{F}_{\beta\gamma} \ . \tag{22.184}$$

Combining the Ward–Takahashi identity (12.190) in QED with the equation above, we find

$$\begin{aligned}\partial_\mu K_{\mu\rho}(\partial_x)\langle \mathbf{0}|T[\boldsymbol{\phi}_\rho(x)\boldsymbol{\phi}_\sigma(y)\boldsymbol{C}_\lambda(z)]|\mathbf{0}\rangle \\ = \frac{1}{2}Z_C^{-1}Z_3\epsilon_{\lambda\alpha\beta\gamma}\frac{\partial}{\partial x_\alpha}\delta^4(x-z)\frac{\partial}{\partial y_\beta}G_{\sigma\gamma}^{(2,0)}(y-z) \ ,\end{aligned} \tag{22.185}$$

where $G^{(2,0)}$ is the Green's function for the electromagnetic field. We note that, using the renormalization condition (22.163),

$$Z_C^{-1}Z_3 = 1 \ , \quad \text{or} \quad Z_C = Z_3 \ , \tag{22.186}$$

and the right-hand side of (22.185) is also simplified using this equation.

22.7.3 Proof of the Adler–Bardeen Theorem Using the Callan–Symanzik Equation

Recall (20.166) in which the mass-insertion terms are introduced for Green's functions. We now do the same thing for QED. For example,

$$\begin{aligned}A^{(n,m)}(w; x\ldots y\ldots z\ldots : K) \\ = \langle \mathbf{0}|T\Big[\boldsymbol{A}(w)\boldsymbol{\phi}(x)\ldots\boldsymbol{\psi}(y)\ldots\bar{\boldsymbol{\psi}}(z)\ldots\exp\Big[-\mathrm{i}Km\int \mathrm{d}^4u\, \boldsymbol{S}(u)\Big]\Big]|\mathbf{0}\rangle .\end{aligned} \tag{22.187}$$

The Callan–Symanzik equation is then

$$\left[\tilde{D}+\gamma^{(n,m)}+\gamma_A\right]A^{(n,m)}(\ldots:K)=0\,. \tag{22.188}$$

Here, γ_A is the anomalous dimension of $\boldsymbol{A}$:

$$\tilde{D}=m\frac{\partial}{\partial m}+\beta(e)\frac{\partial}{\partial e}-2\alpha\gamma_{\text{ph}}(e)\frac{\partial}{\partial\alpha}-\left\{1+\left[1-\gamma_S(e)\right]K\right\}\frac{\partial}{\partial K}\,. \tag{22.189}$$

Here, for $\boldsymbol{A}$, we use $\boldsymbol{S}$, $\boldsymbol{P}$, $\boldsymbol{A}_\lambda$, $\boldsymbol{W}$, and so on. From the definition,

$$\gamma_W=0\,. \tag{22.190}$$

Moreover, for $\boldsymbol{C}_\lambda$, Eq. (20.158) implies

$$\left[\tilde{D}+\gamma^{(n,m)}+\gamma_C\right]C_\lambda^{(n,m)}(\ldots:K)+\gamma_M A_\lambda^{(n,m)}(\ldots:K)=0\,. \tag{22.191}$$

The quantity $W^{(n,m)}$ was originally generated by the equal-time commutation relations between $(\boldsymbol{A}_0-a\boldsymbol{C}_0)$ and $\boldsymbol{\psi}$, $\bar{\boldsymbol{\psi}}$. If there is a mass-insertion term, the commutator of $(\boldsymbol{A}_0-a\boldsymbol{C}_0)$ and $\boldsymbol{S}$ is generated. This is proportional to $\boldsymbol{P}$, and if we take its proportionality coefficient as b, then (22.140) generalizes to

$$\begin{aligned}\partial_\lambda\left[A_\lambda^{(n,m)}(\ldots:K)-aC_\lambda^{(n,m)}(\ldots:K)\right]&\\ =2m(1+bK)P^{(n,m)}(\ldots:K)&+W^{(n,m)}(\ldots:K)\,.\end{aligned} \tag{22.192}$$

Similarly, the generalization of (22.185) is

$$\partial_\mu K_{\mu\rho}(\partial_x)C_{\lambda\rho\sigma}^{(2,0)}(x,y,z:K)=\frac{1}{2}\epsilon_{\lambda\alpha\beta\gamma}\frac{\partial}{\partial x_\alpha}\delta^4(x-z)\frac{\partial}{\partial y_\beta}G_{\sigma\gamma}^{(2,0)}(y-z:K)\,. \tag{22.193}$$

Multiplying (22.192) by $(\tilde{\mathscr{D}}+\gamma^{(n,m)})$, a linear combination of $\{\partial_\lambda A_\lambda\}$, $\{\partial_\lambda C_\lambda\}$, and $\{P\}$ is obtained once again, and $\{W\}$ disappears. Without $\{W\}$, the remaining three sets are linearly independent, so all coefficients turn out to disappear. From this, we obtain

$$\left(\beta\frac{\mathrm{d}}{\mathrm{d}e}-\gamma_C\right)a=0\,, \tag{22.194}$$

$$\gamma_A=a\gamma_M\,, \tag{22.195}$$

$$b=1-\gamma_P\,, \tag{22.196}$$

$$\left(\beta\frac{\mathrm{d}}{\mathrm{d}e}+\gamma_S-\gamma_P\right)b=0\,. \tag{22.197}$$

In particular, combining the last two equations, (20.164) is reproduced. Alternatively,

$$\beta\frac{\mathrm{d}}{\mathrm{d}e}\gamma_P = (1-\gamma_P)(\gamma_S-\gamma_P)\,. \tag{22.198}$$

We combine (22.193) with the two equations

$$(\tilde{\mathscr{D}}+2\gamma_{\mathrm{ph}})G^{(2,0)}_{\sigma\gamma}(y-z:K)=0 \tag{22.199}$$

and

$$(\tilde{\mathscr{D}}+2\gamma_{\mathrm{ph}}+\gamma_C)C^{(2,0)}_{\lambda\rho\sigma}(x,y,z:K)+\gamma_M A^{(2,0)}_{\lambda\rho\sigma}(x,y,z:K)=0\,. \tag{22.200}$$

We thus multiply (22.200) by the differential operator

$$\partial_\mu K_{\mu\rho}(\partial)=\frac{1}{\alpha}\Box\partial_\rho\,. \tag{22.201}$$

Hence,

$$(\tilde{\mathscr{D}}+\gamma_C)\partial_\mu K_{\mu\rho}(\partial_x)C^{(2,0)}_{\lambda\rho\sigma}(x,y,z:K)=0\,. \tag{22.202}$$

According to (22.193), since this should be equal to (22.199), we have

$$\gamma_C = 2\gamma_{\mathrm{ph}}\,. \tag{22.203}$$

Hence, by (20.152), we have $\beta = e\gamma_{\mathrm{ph}}$, and (22.194) simplifies to

$$\left(e\frac{d}{de}-2\right)a=0\,. \tag{22.204}$$

Therefore, the value of a is strictly given by the lowest order in perturbation theory:

$$a=-\frac{e^2}{2\pi^2}\,, \tag{22.205}$$

and all higher order corrections disappear. This is the Adler–Bardeen theorem. This proof was given by Higashijima et al. [198].

As we saw above, divergences and indefiniteness do not appear explicitly in the dispersion theory, so it is suitable for treating the delicate issue of anomalous terms.

Regarding these terms, an extremely powerful method using the path integral method was proposed by Fujikawa [199]. This method determines transformations of the path integral measure for fermions under various types of transformations using suitable regularization procedures. The reader is referred to the original work in [199].

Postface

In this book, we have treated various aspects, focusing on well-established methods in field theory. However, many issues remain unaddressed. These include the Nambu–Bethe–Salpeter equation [200, 201] and the dispersion theory for the S-matrix and form factors, which are rarely used nowadays. For these subjects, the reader is referred to my book, "Fields and Particles" [2]. Regarding the renormalisation theory for higher order corrections in the Feynman–Dyson theory and the issue of infrared divergences, the reader is referred to the book "Quantum Field Theory" by Noboru Nakanishi [83].

K. Nishijima, *Quantum Field Theory*,
https://doi.org/10.1007/978-94-024-2190-3

References

1. K. Nishijima, *Fundamental Particles* (Benjamin, New York, 1963)
2. K. Nishijima, *Fields and Particles* (Benjamin, New York, 1969)
3. P.A.M. Dirac, Quantum theory of emission and absorption of radiation, Proc. R. Soc. Lond. A **114**, 243 (1927)
4. P.A.M. Dirac, The quantum theory of the electron. Part 2, Proc. R. Soc. Lond. A **118**, 351 (1928)
5. W. Heisenberg, W. Pauli, On quantum field theory (In German). Z. Phys. **56**, 1 (1929)
6. W. Heisenberg, W. Pauli, On quantum field theory. 2 (In German), Z. Phys. **59**, 168 (1930)
7. C.D. Anderson, The positive electron. Phys. Rev. **43**, 491 (1933)
8. O. Chamberlain, E. Segrè, C. Wiegand, T. Ypsilantis, Observation of antiprotons, UCRL-3172 (1995), ACC0411, W-7405-eng-48, printed for the US Atomic Energy Commission
9. J. Chadwick, Possible existence of a neutron. Nature **129**, 312 (1932)
10. J. Chadwick, The existence of a neutron. Proc. R. Soc. A **136**, 692 (1932)
11. W. Pauli, Dear radioactive ladies and gentlemen, Phys. Today **31N9**, 27 (1978)
12. W. Pauli, Discussions, in Septième Conseil de Physique Solvay, Noyaux Atomiques, Bruxelles (1933), p. 324
13. E. Fermi, Attempts to produce a theory of beta radiation (In Italian). Nuovo Cim. **11**, 1 (1934)
14. D. Iwanenko, Interaction of neutrons and protons. Nature **133**, 981 (1934)
15. I.G. Tamm, Exchange forces between neutrons and protons, and Fermi's theory. Nature **133**, 981 (1934)
16. I. Tamm, Interaction of neutrons and protons. Nature **134**, 1010 (1934)
17. H. Yukawa, On the interaction of elementary particles, Proc. Phys. Math. Soc. Jap. **17**, 48 (1935)
18. O. Klein, International Institute of Intellectual Cooperation, Paris (1938), in *Proceedings of the Conference New Theories in Physics in Warsaw* (1938)
19. J.C. Street, E.C. Stevenson, New evidence for the existence of a particle of mass intermediate between the proton and electron. Phys. Rev. **52**, 1003 (1937)
20. M. Conversi, E. Pancini, O. Piccioni, On the decay process of positive and negative mesons. Phys. Rev. **68**, 232 (1945)
21. C.M.G. Lattes, H. Muirhead, G.P.S. Occhialini, C.F. Powell, Processes involving charged mesons. Nature **159**, 694 (1947)
22. S. Sakata, T. Inoue, On the correlations between mesons and Yukawa particles. Prog. Theo. Phys. **1**, 143 (1946)
23. Y. Tanikawa, On the cosmic-ray meson and the nuclear meson. Prog. Theor. Phys. **2**, 220 (1947)

K. Nishijima, *Quantum Field Theory*,
https://doi.org/10.1007/978-94-024-2190-3

24. R.E. Marshak, H.A. Bethe, On the two-meson hypothesis. Phys. Rev. **72**, 506 (1947)
25. E. Gardner, C.M.G. Lattes, Production of mesons by the 184-inch Berkeley cyclotron, in *Topics on Cosmic Rays*, ed. by J. Bellandi, vol. 2 (1987), pp. 110–111. [Science **107** (1948) 270–271]
26. H.L. Anderson, E. Fermi, E.A. Long, D.E. Nagle, Total cross-sections of positive pions in hydrogen. Phys. Rev. **85**, 936 (1952)
27. N. Kemmer, The charge-dependence of nuclear forces. Proc. Camb. Philos. Soc. **34**, 354 (1938)
28. G.F. Chew, F.E. Low, Effective range approach to the low-energy p wave pion–nucleon interaction. Phys. Rev. **101**, 1570 (1956)
29. S. Mandelstam, Determination of the pion–nucleon scattering amplitude from dispersion relations and unitarity. General theory. Phys. Rev. **112**, 1344 (1958)
30. T. Regge, Introduction to complex orbital momenta. Nuovo Cim. **14**, 951 (1959)
31. G.F. Chew, S.C. Frautschi, Regge trajectories and the principle of maximum strength for strong interactions. Phys. Rev. Lett. **8**, 41 (1962)
32. G.D. Rochester, C.C. Butler, Evidence for the existence of new unstable elementary particles. Nature **160**, 855 (1947)
33. A.J. Seriff, R.B. Leighton, C. Hsiao, E.W. Cowan, C.D. Anderson, Cloud-chamber observations of the new unstable cosmic ray. Phys. Rev. **78**, 209 (1950)
34. R. Armenteros, K.H. Barker, C.C. Butler, A. Cachon, A.H. Chapman, Decay of V-particles. Nature **167**, 501 (1951)
35. Y. Nambu, K. Nishijima, Y. Yamaguchi, On the nature of V-particles. I. Prog. Theor. Phys. **6**, 615–619 (1951). In Eguchi, T. and Nishijima, K. (eds.): Broken symmetry 47–51
36. Y. Nambu, K. Nishijima, Y. Yamaguchi, On the nature of V-particles. II. Prog. Theor. Phys. **6**, 619–622 (1951). In Eguchi, T. and Nishijima, K. (eds.): Broken symmetry 51–54
37. S. Oneda, About V-particles and τ-mesons, in Japanese. Soryushiron Kenkyu **3**(4), 271 (1951)
38. H. Miyazawa, A model of V-particles. Prog. Theor. Phys. **6**, 631 (1951)
39. A. Pais, Some remarks on the V-particles. Phys. Rev. **86**, 663 (1952)
40. W.B. Fowler, R.P. Shutt, A.M. Thorndike, W.L. Whittemore, Production of V_1^0 particles by negative pions in hydrogen. Phys. Rev. **91**, 1287 (1953)
41. T. Nakano, K. Nishijima, Charge independence for V-particles. Prog. Theor. Phys. **10**, 581 (1953)
42. M. Gell-Mann, Isotopic spin and new unstable particles. Phys. Rev. **92**, 833 (1953)
43. R.H. Dalitz, On the analysis of τ-meson data and the nature of the τ-meson. Philos. Mag. **44**, 1068 (1953)
44. E. Fabri, A study of tau-meson decay. Nuovo Cim. **11**, 479 (1954)
45. T.D. Lee, C.N. Yang, Question of parity conservation in weak interactions. Phys. Rev. **104**, 254 (1956)
46. C.S. Wu, E. Ambler, R.W. Hayward, D.D. Hoppes, R.P. Hudson, Experimental test of parity conservation in beta decay. Phys. Rev. **105**, 1413 (1957)
47. T.D. Lee, C.N. Yang, Parity nonconservation and a two-component theory of the neutrino. Phys. Rev. **105**, 1671 (1957)
48. E.C.G. Sudarshan, R.E. Marshak, The nature of the four-fermion interaction, in *Proc. of the Conference on Mesons and Newly-Discovered Particles, Padua-Venice, September* (1957)
49. R.P. Feynman, M. Gell-Mann, Theory of the Fermi interaction. Phys. Rev. **109**, 193 (1958)
50. J.J. Sakurai, Mass reversal and weak interactions. Nuovo Cim. **7**, 649 (1958)
51. S.S. Gershtein, Y.B. Zeldovich, Meson corrections in the theory of beta decay. Zh. Eksp. Teor. Fiz. **29**, 698 (1955)
52. M. Gell-Mann, Symmetries of baryons and mesons. Phys. Rev. **125**, 1067 (1962)
53. M. Gell-Mann, The symmetry group of vector and axial vector currents. Physics **1**, 63 (1964)
54. S.L. Adler, Calculation of the axial vector coupling constant renormalization in beta decay. Phys. Rev. Lett. **14**, 1051 (1965)
55. W.I. Weisberger, Renormalization of the weak axial vector coupling constant. Phys. Rev. Lett. **14**, 1047 (1965)

56. W.I. Weisberger, Unsubtracted dispersion relations and the renormalization of the weak axial vector coupling constants. Phys. Rev. **143**, 1302 (1966)
57. N. Cabibbo, Unitary symmetry and leptonic decays. Phys. Rev. Lett. **10**, 531 (1963)
58. C.L. Cowan, F. Reines, F.B. Harrison, H.W. Kruse, A.D. McGuire, Detection of the free neutrino: A confirmation. Science **124**, 103 (1956)
59. E.J. Konopinski, H.M. Mahmoud, The universal Fermi interaction. Phys. Rev. **92**, 1045 (1953)
60. K. Nishijima, Vanishing of the neutrino rest mass. Phys. Rev. **108**, 907 (1957)
61. J.S. Schwinger, A theory of the fundamental interactions. Ann. Phys. **2**, 407 (1957)
62. G. Danby, J.M. Gaillard, K.A. Goulianos, L.M. Lederman, N.B. Mistry, M. Schwartz, J. Steinberger, Observation of high-energy neutrino reactions and the existence of two kinds of neutrinos. Phys. Rev. Lett. **9**, 36 (1962)
63. E. Fermi, C.N. Yang, Are mesons elementary particles? Phys. Rev. **76**, 1739 (1949)
64. S. Sakata, On a composite model for the new particles, Prog. Theor. Phys. **16**, 686 (1956)
65. M. Ikeda, S. Ogawa, Y. Ohnuki, A possible symmetry in Sakata's model for bosons-baryons system. Prog. Theor. Phys. **22**, 715 (1959)
66. A. Pevsner et al., Evidence for a three pion resonance near 550 MeV. Phys. Rev. Lett. **7**, 421 (1961)
67. M. Gell-Mann, The Eightfold Way: A theory of strong interaction symmetry, CTSL-20, TID-12608
68. Y. Ne'eman, Derivation of strong interactions from a gauge invariance. Nucl. Phys. **26**, 222 (1961)
69. M. Gell-Mann, A theory of strong interaction symmetry, Caltech Report CTLS-20 (1961)
70. S. Okubo, Note on unitary symmetry in strong interactions. Prog. Theor. Phys. **27**, 949 (1962)
71. V.E. Barnes et al., Observation of a hyperon with strangeness minus three. Phys. Rev. Lett. **12**, 204 (1964)
72. G.F. Chew, S.C. Frautschi, Principle of equivalence for all strongly interacting particles within the S matrix framework. Phys. Rev. Lett. **7**, 394 (1961)
73. G.F. Chew, Nuclear democracy and bootstrap dynamics, UCRL-11163
74. M. Gell-Mann, A schematic model of baryons and mesons. Phys. Lett. **8**, 214 (1964)
75. G. Zweig, An $SU(3)$ model for strong interaction symmetry and its breaking. Version 1, CERN-TH-401
76. M.Y. Han, Y. Nambu, Three triplet model with double $SU(3)$ symmetry. Phys. Rev. **139**, B1006 (1965)
77. Y. Nambu, A systematics of hadrons in subnuclear physics, in *Preludes in Theoretical Physics*, ed. by A. de Shalit, H. Feshbach, L. Van Hove (North Holland, Amsterdam, 1966)
78. K. Nishijima, *Relativistic Quantum Mechanics* (New Physics Series 13) (Baifukan, 1973) (in Japanese)
79. M. Goldhaber, L. Grodzins, A.W. Sunyar, Helicity of neutrinos. Phys. Rev. **109**, 1015 (1958)
80. W. Pauli, P. Jordan, Zur Quantenelektrodynamik ladungsfreier Felder. Zeits. Phys. **47**, 151 (1928)
81. M. Fierz, Force-free particles with any spin. Helv. Phys. Acta **12**, 3 (1939)
82. W. Pauli, The connection between spin and statistics. Phys. Rev. **58**, 716 (1940)
83. N. Nakanishi, *Quantum Field Theory* (New Physics Series 19) (Baifukan, 1975) (in Japanese)
84. P.A.M. Dirac, Bakerian lecture. The physical interpretation of quantum mechanics. Proc. R. Soc. A **180**, 1 (1942)
85. S.N. Gupta, Theory of longitudinal photons in quantum electrodynamics. Proc. Phys. Soc. A **63**, 681 (1950)
86. K. Bleuler, A new method of treatment of the longitudinal and scalar photons. Helv. Phys. Acta **23**, 567 (1950)
87. S. Tomonaga, On a relativistically invariant formulation of the quantum theory of wave fields. Prog. Theor. Phys. **1**, 27 (1946)
88. J.S. Schwinger, Quantum electrodynamics. I. A covariant formulation. Phys. Rev. **74**, 1439 (1948)
89. P.A.M. Dirac, Relativistic quantum mechanics. Proc. R. Soc. A **136**, 453 (1932)

90. C.N. Yang, D. Feldman, The S matrix in the Heisenberg representation. Phys. Rev. **79**, 972 (1950)
91. Y. Takahashi, *Introduction to Analytical Mechanics for Learning Quantum Mechanics*, Enlarged second edition, (Kodansha Scientific, 2000) (in Japanese)
92. F.J. Belinfante, On the current and the density of the electric charge, the energy, the linear momentum and the angular momentum of arbitrary fields. Physica **7**, 449 (1940)
93. L. Rosenfeld, Sur le tenseur dimpulsion-énergie. Acad. R. Belg. Mémoirs de classes de Science **18** (1940)
94. W. Pauli, On the conservation of the lepton charge. Nuovo Cimento **6**, 204 (1957)
95. F. Gürsey, Relation of charge independence and baryon conservation to Pauli's transformation. Nuovo Cimento **7**, 411 (1958)
96. R.P. Feynman, Space-time approach to nonrelativistic quantum mechanics. Rev. Mod. Phys. **20**, 367 (1948)
97. F.J. Dyson, The Radiation theories of Tomonaga, Schwinger, and Feynman. Phys. Rev. **75**, 486 (1949)
98. F.J. Dyson, The S matrix in quantum electrodynamics. Phys. Rev. **75**, 1736 (1949)
99. W.H. Furry, A symmetry theorem in the positron theory. Phys. Rev. **51**, 125 (1937)
100. R.P. Feynman, The theory of positrons. Phys. Rev. **76**, 749 (1949)
101. L.D. Landau, On the angular momentum of a two-photon system. Dokl. Akad. Nauk Ser. Fiz. **60**, 207 (1948)
102. C.N. Yang, Possible experimental determination of whether the neutral meson is scalar or pseudoscalar. Phys. Rev. **77**, 722 (1950)
103. C. Møller, Zur Theorie des Durchgangs schneller Elektronen durch Materie. Ann. Physik **14**, 531 (1932)
104. O. Klein, Y. Nishina, Über die Streuung von Strahlung durch freie Elektronen nach der neuen relativistischen Quantendynamik von Dirac. Zeitsch. Physik **52**, 853 (1929)
105. J.D. Bjorken, Asymptotic sum rules at infinite momentum. Phys. Rev. **179**, 1547 (1969)
106. L. Michel, Interaction between four half-spin particles and the decay of the μ-meson. Proc. Phys. Soc. A **63**, 514 (1950)
107. R.H. Dalitz, Decay of τ mesons of known charge. Phys. Rev. **94**, 1046 (1954)
108. J.H. Christenson, J.W. Cronin, V.L. Fitch, R. Turlay, Evidence for the 2π decay of the K_2^0 meson. Phys. Rev. Lett. **13**, 138 (1964)
109. G. Lüders, On the equivalence of invariance under time reversal and under particle–antiparticle conjugation for relativistic field theories. Kong. Dan. Vid. Sel. Mat. Fys. Med. **28N5**, 1 (1954)
110. W. Pauli, *Niels Bohr and the Development of Physics*. Essays Dedicated to Niels Bohr on the Occasion of his Seventieth birthday (Pergamon Press, London, 1955)
111. M. Gell-Mann, F. Low, Bound states in quantum field theory. Phys. Rev. **84**, 350 (1951)
112. P.T. Matthews, The application of Dyson's methods to meson interactions. Phys. Rev. **76**, 684 (1949)
113. H. Lehmann, K. Symanzik, W. Zimmermann, On the formulation of quantized field theories. Nuovo Cim. **1**, 205 (1955)
114. B.A. Lippmann, J. Schwinger, Variational principles for scattering processes. I. Phys. Rev. **79**, 469 (1950)
115. G.F. Chew, M.L. Goldberger, The scattering of elementary particles by complex nuclei. A generalization of the impulse approximation. Phys. Rev. **87**, 778 (1952)
116. J.S. Schwinger, On quantum electrodynamics and the magnetic moment of the electron. Phys. Rev. **73**, 416 (1948)
117. H. Fukuda, Y. Miyamoto, S. Tomonaga, A self-consistent subtraction method in the quantum field theory. II. Prog. Theor. Phys. **2**, 121 (1949)
118. J.C. Ward, An identity in quantum electrodynamics. Phys. Rev. **78**, 182 (1950)
119. Y. Takahashi, On the generalized Ward identity. Nuovo Cim. **6**, 371 (1957)
120. K. Nishijima, Asymptotic conditions and perturbation theory. Phys. Rev. **119**, 485 (1960)

121. H. Umezawa, S. Kamefuchi, The vacuum in quantum electrodynamics. Prog. Theor. Phys. **6**, 543 (1951)
122. G. Källén, On the definition of the renormalization constants in quantum electrodynamics. Helv. Phys. Acta **25**(4), 417 (1952)
123. H. Lehmann, Über Eigenschatten von Ausbreitungsfunktionen und Renormierungskonsten quantisierter Felder. Nuovo Cim. **11**, 342 (1954)
124. T. Goto, T. Imamura, Note on the non-perturbation-approach to quantum field theory. Prog. Theor. Phys. **14**, 396 (1955)
125. J.S. Schwinger, Field theory commutators. Phys. Rev. Lett. **3**, 296 (1959)
126. M.L. Perl et al., Evidence for anomalous lepton production in e^+–e^- annihilation. Phys. Rev. Lett. **35**, 1489 (1975)
127. H. Frauenfelder et al., Parity and the polarization of electrons from ^{60}Co. Phys. Rev. **106**, 386 (1957)
128. H. Frauenfelder et al., Parity and electron polarization: Møller scattering. Phys. Rev. **107**, 643 (1957)
129. M. Gell-Mann, M. Levy, The axial vector current in beta decay. Nuovo Cim. **16**, 705 (1960)
130. S.L. Glashow, J. Iliopoulos, L. Maiani, Weak interactions with lepton–hadron symmetry. Phys. Rev. D **2**, 1285 (1970)
131. M. Kobayashi, T. Maskawa, CP violation in the renormalizable theory of weak interaction. Prog. Theor. Phys. **49**, 652 (1973)
132. J.J. Aubert et al. [E598 Collaboration], Experimental observation of a heavy particle. J. Phys. Rev. Lett. **33**, 1404 (1974)
133. J.E. Augustin et al. [SLAC-SP-017 Collaboration], Discovery of a narrow resonance in e^+–e^- annihilation. Phys. Rev. Lett. **33**, 1406 (1974). [Adv. Exp. Phys. **5** (1976) 141]
134. Z. Maki, The fourth baryon, Sakata model and modified $B - L$ symmetry. 1. Prog. Theor. Phys. **31**, 331 (1964)
135. Y. Hara, Unitary triplets and the eightfold way. Phys. Rev. **134**, B701 (1964)
136. S.W. Herb et al., Observation of a dimuon resonance at 9.5 GeV in 400 GeV proton–nucleus collisions. Phys. Rev. Lett. **39**, 252 (1977)
137. R.P. Feynman, The behavior of hadron collisions at extreme energies. Conf. Proc. C **690905**, 237 (1969)
138. C.G. Callan Jr., D.J. Gross, High-energy electroproduction and the constitution of the electric current. Phys. Rev. Lett. **22**, 156 (1969)
139. C.N. Yang, R.L. Mills, Conservation of isotopic spin and isotopic gauge invariance. Phys. Rev. **96**, 191 (1954)
140. R. Utiyama, Invariant theoretical interpretation of interaction. Phys. Rev. **101**, 1597 (1956)
141. Y. Nambu, G. Jona-Lasinio, Dynamical model of elementary particles based on an analogy with superconductivity. I. Phys. Rev. **122**, 345 (1961)
142. Y. Nambu, G. Jona-Lasinio, Dynamical model of elementary particles based on an analogy with superconductivity. II. Phys. Rev. **124**, 246 (1961)
143. J. Goldstone, Field theories with superconductor solutions. Nuovo Cim. **19**, 154 (1961)
144. F. Englert, R. Brout, Broken symmetry and the mass of gauge vector mesons. Phys. Rev. Lett. **13**, 321 (1964)
145. P.W. Higgs, Broken symmetries, massless particles and gauge fields. Phys. Lett. **12**, 132 (1964)
146. P.W. Higgs, Broken symmetries and the masses of gauge bosons. Phys. Rev. Lett. **13**, 508 (1964)
147. P.W. Higgs, Spontaneous symmetry breakdown without massless bosons. Phys. Rev. **145**, 1156 (1966)
148. N. Nakanishi, Indefinite-metric quantum theory of genuine and Higgs-type massive vector fields. Prog. Theor. Phys. **49**, 640 (1973)
149. P.W. Anderson, Plasmons, gauge invariance, and mass. Phys. Rev. **130**, 439 (1963)
150. T.W.B. Kibble, Symmetry breaking in non-Abelian gauge theories. Phys. Rev. **155**, 1554 (1967)

151. S. Weinberg, A model of leptons. Phys. Rev. Lett. **19**, 1264 (1967)
152. A. Salam, Weak and electromagnetic interactions. Conf. Proc. C **680519**, 367 (1968)
153. S. Glashow, Partial-symmetries of weak interactions. Nucl. Phys. **22**(4), 579–588 (1961)
154. G. Arnison et al. [UA1 Collaboration], Experimental observation of isolated large transverse energy electrons with associated missing energy at s(1/2) = 540 GeV. Phys. Lett. B **122**, 103 (1983)
155. M. Banner et al. [UA2 Collaboration], Observation of single isolated electrons of high transverse momentum in events with missing transverse energy at the CERN anti-p p collider. Phys. Lett. B **122**, 476 (1983)
156. G. Arnison et al. [UA1 Collaboration], Experimental observation of lepton pairs of invariant mass around 95 GeV/c^2 at the CERN SPS collider. Phys. Lett. B **126**, 398 (1983)
157. N.S. Craigie, K. Hidaka, P. Ratcliffe, The role helicity asymmetries could play in the search for supersymmetric interactions. Phys. Lett. B **129**, 310 (1983)
158. F.J. Hasert et al. [Gargamelle Neutrino Collaboration], Observation of neutrino-like interactions without muon or electron in the Gargamelle neutrino experiment. Phys. Lett. B **46**, 138 (1973)
159. S.L. Adler, Axial vector vertex in spinor electrodynamics. Phys. Rev. **177**, 2426 (1969)
160. J.S. Bell, R. Jackiw, A PCAC puzzle: $\pi^0 \to \gamma\gamma$ in the sigma model. Nuovo Cim. A **60**, 47 (1969)
161. W.A. Bardeen, Anomalous Ward identities in spinor field theories. Phys. Rev. **184**, 1848 (1969)
162. H. Georgi, S.L. Glashow, Gauge theories without anomalies. Phys. Rev. D **6**, 429 (1972)
163. H. Georgi, S.L. Glashow, Unity of all elementary particle forces. Phys. Rev. Lett. **32**, 438 (1974)
164. L.D. Faddeev, V.N. Popov, Feynman diagrams for the Yang–Mills field. Phys. Lett. B **25**, 29 (1967)
165. G. 't Hooft, Renormalization of massless Yang–Mills fields. Nucl. Phys. B **33**, 173 (1971)
166. T. Kugo, I. Ojima, Manifestly covariant canonical formulation of Yang–Mills field theories. 1. The case of Yang–Mills fields of Higgs–Kibble type in the Landau gauge. Prog. Theor. Phys. **60**, 1869 (1978)
167. C. Becchi, A. Rouet, R. Stora, Renormalization of gauge theories. Ann. Phys. **98**, 287 (1976)
168. G. Curci, R. Ferrari, Slavnov transformations and supersymmetry. Phys. Lett. B **63**, 91 (1976)
169. I. Ojima, Another BRS transformation. Prog. Theor. Phys. **64**, 625 (1980)
170. L. Bonora, M. Tonin, Superfield formulation of extended BRS symmetry. Phys. Lett. B **98**, 48 (1981)
171. A.A. Slavnov, Invariant regularization of gauge theories. Teor. Mat. Fiz. **13**, 174 (1972)
172. J.C. Taylor, Ward identities and charge renormalization of the Yang–Mills field. Nucl. Phys. B **33**, 436 (1971)
173. T. Kugo, I. Ojima, Local covariant operator formalism of nonabelian gauge theories and quark confinement problem. Prog. Theor. Phys. Suppl. **66**, 1 (1979)
174. T. Kugo, I. Ojima, Manifestly covariant canonical formulation of Yang–Mills field theories: Physical state subsidiary conditions and physical S matrix unitarity. Phys. Lett. B **73**, 459 (1978)
175. N. Nakanishi, I. Ojima, Superalgebras of nonabelian gauge theories in the manifestly covariant canonical formalism. Z. Phys. C **6**, 155 (1980). [Z. Phys. C **8** (1981) 94]
176. E.C.G. Stueckelberg, A. Petermann, La normalisation des constantes dans la théorie des quanta (Normalization of constants in the quantum theory). Helv. Phys. Acta **26**, 499 (1953)
177. M. Gell-Mann, F.E. Low, Quantum electrodynamics at small distances. Phys. Rev. **95**, 1300 (1954)
178. N.N. Bogolyubov, D.V. Shirkov, Introduction to the theory of quantized fields. Intersci. Monogr. Phys. Astron. **3**, 1 (1959)
179. C.G. Callan Jr., Broken scale invariance in scalar field theory. Phys. Rev. D **2**, 1541 (1970)
180. K. Symanzik, Small distance behavior in field theory and power counting. Commun. Math. Phys. **18**, 227 (1970)

181. K. Symanzik, Small distance behavior analysis and Wilson expansion. Commun. Math. Phys. **23**, 49 (1971)
182. K. Higashijima, K. Nishijima, Renormalization groups of Gell-Mann–Low and of Callan–Symanzik. Prog. Theor. Phys. **64**, 2179 (1980)
183. D.J. Gross, F. Wilczek, Ultraviolet behavior of nonabelian gauge theories. Phys. Rev. Lett. **30**, 1343 (1973)
184. H.D. Politzer, Reliable perturbative results for strong interactions? Phys. Rev. Lett. **30**, 1346 (1973)
185. K.G. Wilson, Confinement of quarks. Phys. Rev. D **10**, 2445 (1974)
186. K. Nishijima, BRS transformation and color confinement. Phys. Lett. B **116**, 295 (1982)
187. R. Haag, Quantum field theories with composite particles and asymptotic conditions. Phys. Rev. **112**, 669 (1958)
188. W. Zimmermann, On the bound state problem in quantum field theory. Nuovo Cim. **10**, 597 (1958)
189. K. Nishijima, Formulation of field theories of composite particles. Phys. Rev. **111**, 995 (1958)
190. M. Chaichian, K. Nishijima, Renormalization constant of the color gauge field as a probe of confinement. Eur. Phys. J. C **22**, 463 (2001)
191. K. Nishijima, A. Tureanu, Gauge-dependence of Green's functions in QCD and QED. Eur. Phys. J. C **53**, 649 (2008)
192. H. Fukuda, Y. Miyamoto, On the γ-decay of neutral meson. Prog. Theor. Phys. **4**, 374 (1949)
193. J. Steinberger, On the use of subtraction fields and the lifetimes of some types of meson decay. Phys. Rev. **76**, 1180 (1949)
194. J.S. Schwinger, On gauge invariance and vacuum polarization. Phys. Rev. **82**, 664 (1951)
195. C.G. Callan Jr., S.R. Coleman, R. Jackiw, A new improved energy–momentum tensor. Ann. Phys. **59**, 42 (1970)
196. S.L. Adler, W.A. Bardeen, Absence of higher order corrections in the anomalous axial vector divergence equation. Phys. Rev. **182**, 1517 (1969)
197. N.M. Kroll, M.A. Ruderman, A theorem on photomeson production near threshold and the suppression of pairs in pseudoscalar meson theory. Phys. Rev. **93**, 233 (1954)
198. K. Higashijima, K. Nishijima, M. Okawa, The Adler–Bardeen theorem in quantum electrodynamics. Prog. Theor. Phys. **67**, 668 (1982)
199. K. Fujikawa, Path integral measure for gauge invariant fermion theories. Phys. Rev. Lett. **42**, 1195 (1979)
200. E.E. Salpeter, H.A. Bethe, A relativistic equation for bound-state problems. Phys. Rev. **84**, 1232 (1951)
201. Y. Nambu, Force potentials in quantum field theory. Prog. Theor. Phys. **5**, 614–633 (1950). In Eguchi, T. and Nishijima, K. (eds.): Broken symmetry 27–46; Path integral for gauge theories with fermions, Phys. Rev. D **21** (1980) 2848 [Phys. Rev. D **22** (1980) 1499]

Index

K. Nishijima, *Quantum Field Theory*,
https://doi.org/10.1007/978-94-024-2190-3

Printed in the United States
by Baker & Taylor Publisher Services